Lecture Notes in Computer Science 15669

The series Lecture Notes in Computer Science (LNCS), including its subseries Lecture Notes in Artificial Intelligence (LNAI) and Lecture Notes in Bioinformatics (LNBI), has established itself as a medium for the publication of new developments in computer science and information technology research, teaching, and education.

LNCS enjoys close cooperation with the computer science R & D community, the series counts many renowned academics among its volume editors and paper authors, and collaborates with prestigious societies. Its mission is to serve this international community by providing an invaluable service, mainly focused on the publication of conference and workshop proceedings and postproceedings. LNCS commenced publication in 1973.

Georg Krempl · Kai Puolamäki · Ioanna Miliou
Editors

Advances in Intelligent Data Analysis XXIII

23rd International Symposium on Intelligent Data Analysis, IDA 2025
Konstanz, Germany, May 7–9, 2025
Proceedings

Springer

Editors
Georg Krempl
Utrecht University
Utrecht, Utrecht, The Netherlands

Kai Puolamäki
University of Helsinki
Helsinki, Finland

Ioanna Miliou
Stockholm University
Kista, Sweden

ISSN 0302-9743 ISSN 1611-3349 (electronic)
Lecture Notes in Computer Science
ISBN 978-3-031-91397-6 ISBN 978-3-031-91398-3 (eBook)
https://doi.org/10.1007/978-3-031-91398-3

This Springer imprint is published by the registered company Springer Nature Switzerland AG
The registered company address is: Gewerbestrasse 11, 6330 Cham, Switzerland

Preface

We are delighted to introduce the proceedings of the 23rd International Symposium on Intelligent Data Analysis (IDA), held from 7 to 9 May 2025 in Konstanz, Germany. This 23rd edition also marked IDA's 30th anniversary. The symposium was founded in 1995, with its first meeting in Baden-Baden, Germany. IDA was initially a biennial event, but since 2010, it has been held annually, shifting its focus towards encouraging submissions presenting groundbreaking and innovative ideas, even if they are not yet fully developed. IDA 2025 continued this tradition, encouraging submissions at all stages of development, including those that, while preliminary in other contexts, promise significant research potential.

The IDA Symposium brings together a wide range of modeling and analysis methodologies, fostering interdisciplinary discussions on intelligent data analysis. It welcomes contributions that address intelligent support for modeling and analyzing data from complex, dynamic systems, encouraging novel approaches and innovative perspectives. In this context, IDA 2025 extended its support for data analysis beyond the conventional algorithmic solutions typically discussed in academic literature.

The traditional review process often favors minor, incremental improvements over true innovation. To address this, reviewers and senior PC advisors were encouraged to favor innovative ideas over complex solutions and extensive experimental evaluations. The outcome was a diverse and engaging program. We received 91 full submissions, of which 35 were accepted as regular papers. Each submission underwent a rigorous single-blind review by at least two members of the program committee and one senior PC advisor.

In addition, this year's edition featured a special event, "IDA from 1995 to today: A 30-Year Retrospective," celebrating the 30th anniversary of the symposium. This dedicated session provided a platform for insightful discussions. It featured lively discussions on past IDA research that has stood the test of time or significantly influenced the field. Finally, we were honored to announce the following invited speakers:

- Katharina Morik, TU Dortmund, Germany on the topic "Embedding Machine Learning into Real-World Processes – A Time Ride through AI Applications"
- Jessica Lin, George Mason University, USA on the topic "From Repetition to Insight: Shaping a More Interpretable Future through Time Series Motif Discovery"

We sincerely thank all authors for their contributions, as well as the Program Committee members, Advisors, and additional reviewers for their effort in reviewing, discussing, and providing feedback on the submitted papers. Special thanks go to Francesca Pratesi for managing the PhD Forum Track, Allan Tucker as History Chair for organizing the "IDA from 1995 to today: A 30-Year Retrospective" event, Panagiotis Papapetrou as Frontier Prize Chair, and the members of the IDA Steering Committee for their ongoing guidance and support. We are also grateful to Nicole Stockwell, the local chair, and the volunteer team from KNIME for local organization.

We extend our gratitude to our advisory chair, Michael Berthold, and the general chair of the IDA society, Matthijs van Leeuwen, for their invaluable guidance during the preparation of IDA 2025. We also thank Maria Bampa, the Social Media Chair, who, together with Ioanna Miliou, handled all strategic social media communications related to IDA. Finally, we thank Zed Lee and Ioanna Miliou for running the conference website. We thank our sponsors, KNIME and Springer, with special appreciation to KNIME for funding the IDA Frontier Prize, which was awarded for the most visionary contribution.

May 2025

Georg Krempl
Kai Puolamäki
Ioanna Miliou

Organization

General Chair

Ioanna Miliou	Stockholm University, Sweden

Program Committee Chairs

Kai Puolamäki	University of Helsinki, Finland
Georg Krempl	Utrecht University, The Netherlands

Steering Committee: IDA Council

Michael Berthold	Konstanz University, Germany
Elizabeth Bradley	University of Colorado Boulder, USA
Tijl De Bie	Ghent University, Belgium
Elisa Fromont	University of Rennes 1, France
Jaakko Hollmen	Aalto University, Finland
Frank Höppner	Ostfalia University of Applied Sciences, Germany
Joost Kok	University of Twente, The Netherlands
Matthijs van Leeuwen	Leiden University, The Netherlands
Xiaohui Liu	Brunel University London, UK
Arno Siebes	Utrecht University, The Netherlands
Allan Tucker	Brunel University London, UK

Senior Program Committee

Hendrik Blockeel	KU Leuven, Belgium
Henrik Boström	KTH Royal Institute of Technology, Sweden
Peggy Cellier	INSA Rennes, IRISA, France
Bruno Cremilleux	Université de Caen Normandie, France
Wouter Duivesteijn	TU Eindhoven, Netherlands
Ad Feelders	Universiteit Utrecht, Netherlands
Benoît Frénay	University of Namur, Belgium
Esther Galbrun	University of Eastern Finland, Finland
Joao Gama	INESC TEC - LIAAD, Portugal

Barbara Hammer	CITEC, Bielefeld University, Germany
Sibylle Hess	TU Eindhoven, Netherlands
Frank Hoeppner	Ostfalia University of Applied Science, Germany
Arno Knobbe	Leiden University, Netherlands
Charlotte Laclau	Télécom Paris, France
Nada Lavrač	Jožef Stefan Institute, Slovenia
Siegfried Nijssen	Université Catholique de Louvain, Belgium
Nico Piatkowski	Fraunhofer IAIS, Germany
Celine Robardet	INSA Lyon, France
Arno Siebes	Universiteit Utrecht, Netherlands
Myra Spiliopoulou	Otto-von-Guericke-University Magdeburg, Germany
Jerzy Stefanowski	Poznań University of Technology, Poland
Stephen Swift	Brunel University London, UK
Maryam Tavakol	TU Eindhoven, Netherlands
Allan Tucker	Brunel University London, UK
Matthijs van Leeuwen	Leiden University, Netherlands
Veronica Vinciotti	University of Trento, Italy
David Weston	Birkbeck, University of London, UK
Albrecht Zimmermann	Université de Caen Normandie, France

Program Committee

Pedro Henriques Abreu	CISUC, Portugal
Gael Aglin	UCLouvain, Belgium
Thiago Andrade	INESC TEC - LIAAD, Portugal
Fabrizio Angiulli	University of Calabria, Italy
Sahar Asadi	King (part of Microsoft), Sweden
Ali Ayadi	University of Strasbourg, France
Paulo J. Azevedo	Universidade do Minho, Portugal
José Luis Balcázar	UPC, Spain
Maria Bampa	Stockholm University, Sweden
Anton Björklund	University of Oxford, UK
Veselka Boeva	Blekinge Institute of Technology, Sweden
Giacomo Boracchi	Politecnico di Milano, Italy
José L. Borges	FEUP/INESC-TEC, Portugal
Tassadit Bouadi	Université de Rennes, France
Paula Brito	Universidade do Porto & LIAAD-INESC TEC, Portugal
Dariusz Brzezinski	Poznań University of Technology, Poland
Sebastian Buschjäger	Lamarr Institute for ML and AI, Germany

Paulo Cortez	University of Minho, Portugal
Bertrand Cuissart	University of Caen Normandy, France
Thi-Bich-Hanh Dao	University of Orléans, France
Nuno Escudeiro	Instituto Politécnico do Porto, Portugal
Brigida Monica Faria	ESS-P.PORTO and LIACC, Portugal
Sébastien Ferré	Université de Rennes, CNRS, Inria, IRISA, France
Carlos Ferreira	INESC TEC, Portugal
Françoise Fessant	Orange, France
Douglas H. Fisher	Vanderbilt University, USA
Mikel Galar	Universidad Pública de Navarra, Spain
Rui Jorge Gomes	Universidade Portucalense Infante D. Henrique, Portugal
Maciej Grzenda	Warsaw University of Technology, Poland
Riccardo Guidotti	University of Pisa, Italy
Alberto Fernandez Hilario	University of Granada, Spain
Martin Holena	Institute of Computer Science, Czech Republic
Tomáš Horváth	Edinburgh Napier University, UK
Dino Ienco	Irstea Institute, France
Sasan Jafarnejad	University of Luxembourg, Luxembourg
Baptiste Jeudy	Université de Saint-Étienne, France
Ulf Johansson	Jönköping University, Sweden
Alipio M. G. Jorge	INESC TEC/University of Porto, Portugal
Andrej Kastrin	University of Ljubljana, Slovenia
Frank Klawonn	Ostfalia University of Applied Sciences, Germany
Jiri Klema	Czech Technical University, Czech Republic
Alejandro Kuratomi	Stockholm University, Sweden
Christine Largeron	LabHC, Lyon University, France
Mark Last	Ben-Gurion University of the Negev, Israel
Jefrey Lijffijt	Ghent University, Belgium
Sindri Magnusson	Stockholm University, Sweden
George Magoulas	Birkbeck College, University of London, UK
Alessio Malizia	University of Pisa, Italy
Joao Mendes-Moreira	University of Porto, Portugal
Mohamed Nadif	Centre Borelli - Université Paris Cité, France
Ana Rita Nogueira	INESCTEC, Portugal
Slawomir Nowaczyk	Halmstad University, Sweden
Andreas Nürnberger	Otto-von-Guericke-Universität Magdeburg, Germany
Loren Nuyts	KU Leuven, Belgium

Kaustubh Patil	Institute of Neuroscience and Medicine, Brain & Behaviour (INM-7), Research Centre Jülich, Germany
Ruggero G. Pensa	University of Turin, Italy
João P. Pimentel	Dalhousie University, Canada
Marc Plantevit	EPITA, France
Senja Pollak	Jožef Stefan Institute, Slovenia
Luboš Popelínský	Masaryk University, Czechia
Filipe Portela	University of Minho, Portugal
Miguel A. Prada	Universidad de León, Spain
Jose Puerta	Universidad de Castilla-La Mancha, Spain
Justine Reynaud	Université de Caen Normandie, CNRS, France
Rita P. Ribeiro	University of Porto & INESC TEC, Portugal
Salvatore Rinzivillo	KDDLab - ISTI - CNR, Italy
Duncan Ruiz	Pontifícia Universidade Católica do Rio Grande do Sul, Brazil
Amal Saadallah	TU Dortmund, Germany
Miriam Santos	University of Porto, Portugal
Akrati Saxena	Leiden University, Netherlands
Jörg Schlötterer	University of Marburg, University of Mannheim, Germany
Johannes Schneider	University of Liechtenstein, Liechtenstein
Roberta Siciliano	University of Naples Federico II, Italy
Paula Silva	INESC TEC - LIAAD, Portugal
Carlos Soares	University of Porto/Fraunhofer AICOS Portugal, Portugal
Amina Souag	Canterbury Christ Church University, UK
Arnaud Soulet	University of Tours, France
Francesco Spinnato	University of Pisa, Italy
Shazia Tabassum	INESCTEC, Portugal
Maguelonne Teisseire	INRAE - UMR Tetis, France
Sónia Teixeira	INESC TEC, Portugal
Andreas Theissler	Aalen University of Applied Sciences, Germany
Alicia Troncoso	Pablo de Olavide University, Spain
Elena Tsiporkova	Sirris, Belgium
Peter van der Putten	Leiden University, Netherlands
Jan N. Van Rijn	LIACS, Leiden University, Netherlands
Cor Veenman	TNO/Leiden University, Netherlands
Bruno Veloso	INESC TEC & University of Porto, Portugal
Rakesh Verma	University of Houston, USA
Hilde J. P. Weerts	Eindhoven University of Technology, Netherlands
Indre Zliobaite	University of Helsinki, Finland

Sponsor

We thank KNIME Germany as our principal sponsor.

Contents

Applications of Data Science

Credal Knowledge Tracing for Imprecise and Uncertain MCQ 3
Dorra Sassi, Constance Thierry, and David Gross-Amblard

Development of Models to Quantify Training Load in Outdoor Running Using Inertial Sensors 17
Bouke L. Scheltinga, Jasper Reenalda, Jaap H. Buurke, and Joost N. Kok

Estimating the Learning Capacity of Bacterial Metabolic Networks 28
Bastien Mollet, Paul Ahavi, Antoine Cornuéjols, Jean-Loup Faulon, Evelyne Lutton, and Alberto Tonda

Semi-supervised Learning with Pairwise Instance Comparisons for Medical Instance Classification 41
Anne Rother, Till Ittermann, and Myra Spiliopoulou

Local-Global Data Augmentation for Contrastive Learning in Static Sign Language Recognition 54
Ariel Basso Madjoukeng, Edith Bélise Kenmogne, Pierre Poitier, Benoît Frénay, and Jérôme Fink

SiamCircle: Trajectory Representation Learning in Free Settings 67
Maedeh Nasri, Mitra Baratchi, Alexander Koutamanis, and Carolien Rieffe

Synthetic Tabular Data Detection in the Wild 81
G. Charbel N. Kindji, Elisa Fromont, Lina M. Rojas-Barahona, and Tanguy Urvoy

Assessing the Graph Structure Learning in Graph Deviation Networks 97
Canberk Ozen, Sławomir Nowaczyk, Prayag Tiwari, and Sepideh Pashami

Foundations of Data Science

The When and How of Target Variable Transformations 113
Loren Nuyts and Jesse Davis

Balancing Performance and Scalability of Demand Forecasting ML Models ... 127
Mateusz Żarski and Sławomir Nowaczyk

Balancing Global Importance and Source Proximity for Personalized Recommendations Using Random Walk Length 141
Tsuyoshi Yamashita and Kunitake Kaneko

Counterintuitive Behavior of Clustering Quality: Findings for K-Means on Synthetic and Real Data 154
Marco Loog, Jesse H. Krijthe, and Manuele Bicego

BOWSA: A Contribution of Sensitivity Analysis to Improve Bayesian Optimization for Parameter Tuning 167
Lise Kastner, Bertrand Cuissart, and Jean-Luc Lamotte

Overfitting in Combined Algorithm Selection and Hyperparameter Optimization 181
Sietse Schröder, Mitra Baratchi, and Jan N. van Rijn

Local Subgroup Discovery on Attributed Network Graphs 195
Carl Vico Heinrich, Tommie Lombarts, Jules Mallens, Luc Tortike, David Wolf, and Wouter Duivesteijn

Imposing Constraints in Probabilistic Circuits via Gradient Optimization 209
Soroush Ghandi, Benjamin Quost, and Cassio de Campos

Natural Language Processing

Improving Next Tokens via Second-to-Last Predictions with Generate and Refine 223
Johannes Schneider

Detection of Large Language Model Contamination with Tabular Data 234
Benoît Ronval, Pierre Dupont, and Siegfried Nijssen

Imbalanced Data Clustering via Targeted Data Augmentation Using GMM and LLM 246
Noor Khalal, Abdallah Alaa-Eddine Djamai, Imed Keraghel, and Mohamed Nadif

Make Literature-Based Discovery Great Again Through Reproducible Pipelines 261
Bojan Cestnik, Andrej Kastrin, Boshko Koloski, and Nada Lavrač

Extracting Information in a Low-Resource Setting: Case Study on Bioinformatics Workflows 274
Clémence Sebe, Sarah Cohen-Boulakia, Olivier Ferret, and Aurélie Névéol

Vocabulary Quality in NLP Datasets: An Autoencoder-Based Framework Across Domains and Languages ... 288
Vu Minh Hoang Dang and Rakesh M. Verma

Temporal and Streaming Data

Expertise Prediction of Tetris Players Using Eye Tracking Information ... 305
Stijn J. Rotman, Gianluca Guglielmo, Boris Cule, and Michal Klincewicz

Integrating Inverse and Forward Modeling for Sparse Temporal Data from Sensor Networks ... 318
Julian Vexler, Björn Vieten, Martin Nelke, and Stefan Kramer

Bridging Spatial and Temporal Contexts: Sparse Transfer Learning ... 330
Daniel Persson, William Wahlberg, Anna Vettoruzzo, and Sławomir Nowaczyk

Meta-learning and Data Augmentation for Stress Testing Forecasting Models ... 343
Ricardo Inácio, Vitor Cerqueira, Marília Barandas, and Carlos Soares

Pragmatic Paradigm for Multi-stream Regression ... 358
Nuwan Gunasekara, Sławomir Nowaczyk, and Sepideh Pashami

Two-in-One Models for Event Prediction and Time Series Forecasting. Comparison of Four Deep Learning Approaches to Simulate a Digital Patient Under Anesthesia ... 373
Quentin Victor, Ianis Clavier, Hugo Boisaubert, Fabien Picarougne, Corinne Lejus-Bourdeau, and Christine Sinoquet

An Analysis of Temporal Dropout in Earth Observation Time Series for Regression Tasks ... 389
Miro Miranda, Francisco Mena, and Andreas Dengel

Performative Drift Resistant Classification Using Generative Domain Adversarial Networks ... 403
Maciej Makowski, Brandon Gower-Winter, and Georg Krempl

Explainable and Interpretable Data Science

Extracting Moore Machines from Transformers Using Queries and Counterexamples ... 419
Rik Adriaensen and Jaron Maene

Obtaining Example-Based Explanations from Deep Neural Networks 432
Genghua Dong, Henrik Boström, Michalis Vazirgiannis, and Roman Bresson

Relevance-Aware Algorithmic Recourse 444
Dongwhi Kim and Nuno Moniz

Expanding Polynomial Kernels for Global and Local Explanations of Support Vector Machines 456
Rikard Vinge, Stefan Byttner, and Jens Lundström

A Constrained Declarative Based Approach for Explainable Clustering 469
Mathieu Guilbert, Christel Vrain, and Thi-Bich-Hanh Dao

Author Index 485

Applications of Data Science

Credal Knowledge Tracing for Imprecise and Uncertain MCQ

Dorra Sassi(✉), Constance Thierry, and David Gross-Amblard

University of Rennes, IRISA, CNRS, Rennes, France
dorra.sassi@irisa.fr

Abstract. Online quizzes constitute a classical building block of digital learning environments. Their answers can be aggregated using Bayesian Knowledge Tracing (BKT), a widely used model for student knowledge evaluation. But traditional quiz systems force the user to provide precise answers, limiting the expression of her doubt, uncertainty or imprecision. To address these limitations, we introduce a novel approach called Credal Knowledge Tracing (CrKT), based on the Dempster-Shafer Theory. In this setting, a doubting learner is not constrained to choose randomly an answer but can choose different ones simultaneously, and can weight each of them with a level of certainty. In this work, we provide a definition of CrKT and compare it with alternative approaches. Using a multi-attempt simulation framework, our experiments demonstrate that CrKT converges more rapidly to knowledge mastery than BKT, highlighting CrKT's potential to enhance intelligent tutoring systems.

Keywords: Knowledge Tracing · Bayesian Knowledge Tracing · Credal Knowledge Tracing · Dempster-Shafer Theory · Educational Technology

1 Introduction

With the rapid development of online education and digital learning, the need for sophisticated tools to assess and track students' understanding has generated increasing interest in Knowledge Tracing (KT) models. KT aims to estimate students' mastery of specific knowledge components by analyzing their performance on various exercises over time, predicting future success on learning tasks, and guiding instructional interventions based on students' progress [1]. Among KT models [9,10,12,15], Bayesian Knowledge Tracing (BKT) is one of the most studied and widely used.

In BKT, each knowledge element is modeled using two hidden states: *mastered* and *not mastered*. These states are inferred from student scores (correct/incorrect) on exercises, with slip and guess parameters accounting for occasional errors or lucky guesses. While BKT has proven effective and interpretable, its reliance on binary evaluation of answers significantly limits its applicability in learning environments where students might provide imprecise or ambiguous responses. For example, a student in doubt may choose to give multiple and

G. Krempl et al. (Eds.): IDA 2025, LNCS 15669, pp. 3–16, 2025.
https://doi.org/10.1007/978-3-031-91398-3_1

uncertain answers rather than committing to a single one, which BKT cannot effectively handle. Liu et al. [6] addressed this limitation by introducing Fuzzy Bayesian Knowledge Tracing (FBKT) and Type-2 Fuzzy Bayesian Knowledge Tracing (T2FBKT). These models leverage fuzzy sets to represent continuous cognitive states, such as *partially mastered*, providing finer granularity. However, they primarily focus on modeling uncertainty in the relationships between cognitive states and responses and do not explicitly handle imprecise or ambiguous answers. Other advancements, such as Deep Knowledge Tracing (DKT) [10] and Attentive Knowledge Tracing (AKT) [4], utilize neural networks and attention mechanisms to enhance prediction accuracy. However, these approaches lack mechanisms to explicitly manage uncertainty or ignorance in student responses, focusing instead on modeling sequential dependencies. While fuzzy models [6] represent progress, they fall short of addressing the full spectrum of uncertainty and imprecision seen in real-world learning scenarios. This gap highlights the need for a framework which explicitly manages uncertainty, imprecision, and ambiguous responses.

In this work, we propose a novel Credal Knowledge Tracing (CrKT) approach, based on belief functions theory, to overcome these limitations. Belief Functions Theory, introduced by Dempster in 1967 [2] and formalized by Shafer in 1976 [11], offers a powerful approach for managing uncertainty and imprecision in data interpretation. Unlike traditional probability, which enforce precise probability values for individual events, belief functions provide a range of confidence levels, allowing for the representation of both uncertainty (lack of complete knowledge) and imprecision (ambiguity in information). This flexibility is particularly useful in educational contexts where student responses often reflect partial knowledge or uncertain understanding. Belief functions have been successfully applied in fields such as artificial intelligence and machine learning for tasks requiring robust handling of incomplete information [3,7]. In addition, they are particularly well-suited for modeling human responses in questionnaires, as they can naturally represent varying levels of certainty, hesitation, and ambiguity in the answers provided by individuals [13,14]. This ability to handle partial and uncertain information makes belief functions an appropriate framework for capturing the complexity of student responses in educational assessments. However, to our knowledge, they have yet to be applied to Knowledge Tracing.

The CrKT model we propose extends the BKT framework to allow for imprecise and uncertain responses from students, thus enabling a more accurate assessment of their mastery of a concept. To validate our approach, we compare the performance of CrKT with the BKT model on numerous simulations. We evaluate both models in terms of prediction accuracy for knowledge mastery and robustness to uncertainty in student responses. In summary, the contributions of this work are as follows:

- An extension of the BKT framework using belief functions to capture both uncertainty and imprecision in student responses;
- An extensive evaluation comparing CrKT with BKT in terms of prediction accuracy and robustness to uncertain responses on synthetic data.

The structure of this paper is as follows: Sect. 2 provides an overview of the related work, including traditional KT models and fuzzy approaches. Section 3 presents our CrKT approach, detailing the integration of belief functions into KT. Section 4 describes the experimental setup and results. Finally, Sect. 5 summarizes the implications of our findings and offers concluding remarks.

2 Related Work

Knowledge Tracing (KT) has become essential in educational technology for monitoring students' learning progress and predicting their performance. Various KT models have been proposed to address challenges in accurately capturing and modeling students' cognitive states. Bayesian Knowledge Tracing (BKT), introduced by Corbett and Anderson [1], laid the foundation for modeling students' mastery of knowledge components through a binary framework of mastery and non-mastery states. The Hidden Markov Model (HMM) in BKT uses observable responses (correct or incorrect) to estimate the latent cognitive state. While BKT is interpretable and computationally efficient, it is constrained by the binary evaluation of responses, which limits its application in scenarios requiring nuanced understanding of student cognition.

To address this limitation, Liu et al. [6] proposed Fuzzy Bayesian Knowledge Tracing (FBKT), which introduces fuzzy sets to capture continuous cognitive states, such as "partially mastered." FBKT improves granularity and representational richness in KT tasks. Furthermore, the Type-2 Fuzzy Bayesian Knowledge Tracing (T2FBKT) extends FBKT by addressing model uncertainty through type-2 fuzzy theory, which accounts for uncertainty in fuzzy relations between cognitive states and responses. These advancements have shown promising results in continuous score scenarios, demonstrating their ability to model fine-grained cognitive changes and manage uncertainty more effectively than traditional BKT. However, this approach, while more flexible than BKT, lacks the ability to represent ambiguity or ignorance effectively, particularly in cases where students provide incomplete or contradictory answers.

Recent hybrid KT models have attempted to enhance predictive accuracy by integrating external factors, such as temporal features, attention mechanisms, and convolutional neural networks (CNNs). For instance, Nagatani et al. [8] incorporated forgetting effects into KT models to align with cognitive theories, while Attentive Knowledge Tracing (AKT) [4] leverages attention mechanisms to dynamically capture relationships between questions and concepts. Similarly, Convolutional Knowledge Tracing (CKT) [12] combines CNNs with attention mechanisms to model temporal patterns in students' performance sequences more effectively. These hybrid approaches broaden KT's scope, but they still fall short in explicitly managing imprecision or uncertainty in responses. When students provide incomplete or ambiguous answers, these models lack the explicit ability to handle such responses, which affects their accuracy and robustness.

Despite the advancements made by fuzzy and hybrid models, most KT models focus on representing continuous states or mitigating parameter uncertainty but do not explicitly model ignorance or handle ambiguous responses. The Credal Knowledge Tracing (CrKT) model addresses these gaps by integrating belief functions theory from Dempster-Shafer [11]. Unlike fuzzy approaches that rely on membership values, CrKT allows for the allocation of belief to specific responses, combinations of responses, or the entire set of possible responses, enabling robust management of both uncertainty and ignorance. This capacity to represent and reconcile conflicting information distinguishes CrKT from its predecessors, making it a promising advancement in the field of Knowledge Tracing.

3 Our Proposal: Credal Knowledge Tracing

In this section we present our model for Credal Knowledge Tracing. We first explain how to model uncertain and imprecise answer to a multiple answer questionnaire. In a second part we extend the classical BKT model to incorporate these answers in knowledge tracing.

3.1 Modeling Imprecise and Uncertain Answers

We consider a learner facing a multiple answer question. We call $\Omega = \{\omega_1, \omega_2, \dots\}$ our *frame of discernment*, the finite set of possible individual answers ω_i. In our setting, a learner's answer can be imprecise, i.e. the answer can be $\omega_1 \wedge \omega_3$, meaning that answers ω_1 and ω_3 are chosen altogether. Hence, the set 2^Ω denotes all the possible imprecise answers to the quiz. Then, in order to express the belief of a learner on her answer, we defined a *mass function* m_Ω on 2^Ω that takes values in the range $[0, 1]$. Mass functions satisfy the normalization condition:

$$\sum_{X \in 2^\Omega} m_\Omega(X) = 1. \tag{1}$$

The mass $m_\Omega(X)$ represents the degree of belief assigned by a source of information S to the element $X \in 2^\Omega$. The higher the mass, the stronger the belief. Valuation $m_\Omega(\emptyset) = 0$ means that the world outside the frame of discernment is not considered, known as a "closed world" assumption. An element $X \in 2^\Omega$ where $m_\Omega(X) > 0$ is called a *focal element*, and the union of the focal elements constitutes the *kernel*. If only the singletons of Ω are focal elements, then m_Ω is a probability function and is referred to as a *Bayesian mass function*. We will consider specific mass functions such as:

- **Simple Support Mass Functions:** This type of mass function represents partial belief, denoted as α_{cq}, in the set X, which may be imprecise. When the source S is not fully certain, a portion of the mass $(1 - \alpha_{cq})$ is assigned to the ignorance represented by Ω. The simple support mass function is defined as follows in Eq. 2:

$$m_\Omega^{cq}(X) = \begin{cases} \alpha_{cq} & \text{if } X \in 2^\Omega \setminus \Omega, \text{ and } \alpha_{cq} \in [0, 1], \\ 1 - \alpha_{cq} & \text{if } X = \Omega, \\ 0 & \text{otherwise.} \end{cases} \tag{2}$$

Here, α_{cq} represents the degree of belief in the subset X, where c refers to the contributor (source of information) and q indicates the specific question or context. If $\alpha_{cq} = 1$, the belief is entirely focused on X, indicating no uncertainty. However, if $\alpha_{cq} < 1$, the remaining mass $(1 - \alpha_{cq})$ is distributed to the entire frame of discernment Ω, reflecting a lack of complete certainty or a state of ignorance.

- **Categorical Mass Functions:** This mass function reflects a situation where all belief is assigned to a single focal element, thus expressing total certainty, whether precise or imprecise answer. It is defined by Eq. 3:

$$m_{\Omega}(X) = \begin{cases} 1 & \text{if } X = A, A \in 2^{\Omega}, \\ 0 & \text{otherwise.} \end{cases} \tag{3}$$

  In this equation, the set A is the unique focal element, and all belief mass is assigned to it. Thus, the categorical mass function expresses complete certainty. This type of function is often used in contexts where fully reliable information is available but may vary in granularity based on the nature of the focal set.
- **Consonant Mass Functions:** The consonant mass function structures belief by assigning masses to sets that are progressively nested within each other (e.g. $\{\omega_1\} \subset \{\omega_1, \omega_2\} \subset \ldots$). This models answers such that "X, or maybe X and Y, or maybe X,Y,and Z".

Example 1. Consider a learner answering a multiple-choice question about birds, where $\Omega = \{\text{sparrow, robin, pigeon}\}$ represents the possible answers. Table 1 lists some typical mass functions and their meaning.

Table 1. Examples of Different Mass Functions

Mass Function Type	Learner	Belief Distribution
Simple Support	Learner is fairly certain the answer is *sparrow*, but acknowledges uncertainty.	$m(\{\text{sparrow}\}) = 0.8$, $m(\Omega) = 0.2$
Categorical	Learner is entirely certain the answer is *sparrow*.	$m(\{\text{sparrow}\}) = 1$,
Consonant	Learner believes the answer is likely *sparrow*, but considers the possibility of *robin* or *pigeon*.	$m(\{\text{sparrow}\}) = 0.6$, $m(\{\text{sparrow, robin}\}) = 0.3$, $m(\{\text{sparrow, robin, pigeon}\}) = 0.1$
General	Learner believes the answer is either a *sparrow* or *pigeon*, if not perhaps a *robin*, with uncertainty.	$m(\{\text{sparrow, pigeon}\}) = 0.7$, $m(\{\text{robin}\}) = 0.2$, $m(\Omega) = 0.1$

To apply mass functions to real student responses, a *Likert scale* for self-reported confidence [14] can be used. Students select an answer and rate their certainty for example from "totally uncertain" to "totally certain". Then this certainty level will be transformed into a equivalent belief (e.g. for a seven level Likert scale : "Totally uncertain" $\alpha_{cq} = 0$, "Uncertain" $\alpha_{cq} = \frac{1}{6}$, "Rather uncertain" $\alpha = \frac{2}{6}$, "Neutral" $\alpha_{cq} = \frac{3}{6}$, "Rather certain" $\alpha_{cq} = \frac{4}{6}$, "Certain" $\alpha_{cq} = \frac{5}{6}$, "Totally Certain" $\alpha_{cq} = 1$).

Then to obtain a mass function respecting the normalisation condition (1) we can for example do:

$$m(X) = \frac{1}{\sum_{Y \in 2^{\Omega}} \alpha_{cq}(Y)} \times \alpha_{cq}(X)$$

.

For instance, if a student selects "sparrow" with "Rather certain", "sparrow or robin" with an highest certainty ("Certain") but keep an hesitation (Ω):

$$m(\{\text{sparrow}\}) = \frac{1}{2} \times \frac{4}{6}, \quad m(\{\text{sparrow}, \text{robin}\}) = \frac{1}{2} \times \frac{5}{6}, \quad m(\Omega) = \frac{1}{2} \times \frac{3}{6}.$$

3.2 Computing Distance Between Answers

Let us consider the mass function $m_{learner}$ of a learner's answer. Let $m_{correct}$ be the mass function of the correct answer. In order to compare them, we use the Jousselme distance [5] d_J, which was designed to measure the dissimilarity between two mass functions. It is defined over the interval $[0, 1]$, where:

- $d_J = 0$: the mass functions are identical.
- $d_J = 1$: the mass functions are as dissimilar as possible.

This distance is based on the Jaccard index, which quantifies the similarity between sets with binary attributes. For two mass functions m_{learner} and m_{correct}, the Jousselme distance is calculated as follows:

$$d_J(m_{\text{learner}}, m_{\text{correct}}) = \sqrt{\frac{1}{2} \cdot (m_{\text{learner}} - m_{\text{correct}})^T \cdot D \cdot (m_{\text{learner}} - m_{\text{correct}})} \tag{4}$$

where D is a similarity matrix based on the Jaccard index. The similarity matrix D is defined as:

$$D_{ij} = \frac{|A_i \cap B_j|}{|A_i \cup B_j|} \tag{5}$$

where $A_i, B_j \in 2^{\Omega}$, and $|A_i \cap B_j|$ and $|A_i \cup B_j|$ are the cardinalities of their intersection and union, respectively. Hence,D_{ij} measures the degree of similarity between subsets A_i and B_j:

- $D_{ij} = 1$: $A_i = B_j$, indicating a perfect match.
- $D_{ij} = 0$: $A_i \cap B_j = \emptyset$, indicating no overlap.

This distance is specifically designed to compare mass functions within the framework of Dempster-Shafer belief theory. Unlike generic metrics such as Euclidean distance or Kullback-Leibler divergence, d_J accounts for both the cardinality of focal sets (imprecision) and the mass values (certainty) associated with these sets. This dual consideration makes it particularly suitable for the CrKT model, where responses are inherently uncertain and may involve overlapping or imprecise knowledge states. Furthermore, the Jousselme distance incorporates the Jaccard index to measure the similarity between subsets, providing a mathematically grounded way to handle the relationships between hypotheses represented in the power set 2^{Ω}. This allows d_J to reflect not only the exact matches between responses but also partial overlaps, ensuring a nuanced evaluation of correctness in student answers.

Empirically, we observed that d_J outperformed other tested similarity measures, such as cosine distance, in maintaining the robustness and accuracy of mastery predictions. Its alignment with the theoretical foundations of belief functions and its proven efficacy in capturing both certainty and imprecision strongly justify its selection as the preferred metric for CrKT.

With this in mind, a response is considered correct if the Jousselme distance between m_{learner} and m_{correct} is below a predefined threshold ϵ:

$$d_J < \epsilon \tag{6}$$

This threshold, typically set within $[0, 1]$, is chosen based on the desired tolerance for discrepancies between the learner's beliefs and the correct answer. A lower d_J value indicates greater similarity between the belief functions, ensuring that near-correct responses are appropriately recognized. We discuss its value in Sect. 4. Once correct answers are defined, we can now present our overall procedure.

3.3 The CrKT Procedure

For a given knowledge element to be learned, the classical Bayesian Knowledge Tracing (BKT) model estimates a student mastery after n quiz questions as a probability $p(L_n)$. This model has several parameters:

- $p(L_0) \in [0, 1]$: The initial probability of mastery;
- $p(T) \in [0, 1]$: the transition probability from a non-mastered to a mastered state;
- $p(S) \in [0, 1]$: the slip probability, representing the likelihood of an error even when mastery is achieved;
- $p(G) \in [0, 1]$: the guess probability, representing the likelihood of a correct answer when the knowledge element is not mastered.

Our method extends the BKT model by adding a credibilist decision procedure on mass functions. We consider the two complementary parameters:

- $\epsilon \in [0, 1]$: the threshold for the Jousselme distance;
- Ω: The set of possible individual answers (frame of discernment).

Consider a quiz at step $n > 1$. Let m_{answer} and $m_{correct}$ be the respective mass function of the learner's answer and the awaited correct one. Recall that a response is considered correct if the Jousselme distance is below the ϵ threshold.

$$\text{If } d_J(m_{\text{learner}}, m_{\text{correct}}) < \epsilon, \text{ then the response is correct.} \tag{7}$$

According to the answer, we can compute the Posterior probabilities:

- If the response is correct:

$$\text{Posterior}(L_{n-1}) = \frac{p(L_{n-1}) \times (1 - p(S))}{p(L_{n-1}) \times (1 - p(S)) + (1 - p(L_{n-1})) \times p(G)} \tag{8}$$

- If the response is incorrect:

$$\text{Posterior}(L_{n-1}) = \frac{p(L_{n-1}) \times p(S)}{p(L_{n-1}) \times p(S) + (1 - p(L_{n-1})) \times (1 - p(G))} \tag{9}$$

Finally, the mastery probability at step n is

$$p(L_n) = \text{Posterior}(L_{n-1}) + (1 - \text{Posterior}(L_{n-1})) \times p(T) \tag{10}$$

In the next section we assess experimentally this procedure.

4 Experimental Evaluation and Discussion

4.1 Simulation Setup

The aim of this experiment is to evaluate the performance of the proposed **CrKT** approach, by simulating the responses of students. These simulated responses are modeled using various mass functions to capture the uncertainty and imprecision present in real-world learning environments. The results are compared with the traditional **BKT** model.

Experimental Setting. Simulations were conducted using Python on a standard computational setup, ensuring reproducibility across different environments. All scripts, notebooks, and associated resources are made publicly available for transparency and replication purposes[1]. As simulation parameters, we use 100 simulated learners, quizzes with 100 questions, each multiple-choice question having 3 options $\Omega = \{\omega_1, \omega_2, \omega_3\}$, with only one unique correct answer, say ω_1.

In each round of the simulation, the learner's answers are modeled by different belief mass functions that represent varying levels of certainty and precision in their answers. The learning process results were averaged across all learners to produce the overall mastery progression curves. The belief mass functions used were: **Consonant**, **Simple Support**, **General**, and **Categorical**.

The parameter values for the CrKT model were estimated using a numerical optimization approach. We generated synthetic response data for multiple simulated learners, based on predefined parameters: initial knowledge probability ($p(L_0)$), learning probability ($p(T)$), slip probability ($p(S)$), and guess probability ($p(G)$). These simulated responses were then used to optimize the parameters

[1] https://gitlab.inria.fr/dsassi/crkt.git.

by minimizing the error between the predicted and observed probabilities of correct responses. The optimization process was carried out using the L-BFGS-B algorithm, implemented in `scipy.optimize.minimize`, which is well-suited for bounded numerical optimization. The algorithm iteratively adjusted the parameter estimates to minimize the total squared error, leveraging gradient-based techniques while respecting the constraints $p(L_0), p(T), p(S), p(G) \in [0, 1]$. This approach ensures the parameter values are not only accurate but also realistic for modeling student learning behaviors.

4.2 Experimental Methodology of Mass Function Generation

This section explains the generation of belief mass functions used in the simulation. Each type of mass function is designed to represent distinct patterns of uncertainty, precision, and belief allocation.

Consonant Mass Function. The consonant mass function ensures that its focal elements are nested, meaning subsets are entirely included within larger subsets.

- **Selection of inclusion chains:** From the power set 2^{Ω}, non-empty subsets are selected. Chains of nested subsets (e.g., $\{\omega_1\} \subset \{\omega_1, \omega_2\} \subset \Omega$) are generated.
- **Random selection:** A chain is randomly chosen from the set of all possible inclusion chains.
- **Mass assignment:** A Dirichlet distribution is used to assign normalized belief masses to the subsets within the chosen chain. The Dirichlet distribution ensures that the total belief sum equals 1.

This function reflects scenarios where uncertainty is concentrated progressively within larger subsets.

Simple Support Mass Function. The simple support mass function assigns belief mass primarily to:

- A single subset A, representing a specific hypothesis.
- The frame of discernment Ω, representing ignorance.

- **Random subset selection:** A subset A is randomly selected from 2^{Ω}, excluding the empty set.
- **Mass distribution:** A random weight $\omega_{cq} \in [0, 1]$ is generated. Belief is assigned as:

$$m(A) = \omega_{cq}, \quad m(\Omega) = 1 - \omega_{cq}.$$

This function represents a focused belief with some allowance for ignorance.

General Mass Function. The general mass function provides maximum flexibility by allowing belief to be distributed across multiple focal elements without structural constraints.

- **Subset selection:** Randomly selects k focal elements from 2^{Ω}, excluding the empty set. The number k is also chosen randomly.
- **Mass assignment:** A Dirichlet distribution assigns belief masses to the chosen focal elements. Additionally, a small mass is allocated to Ω to account for residual ignorance.

This function captures highly variable scenarios where uncertainty and precision fluctuate across subsets.

Categorical Mass Function. The categorical mass function represents the most deterministic case:

- Belief is entirely assigned to a single subset A, chosen randomly from $2^{\Omega} \setminus \{\emptyset\}$.

This function models scenarios where there is complete certainty in a single hypothesis (i.e. traditional quiz with unique answer).

4.3 Model Parameters

To ensure comparability, the initial parameters for the CrKT and BKT models in experimentation were standardized as follows:

- **Initial Mastery Probability** ($p(L_0)$): Set to 0.2, indicating a low initial probability of mastery.
- **Transition Probability** ($p(T)$): Fixed at 0.1, reflecting incremental mastery after a successful attempt.
- **Slip Probability** ($p(S)$): Set at 0.3, capturing the likelihood of incorrect responses despite mastery.
- **Guess Probability** ($p(G)$): Set to 0 for CrKT to prioritize credal reasoning, while BKT uses 0.333 to account for one random correct responses within $\Omega = \{\omega_1, \omega_2, \omega_3\}$.

4.4 Simulation Results

The results of our simulation are presented in Fig. 1, which illustrates the comparison between different mass functions and the BKT model. The figure displays the average mastery probability across all learners, comparing the performance of **Consonant Mass Function**, **Simple Support Mass Function**, **General Mass Function**, **Categorical Mass Function**, and the **BKT baseline**. The following findings were observed from the simulation results:

- **Consonant Mass Function:** this function showed the slowest progression toward mastery. This function's slower progression is beneficial in cases where gradual knowledge consolidation is preferred. It may also be suitable for learners who need additional time to assimilate knowledge.

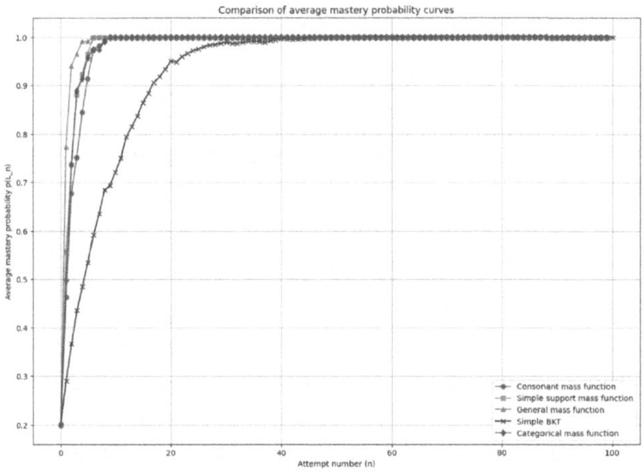

Fig. 1. Comparison of average mastery probability curves across different mass functions and the BKT model, showing faster convergence to mastery.

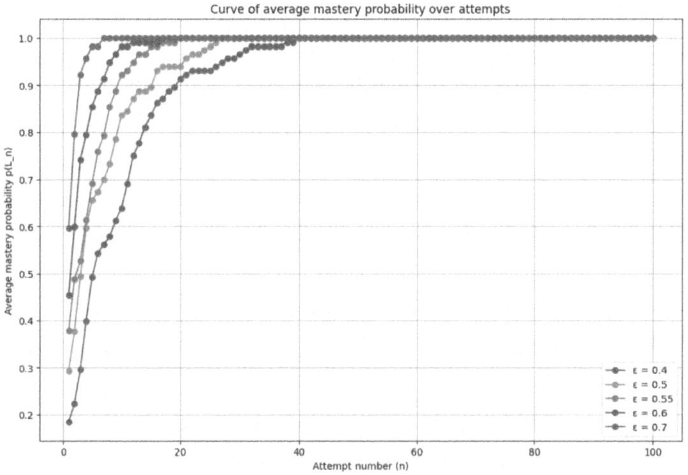

Fig. 2. Tradeoff for ϵ threshold: increasing ϵ speeds up learning, but allows permissive answers.

- **Simple Support Mass Function:** demonstrated rapid convergence to mastery, akin to the BKT model. However, the Simple Support Mass Function allows for faster adjustment and is useful when learners have partial knowledge that can be easily strengthened over time.
- **General Mass Function:** showed a mastery progression similar to the Simple Support Mass Function. Its flexibility in belief assignment makes it useful for learners who exhibit fluctuating understanding across attempts.
- **Categorical Mass Function:** achieved the fastest convergence to mastery, indicating the effectiveness of a function that commits to certainty once suffi-

cient evidence is gathered. This function is ideal for learners who demonstrate quick mastery once key concepts are well understood.
- **Baseline BKT Model:** the BKT model exhibited the slowest progression to mastery, as only exact correct answers are accepted. This slower progression highlights BKT's limitation in handling uncertainty and intermediate stages of learning effectively.

Figure 2 shows the impact of threshold ϵ on the convergence rate. The choice of ϵ, the threshold for determining correctness based on the Jousselme distance, is critical in balancing the strictness and flexibility of the model. A lower ϵ enforces stricter criteria for classifying responses as correct, which slows the learner's perceived progress but ensures higher precision. Conversely, a higher ϵ allows more responses to be considered correct, leading to faster mastery updates but potentially at the cost of accuracy. After testing various ϵ values, we observed that overly strict thresholds could hinder meaningful progress, while overly lenient ones might fail to reflect actual mastery. To address this trade-off, we selected a moderate ϵ value that balances the convergence rate and correctness precision. Furthermore, instead of assigning the belief mass exclusively to the correct answer ($m_{\text{correct}}(\{\omega_1\}) = 1$), we opted for an imprecise belief mass function to better represent uncertainty and ambiguity in learners' responses. Specifically, we distributed the mass across plausible hypotheses as follows:

$$m(\{\omega_1\}) = 0.7, \quad m(\{\omega_1, \omega_2\}) = 0.15, \quad m(\{\omega_1, \omega_3\}) = 0.15.$$

This adjustment acknowledges that learners may hold partial or uncertain knowledge, enhancing the model's robustness and realism in capturing nuanced cognitive states.

5 Conclusion and Future Work

Our approach combines belief theory with the BKT model, enabling nuanced assessments of learner mastery. The use of diverse mass functions highlights the flexibility of credal reasoning in adaptive learning contexts. Simulation results confirm the effectiveness of *CrKT* in handling uncertainty and imprecision in student responses. Unlike *BKT*, which operates on binary responses (correct/incorrect), *CrKT* leverages belief functions to represent a broader spectrum of cognitive states, including partial knowledge and uncertain understanding. This flexibility allows *CrKT* to handle real-world learning scenarios more effectively, where students' knowledge is often incomplete or uncertain.

In particular:

- *CrKT* demonstrated robustness in scenarios with imprecise responses, offering a nuanced way to represent both certainty and plausibility.
- The *CrKT* model avoids the guessing mechanism present in BKT, which is a significant advantage when dealing with uncertain responses.
- *CrKT* provides a finer granularity in tracking student progress by representing a continuous spectrum of mastery levels.

Implementing CrKT in real-world learning systems could provide practical insights into its effectiveness and potential for enhancing learner personalization. As a future work, we plan to evaluate our method on real students both in a controlled setting and on crowdsourcing platforms. In future real-world experiments, we plan to address the challenge of determining optimal parameter values through a combination of cold-start strategies and co-evolutionary approaches. By conducting iterative experiments and analyzing learner interactions, we aim to refine the calibration of key parameters such as ϵ, ensuring better adaptation to real educational settings. We also envision equipping the frame of discernment Ω with a semantic distance, to better capture the learned concepts.

Disclosure of Interests. The authors have no competing interests to declare that are relevant to the content of this article.

References

1. Corbett, A.T., Anderson, J.R.: Knowledge tracing: modeling the acquisition of procedural knowledge. User Model. User-Adap. Inter. **4**, 253–278 (1994)
2. Dempster, A.P.: Upper and lower probability inferences based on a sample from a finite univariate population. Biometrika **54**(3–4), 515–528 (1967)
3. Ellouze, A., Kahouli, O., Ksantini, M., Alsaif, H., Aloui, A., Kahouli, B.: Artificial intelligence-based diabetes diagnosis with belief functions theory. Symmetry **14**(10), 2197 (2022)
4. Ghosh, A., Heffernan, N., Lan, A.S.: Context-aware attentive knowledge tracing. In: Proceedings of the 26th ACM SIGKDD International Conference on Knowledge Discovery & Data Mining, pp. 2330–2339 (2020)
5. Jousselme, A.L., Grenier, D., Bossé, É.: A new distance between two bodies of evidence. Inf. Fusion **2**(2), 91–101 (2001)
6. Liu, F., Hu, X., Bu, C., Yu, K.: Fuzzy bayesian knowledge tracing. IEEE Trans. Fuzzy Syst. **30**(7), 2412–2425 (2021)
7. Liu, Z., Letchmunan, S.: Representing uncertainty and imprecision in machine learning: a survey on belief functions. J. King Saud Univ.-Comput. Inf. Sci., 101904 (2024)
8. Nagatani, K., Zhang, Q., Sato, M., Chen, Y.Y., Chen, F., Ohkuma, T.: Augmenting knowledge tracing by considering forgetting behavior. In: The World Wide Web Conference, pp. 3101–3107 (2019)
9. Pandey, S., Karypis, G.: A self-attentive model for knowledge tracing. arXiv preprint arXiv:1907.06837 (2019)
10. Piech, C., et al.: Deep knowledge tracing. Adv. Neural Inf. Process. Syst. **28** (2015)
11. Shafer, G.: A Mathematical Theory of Evidence. Princeton University Press (1976)
12. Shen, S., et al.: Convolutional knowledge tracing: modeling individualization in student learning process. In: Proceedings of the 43rd International ACM SIGIR Conference on Research and Development in Information Retrieval, pp. 1857–1860 (2020)
13. Thierry, C., Martin, A., Dubois, J.C., Gall, Y.L.: Validation of smets' hypothesis in the crowdsourcing environment. In: Belief Functions: Theory and Applications: 6th International Conference, BELIEF 2021, Shanghai, China, 15–19 October 2021 (2021) Proceedings 6. pp. 259–268. Springer (2021)

14. Thierry, C., Martin, A., Le Gall, Y., Dubois, J.C.: Modeling evolutionary responses in crowdsourcing mcq using belief function theory. Procedia Comput. Sci. **225**, 2575–2584 (2023)
15. Tong, S., et al.: Structure-based knowledge tracing: an influence propagation view. In: 2020 IEEE International Conference on Data Mining (ICDM), pp. 541–550. IEEE (2020)

Development of Models to Quantify Training Load in Outdoor Running Using Inertial Sensors

Bouke L. Scheltinga[1(✉)], Jasper Reenalda[1], Jaap H. Buurke[1], and Joost N. Kok[2]

[1] University of Twente, Biomedical Signals and Systems group, Enschede, The Netherlands
b.l.scheltinga@utwente.nl

[2] Eindhoven University of Technology, Eindhoven, The Netherlands

Abstract. Running is a popular sport that offers health benefits but also poses a high risk of overuse injuries, often leading to a temporary or permanent cessation of running. These injuries are often caused by repetitive mechanical loading, emphasizing the need to quantify training loads to understand injury development. While such quantification is feasible in controlled settings like gait labs, capturing accurate data in natural outdoor conditions is challenging. This requires a practical sensor setup that is feasible for daily use and robust modelling approaches to relate sensor-derived data to mechanical load.

We address these challenges by quantifying biomechanical loads during outdoor running using a minimal sensor setup. A key achievement is a model that uses inertial sensors to estimate vertical ground reaction forces (GRFs) used to estimate load. Moreover, we show that runners' perceived exertion is often misaligned with sensor-derived load measurements, highlighting the importance of combining subjective and objective metrics for a holistic view of training loads. The discrepancy also underscores the need to quantify structure-specific loads, such as forces on the lower legs, to better understand their potential role in mechanical fatigue and injury risk. We present a neural network model to estimate 3D GRFs needed to quantify these structure-specific loads accurately. These results advance biomechanics with novel data methods for quantifying training loads in outdoor environments with a feasible setup for many runners, enabling large future studies to better understand injury mechanisms. The study highlights the difficulty of data analysis when moving from a controlled situation in the laboratory to a large study outside the laboratory.

Keywords: Running injuries · Training Load Quantification · Sensor Technology

1 Introduction

Running is one of the most popular sports globally. In the Netherlands, it ranks as the second most popular physical activity, with a significant portion of the

G. Krempl et al. (Eds.): IDA 2025, LNCS 15669, pp. 17–27, 2025.
https://doi.org/10.1007/978-3-031-91398-3_2

population engaging in regular running activities. However, despite its health benefits, running is often associated with a high incidence of injuries, particularly overuse injuries [16]. These injuries, usually the result of repetitive strain on musculoskeletal tissues, can temporarily force runners to stop running or even cause individuals to quit the sport [10].

Beyond the physical burden, running-related injuries also impose significant economic costs. The number of registered sports injuries in the Netherlands was 4.3 million in 2021. In 2022, this number increased to 5.1 million. This resulted in 220 million euros (an average of 2100 euros per injury) in hospital costs. In addition, the associated costs of absenteeism in 2022 increased to an additional 270 million euros [7].

Beyond the direct costs of running-related injuries, it is essential to consider the broader health benefits of running. Regular physical activity, including running, is crucial in preventing lifestyle-related diseases such as obesity, type 2 diabetes, and cardiovascular diseases [14]. By encouraging safe and sustainable running practices, injury prevention efforts reduce the financial burden associated with sports injuries and contribute to overall improvements in public health. Reducing running injuries allows people to maintain an active lifestyle, which helps mitigate healthcare costs related to chronic conditions. Therefore, promoting safe running habits has a dual benefit: it minimizes injury-related expenses while supporting long-term health and well-being.

Unfortunately, the exact mechanisms underlying running-related injuries remain poorly understood [4]. These injuries are multifactorial, influenced by both non-modifiable and modifiable risk factors [24]. Non-modifiable factors include individual biomechanical characteristics, prior injury history, and sex, while modifiable factors-such as running technique, muscle strength and bone density adapt over time. Despite extensive research, no definitive predictive model exists that reliably links specific biomechanical loads to injury risk. This gap highlights the need for real-world approach to monitoring biomechanical loads and their contribution to injury risk. The complex interplay between biomechanical loads, tissue adaptation, and risk factors over time remains a significant challenge in understanding injury development. Bridging this gap requires a deeper exploration of real-world running biomechanics and long-term load monitoring outside of controlled laboratory settings.

Overuse injuries can be seen as a mechanical fatigue phenomenon, where (biological) tissue damage occurs due to repetitive loading of the structure caused by the repetitive origin of running [8]. In the gait laboratory, GRF can be accurately measured using a force plate embedded in the floor or treadmill. During each running step, GRF peaks at approximately 2.5 to 2.8 times body weight [5]. These forces are fundamental to the dynamics of running, determining the running direction and velocity. Although not always consistent [6,15], GRF, or derivatives, can be linked to injury risk [11], or used to estimate load [12].

To increase understanding of the relationship between load and injuries, it is important to study the relationship between load and injuries over a longer period [3]. Therefore, the challenge lies in accurately capturing the training load in the natural outdoor environments where runners train and compete [2].

While objective biomechanical measurements provide quantifiable data on running loads, it is unclear whether runners' perception of load aligns with these objective metrics. Understanding this relationship is crucial, as training decisions are often based on subjective sensations rather than biomechanical data, which may increase the risk of overuse injuries if actual mechanical loads are underestimated or misinterpreted.

In this paper, we propose moving from indoor to outdoor load monitoring in runners to increase understanding of injury mechanisms.

First, we fit a physical model that estimates GRFs in the vertical direction using three inertial sensors. This setup can estimate the GRF during outdoor running with good repeatability. However, the estimated load does not match the load experience of the runner.

Then, we estimate the GRF in three dimensions by constructing a novel neural network model with data from only three inertial sensors as input. With a three-dimensional GRF estimation model. The proposed quantification of 3-dimensional GRF during running will enable future studies to relate these forces to different injuries. Furthermore, this 3D GRF model can be combined with musculoskeletal models to further investigate tissue loading and injury mechanisms in greater detail.

The remainder of the paper is structured as follows: Sect. 2 introduces a physical model for estimating vertical GRF using a minimal sensor setup and assesses its accuracy and repeatability in outdoor environments. Section 3 compares objective and subjective training load measures, exploring their implications for injury prevention and optimization. Section 4 describes developing a three-dimensional neural network model to estimate GRFs in all directions and evaluates its performance in real-world applications. Finally, Sect. 5 concludes the study, discusses the implications of the findings, and outlines future research directions.

2 Outdoor Estimation of the External Load Using a One-Dimensional Physical Model

To develop a model for estimating GRF, we first collected data. Twelve experienced runners participated, equipped with Xsens inertial sensors (sampling rate of 240 Hz) and tested on a force plate treadmill (2048 Hz). Each participant completed 90-s trials at three different running speeds (10, 12, and 14 km/h) and three stride frequencies (90%, 100%, and 110% of their preferred stride frequency) [22].

Inertial sensors record accelerations, which can be used to derive forces for rigid bodies using Newton's second law. For human motion, this requires knowledge of all body segments accelerations and masses:

$$eGRF = (m_b \times g) + \sum_{i=1}^{N} m_b \times WF_i \times (a_{z,i}), \tag{1}$$

where m_b is the subject's body mass, g is gravitational acceleration, WF_i is the weight factor for the sensor i, and $a_{z,i}$ is low-pass filtered sensor acceleration in the global vertical direction, without a gravity component.

Various three-sensor configurations, with sensors on different locations on the body, were tested. For each configuration, the cutoff frequency of the filter and the weight factor for the corresponding acceleration were optimized. For this optimization, root mean squared error (RMSE) between estimated and measured GRF during ground contact was used as the loss function.

The configuration with sensors on the lower legs (tibias) and rear side of the hip (pelvis) performed best, with an RMSE of 0.179 BW and a strong correlation (Pearson's r=0.98) with the measured force. The model achieves greater accuracy than previous work (e.g., 0.27 BW [19,25]). Estimated and measured GRF profiles are shown in Fig. 1. The three-sensor method achieves a low RMSE and a strong correlation with the measured GRF, demonstrating its accuracy and precision.

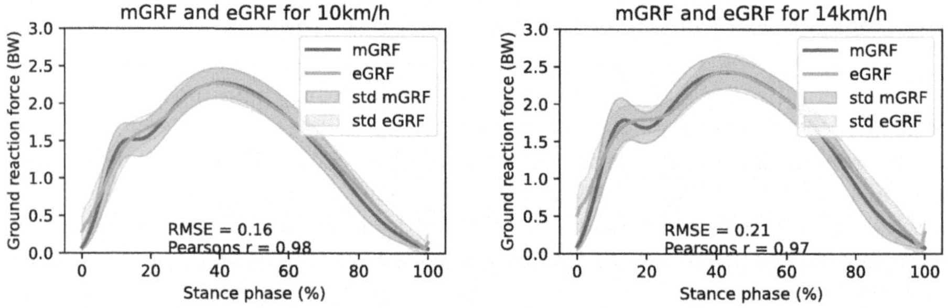

(a) 10 km/h (b) 14km/h

Fig. 1. Measured (blue) and estimated (orange) ground reaction force (GRF) for 10 and 14 km/h. Estimations based on the pelvis-tibias configuration averaged over all subjects. (Color figure online)

Next, the repeatability of the GRF estimation algorithm during outdoor running was investigated. Twelve well-trained participants completed three running sessions on different days on an athletics track, instrumented with inertial sensors on the lower legs and pelvis. The goal was to assess the algorithm's repeatability across three real-world sessions to bridge the gap between laboratory and outdoor measurements.

Results demonstrated high repeatability, with an intra-class correlation coefficient (ICC2,k) of 0.86 for peak GRF. To assess the consistency of waveform shape across days, we calculated the ICC for skewness (0.92) and kurtosis (0.94), confirming strong agreement. The root mean squared difference (RMSD) between days was 0.08 body weight, and Pearson correlation coefficients exceeded 0.99, indicating strong consistency in GRF estimates across sessions.

The GRF profiles measured over different days are shown in Fig. 2 for two subjects. To our knowledge, this is the first study looking into the day-to-day outdoor repeatability of the estimation of GRF, showing the potential to use this algorithm daily outdoors [20].

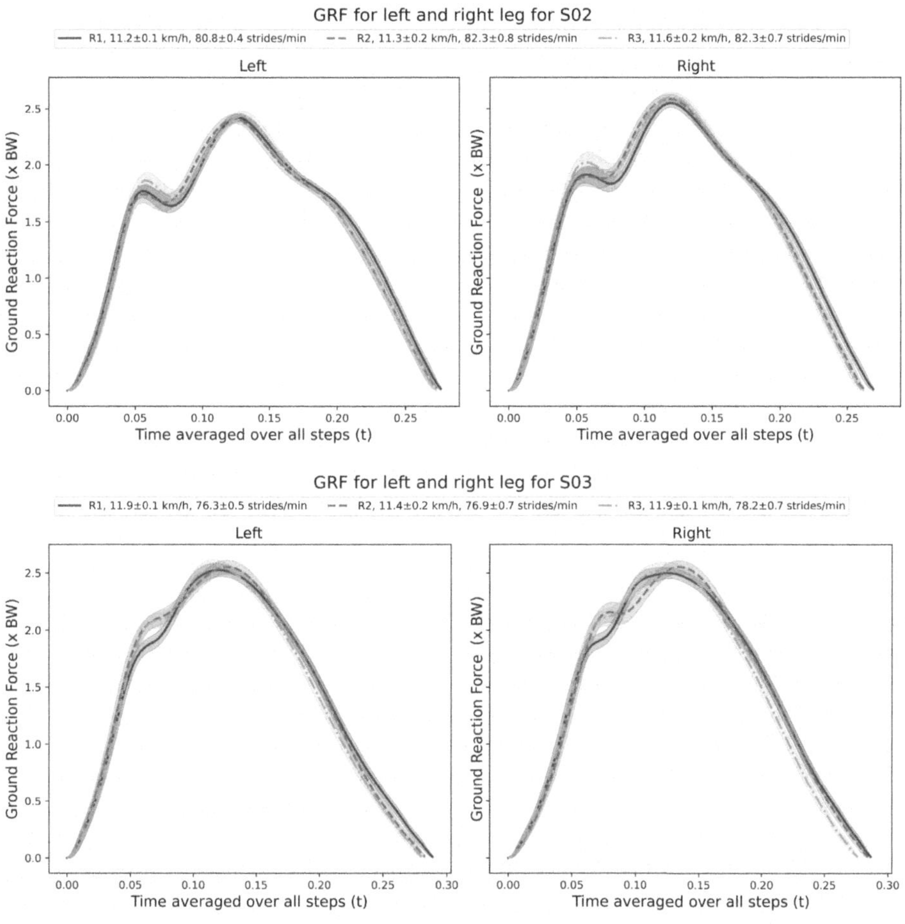

Fig. 2. Estimated ground reaction force (GRF) waveform for S02 (top) and S03 (bottom) for left and right leg. These participants had the lowest root mean squared difference. R1, R2 and R3 represent the runs on the three different days.

While these findings demonstrate the feasibility of estimating external load in outdoor conditions, they do not account for how runners perceive training load. The following section examines the relationship between biomechanical load quantification and subjective effort perception, exploring whether runners' self-reported exertion aligns with objective biomechanical data.

3 Relating Objective and Subjective Load

While biomechanical models provide an objective estimate of external load, runners often rely on total running distance and subjective measures such as perceived exertion to guide training decisions. The relationship between these subjective and objective measures remains unclear. In this section, we investigate whether the estimated biomechanical load correlates with the self-reported perceptions of exertion by runners.

Rate of Perceived Exertion (RPE) is a widely used subjective measure of exercise intensity, where individuals rate their exertion level based on physical sensations such as breathing effort, muscle fatigue, and overall discomfort [23]. The RPE scale used in this work ranges from 1 (no exertion) to 10 (maximal exertion) and correlates with physiological markers such as heart rate [9]. RPE is frequently used in training load management due to its simplicity and accessibility; however, it remains subjective and influenced by factors such as motivation, psychological state, and external conditions [17]. While RPE provides valuable insights into an athlete's perceived workload, its relationship with objective biomechanical load remains unclear. This raises questions about its reliability for injury prevention and performance optimization.

So far, only one study has investigated the effect of estimated GRF peak force on running injuries over an extended period [12]. Kiernan and colleagues applied the following function to estimate cumulative damage (D):

$$D \propto n\sigma^b, \tag{2}$$

where D is proportional to the number of times (n) the vertical peak GRF (σ) is applied, raised to the power of a constant (b). This function is based on the Palmgren-Miner rule for mechanical fatigue, in which b represents the slope of the stress-life curve [8]. Due to the exponential nature of this relationship, the choice of b has a significant impact on the estimated damage. Although a value of b $=$ 9 was derived from in vivo measurements [1], its suitability to model running-related loading remains uncertain and requires further validation.

Our study aims to estimate cumulative damage (D) across three different training sessions in real-world running conditions and examine its relationship with RPE. By comparing sensor-derived biomechanical load with self-reported exertion, we seek to determine whether runners accurately perceive the mechanical stress their bodies experience during training.

On three different days, participants performed an endurance run, a submaximal effort and an interval training. Cumulative load was calculated using Eq. (2) with parameter b $=$ 9, in accordance with Kiernan et al. [12]. After each run, participants reported their RPE.

Despite variations in RPE across the sessions, no significant differences in cumulative damage were detected using inertial sensors. This suggests that subjective exertion does not strongly correlate with objective biomechanical load. This could mean that (1) runners are not capable of estimating training load, (2) quantified training load is not related to perceived load, or (3) the method to quantify load using GRF is not suitable. Relating objective load and subjective load remains an interesting topic for future research, as with our experiment we could not relate the two.

These findings highlight the complexity of assessing training load and suggest that a combined approach, integrating both objective and subjective measures, may be needed to develop a more accurate view of injury risk. The weak correlation between perceived exertion and objective biomechanical load suggests that single-metric assessments may not fully capture training stress. A more detailed approach is needed to better quantify biomechanical load in real-world conditions. To improve external load estimation, we introduce a novel neural network model capable of estimating three-dimensional GRFs during outdoor running, allowing for a more comprehensive evaluation of structure-specific loads and injury risk.

4 Outdoor Estimation of the External Load Using a Three-Dimensional Neural Network Model

Current inertial sensor-based GRF estimation methods primarily focus on vertical forces, limiting their ability to capture the full complexity of human movement. However, running occurs in three dimensions, with forces acting in the vertical, anterior-posterior (AP), and mediolateral (ML) directions. Accurately estimating 3D GRFs is essential for assessing structure-specific loads and understanding injury mechanisms [18].

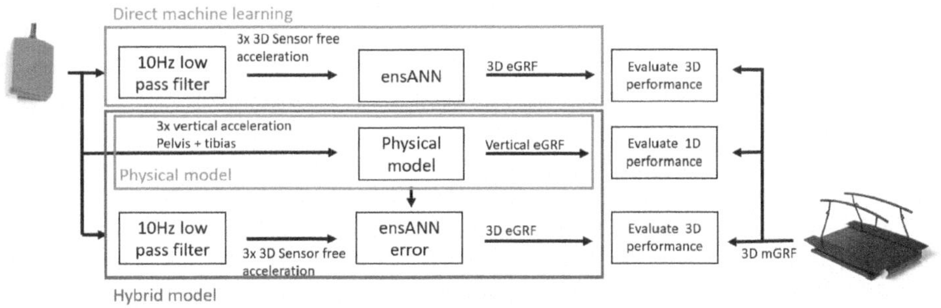

Fig. 3. Overview of the different models. An ensemble MLP (ensANN) was used to estimate 3-dimensional GRF directly. A hybrid model was created by adding the physical estimate as additional input for the MLP. The models' estimates were compared with measured GRF (mGRF) to evaluate the performance.

Twelve experienced runners were equipped with Xsens inertial sensors (sampling at 240 Hz) and tested on a force plate treadmill (2048 Hz). Each participant completed 90-s trials at three different running speeds (10, 12, and 14 km/h) and three stride frequencies (90%, 100%, and 110% of their preferred stride frequency).

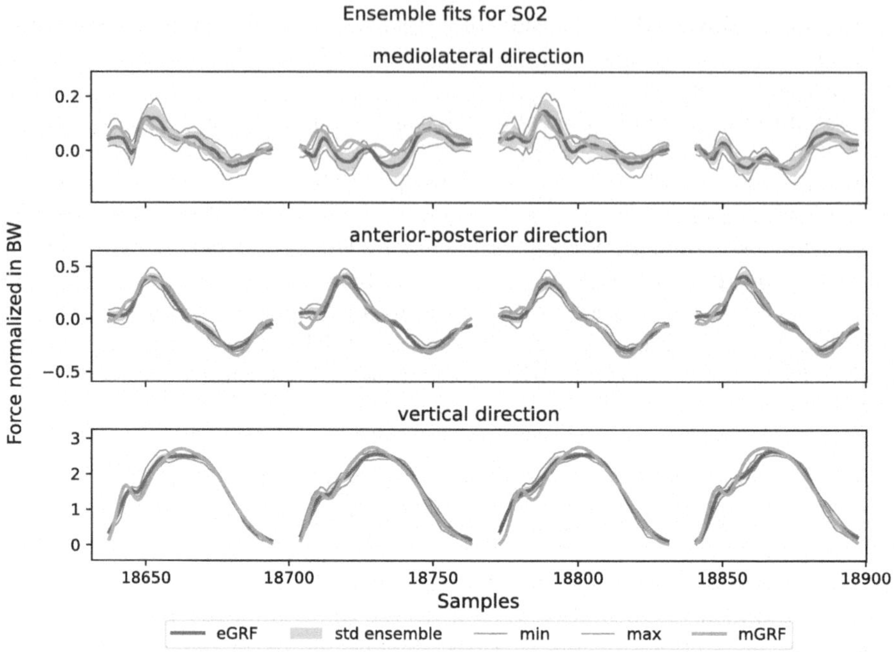

Fig. 4. The estimated ground reaction curves for the different axes during stance from multilayer perceptron at 12 km/h preferred stride frequency. The shading indicates the average over the ensemble members plus/minus one standard deviation. Also, the maximum and minimum values are plotted. A fixed gap in the data replaces the stance phase. Note that the y-axis is not the same scale for the different sub-plots.

Next, acceleration data from the inertial sensors placed on the pelvis and lower legs were used as input for multilayer perceptrons (MLPs) to estimate 3D GRFs. The sensor-free acceleration data were filtered using a 3rd order bidirectional Butterworth low-pass filter with a cutoff frequency of 10 Hz.

The MLPs consisted of two fully connected layers with 100 neurons and ReLU activation functions. This architecture was chosen based on previous findings demonstrating its effectiveness in estimating GRF from inertial sensor orientation data [25], while remaining computationally efficient. Mean squared error was used as the loss function, and adaptive moment estimation (Adam) was used as the update rule [13]. Models were trained for 1,000 epochs, with a batch size of 250. An early stopping criterium was implemented as the validation loss did not decrease for more than 100 epochs.

To improve estimation accuracy, we developed a hybrid model by incorporating vertical GRF estimates from a physics-based model as an additional input feature (Fig. 3). Furthermore, we implemented an ensemble learning approach, training multiple MLPs on different validation and training splits. The final ensemble estimation was obtained by averaging the predictions of all individual MLP models [21].

Incorporating the estimate from a physical model, the hybrid model demonstrated superior performance with a relative root mean square error (rRMSE) of 6.8% in the vertical direction, slightly better than the direct model's 7.3%. This accuracy was confirmed by a strong Pearson's correlation coefficient of 0.97 with measured GRF, indicating the hybrid model's higher precision for estimating vertical GRFs. Despite these advancements, the models showed limitations in accurately estimating mediolateral forces (Fig. 4).

This model represents a significant advancement in applying wearable sensor technology for biomechanical load assessment in real-world environments. By leveraging ensemble learning, this approach improves generalization and robustness, paving the way for more accurate injury risk assessment and performance monitoring outside laboratory settings.

5 Conclusions and Future Work

The findings of this study contribute to biomechanics by introducing novel data-driven methods for quantifying training loads in outdoor environments using a practical and feasible sensor setup. This approach enables large-scale studies to better understand injury mechanisms in real-world running conditions. Additionally, our results highlight the challenges of transitioning from controlled laboratory experiments to field-based studies, emphasizing the need for robust data processing techniques and model validation in diverse environments.

While external load quantification provides valuable insight into injury mechanisms, individual susceptibility to injuries remains highly variable. Factors such as biomechanical differences, running experience, training history, genetic predispositions, and recovery capacity play a significant role in determining how an individual responds to mechanical stress. Two runners exposed to the same external load may experience different physiological adaptations, with one remaining injury-free and the other developing an overuse injury. Understanding this variability requires integrating personalized load monitoring that accounts for individual factors.

Future research should focus on refining models to estimate load, to better capture the nuances of different running styles, footwear, terrain and environmental conditions, thereby enhancing their applicability across different athletic populations and clinical settings. Expanding these models to other sports and activities could further broaden their impact. In addition, integrating individual risk factors will be critical for making training load assessments more personalized and clinically relevant.

To establish strong predictive relationships between workload, injury risk, and other biomechanical factors, large-scale longitudinal data collection is essential. Running events present an ideal opportunity for this data acquisition, as they bring together diverse groups of athletes with different levels of experience, training regimens, and injury histories. By collaborating with event organizers, it is possible to collect extensive real-world data, capturing workload metrics, training habits, environmental influences, and injury occurrences on a scale not easily achieved in controlled studies.

Using such large data sets will enable the development of more precise injury risk models, ultimately supporting targeted interventions and personalized injury prevention strategies. We are actively engaged with organizers of events to explore these possibilities and initiate broader data collection efforts.

Acknowledgments. This work is part of the research project Perfect Fit, which is supported by several funders organized by the Netherlands Organization for Scientific Research (NWO), program Commit2Data - Big Data & Health (project number 628.011.211). Besides NWO, the funders include the Netherlands Organization for Health Research and Development (ZonMw), Hartstichting, the Ministry of Health, Welfare and Sport (VWS), Health Holland, and the Netherlands eScience Center.

Disclosure of Interests. The authors have no competing interests to declare that are relevant to the content of this article.

References

1. Baggaley, M., Brent Edwards, W.: Effect of running speed on achilles tendon injury potential. Med. Sci. Sports Exerc. **49**(5S), 139 (2017). https://doi.org/10.1249/01.mss.0000517210.29165.6c
2. Benson, L.C., Räisänen, A.M., Clermont, C.A., Ferber, R.: Is this the real life, or is this just laboratory? A scoping review of IMU-based running gait analysis. Sensors **22**(5), 1722 (2022). https://doi.org/10.3390/S22051722. https://www.mdpi.com/1424-8220/22/5/1722
3. Bertelsen, M.L., et al.: A framework for the etiology of running-related injuries. Scand. J. Med. Sci. Sports **27**(11), 1170–1180 (2017). https://doi.org/10.1111/sms.12883
4. Bittencourt, N.F., Meeuwisse, W.H., Mendonça, L.D., Nettel-Aguirre, A., Ocarino, J.M., Fonseca, S.T.: Complex systems approach for sports injuries: moving from risk factor identification to injury pattern recognition - Narrative review and new concept. Br. J. Sports Med. **50**(21), 1309–1314) (2016). https://doi.org/10.1136/bjsports-2015-095850
5. Cavanagh, P.R., Lafortune, M.A.: Ground reaction forces in distance running. J. Biomech. **13**(5), 397–406 (1980). https://doi.org/10.1016/0021-9290(80)90033-0
6. Ceyssens, L., Vanelderen, R., Barton, C., Malliaras, P., Dingenen, B.: Biomechanical risk factors associated with running - related injuries: a systematic review. Sports Med. **49**(7), 1095–1115 (2019). https://doi.org/10.1007/s40279-019-01110-z
7. Van der Does, H., Stam, C., Valkenberg, H.: Sportblessures in Nederland Cijfers 2022. Technical report, VeiligheidNL (2023)
8. Edwards, W.B.: Modeling overuse injuries in sport as a mechanical fatigue phenomenon. Exerc. Sport Sci. Rev. **46**(4), 224–231 (2018). https://doi.org/10.1249/JES.0000000000000163. https://journals.lww.com/00003677-201810000-00005
9. van Erp, T., Foster, C., de Koning, J.J.: Relationship between various training-load measures in elite cyclists during training, road races, and time trials. Int. J. Sports Physiol. Perform. **14**(4), 493–500 (2019). https://doi.org/10.1123/ijspp.2017-0722
10. Fokkema, T., et al.: Reasons and predictors of discontinuation of running after a running program for novice runners. J. Sci. Med. Sport **22**(1), 106–111 (2019)

11. Johnson, C.D., Tenforde, A.S., Outerleys, J., Reilly, J., Davis, I.S.: Impact-related ground reaction forces are more strongly associated with some running injuries than others. Am. J. Sports Med. **48**(12), 3072–3080 (2020). https://doi.org/10.1177/0363546520950731
12. Kiernan, D., et al.: Accelerometer-based prediction of running injury in National Collegiate Athletic Association track athletes. J. Biomech. **73**, 201–209 (2018). https://doi.org/10.1016/j.jbiomech.2018.04.001
13. Kingma, D.P., Ba, J.L.: Adam: a method for stochastic optimization. In: 3rd International Conference on Learning Representations, ICLR 2015 - Conference Track Proceedings. International Conference on Learning Representations, ICLR (2015)
14. Lee, D.C., Pate, R.R., Lavie, C.J., Sui, X., Church, T.S., Blair, S.N.: Leisure-time running reduces all-cause and cardiovascular mortality risk. J. Am. Coll. Cardiol. **64**(5), 472–481 (2014). https://doi.org/10.1016/j.jacc.2014.04.058
15. Matijevich, E.S., Branscombe, L.M., Scott, L.R., Zelik, K.E.: Ground reaction force metrics are not strongly correlated with tibial bone load when running across speeds and slopes: Implications for science, sport and wearable tech. PLoS ONE **14**(1), 1–19 (2019). https://doi.org/10.1371/journal.pone.0210000
16. Nielsen, R., Ramskov, D., Blacket, C.T., Malisoux, L.: Running-related injuries among more than 7000 runners in 87 different countries: the garmin-RUNSAFE running health study. J. Orthopaedic Sports Phys. Therapy **54**(2), 133–141 (2024). https://doi.org/10.2519/JOSPT.2023.11959
17. Noakes, T.D.: Fatigue is a brain-derived emotion that regulates the exercise behavior to ensure the protection of whole body homeostasis. Front. Physiol. **3**(April), 1–13 (2012). https://doi.org/10.3389/fphys.2012.00082
18. Novacheck, T.: The biomechanics of running. Gait Post. **7**(1), 77–95 (1998). https://doi.org/10.1016/s0966-6362(97)00038-6
19. Ohtaki, Y., Sagawa, K., Inooka, H.: A method for gait analysis in a daily living environment by body-mounted instruments. JSME Int J., Ser. C **44**(4), 1125–1132 (2001). https://doi.org/10.1299/jsmec.44.1125
20. Scheltinga, B.L., Buurke, J.H., Kok, J.N., Reenalda, J.: Repeatability of vertical ground reaction force estimation during running on the athletics track on 3 different days. J. Appl. Biomech. **-1**(aop), 1–12 (2025). https://doi.org/10.1123/JAB.2024-0126
21. Scheltinga, B.L., Kok, J.N., Buurke, J.H., Reenalda, J.: Estimating 3D ground reaction forces in running using three inertial measurement units. Front. Sports Active Living **5**, 1176466 (2023). https://doi.org/10.3389/fspor.2023.1176466
22. Scheltinga, B.L., Usta, H., Reenalda, J., Buurke, J.H.: Estimating vertical ground reaction force during running with 3 inertial measurement units. J. Biomed. Eng. Biosci. **9**, 31–38 (2022). https://doi.org/10.11159/jbeb.2022.006
23. Williams, N.: The borg rating of perceived exertion (rpe) scale. Occup. Med. **67**(5), 404–405 (2017). https://doi.org/10.1093/occmed/kqx063
24. Windt, J., Gabbett, T.J.: How do training and competition workloads relate to injury? The workload - injury aetiology model. Br. J. Sports Med. **51**(5), 428–435 (2017). https://doi.org/10.1136/bjsports-2016-096040
25. Wouda, F.J., et al.: Estimation of vertical ground reaction forces and sagittal knee kinematics during running using three inertial sensors. Front. Physiol. **9**, 1–14 (2018). https://doi.org/10.3389/fphys.2018.00218

Estimating the Learning Capacity of Bacterial Metabolic Networks

Bastien Mollet[1(✉)], Paul Ahavi[2], Antoine Cornuéjols[1], Jean-Loup Faulon[2], Evelyne Lutton[1,3], and Alberto Tonda[1,3]

[1] EKINOCS, UMR 518 MIA-PS, INRAE, AgroParisTech, University of Paris-Saclay, 91120 Palaiseau, France
bastien.mollet@inrae.fr

[2] MICALIS Institute, INRAE, AgroParisTech, University of Paris-Saclay, 78350 Jouy-en-Josas, France

[3] UAR 3611 ISC-PIF, CNRS, Paris, France

Abstract. Biocomputing has emerged as a promising field with the potential for energy-efficient computation, but standardized methods to evaluate living systems' problem-solving capabilities remain underdeveloped. To address this gap, we focus on bacterial systems and their metabolic adaptation. Using an *in-silico* model of bacterial behavior in varying environments, we propose a new framework for transforming supervised machine learning (ML) problems into a format solvable by biological systems. We then evaluate the framework's performance against other ML algorithms on standard classification and regression benchmarks. Experimental results show that bacterial metabolic networks often outperform linear methods and rival boosted trees, which are considered state-of-the-art for tabular data. A final ablation study suggests that the system's computational capacity may stem from its biological components rather than the translation tools used for the learning problem.

Keywords: Biocomputing · Leaning Capacity · Metabolic Networks

1 Introduction

Living organisms can be viewed as information-processing systems that integrate input signals, perform computations, and produce results essential for survival. Since the 1930s and 1940s, the information processing metaphor has been used to understand biological functions across various scales, from molecular components to ecosystems [1,9,11]. Recently, there has been growing interest in harnessing the computing power of living organisms for practical applications in medicine, industry, and the environment [8,12,15]. Reviews highlight the potential of engineered bacteria for delivering tumor-specific drugs [3].

To assess the usability of living organisms for computations, their performance must be evaluated and quantified. This requires adopting an information-

G. Krempl et al. (Eds.): IDA 2025, LNCS 15669, pp. 28–40, 2025.
https://doi.org/10.1007/978-3-031-91398-3_3

processing perspective, viewing organisms like bacteria as functions that associate inputs with outputs. Inputs can include chemical concentrations, temperature, and light, while outputs may involve changes in protein conformation, gene expression, cell growth, and motility.

Researchers have turned to the machine-learning concept of "reservoir computing" (RC) to explore whether biological and physical systems can perform complex, time-dependent computations. In this approach, a system associates temporal inputs with outputs, performing a computation. To use a "reservoir" effectively, two key steps are required: (i) mapping the desired inputs to the system's actual inputs, and (ii) learning how to interpret the system's output. The goal is to explore the types of input-output relationships achievable through RC and their complexity. For example, studies [7] have asked if a "Liquid State Machine" (a type of reservoir) exists in Escherichia coli (*E. coli*). Other [10] and [13] explored the computational abilities of soft machines like plants used as reservoirs to compute complex time-dependent input-output relationships.

While previous work focuses on assessing how physical or biological systems can approximate specific target relationships within a reservoir computing framework, this paper takes inspiration from the Machine Learning framework. Specifically, we **evaluate the learning capacity of biological systems** using classical benchmark datasets for regression and classification tasks. We measure the complexity required for an in-silico learning algorithm to achieve the same predictive performance. For example, we compare the system's performance to that of linear methods (e.g., linear regression or perceptrons) or more sophisticated, non-linear methods, using ensembles of boosted trees like XGBoost [2]. This work expands on questions raised in [4], with technical differences in the framework and a deeper experimental analysis of the bacterium's contribution.

Our contribution in this paper is fourfold. (*i*) We propose to employ bacteria as part of supervised ML systems to perform regression and classification tasks; This entails performing back-propagation through the bacterium's surrogate model -here an Artificial Metabolic Network (AMN) [5]-, and we show how to do this (Sect. 2). (*ii*) We describe a new method to evaluate the capability of a learning algorithm (Sect. 3). (*iii*) We assess the learning capacity of a system combining a simple neural network to translate input data into a form readable by the AMN, with the AMN acting as a reservoir (Sects. 3.1 and 3.4). (*iv*) Using an ablation/substitution approach, we evaluate the AMN's contribution to the problem-solving process and its potential for future computational devices (Sect. 4). In the conclusion section, we draw lessons from this new method for assessing the computational capacity of living organisms (indeed any learning algorithm), particularly its empirical results when applied to *E. coli*, and highlight future research directions.

2 Bio-Computing with *E. coli*

Bio-computing aims to explore new substrates for computation by leveraging biological components, such as genes, that receive inputs (either endogenous or

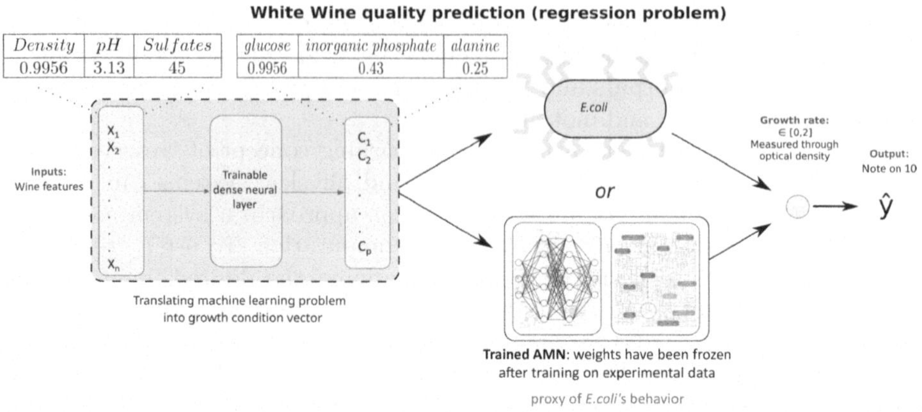

Fig. 1. Example of a bacteria used as a ML regressor. The values of input features (`Density`, `pH`, and `Sulfates`) are translated into concentrations of glucose, inorganic phosphate, and alanine, nutrients for the bacteria, using a trainable neural network. *E. coli*'s resulting growth rate is then interpreted as an output value for the regression problem. For full *in-silico* experiments, the bacteria can be replaced by an Artificial Metabolic Network modeling its behavior.

exogenous), transform those inputs, and produce outputs determined by their internal "rules." From this perspective, biological systems like *E. coli*, which take complex decisions based on environmental signals, can be viewed as signal-processing and computing units. This concept reveals the potential for using bacteria in computational tasks.

As illustrated in Fig. 1, *E. coli* can be integrated into hybrid systems: an Artificial Neural Network (ANN) performs pre-processing, translating problem features into nutrients that serve as input signals for the bacterium. A second ANN is used for post-processing, converting the bacterium's output—such as growth rate or metabolite fluxes—into results suitable for tasks like regression or classification.

However, conducting experiments with live bacteria in *in vivo* settings can be impractical. To overcome this, we use an *in silico* surrogate: a genome-scale metabolic model (GEM) of *E. coli*, which makes it possible to simulate the bacterium's cellular metabolism for our computational experiments.

2.1 Using an *in-Silico* Surrogate of *E. coli*

In this study, we propose a novel hybrid approach that combines Machine Learning (ML) with Mechanistic Modeling (MM) to model the bacteria's metabolism. This leverages the complementary strengths of both methods. Mechanistic models are designed to explain biological phenomena by incorporating detailed physical and biochemical processes, but they face challenges in handling complex systems. Conversely, ML models can accurately predict complex biological outcomes even without a mechanistic understanding; however, they require large, representative datasets for training. This complementarity has led to hybrid approaches, such as the Physics-Informed Neural Network (PINN) framework [14]. One hybrid model resulting from these considerations is the AMN model recently presented in [5]. One key property of this hybrid model being that *it makes it possible to backpropagate errors through the whole system*, which is crucial to be able to train the pre-processing ANN that translates the problem features into a signal for *E. coli*. Therefore, for the rest of this paper, we use the AMN as trained and described in [5] in place of *E. coli*, as in Fig. 1. For transparency and reproducibility, the code is available on GitHub[1]

2.2 An *E. coli* Based Processing Pipeline for ML Tasks

In order to use *E.coli* – replaced for *in-silico* experiments by an AMN, as described in Sect. 2.1 – we designed a processing pipeline (Fig. 1) that associates an input for the ML task with an output used for prediction. Figure 1 illustrates this for the problem of predicting the rating of a wine described by a set of features $X_1, X_2, \ldots, X_n$. Since bacteria react to a set of average concentrations $C_1, C_2, \ldots, C_p$, we need to translate the features describing an example into a set of average concentrations that the bacteria can perceive and react to. In our approach, this is done in three steps: (*i*) First, the input data are normalized using the MinMaxScaler function of the sklearn Python library on the training set. (*ii*) Next, if the input space dimensionality exceeds that of the bacteria ($n > p$), PCA is applied to retain the first p principal components; otherwise, the inputs are used as is in the next step. (*iii*) Finally, a single-hidden-layer ANN is trained to translate the problem inputs into inputs for the bacteria in such a way that the training error is minimized. In this study, we use a hidden layer with 200 neurons and a Relu activation function. The principal trainable part of the overall system is therefore located *before* the AMN, whereas in the classical RC framework, the trainable part is located *after* the reservoir. In this framework, a final trainable layer converts the output signal i.e. growth rate into the predicted output $\hat{y}$, this layer has only one node. Its has no activation function for regression tasks (corresponding to a rescaling) and for classification task the activation function was sigmoid, sharpened with a factor $10e4$. This is possible because the output error can be backpropagated through the AMN.

3 Learning Capacity of the *E. coli* pipeline

3.1 Dataset Selection

The datasets used in our experiments are taken from the OpenML [17] benchmark suites OpenML-CC18 (classification) and OpenML-CTR23 (regression) [6]. Starting from the full suites (72 classification and 40 regression problems), we further selected datasets containing only real-valued features, and for classification we considered only binary classification tasks, with a final selection of 24

[1] https://github.com/bmollet01/LCBMN.

regression and 20 classification datasets. To illustrate possible device behaviors, from the datasets that met the conditions, three were selected for classification on the one hand, and for regression on the other. They were chosen on the basis of the size of the dataset, neither too small to be too easy, nor too large to be too costly in terms of time complexity, and on the basis of preliminary results to exclude learning problems that were too easy or those requiring multiple outputs. More specifically, the `energy_efficiency` and `phoneme` datasets were selected to be challenging for the approach.

3.2 Methodology

The approach is as follows: for each ML task, we measure the predictive performance of the full pipeline after training (details below) and then attempt to achieve *same average performance* using either a linear model or XGBoost, with parameters such as the number and depth of trees controlled (see Fig. 2). To simplify computational capacity evaluation, we use single-node trees ("decision stumps") in XGBoost, adjusting the number of trees to match the pipeline's performance. Section 4 further discusses methods for evaluating the respective roles of the translation component and of the bacterium.

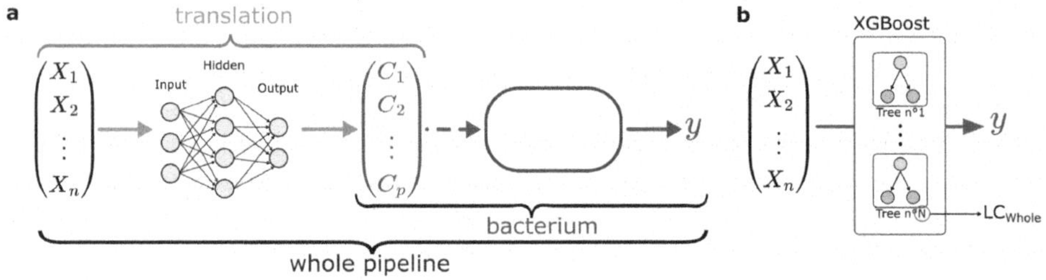

Fig. 2. (**a**) The proposed pipeline is composed of a first stage with a trainable NN, translating the problem's feature values into concentrations of nutrients for the bacteria, whose growth rate is then interpreted as the output. (**b**) XGBoost can be used to estimate the learning capacity of another ML method, increasing its number of trees until a target performance is reached.

3.3 Capacity Measurement

In order to measure the learning capacity of a given system on a given task, we first measure its predictive performance on that task using a 5-fold cross validation. We then iteratively apply XGBoost increasing the number N of single-node trees until it reaches the same predictive performance, again using a 5-fold cross validation.

In this section, we focus on the learning capacity of the whole pipeline, thus considering *LC–whole* (Fig. 2(b)). As shown in Fig. 3, where the x-axis stands

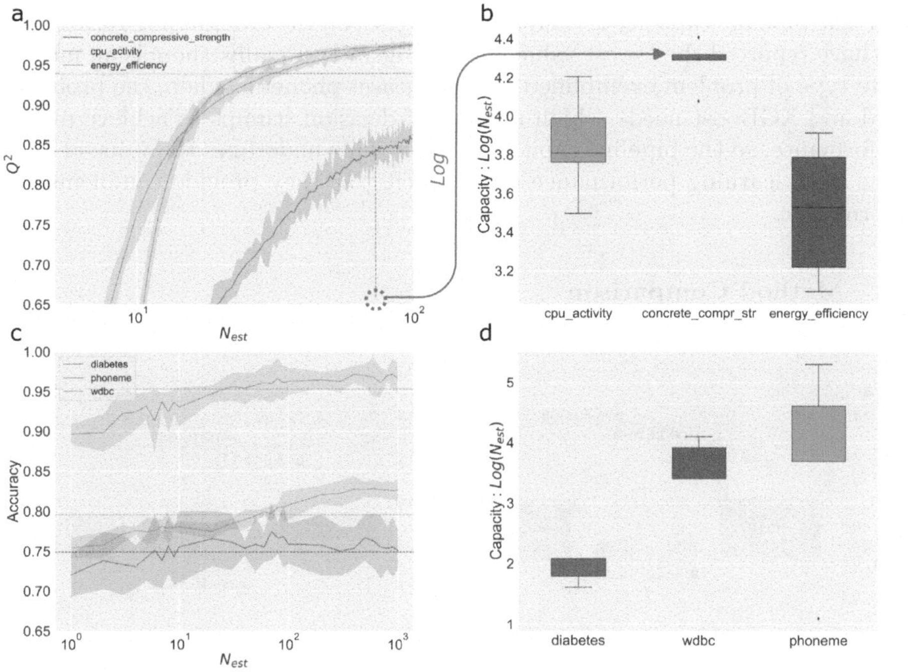

Fig. 3. Evaluation of the learning capacity of the pipeline with *E. coli* on three regression tasks (a & b) and on three classification tasks (c & d).

for the number of node trees (here in log scale) and the y-axis stands for the error rate (for classification tasks) or Q^2 score (for regression tasks). It appears from Fig. 3 that the number N_{est} of XGBoost stumps (in a log scale), tends to plateau as the number of decision or regression trees increases.

In Fig. 3, (a) and (c), the *dotted horizontal lines* indicate the performance of the pipeline for each learning tasks. Thus, the number N_{est} of XGBoost stumps (in a log scale), required to reach a given level of performance can be read on the x-axis. The learning capacity of the pipeline is displayed in (b) for all repetitions and for the three *regression tasks* and in (d) for the classification tasks. For instance, in (a), $\sim$ 75 stumps are needed in order to attain the performance of the pipeline on the `concrete_compressive_strength` task (i.e. $Q^2 = 0.84 \pm 9.4e{-}5$), corresponding to a capacity of $\log 75 \approx 4.3$ reported in (b) with standard deviation.

Looking at *classification tasks*, we observe three interesting behaviors. In one, `wdbc`, both the pipeline and XGBoost achieve high accuracy ($\geq$ 0.97), and XGBoost needs a large number of decision stumps to reach this level. The pipeline's learning capacity is therefore high. For the `diabetes` dataset, we see a completely different case, where the classification problem seems difficult for both the pipeline and XGBoost, with an accuracy that remains around 0.75 for both and whatever the number of decision stumps in XGBoost between $\approx$ 10 and

10^3. The pipeline capacity is therefore imprecise, on the order of [log 10, log 1000]. We have reported the lowest value here in Fig. 3(d). Finally, there is an intermediate type of problem exemplified by the dataset `phoneme`, where the problem is hard and XGBoost needs a high number of decision stumps to achieve pipeline performance, so the pipeline capacity is high. This underlines the value of examining both learning performance and capacity, as they provide complementary information.

3.4 Method Comparison

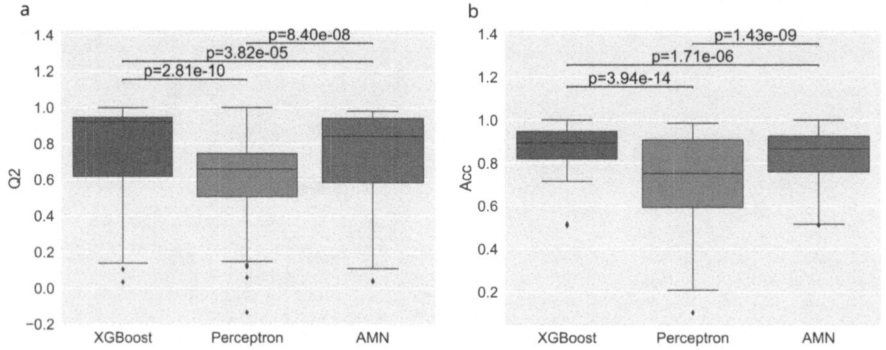

Fig. 4. Method comparison between AMNs, linear perceptron and XGBoost using Wilcoxon Signed rank test. (**a**) Q^2 scores of different methods on 10 different regression problems with $n = 5$ replicates. Here the Q^2 score corresponds to R^2 performed on test data so it is a metric of models' predictive power.(**b**) Accuracy scores of AMNs, linear perceptron and XGBoost classifier on 20 classification problems with $n = 5$ replicates.

Figure 4 sums up the results for 10 datasets of the *regression tasks* (a), and for 20 datasets for classification tasks (b). On the regression tasks (a), the pipeline with *E. coli* reaches a performance in between the one of linear regression and using XGBoost with 50 regression trees of depth 30. This setup for XGBoost has been set through hyperparameter optimization to maximize average performance on all datasets. The methods were compared using a Wilcoxon signed-rand text with two-sided hypothesis and displaying the *p-value*.

For some specific datasets such as `energy_efficiency` (a regression task), the pipeline has a lower performance than linear regression. However, this is not surprising since it is known that this dataset is well described by linear relations between only 3 of its 8 features.

On the 20 datasets for *classification* (Fig. 4(b)), the same trend is observed, with the pipeline having a performance that lies between linear classification and XGBoost. From this analysis, we can tentatively conclude that the *E. coli* based pipeline **outperforms linear methods for nonlinear problems**, and

underperforms a high-performance nonlinear method like XGBoost given enough trees.

4 Evaluating the Contribution of the Bacterium

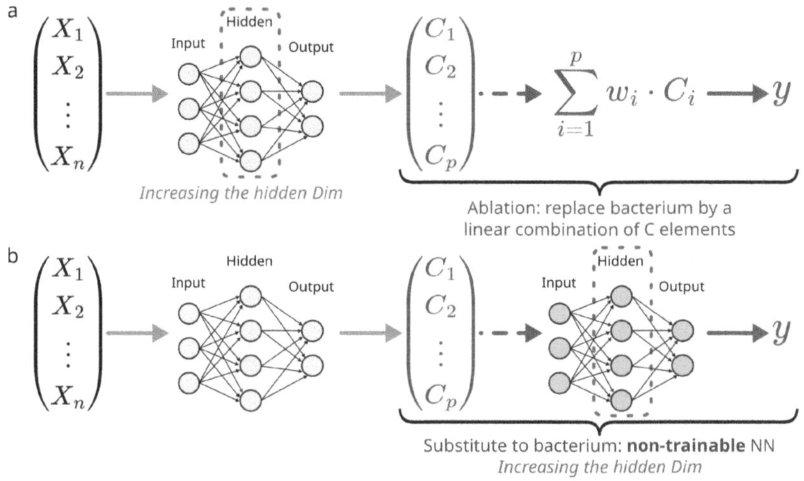

Fig. 5. (**a**) Architecture used for the **ablation study** where the bacterium is replaced with a simple linear transformation. Note that the trainable ANN has the same number of trainable weights than the translation ANN of the whole pipeline. (**b**) Architecture used for the **substitution study** where the bacterium is replaced by a one hidden layer ANN with frozen randomly initialized weights and 200 neurons in the hidden layer of the translation ANN (see Sect. 4). (Color figure online)

As illustrated in Fig. 2, the bacterium is but one element in the computational pipeline. A critically important question is to evaluate its contribution to the learning process. We do this in two ways.

First, we *evaluate what is required of the translation step*, performed by a trainable one hidden layer ANN, in order to solve the learning problem when the bacterium is replaced by a simple and fixed *linear function* $\mathbb{R}^p \rightarrow \mathbb{R}$ (Fig. 5(a)). We call this the **ablation study**. The idea is that, the larger the number of hidden neurons required to solve the task on the translation side, the higher is the contribution of the bacterium since this shows that it cannot easily be replaced with just a linear function.

Second, we set the size of the translation ANN to 200, as this gives a performance close to the maximum achievable value (Fig. 6). While this ANN is still learnt for optimizing the performance on each problem, we now *use a **frozen** one hidden layer ANN as a substitute for the bacterium* so that it can implement non linear functions (Fig. 5(b)). We call this the **substitution study**. This time, we *vary the size of the hidden layer* of the substitute so as to evaluate its required capacity for the learning problem to be solved. It is assumed that the larger this size, the bigger is the role of the bacterium.

It is expected that (i) increasing the number of hidden neurons in the bacterial substitute will lead to an increase in performance, which could reach that of the entire pipeline. If so, this would give us an estimate of the bacterium's capacity itself (i.e. excluding the translation layer) and shows that it does the work of a one hidden layer ANN. It is also possible that (ii) even by increasing the number of hidden neurons, the performance of the whole pipeline is never matched, in which case we could hypothesize that bacteria have a higher capacity than a single-hidden-neuron ANN using Relu functions. A third case is (iii) where the neural substitute enables the system to outperform the entire pipeline. In this case, the bacterium can be said to have a lower capacity than that of a single hidden layer ANN.

In both the ablation and substitution studies, the trainable part lies in the translation component, while the bacterium is replaced either by a fixed linear function (ablation) or by a fixed one hidden layer ANN with the Relu activation function (substitution). $n = 20$ replicates were randomly generated for each substitute size. For each replicate, weights were drawn from a uniform distribution in $\left[-\sqrt{\frac{6}{p_{in}+p_{out}}}, \sqrt{\frac{6}{p_{in}+p_{out}}}\right]$, with p_{in} the number of inputs and p_{out} the number of outputs (default initialisation of the keras python library). For the replicates of the linear function, the weights were drawn from a uniform distribution in $[-1, 1]$.

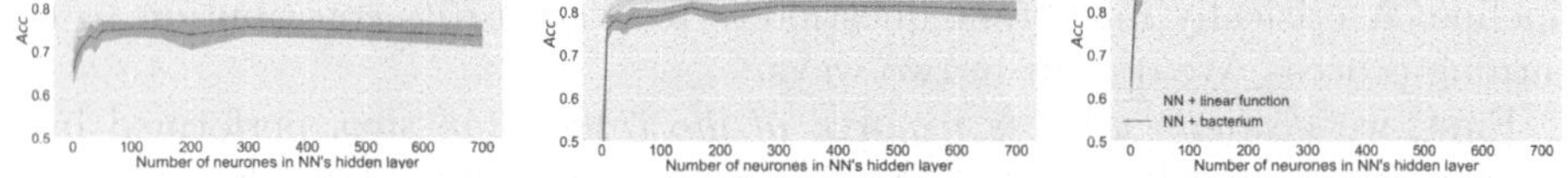

Fig. 6. Ablation study on three regression tasks (a-c) and three classification tasks (d-f). The performances of our pipeline (blue lines) with increasing number of neurons in the translation hidden layer (ranging from 1 to 700) were compared to the performances of the pipeline where *the bacterium is replaced by a simple linear transformation*, as presented in Fig. 5(a). Mean and variance were obtained with $n = 20$ random initializations of the linear layer's weights. (Color figure online)

4.1 Results of the Ablation Study

The question is: does the learning capacity can be explained entirely by the translation component?

For *regression*, we base our observations on the three tasks reported in Fig. 6(a, b, c). In all cases, it appears that either it is impossible for the translation component to overcome the absence of the bacterium for all numbers of hidden neurons, leading to impoverished performance (as is apparent for the `energy_efficiency` and the `cpu_activity` datasets), or it can equal the whole pipeline performance as for the `concrete_compressive_strength` dataset.

For *classification* tasks, the results in Fig. 6(d, e, f) on three datasets show a somewhat different picture. Here, the translation component can manage to learn as well as or better than the whole pipeline.

Thus, regarding the `cpu_activity` dataset, for regression, the pipeline performs much better than a linear model, and the ablation study shows that this cannot be explained by the performance of the translation component. Conversely, on the `diabetes` dataset related to a classification task, the good performance of the pipeline could be due to the translation component.

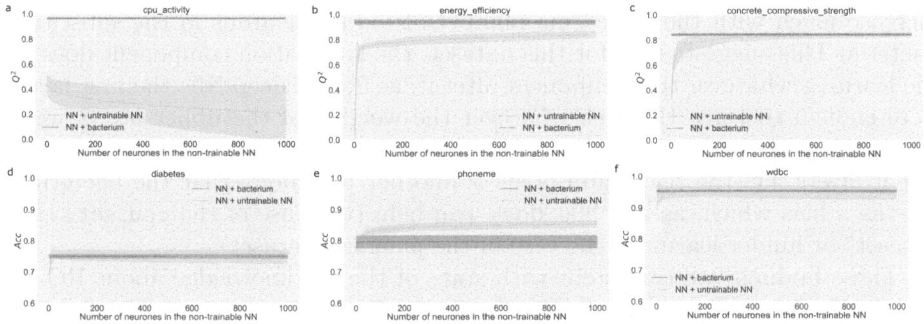

Fig. 7. Substitution study on three regression tasks (a–c) and three classification tasks (d–f). The performances of the whole pipeline (Fig. 4), here represented as dashed blue lines, were compared to the performances of the framework presented in Fig. 5(b) where *the bacterium is replaced by a one hidden layer ANN.* The mean and variance, were obtained with $n = 20$ random initializations of the non-trainable ANN with its hidden layer dimension ranging in [1–1000].

4.2 Results of the Substitution Study

Here, the question is: what is the level of non linearity and expressivness of the bacterium? This being measured by the number of neurons in a one hidden layer neural network that reaches the same level of perfomance if possible. (Fig. 5 (b)).

The experimental results for the substitution study (Fig. 7) confirm the difference between regression problems and classification ones found in the ablation study.

On *regression tasks*, the bacterium based pipeline exhibits greater or equal performance than the substitute based bacterium, for all number of hidden neurons that were tested. For instance, the examination of the `cpu_activity` dataset shows that the performance of the bacterium-based pipeline could not be matched by any of the one hidden layer ANN with Relu activation functions tested. This suggests that the bacterium performs a computation that requires more than what a one hidden layer ANN can compute. This analysis also applies, to a lesser extent, to the case of `energy_efficiency` dataset.

The situation is reversed for *classification tasks*. On the `wdbc` dataset, the performance of both the bacterium-based pipeline and of the substitute based pipeline are very good, This is true whatever the number of hidden neurons in the translation component (ablation study) or in the substitute based pipeline (substitution study). It may be hypothesized, also using Fig. 4, that a simple model captures all there is to capture in this dataset and that more powerful models are not necessary.

In the case of the `phoneme` dataset, the performance of the pipeline with substitute bacteria is superior to that of the pipeline with bacteria. This is not surprising, since it was already the case with a simple linear transformation instead of the bacteria. What is more surprising is that performance does not increase much with the increasing number of hidden neurons in the substitute bacteria. This suggests that for this dataset, the translation component does all the learning whatever the component after it, as if, in a deep NN, the first layers were enough to learn the task whatever the weights of the upper layers were. Except that here, the first layers cannot overcome the bias introduced by the upper layers, i.e. the bacterium. This is another testimony that the bacterium acts as a bias which, as any bias does, can help (the case of the `cpu_activity` dataset) or hinder learning (the case of the `phoneme` dataset).

These findings are consistent with state of the art knowledge about RC. In their 2019 review, Tanaka et al. [16] stated that (*i*) preprocessing is a crucial step for RC, and (*ii*) that RC performances are subject to hyperparameters optimization. We did not conduct any optimization here since we took the bacterium in one specific condition, but actually bacteria offers a wide variety of optimizable variables ranging from environmental conditions to genetic modifications. It is therefore entirely conceivable to seek to optimize the functioning of a bacterium according to the problem under study.

5 Conclusions

Given the energetic efficiency of bacteria in transforming input signals into output, and therefore in performing calculations, and the wealth of applications, particularly in the healthcare field, where they could be used, it is important to assess the extent to which they can help in learning tasks. For example, in the detection of a subject's pathology using a probe applied directly to the skin or to a blood sample.

In this paper, to answer the general question of whether the metabolism of living organisms such as bacteria can be used for data processing, we presented a **method** based on Machine Learning evaluation protocols. We proposed to assess a system's learning capacity by considering a parameterizable model that can equal its performance. Here, we have considered the XGBoost method, with the number of trees (for regression) or the number of decision stumps (for classification) as the hyper-parameter. We thus evaluated the learning capacity of a pipeline using a bacterial metabolic network for *E. coli.* A series of experiments with datasets corresponding to regression and classification tasks led to the conclusion that using *E. coli* as a computational device provides *a learning capability between linear methods and the XGBoost* algorithm known for its high level of performance for non-linear problems.

In order to evaluate the contribution of the bacteria to the learning task, we carried out an *ablation* study, replacing it with the simplest transformation between the output of the bacteria and the output corresponding to the learning task, i.e. a linear function. At the same time, we also introduced the principle of a *substitution* study to assess the degree of non-linearity and expressiveness of the organism in the learning pipeline. We found that, for regression tasks, E.coli behaved as a positive bias and exhibited a high degree of expressivity, whereas, for classification tasks, this was less so.

This study brings a first characterization of the data processing potential of *E. coli.* Further work, currently in progress, includes (*i*) building a broader overview of the types of learning problems that can benefit from the use of bacteria in a learning pipeline, (*ii*) replicating *in vivo* the experiments detailed in this article and (*iii*) carrying out experiments in which multiple outputs from bacteria are taken into account.

References

1. Balasubramaniam, S., Somathilaka, S., Sun, S., Ratwatte, A., Pierobon, M.: Realizing molecular machine learning through communications for biological AI. IEEE Nanotechnol. Maga. **17**, 10–20 (2023)
2. Chen, T., Guestrin, C.: XGBoost: a scalable tree boosting system. In: Proceedings of the 22nd ACM SIGKDD International Conference on Knowledge Discovery and Data Mining, pp. 785–794. ACM (2016). https://doi.org/10.1145/2939672.2939785
3. Chowdhury, S., Castro, S., Coker, C., Hinchliffe, T.E., Arpaia, N., Danino, T.: Programmable bacteria induce durable tumor regression and systemic antitumor immunity. Nat. Med. **25**(7), 1057–1063 (2019). https://doi.org/10.1038/s41591-019-0498-z
4. Faulon, J.L., Ahavi, P., Hoang, T.: Reservoir computing with bacteria. bioRxiv (2024). https://doi.org/10.1101/2024.09.12.612674
5. Faure, L., Mollet, B., Liebermeister, W., Faulon, J.L.: A neural-mechanistic hybrid approach improving the predictive power of genome-scale metabolic models. Nat. Commun. **14**(1), 4669 (2023). https://doi.org/10.1038/s41467-023-40380-0
6. Fischer, S.F., Feurer, M., Bischl, B.: OpenML-CTR23 – a curated tabular regression benchmarking suite. In: AutoML Conference 2023 (Workshop) (2023)
7. Jones, B., Stekel, D., Rowe, J., Fernando, C.: Is there a liquid state machine in the bacterium escherichia coli? In: 2007 IEEE Symposium on Artificial Life. IEEE (2007). https://doi.org/10.1109/alife.2007.367795

8. Kieffer, C., Genot, A.J., Rondelez, Y., Gines, G.: Molecular computation for molecular classification. Adv. Biol. **7**(3), 2200203 (2023)
9. Monod, J.: On Chance and Necessity. Springer, Heidelberg (1974)
10. Nakajima, K., Hauser, H., Li, T., Pfeifer, R.: Information processing via physical soft body. Sci. Rep. **5**(1), 10487 (2015)
11. Nicholson, D.J.: Is the cell really a machine? J. Theor. Biol. **477**, 108–126 (2019)
12. Pandi, A., et al.: Metabolic perceptrons for neural computing in biological systems. Nat. Commun. **10**(1), 3880 (2019)
13. Pieters, O., De Swaef, T., Stock, M., Wyffels, F.: Leveraging plant physiological dynamics using physical reservoir computing. Sci. Rep. **12**(1), 12594 (2022)
14. Raissi, M., Perdikaris, P., Karniadakis, G.E.: Physics-informed neural networks: a deep learning framework for solving forward and inverse problems involving nonlinear partial differential equations. J. Comput. Phys. **378**, 686–707 (2019)
15. Rizik, L., Danial, L., Habib, M., Weiss, R., Daniel, R.: Synthetic neuromorphic computing in living cells. Nat. Commun. **13**(1), 5602 (2022)
16. Tanaka, G., et al.: Recent advances in physical reservoir computing: a review. Neural Netw. **115**, 100–123 (2019)
17. Vanschoren, J., van Rijn, J.N., Bischl, B., Torgo, L.: OpenML: networked science in machine learning (2014). https://doi.org/10.1145/2641190.2641198

Semi-supervised Learning with Pairwise Instance Comparisons for Medical Instance Classification

Anne Rother[1(✉)], Till Ittermann[2], and Myra Spiliopoulou[1]

[1] Faculty of Computer Science, Otto von Guericke University Magdeburg, Magdeburg, Germany
anne.rother@ovgu.de

[2] Institute for Community Medicine, University Medicine Greifswald, Greifswald, Germany

Abstract. In semi-supervised and active learning, it is assumed that human experts can provide a set of reliably labeled instances. Crowd-working is used increasingly as a business model to acquire such labels by exploiting the wisdom of the crowd. A less investigated application area is a posteriori labeling of medical archives and of entries from population-based studies with diagnoses of diseases that were not recorded when the studies were conducted. Such labeling would make it possible for epidemiologists and clinical researchers to learn about the onset and evolution of diseases from earlier data collections. Data labeling is a complex process, especially for humans. Labels are still needed, but a completely automated approach to data acquisition demands identification of hallucinations, and this can only be done by humans. Methods such as few-shot learning have limitations, e.g. poor data at the beginning to learn the model. We investigate the potential of acquiring labels through the pairwise comparison of instances, and then propagating these labels in a semi-supervised way. For this purpose, we use structured data with hepatic steatosis as outcome and show that its possible, based on very few data to classify this data correctly.

Keywords: semi-supervised learning · structured data · triplets

1 Introduction

Semi-supervised learning (SSL) is a machine learning approach that leverages both labeled and unlabeled data to improve model performance. One of the key advantages of SSL is its ability to work effectively with smaller labeled datasets. Labeling data is often a time-consuming and expensive process, especially in domains requiring expert knowledge, such as medical imaging or legal document classification. By using SSL, we can significantly reduce the amount of labeled data needed, thereby cutting down on costs and resource requirements. Furthermore smaller labeled datasets, when combined with a larger pool of unlabeled

G. Krempl et al. (Eds.): IDA 2025, LNCS 15669, pp. 41–53, 2025.
https://doi.org/10.1007/978-3-031-91398-3_4

data, can help models generalize better. SSL techniques use the structure and patterns in the unlabeled data to enhance the learning process, leading to models that perform well on unseen data. This is particularly useful in scenarios where obtaining a large labeled dataset is impractical.

Artificial intelligence (AI) has become especially popular as a topic in the last few years. There are risks associated with autonomously executed actions and humans have skills that are very difficult to learn for machines. Approaches such as human-in-the-loop (HITL), crowdsourcing or semi-supervised learning can support the decision.

The potential of *pairwise comparison of instances* for the acquisition of labeled datasets and to build semi-supervised mechanisms that demand minimal labeled data is being investigated in a new approach, which considers uncertainty as quantity. AI in medicine can help to improve healthcare, make diagnosis more precise (the final decision on a diagnosis is made by the physician) and reduce costs.

More precisely, the following scientific questions will be investigated:

Q1: Given a very small set of labeled instances, how can a combination of pairwise comparisons with semi-supervised label propagation help in expanding it and what is the effect on classification quality?
Q2: How can we identify the variables that contribute in separating between the classes for most of the pairs?

We postulate that a small number of pairwise comparisons is sufficient for a reliable label propagation. In the following, we use the term 'triplet' to refer to two pairwise comparisons between an unlabeled instance, once with the positive class and once with the negative class for a binary classification task. To address this task, we propose ATLAS to optimize the labeling procedure and the number of labeled instances.

2 Related Work

Pairwise comparisons have been studied intensively from the machine learning perspective, see e.g. [9], where the objective is to induce a distance function over the data space. Insights on exploiting pairwise comparisons by human annotators for machine learning do exist, but are mostly limited to the comparison of images [1]. Arguably, pairwise comparison of structured instances, such as medical records, is different from the comparison of images.

Yao et al., [19], used pairwise comparisons for treatment effect estimation on non-image data: they chose three pairs of instances, one consisting of the most proximal target instance x_i and control instance x_j, one consisting of the most remote target instance with respect to x_i, and one consisting of the most remote control with respect to x_j. They then introduced two counteracting metrics on the basis of loss functions, intended to bring similar instances close to each other but not too close in the representation space.

Semi-supervised learning and especially few-shot learning are becoming increasingly important for generating labels. The selection of instances plays a decisive role here, as Jakubik et al., [7], shows. In addition to the human-in-the-loop approach, they use prototypical neural networks. Crowdworking, where the wisdom of the crowd is used to aquire comparatively "cheap" labels, can lead to misclassification by non-experts in the field [15]. One possibility is to involve experts in the human-in-the-loop process in order to increase the accuracy of the labels [8]. The improvement of the results strongly depends on the task difficulty and the data, e.g. Jakubik et al., [8], uses images as a basis and Rother et al., [15], structured data. For few-shot learning classification Li et al. [12] used learning to self-train. Lazarou et al., [11], predict pseudo-label by iterative label cleaning. Cao et al. [3] shows that one pair is sufficient, but use models learned in advance. In [20], the authors found out that semi-supervised learners benefit with few labels. However, these methods have been designed for images and not structured data, as in our approach.

3 Active Triplet LeArner Under Semisupervision ATLAS

We now introduce our learner ATLAS which assigns pseudo-labels to instances on the basis of pairwise comparisons, exploits the most reliable among them for further pseudo-label assignment and thus attempts to build up a reliably labeled dataset from a minimum number of truly labeled instances. In the following, we use the term 'pseudo-labels' for the labels derived by ATLAS and 'labels' for the true labels.

Let D be the set of unlabeled instances for which ATLAS should deliver 'pseudo-labels' and let DL be the much smaller set of labeled instances to be used for pairwise comparisons. We assume a binary classification problem with classes A & C, and we denote the instances of each class as A-instances and C-instances respectively. We aim to minimize the number of DL instances used for pseudo-labeling while maximizing quality, hence ideally D should be 'pseudolabeled' (i.e. acquire a pseudo-label from ATLAS) by using only one A-instance and one C-instance. Some of these instances are labeled by ATLAS with high certainty: these are then placed in a dataset $D+$, and are thereafter used (similarly to the truly labeled instances from DL) for further pairwise comparisons. This scenario corresponds to a combination of active and semisupervised learning, where pseudolabeled instances are exploited for learning, whenever their pseudo-labels are deemed certain (semi-supervision) and instances are drawn from DL whenever no instance can be pseudolabeled with a high degree of certainty (active label acquisition). The components of ATLAS that realize this workflow are described hereafter.

3.1 ATLAS Basic Module for Temporary Pseudo-Label Assignment

This component takes as input a triplet t comprised of a still unlabeled instance from D and two labeled instances from $DL \cup D+$, namely one A-instance and one C-instance. It performs a pairwise comparison of the unlabeled instance to the two labeled ones and extracts the label of the most proximal instance. This corresponds to an invocation of the lazy classifier kNN with $k = 1$, i.e. to 1NN, except that the unlabeled instance is only *temporarily* assigned a pseudo-label; the assignment may be reversed by the ATLAS core described in Subsect. 3.2.

As similarity function for 1NN we use the Heterogeneous Euclidean-Overlap Metric (HEOM), which computes the distance between two vectors over a space of continuous and categorical variables cf. [5]. HEOM essentially aggregates the Euclidean distances over the subspace of numerical variables and the boolean distances over the subspace of categorical variables.

Albeit further distance functions can be used, and despite the advances on proximity computations in a latent space, we opt for HEOM in the *original* feature space of the raw data for following reasons. First, creating a latent space over DL is unreliable, since the cardinality of DL is TWO in the ideal case. Second, ATLAS is an active learner, intended to receive decisions on pairwise comparisons (i.e. on object proximity) from a human oracle, and humans cannot compare well objects in high-dimensional spaces; our experiments with human annotators [15] were limited to 10 features that could be interpreted and compared, and still annotators were mislead by some variables. Finally, the construction and visualization of a projected space over $D \cup DL$, e.g. with UMAP [13], delivers a possibly suggestive visual placement of the instances, which could mislead the human annotator. This being said, we leave the usage of alternative distance functions over the initial data set union $D \cup DL$ as future work.

3.2 ATLAS Core for Pseudo-Label Assignment and Propagation

The label selected by the ATLAS basic module for the unlabeled instance of a triplet t is associated with uncertainty, which we quantify as

$$uncertainty(t) = \frac{smallest\_distance(t)}{largest\_distance(t)} \tag{1}$$

where the two distance values come from the pairwise comparison between the unlabeled instance, and the labeled A-instance $t.A$ and C-instance $t.C$ in the triplet:

$$\begin{aligned} smallest\_distance(t) &= \min\{HEOM(t.A, t.B), HEOM(t.C, t.B)\} \\ largest\_distance(t) &= \max\{HEOM(t.A, t.B), HEOM(t.C, t.B)\} \end{aligned}$$

where $t.B$ denotes the unlabeled instance in t.

In [15], we used the uncertainty quantification of Eq. 1 as basis for a comparison between an artificial labeler and a human labeler. Here, we use the uncertainty to decide whether the pseudo-label should be retained or not. In particular, let τ be an uncertainty threshold: if $uncertainty(t) < \tau$, then $t.B$ acquires the pseudo-label, is removed from D and is added to $D+$; otherwise, $t.B$ is moved back to D.

Thus, the ATLAS core gradually expands the set $D+$ of pseudo-labeled instances. These instances are used together with the truly labeled instances in DL by the algorithm that builds the triplets input to the core.

3.3 Building the Triplets for the ATLAS Core

The ATLAS core is invoked repeatedly until either D is empty or no more instances can be pseudo-labeled with a lower uncertainty than τ. At each invocation, an unlabeled instance is selected randomly from D and is placed in a triplet that contains one A-instance and one C-instance.

For the first invocation of ATLAS, the **Triplet builder** selects the A-instance and the C-instance from DL. We distinguish between following cases:

1. *DL contains exactly two instances, one per class:* The Triplet Builder repeatedly builds triplets by randomly selecting one unlabeled instance at a time and invoking the ATLAS core[1], until the first pseudo-label is accepted and $D+$ starts growing. From this moment on, labeled instances are selected from $DL \cup D+$.
2. *DL contains more than two instances:* The Triplet Builder chooses randomly one pair, ensuring that the two instances constituting the pair do not belong to the same class.

As soon as there are more than two labeled instances in $DL \cup D+$, the Triplet Builder iteratively creates triplets by choosing randomly one A-instance and one C-instance from $DL \cup D+$ and The instances in D_B are compared to this fixed pair of labeled instances as in the first case above. If the batch D_B is set to contain only one unlabeled instance, then the Triplet Builder performs three random choices at each iteration (one for the unlabeled instance and two for the labeled ones). If D_B is larger, then it is attempted to use this single fixed pair for pseudolabeling all of D_B, before choosing further pairs of labeled instances. In our experiments, D_B consists of only one instance.

Instead of a random selection of instances, proximity-aware variants are thinkable: (1) given an unlabeled instance, the closest labeled instances can be chosen; or (2) given a pair of labeled instances, the closest unlabeled one is chosen. However, the proximity is directly linked to our uncertainty function, so that the effects of proximity are captured despite the usage of a random choice. The impact of the uncertainty threshold is investigated in the experiments.

Uncertainty Gauger. The last component of ATLAS is the uncertainty gauger. This component can be configured, so that less but more certain pseudo-labels are acquired.

[1] Obviously, unlabeled instances that are pushed back to D can only be considered again *after* the first pseudo-label is accepted. Comparing them repeatedly to the only existing pair of labeled instances will not improve their chance of being labeled.

4 Evaluation Procedure

The Study of Health in Pomerania (SHIP) is a population-based, epidemiological study in the region of Vorpommern in the German state of Mecklenburg-Vorpommern and focus on the living conditions, environment and health since 1997. The instances of the data set used for the experiments were selected from 852 individuals of the "Study of Health in Pomerania" (SHIP-2), the second follow-up of a population-based study in Northeast Germany. Hepatic steatosis was used as classification outcome [18]. Hepatic steatosis was defined according to liver fat measurements based on magnetic resonance imaging (MRI) [10]. Ninety records were selected based on the MRI-assessed fat fraction in the liver. The experiments used selected variables as shown in Table 1. In addition to the outcome hepatis steatosis, the selected dataset consists of tree subspaces behavioural & sociodemographic, liver biomarkers and metabolism and includes a total of 90 records.

The evaluation is based on the following claim and for this claim the general evaluation focuses on the ground truth for each instance.

Claim

A correct classification whether an instance is to be assigned to ill or healthy, is possible with few triplets and with few variables inside a triplet.

For research question $Q1$, we evaluate performance across following dimensions:

- *Correctness:* Set of correctly classified instances – the larger the better.
- *Number of pseudo-labeled instances:* This corresponds to the size of $D+$ – the larger the better
- *Size of DL:* The set of truly labeled instances is used in its entirety, but we vary this size and investigate how the number of labeled instances made available affects the previous quantities.
- *Uncertainty threshold:* The smaller this threshold, the faster does the $D+$ grow, i.e. the less iterations are needed, but the correctness may be affected negatively. In our experiments, we consider $\tau = 0.4$ and $\tau = 0.5$.
- *Iterations until termination:* The number of ATLAS iterations refers the elapsed time until the maximum possible number of pseudo-labels is created. Since this number is not necessarily equal to the cardinality of D, we specify an upper boundary. In our experiments, this boundary is set to 80.

For $Q2$, we organize the set of variables (i.e. the feature space) into subspaces, and investigate how the usage of less than the complete feature space affects the different dimensions of performance. The set of variables for our dataset are depicted in Table 1.

Table 1. Overview of variables: The first section lists the outcome heapatic steatosis. The second section listed the 10 variables and the variables are ordered alphabetically by subspaces.

Variable	Description	Values margin	Subspaces
livfat_per	liver fat concentration	%	
stea	hepatis steatosis	1 (yes) or 0 (no)	
age	age at examination	years	behavioural & sociodemographic
alkligt	alcohol in g/day	g/day	behavioural & sociodemographic
sex	sex	1 (male) or 2 (female)	behavioural & sociodemographic
smoking	smoking status	0 (never), 1 (former) or 2 (current)	behavioural & sociodemographic
alat_s	alanin-aminotransferase	μ katal/l	liver biomarkers
C07	beta blocker intake	1 (yes) or 0 (no)	metabolism
crp_hs	high-sensitive CRP	mg/l	metabolism
diabetes	type 2 diabetes mellitus	1 (yes) or 0 (no)	metabolism
hyp_sr	hypertension	1 (yes) or 0 (no)	metabolism
ldlch	LDL-cholesterol	mmol/l	metabolism

4.1 Competitors

The competitors consist of supervised approaches such as random forest and K Nearest Neighbors, which include all A, B and C instances. The supervised-baselines includes Random Forest (RF) and K Nearest Neighbors (KNN) and are independented of ATLAS. For both, RF and KNN, a split of $p = 0.75$ is used, hence for 60 instances 14 test-instances and 46 train-instances were used for parameter optimization. Thus, with the same $p = 0.75$ and with 90 instances, test and train instances are distributed as 22 (test) and 68 (train).

5 Results

5.1 Results for Q1: Model Learning on Very Few Labeled Instances

The course of the decisions can be seen in Fig. 1 and Fig. 2. In both subspace "behavioural & sociodemographic" and subspace "metabolism", the decisions increase from run to run. The number of labeled instances used is remarkable: Table 2 and Table 3 compares the instances used for two thresholds. It is worth noting that only one A instance and one C instance are required for the subspaces. After that, it is sufficient for the further assignment to take place through the distance calculation and new assignment of the respective B instance. The number of used (labeled) instances increases in the full feature space to up to five A instances and four C instances.

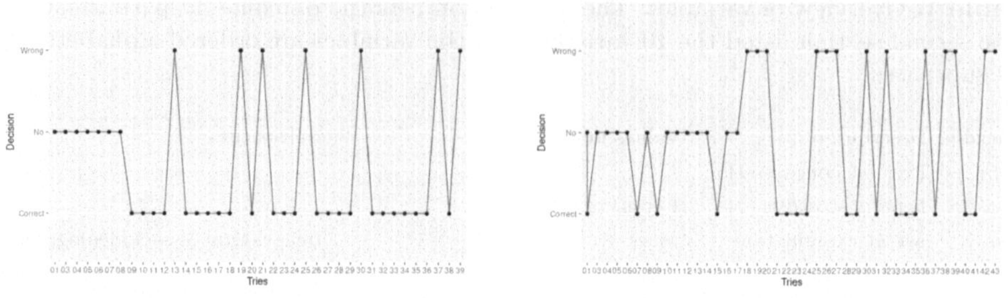

Fig. 1. Lineplot with subspace "behavioural & sociodemographic": Left subfigure: Time course with a threshold value of 0.4 and right subfigure time course with a threshold value of 0.5: All 30 unlabeled B instances can be classified.

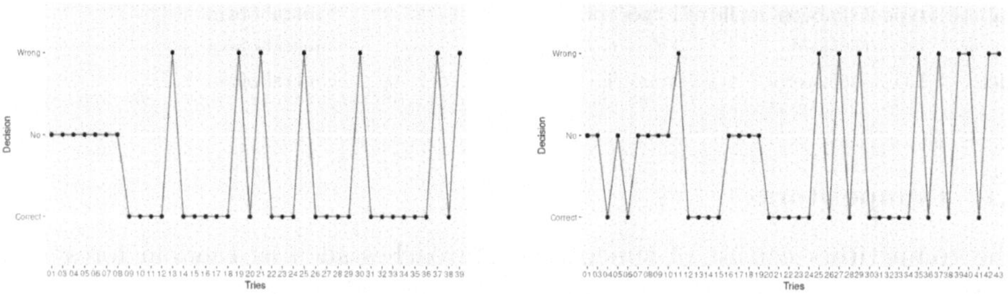

Fig. 2. Lineplot with subspace "metabolism": Left subfigure: Time course with a threshold value of 0.4 and right subfigure time course with a threshold value of 0.5: All B instances can be assigned at both threshold values – the number of misclassified B instances is lower

5.2 Results for Q2: Role of different variables in the pairwise comparisons

This research question, Q2, focuses on the variables involved and the difference to the full feature space.

The Choice of Variables has a Strong Influence on Performance. The analysis indicates the strong impact of the variables in the subspaces. In "behavioural & sociodemographic" are four variables, namely: age, alcohol intake, sex and smoking status. "Metabolism" includes five variables (beta blocker intake, CRP, diabetes melitus, hypertension and LDL-cholesterol. As the results of the subspaces indicate, these variables have a strong influence on the correct classification, especially in the specific subspaces. The results from the "metabolism" subspace are slightly better than in the "behavioural & sociodemographic" subspace (see Table 2). The dimensions also have an influence: the variables which are represented in the subspaces are included in the complete data set. In addition to the uncertainty, this has an influence on the correctness.

Table 2. Summary: Each run requires a different number of tries to classify the unlabeled instances. A stop criterion has been set (max. 80 iterations), which is not always required.

Threshold	run	Tries	Decisions	Correct classifications
0.50	full feature space	80	1	1
0.50	behavioural & sociodemographic	43	30	17 (56.67 %)
0.50	metabolism	43	30	21 (70.00 %)
0.40	full feature space	80	0	0
0.40	behavioural & sociodemographic	39	30	24 (80%)
0.40	metabolism	39	30	24 (80%)

Table 3. Summary with a threshold 0.50: Very few labeled instances are required

Threshold 0.50	subspaces		
Number of...	full feature space	behavioural & sociodemo-graphic	metabolism
used A instances	4	1	1
used C instances	4	1	1
tries until the first assignment	4	2	4

Uncertainty as a Fundamental Decision Criterion Uncertainty should be as low as possible and the choice of variables for the classification also plays a key role, as can be seen in Table 5 and Table 6. In both tables the uncertainty values for the subspaces are lower than in the full feature space. It is remarkable that ATLAS is certain. Values of 0.37 and 0.39 with a threshold of 0.40 are comparatively low. With a threshold of 0.50, the average uncertainty was 0.52 (for behavioural & sociodemographic) and 0.48 (metabolism) (Table 4).

Table 4. ATLAS: Summary with a threshold 0.40: No B instances can be assigned in the full feature space.

Threshold 0.40	subspaces		
Number of...	full feature space	behavioural & sociodemo-graphic	metabolism
used A instances	5	1	1
used C instances	4	1	1
tries until the first assignment	–	9	9

Table 5. ATLAS: Summary of average uncertainty and standard deviation for 0.50

Variables	Average uncertainty	StdDev
full feature space	0.74	0.09
behavioural & sociodemographic	0.52	0.19
metabolism	0.48	0.12

Table 6. ATLAS: Summary of average uncertainty and standard deviation for 0.40

Variables	Average uncertainty	StdDev
full feature space	0.79	0.11
behavioural & sociodemographic	0.37	0.20
metabolism	0.39	0.22

Performance of the Competing Options. The results with the highest accuracy values can be found in Table 8 for 60 instances and in Table 7 for 90 instances.

Table 7. Results for competitors: Random Forest (RF) and K Nearest Neighbors (KNN). Separates between all 10 variables (column "Accuracy all") and three groups (column "Accuracy subspaces") of all As, Bs and Cs, thus 90 instances.

Classifier	Accuracy (all)	Accuracy (subspaces)	
		behavioural & sociodemographic	metabolism
RF	**0.81**	0.54	0.63
KNN	0.81 (k = 6)	0.59 (k = 8)	0.77 (k = 1)

As can be seen in Table 7 and Table 8 the accuracy results of the complete feature space are above 70 % for both RF and KNN. When considering all 90 instances in the subspaces (30 A instances, 30 B instances and 30 C instances), the performance is lower than just in the separation of A and C instances.

6 Discussion

The interplay of correct classification with as few instances as possible was the main objective of the investigations. ATLAS, as a mechanism for classification, serves as the basis for the results. Main findings relate to the small number of A and C instances that are required for a correct classification. With smaller labeled datasets, the training process is generally faster, allowing for quicker iterations and experimentation. This can be particularly advantageous in research and development settings, where rapid prototyping and testing of different models and approaches are essential.

Table 8. Results for competitors: Random Forest (RF) and K Nearest Neighbors (KNN). Separates between all 10 variables (column "Accuracy all") and three groups (column "Accuracy subspaces") of all As and Cs, thus 60 instances.

Classifier	Accuracy (all)	Accuracy (subspaces)	
		behavioural & sociodemographic	metabolism
RF	0.71	**0.78**	0.64
KNN	0.78 (k = 1, 3, 4, 5, 6, 7)	0.71 (k = 3, 4, 6, 7, 8)	0.78 (k = 6, k = 8)

6.1 Discussion of the Results in Comparison to the State of the Art

In [4] the authors implemented an machine learning model, which generates labels for normal liver disease or abnormal disease, including an abstain approach to generate a diagnose. They used 99 liver ultrasound images and show an accuracy of 98%. In contrast *ATLAS* used triplet-based learning and non-images to classify whether an instance is ill or healthy.

Semi-supervised learning was used in [17] to create triplets for audio tagging task. They found out that semi-supervised sampling outperform supervised and unsupervised strategies. Furthermore "the ratio between labeled and unlabeled data can then have a great impact on the final performance". This can be partially confirmed: The full feature space in RF and KNN had a higher accuracy, however, the subspaces in ATLAS were partly stronger.

In [2] the authors investigated logical assumptions to make the aggregation model more transparent. For this, the authors include human experts, which provide opinions with different degrees of confidence. This reflects uncertainty and disagreement in the answers. Disagreement was not investigated, but uncertainty regarding the classification of the machine, ATLAS. The chosen thresholds of uncertainty can be configured by humans.

[16] investigated how quality data classes can be used to benefit decision making under uncertainty. As an approach they compared data to past quality levels. Uncertainty in the decision also has an influence on ATLAS: if a defined threshold value is not reached, no final label is provided.

Compared to previous work in [15], ASBA was developed. Here, triplets were specifically selected with labeled A and C instances and uncertainty and correctness of the selected triplets was investigated. ASBA labeled 18 of 30 triplets correct for the full feature space and was also mostly uncertain. This outperforms ATLAS in the full feature space, but the aim of ATLAS is to minimize the labeled instances and this was shown by including two subspaces.

The authors of [14] used pairwise comparisons to reduces bias and error. They found out that "the comparison approach is less susceptible to inflating bias, which majority voting tends to do". Majority voting, especially by humans, did not play a role in this work. However, pairwise comparisons have an increasing importance and a better understanding of human-in-the-loop can support future improvements of ML algorithms.

6.2 Limitations to Validity

Active learning, few-shot learning, supervised or unsupervised methods can help to improve strategies for selecting instances for the classification. Labeled instances are expensive. Therefore, the number of labeled instances used should be as low as possible. Including the human in the loop is a common strategy [6] and semi-supervised learning can improve the labeling. The results are based on a very limited data set. Therefore, the robustness of the implemented approach against variance or abnormalities in the data needs to be further investigated.

Some of the above limitations can be taken over as lessons learned for further optimizations: (a) verify the results through the execution of a large number of runs (b) intelligent ways to choose instances to reduce uncertainty and adapt for other data sets.

6.3 Outlook

The importance of the investigated topic is growing all the time. The approach of pairwise comparisons in medicine in particular has great potential. *ATLAS* serves as a basis for further improvement to enhance the classification as a tool that can exactly compute distances in a high-dimensional space. Large language models seem an alternative worth considering for semi-supervised label propagation, whereby the prompts needed to acquire relevant information constitute a dataset-specific challenge, while ATLAS is dataset-independent. Investigation of the potential of ATLAS on larger data sets like MIMIC IV would allow for more insights on how large feature spaces with many missing values affect the label propagation process.

Acknowledgments. SHIP is part of the Community Medicine Research net of the University of Greifswald, Germany, supported by the Federal Ministry of Education and Research (grants no. 01ZZ9603, 01ZZ0103, and 01ZZ0403), the Ministry of Cultural Affairs as well as the Social Ministry of the Federal State of Mecklenburg-West Pomerania.

Disclosure of Interests. The authors have no competing interests to declare that are relevant to the content of this article.

References

1. Ahonen, L., Cowley, B., Torniainen, J., Ukkonen, A., Vihavainen, A., Puolamaki, K.: S1: analysis of electrodermal activity recordings in pair programming from 2 dyads. PLoS One (2016). http://journalsplos.org/plosone/article/asset
2. Canfora, G., Troiano, L.: A model for opinion agreement and confidence in multi-expert multi-criteria decision making. Mathware Soft Comput. **11**(2) (2004)
3. Cao, Q., Deng, L.J., Wang, W., Hou, J., Vivone, G.: Zero-shot semi-supervised learning for pansharpening. Inf. Fusion **101**, 102001 (2024)
4. Hamid, K., Asif, A., Abbasi, W., Sabih, D., et al.: Machine learning with abstention for automated liver disease diagnosis. In: 2017 International Conference on Frontiers of Information Technology (FIT), pp. 356–361. IEEE (2017)

5. Hielscher, T., Spiliopoulou, M., Völzke, H., Kühn, J.P.: Using participant similarity for the classification of epidemiological data on hepatic steatosis. In: 2014 IEEE 27th International Symposium on Computer-Based Medical Systems, pp. 1–7. IEEE (2014)
6. Holzinger, A.: Interactive machine learning for health informatics: when do we need the human-in-the-loop? Brain Inf. **3**(2), 119–131 (2016). https://doi.org/10.1007/s40708-016-0042-6
7. Jakubik, J., Blumenstiel, B., Vössing, M., Hemmer, P.: Instance selection mechanisms for human-in-the-loop systems in few-shot learning. arXiv preprint arXiv:2207.06835 (2022)
8. Jakubik, J., Weber, D., Hemmer, P., Vössing, M., Satzger, G.: Improving the efficiency of human-in-the-loop systems: adding artificial to human experts. arXiv preprint arXiv:2307.03003 (2023)
9. Kleindessner, M., von Luxburg, U.: Kernel functions based on triplet comparisons (2017)
10. Kühn, J.P., et al.: Quantitative chemical shift-encoded MRI is an accurate method to quantify hepatic steatosis. J. Magn. Reson. Imaging **39**(6), 1494–1501 (2014)
11. Lazarou, M., Stathaki, T., Avrithis, Y.: Iterative label cleaning for transductive and semi-supervised few-shot learning. In: Proceedings of the IEEE/CVF International Conference on Computer Vision, pp. 8751–8760 (2021)
12. Li, X., et al.: Learning to self-train for semi-supervised few-shot classification. Adv. Neural Inf. Process. Syst. **32** (2019)
13. McInnes, L., Healy, J., Melville, J.: Umap: uniform manifold approximation and projection for dimension reduction. arXiv preprint arXiv:1802.03426 (2018)
14. Narimanzadeh, H., Badie-Modiri, A., Smirnova, I., Chen, T.H.Y.: Crowdsourcing subjective annotations using pairwise comparisons reduces bias and error compared to the majority-vote method. arXiv preprint arXiv:2305.20042 (2023)
15. Rother, A., Niemann, U., Hielscher, T., Völzke, H., Ittermann, T., Spiliopoulou, M.: Assessing the difficulty of annotating medical data in crowdworking with help of experiments. PLoS ONE **16**(7), e0254764 (2021)
16. Simard, V., Rönnqvist, M., Lebel, L., Lehoux, N.: A method to classify data quality for decision making under uncertainty. ACM J. Data Inf. Qual. (2022)
17. Turpault, N., Serizel, R., Vincent, E.: Semi-supervised triplet loss based learning of ambient audio embeddings. In: ICASSP 2019 - 2019 IEEE International Conference on Acoustics, Speech and Signal Processing (ICASSP), pp. 760–764 (2019). https://doi.org/10.1109/ICASSP.2019.8683774
18. Völzke, H., et al.: Cohort profile: the study of health in pomerania. Int. J. Epidemiol. **40**(2), 294–307 (2011)
19. Yao, L., Li, S., Li, Y., Huai, M., Gao, J., Zhang, A.: Representation learning for treatment effect estimation from observational data. Adv. Neural. Inf. Process. Syst. **31**, 2633–2643 (2018)
20. You, Z., Zhong, Y., Bao, F., Sun, J., Li, C., Zhu, J.: Diffusion models and semi-supervised learners benefit mutually with few labels. Adv. Neural Inf. Process. Syst. **36** (2024)

Local-Global Data Augmentation for Contrastive Learning in Static Sign Language Recognition

Ariel Basso Madjoukeng[1(✉)], Edith Bélise Kenmogne[2], Pierre Poitier[1], Benoît Frénay[1], and Jérôme Fink[1]

[1] University of Namur, NADI - PReCISE - HuMaLearn, Namur, Belgium
arielbassodev@gmail.com
[2] University of Dschang - FS, Dschang, Cameroon

Abstract. Sign language (SL) is a visual language used by the Deaf community. Static sign language recognition (SLR) consists of classifying static hand configurations, i.e., signs, present in isolated images. Due to the expertise required for manual annotation, SLR suffers from a data scarcity issue. Recent studies show that contrastive learning is an effective method for addressing this issue by proposing an efficient unsupervised pre-training. Contrastive learning leverages data augmentation techniques applied to entire images (global-global augmentation). However, fine-tuned, contrastive models often rely on irrelevant aspects of those images, like the background, without focusing solely on the regions of interest. Such models are prone to bias that could lead to unreliable predictions. In response, this paper proposes a new local-global data augmentation technique that helps contrastive models focus during the fine-tuning step on regions of interest, i.e., the signer's hands. This approach (i) improves the accuracy of contrastive learning by up to 15% on some SLR datasets, and (ii) help the fine-tuned contrastive models to better focus on relevant regions of images for SLR.

Keywords: Contrastive Learning · Data Augmentation · Sign Language Recognition · Local-Global Augmentation

1 Introduction

Sign language recognition (SLR) studies the translation of sign language into written or oral languages. The lack of people able to understand and translate sign languages (there is no universal sign language, but many of them) slows down data collection. To overcome this shortage in annotation, a solution is to leverage unsupervised methods, enabling models to learn useful features for SLR.

P. Poitier—Ph.D. grant from FRIA (F.R.S.-FNRS).
B. Frénay—Supported by SPWR under grant n°2010235 - ARIAC by DIGITALWALLONIA4.AI.

G. Krempl et al. (Eds.): IDA 2025, LNCS 15669, pp. 54–66, 2025.
https://doi.org/10.1007/978-3-031-91398-3_5

Contrastive learning is a popular and versatile, unsupervised approach that relies on augmentations of images to learn useful representations that are invariant to those augmentations. The resulting model can be fine-tuned on a small proportion of available annotated data of the same nature. The fine-tuned model performances are closely tied to the representations learned during the unsupervised training. In many application scenarios, particularly when the size of the unsupervised training dataset is not substantial, instead of learning representations useful for classification, the model sometimes learns representations based on all the features of the image. During the classification phase, predictions could be based on the background or other irrelevant parts of the image.

The above problem is particularly acute in static SLR where the region of interest (RoI), the signer's hands, contains most of the information needed to classify the sign. The background should play no role in the prediction. For SLR, a model whose predictions often rely on aspects of the image other than the RoI may be both unreliable and irrelevant from a linguistic point of view. To overcome those limitations, this paper introduces a novel augmentation method for contrastive pre-training called *local-global* augmentation. It encourages the model to focus on the RoI in the images. This method is applied to static SLR for datasets of single-hand configurations without dynamic components.

Experiments show that, on small datasets, our method helps the model to primarily focus on the RoI and improves accuracy by up to 15%. For large datasets, it concentrates saliency on the RoI without significantly loss of performance.

The rest of this paper is organized as follows. Section 2 introduces SimCLR and MoCo v2, two popular contrastive learning algorithms used here to assess the proposed augmentation strategy. A review of static SLR is also presented, and the datasets used in our experiments are presented. Section 3 shows issues that occur when using a *global-global* augmentation strategy. Our *local-global* augmentation strategy developed to address those issues is introduced in Sect. 4. Section 5 describes the architecture for the experiments and the meta-parameters for the training. Section 6 shows quantitative and qualitative results to assess our new method. Finally, those results are discussed in Sect. 7.

2 Preliminaries and Related Work

This section introduces SimCLR and MoCo v2, two popular contrastive learning algorithms used in our experiments. It also provides an overview of related work in SLR and discusses what augmentation methods can be used for SLR. Finally, the diverse SLR datasets used in our experiments are presented.

2.1 Contrastive Learning

Contrastive learning enables deep learning architectures to learn self-supervised representations through various data augmentations, without the need for prior annotations. There exist several contrastive algorithms, among which the most popular are SimCLR [1] and MoCo v2 [2]. This section briefly presents them.

SimCLR Algorithm: SimCLR is one of the first successful contrastive algorithms and is still widely used in the literature [3–5]. This algorithm consists of four components. First, an *augmentation module* transforms an input image into a positive pairs of augmented images, i.e., augmentations. The other instances in the same batch act as negative examples. Second, the *encoder network* projects the input data in a latent space. Finally, the *projection head*, commonly a multilayer perceptron, transforms the latent space into lower-dimensional vectors called the projections. The contrastive loss function is applied to the projections. The loss function is used to maximize the similarity between positive pairs and minimize the similarity between negative pairs. Once trained, the projection head is removed and only the latent space is used for inference. The obtained embeddings can be grouped with any clustering algorithm [6].

MoCo v2 Algorithm: SimCLR requires large batches to have enough positive and negative examples to converge. It is not always possible to increase the batch size when training a model due to memory limitations. To tackle this issue, MoCo v2 [2] introduces the concept of *dictionary*: a queue containing embeddings of the data from previous batches. Those embeddings serve as negative examples for the current batch. At each step, MoCo v2 receives two inputs: a query and keys. A query is an augmented image from the dataset, while keys are a set of samples stored in a dictionary. MoCo v2 uses two encoders: a query encoder and a key encoder (commonly called *momentum encoder*). During the training, the query encoder parameters are updated during the backpropagation, while the momentum encoder parameters are updated using a momentum update rule.

2.2 Sign Language Recognition

Sign language recognition (SLR) studies the creation of systems to recognize, classify, or translate sign language. This work focuses on static SLR that aims to recognize static signs in images. Since the advent of deep learning, significant progress has been made in this field [7]. Given the visual nature of the data, convolutional neural networks (CNNs) are often used. Murali et al. [8] introduce a three-layer CNN architecture, trained on the American Sign Language (ASL) dataset [9] with 87k images of the ASL alphabet. The results show that their model, trained from scratch, achieves a remarkable accuracy of 90%.

However, for many medium datasets such as the *Indian Sign Language* [10] or the *Arabic Sign Language* [11] datasets, training from scratch does not lead to satisfying results. To address this issue, researchers have proposed to rely on knowledge transfer. Al-Barham et al. [12] show that transfer learning with a pre-trained ResNet-15 achieves 96.77% of accuracy on the ArASL dataset with 7.8k images. Despite these advances, the scarcity of annotated data and the existence of underrepresented sign languages (i.e., small datasets) hinders the usage of transfer learning. For underrepresented datasets like *Bahasa Isyarat Indonesia* [13], transfer learning alone often obtains poor accuracy.

Zhao et al. [14] conducted a study on self-supervised learning methods and concluded that contrastive learning is more efficient than transfer learning in more general tasks such as hands motion estimation. Accordingly, Mohsin et al. [15] propose a contrastive architecture for SLR, applied to the *American Sign language* (ASL) and *Indian Sign Language* (ISL) datasets. They apply augmentations such as color distortion, Gaussian blur, translation, and flipping globally (i.e., applied to the whole image) to form positive contrastive pairs. These transformed images are then used to train contrastive algorithms such as SimSiam [16], SimCLR, MoCo, and others. With SimCLR, they achieve 74.08% of accuracy on ISL and 74.66% on ASL; SimSiam reaches 74.59% on ISL and 77.41% on ASL, while MoCo achieves 70.21% on ISL and 68.03% on ASL.

To the best of our knowledge, in contrastive learning algorithms, data augmentation is only applied globally to the entire image [1,2,10,16], i.e., all of its regions are affected. However, during training, this poses several issues that are further exacerbated when the unsupervised training set is not sufficiently large. Section 3 provides a detailed discussion on the issues associated with the global application of augmentations in contrastive learning for SLR.

2.3 Image Augmentation for SLR

Data augmentation is a critical step in contrastive learning, it should be designed with care as it plays a critical role in the performance of the final model. The SimCLR paper [1] categorizes image augmentations in two main groups, *geometric and appearance transformations*. Geometric transformations are modifications that affect the geometric properties of an image, such as its shape, size, orientation, etc. Such transformations include rotations, translations, scaling, vertical/horizontal flips, and deformations. Appearance transformations focus on the texture without affecting the geometric nature of the image. Color distortion and Gaussian blur are two common appearance augmentations.

In SLR, rotations and flips could be damageable, as the same hand configuration with a different orientation could have a different meaning. These transformations should thus be avoided for contrastive learning in SLR. Moreover, for our method that only transform parts of the image (see Sect. 4.1 for details), only the appearance transformations will be used, as the geometric transformations are hard to apply to the local parts or objects in the image.

2.4 SLR Datasets

Several SLR datasets are used to illustrate issues of current augmentation techniques and to assess our approach: (i) *Bahasa Isyarat Indonesia* (BISINDO) [13] with 312 images across 26 classes representing the alphabet, (ii) *Arabic Alphabets Sign Language* (ArASL) [11] with 7,8k images from 31 unique letter signs, (iii) *the Indian Sign Language* (ISL) [10] with 42k images from 35 static signs from the Indian SL and (iv) the *American Sign Language* (ASL) [9] dataset with 87k images from 26 gestures signs from the American manual alphabet. These datasets are quite diverse and representative of the challenges in static SLR.

3 Issues of Global-Global Augmentation for Contrastive Learning

Contrastive approaches rely on data augmentation to generate positive pairs for their training process. The positive pairs for an image are obtained by applying two distinct augmentations to all the pixels of the image. Since both augmentations are applied to the entire image, we refer to this approach as *global-global* augmentation. As mentioned in Sect. 2, after generating the positive pairs, learning aims to capture invariances between the augmented versions of the same image. However, it often happens that instead of learning invariances related to the RoI (in our case, the hands of the signers), the model captures invariance in regions of the image that are irrelevant for the downstream task (e.g., clothes). There is no reason for the model to focus only on the RoI during the unsupervised pre-training. This issue is even more acute on small datasets.

To assess the existence and seriousness of this issue, a model is trained using a contrastive loss on all the static SLR datasets introduced in Sect. 2.4. The details of the training and the architecture are available in Sect. 5.

To visualize what regions of the image are used by the model to classify a sign, saliency maps [17] are computed. The saliency map of an image highlights the pixels that are the most prominent for the model to take its decision. Therefore, by looking at those maps, it is possible to observe what parts of the image are the most important for the model. Figure 1 shows, for each SLR dataset, two randomly selected images along with their label and a saliency map.

In Fig. 1, the following cases can be observed. Sometimes, the RoI aligns with the saliency map. This happens when the contrastive model captures relevant similarities in the input. It is common for larger datasets. In other cases, the saliency spreads over the whole image. This occurs when the model fails to capture the relevant invariances between the different augmented views of the same image. Typically, it happens with smaller datasets such as BISINDO and ArASL in our experiments, and sometimes with medium-sized datasets.

In response to observations from our preliminary experiment in Fig. 1, this work proposes a new method to mitigate issues for contrasting pre-training in SLR. This new *local-global* augmentation is presented in the next section.

4 Local-Global Data Augmentation

This section introduces the proposed *local-global* augmentation for contrastive training. The intuition and advantages of the method are presented, and we empirically motivate why only applying the augmentation to the RoI for all positive images (i.e., relying on a local-local augmentation) is not desirable.

4.1 Main Idea and Method

The proposed *local-global* augmentation module for contrastive learning is graphically described in Fig. 2. An augmentation is selected and applied to the input

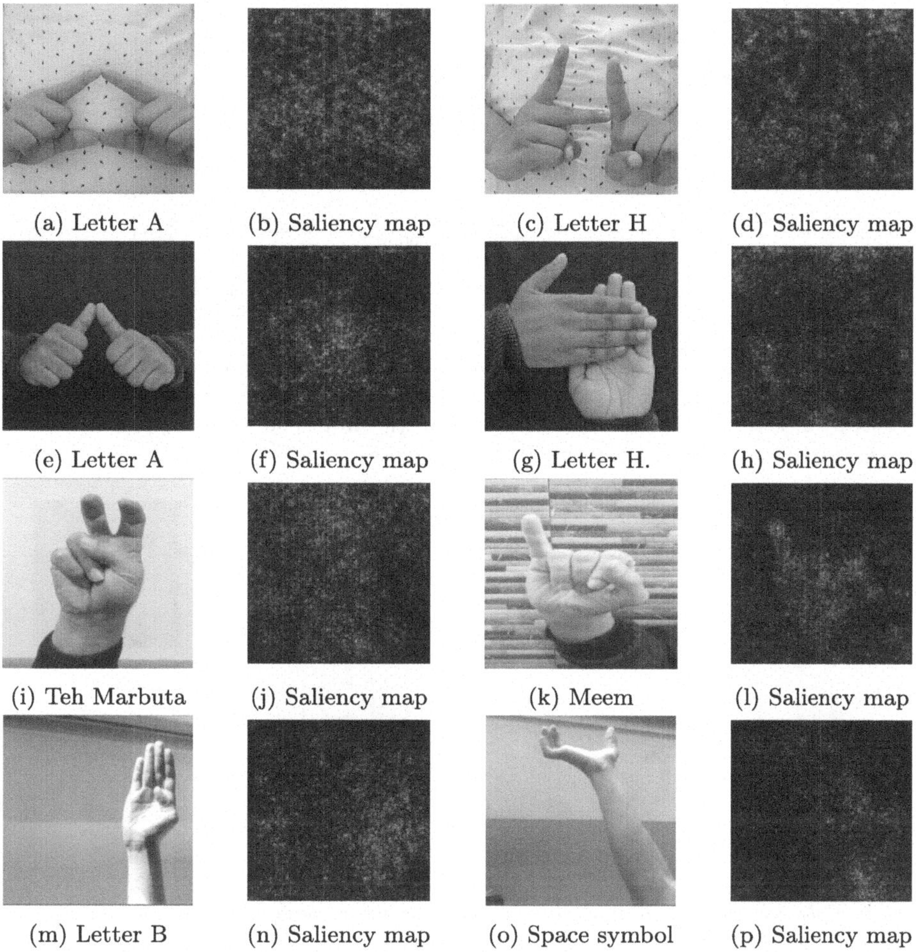

(a) Letter A (b) Saliency map (c) Letter H (d) Saliency map
(e) Letter A (f) Saliency map (g) Letter H. (h) Saliency map
(i) Teh Marbuta (j) Saliency map (k) Meem (l) Saliency map
(m) Letter B (n) Saliency map (o) Space symbol (p) Saliency map

Fig. 1. Images and saliency maps from BISINDO [13] (first line), ISL [10] (second line), ArASL [18] (third line) and ASL [9] (fourth line) datasets. Each image is processed by a model pre-trained using a contrastive loss (see text for details).

image to create a positive pair of images. The key difference with a classical, global-global augmentation module is that both images in the positive pairs are augmented differently in our local-global module. For the first image of the pair, the augmentation is only applied on the RoI. For the second one, the augmentation is applied on the whole image. Figure 3 shows a comparison of a *global-global* and a *local-global* augmentation applied to sign language data.

As already stated in Sect. 3, contrastive algorithms learn invariances between different augmented versions of the same instance. The *local-global* augmentation makes it easier for the model to find similar patterns in the relevant parts of images. On the one hand, maximizing the similarity of the projections for the

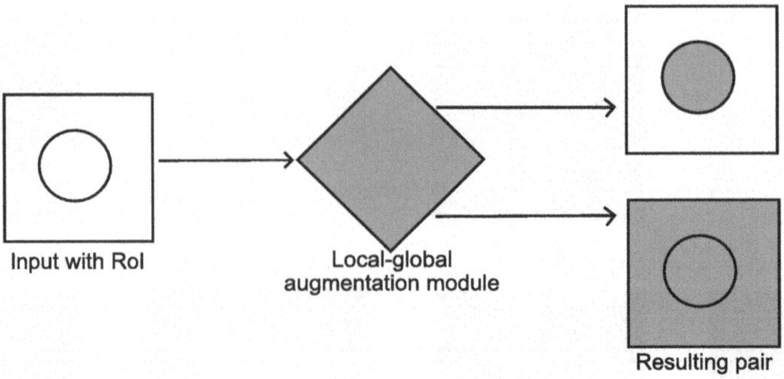

Fig. 2. Overview of the local-global augmentation process. The input with an identified RoI is passed to the augmentation module. It randomly selects a transformation and applies it once to the whole image and once to the RoI. The resulting pair is sent to the contrastive algorithm, here SimCLR or MoCo v2.

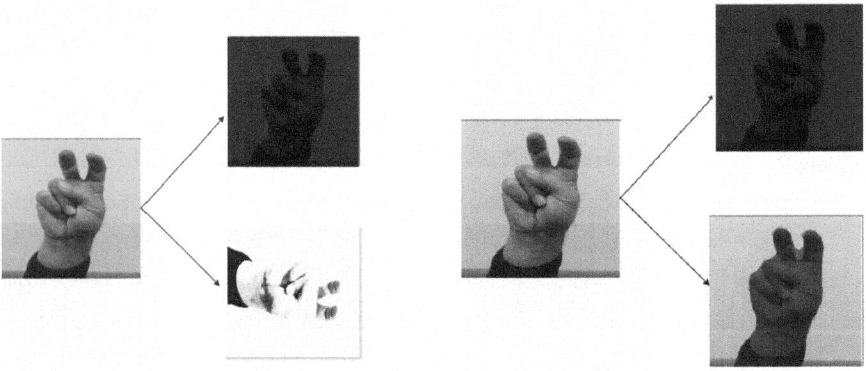

Fig. 3. *global-global* (left) and *local-global* augmentation (right), illustrated on an image from the ArASL dataset [18]. Local augmentation is applied to the RoI.

positive pairs encourages the model to only use the RoI. Indeed, for both images in the positive pairs, *pixels inside the RoI are similar in both positive images as they are altered by the same augmentation*, whereas *pixels outside the RoI are altered in only one of the positive images.* If projections are based only on the RoI, it will be easier to maximize similarity. On the other hand, the negative pairs play the same role as in classical contrastive learning: forcing the model to learn relevant projections to distinguish images. Local-global augmentation makes it easier for the algorithm to find invariances and leads to better results on smaller dataset, as experimentally illustrated in Sect. 6.

A limitation of the *local-global* is the need of RoI annotation to apply the augmentation. For our experiments on static SLR, the RoI is automatically extracted using a network pre-trained for image segmentation called Mask R-CNN [19]. This proved to be effective as we only need to detect hands on a background, and no manual annotation is needed. The *local-global* method also prevents the application of some augmentations impossible to apply locally (e.g., rotation). This explains why our experiments only consider appearance-based augmentation.

4.2 About Local-Local Augmentation

One might wonder why the augmentation is not applied solely on the RoI, leading to a *local-local* augmentation module for contrastive learning. However, this produces the opposite of the desired result: instead of focusing on the RoI, the model seems to divert its focus towards all the other regions of the image, excluding the RoI. This phenomenon suggests that the model, instead of capturing the relevant features of the RoI, ends up capturing the other features of the image. Intuitively, maximizing the similarity of images in the positive pairs no longer encourages contrastive learning to focus on the RoI, *since the background is now what is identical between both images.* Figure 4 presents examples of saliency maps obtained with a contrastive model trained with local-local augmentations. It is clear that the model saliency focuses on everything but the RoI.

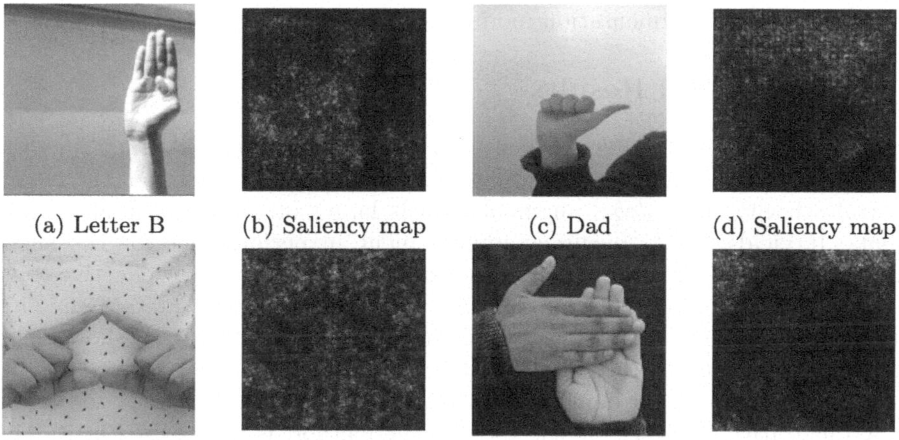

(a) Letter B (b) Saliency map (c) Dad (d) Saliency map

(e) Letter B (f) Saliency map (g) Letter H (h) Saliency map

Fig. 4. Saliency map for images from the ASL [9] (first), ArASL [18] (second), BISINDO (third) and ISL (fourth) datasets with *local-local* augmentation.

5 Architecture and Training Parameters

Let us now introduce the architecture and the meta-parameter used in our experiments. Experiments in Sect. 6 are implemented in PyTorch version 2.4. The model is composed of an encoder and a projection head selected as follows:

- **Encoder**: the encoder is a ResNet-50 pre-trained on ImageNet. This choice is supported by its widespread use [20,21] in transfer learning tasks.

- **Projection head**: the projection head consists of a multilayer perceptron (MLP) with a single hidden layer with an input size of 1000 as it is the size of the ResNet50 embedding and a projection size of 200 followed by a ReLU.

Models are trained using an SGD optimizer with a learning rate of 10^{-3} during 70 epochs and with a batch size of 128 images. The temperature of the contrastive loss is set to 0.1 and, in the case of MoCo v2, the momentum is set to 0.9. Those values match the recommended values reported in the SimCLR and MoCo v2 papers. The contrastive training is unsupervised and does not use the labels provided by the datasets. After the contrastive pre-training, the encoder is fine-tuned with a cross-entropy loss during 70 epochs using the same optimizer, batch size and number of epochs as the contrastive pre-training.

To pre-train, fine-tune and evaluate the contrastive model, three data splits are needed. For the evaluation, the test set is made of 30% of the data. The remaining 70% are used to pre-train and fine-tune the model: 80% of the training set are for the unsupervised training and 20% are for the supervised fine-tuning.

A pre-trained Mask R-CNN [19] with a confidence score value of 0.5 to extracts the RoI: all regions detected by the R-CNN are included in the RoI for the *local-global* augmentation module, whose implementation is available[1].

6 Experimental Results

Tables 1 and 2 show the average accuracies for 10 models trained using SimCLR and MoCo v2 for each dataset. One can observe that the accuracy of the model pre-trained with *local-global* augmentation is larger or equal in all the experiments. It yields a 15% accuracy improvement in performance for the small BISINDO dataset. This suggests that our approach is particularly effective compared to the traditional *global-global* augmentation when data are scarce. This was confirmed for the ASL dataset in an additional experiment (not shown here due to space limitations), where local-global augmentation outperformed global-global augmentation when only 400 instances are used for the supervised part.

A qualitative analysis for the *local-global* method is shown in Fig. 5. The saliency is more concentrated on the RoI, regardless of the dataset size. This proves that the model learns to better focus on the relevant object in the image to learn good representations, whether on a small, medium or large dataset. To complement this qualitative analysis, we computed the average proportion of saliency within the RoI with local-global and global-global augmentation, and the results were respectively: 70% vs. 30% for BISINDO, 63.2% vs. 51.46% for ISL, 83.21% vs. 77.69% for ArASL and 73.91% vs. 72.29% for ASL. This confirms the relocation of saliency.

The limitations of this approach are that it is harder and, sometimes, impossible to apply some augmentation locally. In the case of images, only the appearance-based transformation can be applied locally. This limits the amount of available transformations. It also requires the identification of the RoI in the images, which might not be possible for every use cases.

[1] https://github.com/arielbassodev/local-global-data-Augmentation.

Table 1. Comparison of the accuracies obtained using a *global-global* or *local-global* augmentation module with SimCLR for several static SLR datasets. Accuracy is averaged on 10 repetitions and 95% confidence intervals are given.

Dataset	Augmentation	global-global	local-global
BISINDO	Color Distortion (C_d)	9.72%± 1.27	22.05%±1.70
	Gaussian Blur (G_b)	7.95%±1.43	18.65%±2.45
	G_b and C_d	9.77%±2.06	27.92%±2.60
ISL	Color Distortion (C_d)	54.92%±1.89	62.01%±1.51
	Gaussian Blur (G_b)	52.95%±2.60	54.20%±1.27
	G_b and C_d	57.13%±1.89	65.51%±1.41
ArASL	Color Distortion (C_d)	40.29%±1.04	43.07%±0.75
	Gaussian Bur (G_b)	34.97%±0.69	37.02%±1.03
	G_b and C_d	44.29%±0.86	49.01%±1.17
ASL	Color Distortion (C_d)	67.32%±0.75	67.36%±0.95
	Gaussian Blur (G_b)	64.72%±0.56	64.72%±0.65
	G_b and C_d	69.45%±0.80	70.08%±0.96

Table 2. Comparison of the accuracies obtained using a *global-global* or *local-global* augmentation module with MoCo v2 for several static SLR datasets. Accuracy is averaged on 10 repetitions and 95% confidence intervals are given.

Dataset	Augmentation	global-global	local-global
BISINDO	Color Distortion (C_d)	8.08%±1.36	14.87%±1.51
	Gaussian Blur (G_b)	7.65%±1.09	13.07%±2.15
	G_b and C_d	11.52%±1.51	18.60%±2.30
ISL	Color Distortion (C_d)	57.02%±1.51	59.25%±0.58
	Gaussian Blur (G_b)	54.35%±0.88	55.09%±0.73
	G_b and C_d	59.07%±0.58	62.59%±0.66
ArASL	Color Distortion (C_d)	36.53%±3.02	39.01%±2.89
	Gaussian Blur (G_b)	36.07%±0.27	39.49%±0.59
	G_b and C_d	38.11%±0.43	39.6%±1.04
ASL	Color Distortion (C_d)	64.72%±0.63	64.75%±0.94
	Gaussian Blur (G_b)	61.63%±0.51	62.09%±0.48
	G_b and C_d	65.78%±0.70	66.50%±0.80

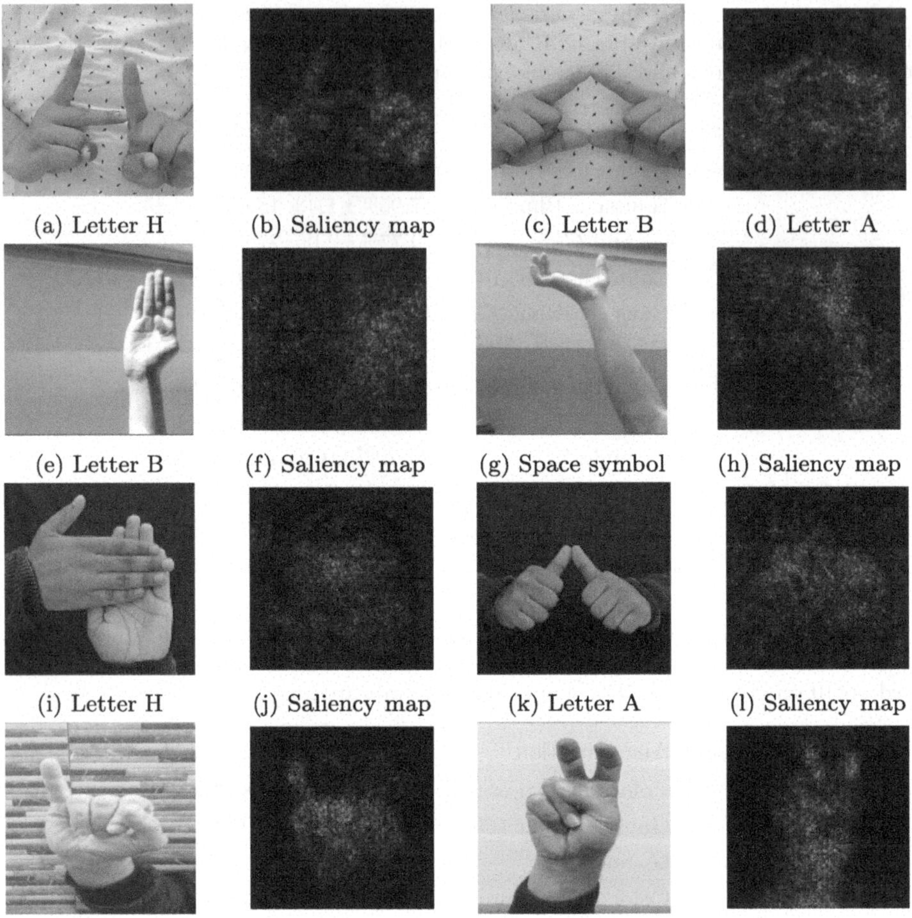

(a) Letter H (b) Saliency map (c) Letter B (d) Letter A

(e) Letter B (f) Saliency map (g) Space symbol (h) Saliency map

(i) Letter H (j) Saliency map (k) Letter A (l) Saliency map

(m) Meem (n) Saliency map (o) Teh Marbuta (p) Saliency map

Fig. 5. Saliency map for the BISINDO (1st row), ASL (2nd row) and ISL (3rd row), and ArASL (4th row) dataset images with local-global augmentation.

7 Conclusion

This work presents a new data augmentation module for contrastive pre-training that forces the model to focus more on regions of interest than on the other parts of the images. The approach is evaluated on the use case of static sign language recognition with four datasets of different sizes. First, a qualitative analysis of a model trained with a classical *global-global* augmentation module shows the pitfalls of this approach. Second, experiments conducted with the proposed local-global augmentation strategy show qualitative improvement. The observation of saliency maps clearly shows that models trained with our *local-global* augmentation module mainly focus on the regions of interest. Those results are quantitatively confirmed by the results obtained by the contrastive model. An improvement of 15% accuracy can be observed on the BISINDO sign language dataset, one of the smallest considered in our SLR experiments.

Although our experiments focus on sign language recognition, they can be transferred to other use cases where patterns to be identified are located in an easily definable region of interest, if masks can be obtained at reasonable cost.

For future work, aiming to further enhance the performance and reliability of data augmentation techniques for sign language recognition, dynamic sign language recognition will be explored by assessing the effectiveness of our method on sign language videos. Local-global augmentation methods for videos will be developed and tested to leverage unannotated sign language data in the wild.

References

1. Chen, T., Kornblith, S., Norouzi, M., Hinton, G.: A simple framework for contrastive learning of visual representations. In: Proceedings of ICML, pp. 1597–1607 (2020)
2. He, K., Fan, H., Wu, Y., Xie, S., Girshick, R.: Momentum contrast for unsupervised visual representation learning. In: Proceedings of CVPR, pp. 9729–9738 (2020)
3. Hojjati, H., Ho, T., Armanfard, N.: Self-supervised anomaly detection in computer vision and beyond: a survey and outlook. Neural Netw. **172**, 106106 (2024)
4. Ma, J., Duan, H., Yang, C., Wang, X., Niu, Y., Han, Y.: Simclr-unet: an ecg feature wave segmentation algorithm based on a self-supervised learning strategy. In: Proceedings of the 2022 4th RICAI, pp. 1354–1359 (2022)
5. Willi, A., Baumann, P., Erb, S., Gröger, F., Zeder, Y., Lionetti, S.: Self-supervised and few-shot learning for robust bioaerosol monitoring. arXiv preprint arXiv:2406.09984 (2024)
6. Madjoukeng, A.B., Kenmogne, E.B., Frenay, B.: Fast k-means with stable instance sets. In: 2024 Proceedings of IJCNN, pp. 1–8. IEEE (2024)
7. Al-Qurishi, M., Khalid, T., Souissi, R.: Deep learning for sign language recognition: current techniques, benchmarks, and open issues. IEEE Access **9**, 126917–126951 (2021)
8. Murali, R., Ramayya, L.D., Santosh, V.A.: Sign language recognition system using convolutional neural network and computer vision. Int. J. Eng. Innov. Adv. Technol. **4**, 138–141 (2022)
9. Nagaraj, A.: ASL alphabet (2018)
10. Arikeri, P.: Indian sign language character level dataset (2021). Accessed 05 May 2021
11. Al-Barham, M.: Rgb arabic alphabets sign language dataset (2023)
12. Al-Barham, M., Alomari, O.A., Elnagar, A.: Arabic sign language alphabet classification via transfer learning. In: ICETAI, pp. 226–237. Springer, Heidelberg (2023). https://doi.org/10.1007/978-3-031-56728-5_20
13. Noer, A.: Bahasa isyarat indonesia (bisindo) alphabets. Kaggle Dataset
14. Zhao, Z., Alzubaidi, L., Zhang, J., Duan, Y., Gu, Y.: A comparison review of transfer learning and self-supervised learning: definitions, applications, advantages and limitations. Expert Syst. Appl., 122807 (2023)
15. Mohsin, S., Salim, B.W., Mohamedsaeed, A.K., Ibrahim, B.F., Zeebaree, S.R.: American sign language recognition based on transfer learning algorithms. IJISAE, 390–399 (2024)

16. Chen, X., He, K.: Exploring simple siamese representation learning. In: Proceedings of CVPR, pp. 15750–15758 (2021)
17. Simonyan, K.: Deep inside convolutional networks: visualising image classification models and saliency maps. arXiv preprint arXiv:1312.6034 (2013)
18. Sayed, A.: Arabic sign language dataset 2022 (2022)
19. He, K., Gkioxari, G., Dollár, P., Girshick, R.: Mask r-cnn. In: Proceedings of CVPR, pp. 2961–2969 (2017)
20. Alsharif, B., Altaher, A.S., Altaher, A., Ilyas, M., Alalwany, E.: Deep learning technology to recognize American sign language alphabet. Sensors, 7970 (2023)
21. Salim, B.W., Zeebaree, S.R.: Kurdish sign language recognition based on transfer learning. In: IJISAE, pp. 232–245 (2023)

SiamCircle: Trajectory Representation Learning in Free Settings

Maedeh Nasri[1], Mitra Baratchi[2], Alexander Koutamanis[3], and Carolien Rieffe[1,4,5](✉)

[1] Department of Developmental Psychology, Leiden University, Leiden, The Netherlands
[2] Leiden Institute of Advanced Computer Science, Leiden University, Leiden, The Netherlands
[3] Faculty of Architecture and the Built Environment, Delft University of Technology, Delft, The Netherlands
[4] Faculty of Electrical Engineering, Mathematics and Computer Science, University of Twente, Enschede, The Netherlands
c.j.rieffe@utwente.nl
[5] Department of Psychology and Human Development, Institute of Education, UCL, London, UK

Abstract. Trajectory representation learning (TRL) is an intermediate step in handling trajectory data to realize various downstream machine-learning tasks. While most previous TRL research focuses on modeling structured movements in large-scale urban spaces (e.g., cars or pedestrians on streets), this paper focuses on a more challenging scenario of modeling free movement in small-scale social spaces (e.g., children playing in a schoolyard). We present a TRL model, SiamCircle, to process raw trajectories without additional feature extraction to prevent information loss. SiamCircle adopts a Siamese network with Circle Loss to learn trajectory embeddings. Furthermore, SiamCircle employs a data augmentation process to enable self-supervised learning and enrich the input data to address the limited access to high-quality data and ground truth. We evaluate the performance of SiamCircle in downstream tasks using trajectory ranking and clustering performance via seven evaluation metrics collectively. Using an ablation study, we explored the impact of different loss functions on the model's performance. Accordingly, we selected a 2-D convolutional design with Circle Loss as the best-performing model. In a comparative study, we compared our model against three other baselines. We observed up to 19% improvements in trajectory ranking tasks and achieved the highest average rank in supervised clustering tasks.

Keywords: Triplet Loss · Clustering · Trajectory Representation Learning

1 Introduction

In the wake of rapid growth in wearable technologies, people generate large amounts of movement data using location-aware devices like sports bands and

G. Krempl et al. (Eds.): IDA 2025, LNCS 15669, pp. 67–80, 2025.
https://doi.org/10.1007/978-3-031-91398-3_6

smartwatches. Various organizations address existing challenges in, for example, traffic forecasting [26] and animal migration patterns [13], by collecting and analyzing this data, known as spatio-temporal movement data or trajectories [6].

In a broader context, this movement data is captured in mainly two settings: *structured settings* and *free settings*. In structured settings, movement is recorded in areas influenced by spatial features or regulations, e.g., road networks or driving policies. This setting often exhibits periodic patterns, such as commuting to work on weekdays. Data from structured settings are valuable for applications like traffic forecasting. Conversely, in free settings, individuals move without restrictions, resulting in unstructured trajectories, which often occur in constrained environments, such as students' movements in a schoolyard or athletes' movements in sportsfields. This data is useful for analyzing micro-level behaviors, including social interactions and group dynamics.

In the context of trajectory analysis in structured settings, trajectory representation learning (TRL) has attracted extensive research attention in various machine learning tasks such as trajectory flow forecasting [6] or pair-wise similarity computation [4,27]. Yet, one largely unsolved challenge is effectively designing a TRL framework for movements occurring in free settings. Developing TRL models in free settings is much more challenging than in structured settings due to (i) irregular movement patterns and movements formed in the absence of typical urban features, (ii) limitations in data acquisition systems leading to imperfect location data and complicating trajectory analysis, and (iii) privacy concerns restricting the collection and sharing of high-quality data.

Recent studies in modeling movements in structured settings implement pre-processing methods such as grid-cell projection [15] and graph embeddings [4]. While the results are promising, their application to trajectories collected in free settings may be limited due to the potential information loss during the pre-processing. Besides, their performance highly relies on the choice of hyper-parameters, e.g., cell size. In a free setting, finding the optimal grid size is more challenging, especially in the presence of imprecisions due to noise.

This paper presents a robust trajectory representation learning framework called SiamCircle, which directly applies to raw trajectory data collected in free movement settings. SiamCircle consists of three main components: (1) a data augmentation process to generate the augmented trajectories accounting for imperfections in trajectories and enriching data sets with limited sample size, (2) a neural network applicable to raw trajectory data that maximally exploits spatial and temporal similarities among similar trajectories while amplifying differences between non-similar ones, eliminating common information loss during feature extraction, (3) a loss function uniquely designed with our learning framework to learn representative features through the training process. To our knowledge, this is the first study proposing a TRL framework using raw trajectories collected in free settings. Overall, this paper makes the following contributions:

- We develop a framework to create augmented trajectories that simulate similar movements and imperfections in data. Moreover, this framework generates a ground-truth dataset that annotates the trajectories based on their simi-

larity class and enriches datasets with a limited sample size. Such annotated dataset can be used in supervised training and supervised evaluation metrics.
- We propose a unique deep neural network model with triplet-based Circle Loss to learn trajectory representations directly from raw trajectories to limit information loss commonly occurring during the feature extraction.
- We demonstrate the performance of the proposed framework by conducting an ablation study and a comparative study on five trajectory datasets collected in a free setting using seven evaluation metrics in downstream tasks using two classes of metrics, namely trajectory ranking, and clustering. We compare the performance of SiamCircle to three other baseline models.

The remaining part of the paper is organized as follows. In Sect. 2, we discuss related work. Section 3 presents the problem statement, and Sect. 4 explains the details of our proposed method. The experimental setup and results are presented in Sect. 5 and Sect. 6, respectively. Finally, we summarize the paper and discuss future research directions in Sect. 7.

2 Related Work

Several studies proposed TRL models primarily in structured settings like urban traffic forecasting and detecting transportation modes [3,11]. While these models show promise, they become impractical for data collected in free settings, especially in the absence of external information.

Moreover, TRL via Contrastive learning has been employed in various machine learning problems in natural language processing [10], image processing [21], and spatio-temporal problems [4,7]. Contrastive learning is especially beneficial when using noisy and imperfect datasets with limited sample sizes. Fan et al. [7] proposes a contrastive learning approach powered by Triplet loss [20] in combination with a sliced encoder network, S-BiLSTM, for a user re-identification problem based on trajectory data. S-BiLSTM learns the robust part of individuals' trajectories (segmented in 24 h). This segmentation is not applicable for movements in free settings where trajectories do not exhibit daily patterns. TrajCL [4] is another contrastive learning model including a dual-feature multi-head self-attention-based encoder with InfoNCE loss [17] using trajectories in urban areas. TrajCL proposed a pointwise trajectory feature enrichment method to extract structural and point features to train their model. The optimization process used in both loss functions, Triplet loss and InfoNCE loss, is rigid, as their loss calculations give pairs with different similarities an equal pre-defined margin.

The present study uses a Siamese network with Circle Loss to learn trajectories' representative features designed explicitly for free movements in constrained environments (i.e., free settings). Our model uses raw trajectories without any feature extraction techniques as opposed to earlier work [4,28] to limit the unnecessary loss of data and oversimplification in free settings. Circle Loss adopts a dynamic penalty scheme based on the degree of similarity between trajectories in a triplet to obtain a sustainable optimization process and high-quality embeddings to address the rigidness problem in earlier works [4,28].

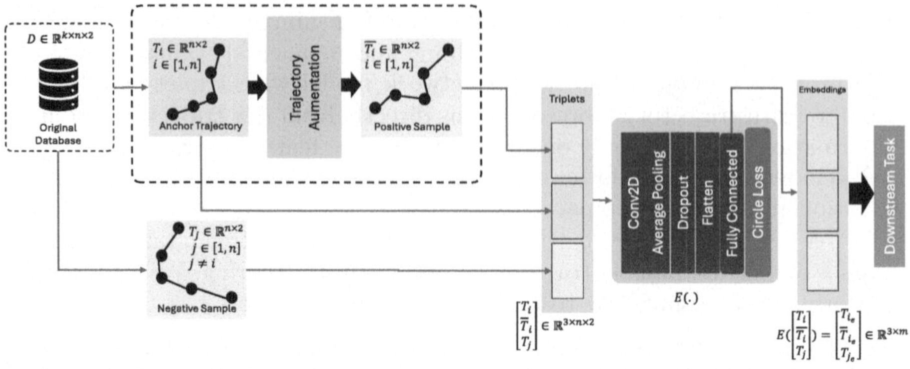

Fig. 1. An overview of SiamCircle: Data Augmentation Process to create the [Anchor, Positive, and Negative] triplets, Siamese Network with Circle Loss to generate the embeddings of the given raw triplets.

3 Problem Definition

Our trajectory modeling problem is based on a given trajectory dataset D, where each trajectory $\{T_i(\mathbf{x}, \mathbf{y}) \in D | (\mathbf{x}, \mathbf{y}) = \{(x_1, y_1), \cdots, (x_n, y_n)\}\}$ is a sequence of two-dimensional points recording the trace of the moving object i over n timestamps.

We are interested in learning an embedding function $E(\cdot)$ that represents trajectory T_i as $E(T_i) = T_{i_e} \in \mathbb{R}^m$ ($m \ll n$ is the dimension of embedding space) such that $f(T_{j_e}, T_{i_e})$ is minimized for any pair of the object i and j moving along the same underlying route, and, conversely, maximized for any pair of objects i and j moving along different underlying routes. While $f(\cdot, \cdot)$ is a distance function, e.g., Euclidean distance.

4 Methodology

This section discusses our proposed method for TRL. First, we explain the trajectory augmentation process. Next, the neural network design is presented. Lastly, we present the loss function used in the proposed framework. An overview of our method is presented in Fig. 1.

4.1 Trajectory Augmentation

In this section, we propose a trajectory augmentation method that adopts meaningful transformations to generate augmented trajectories from original ones. Previously proposed trajectory augmentation methods involve applying predefined transformations to the original data, for instance, by truncating or point masking [4,15]. These transformations often modify the trajectory length, making them only applicable in combination with manual feature extraction methods to create fixed-length sequential data for neural network models. However, as we aim to utilize raw trajectories with no feature extraction, we propose a unique trajectory augmentation approach where the trajectory length remains unchanged while meaningful transformations are applied. Our trajectory augmentation process serves three main purposes: (1) simulating similar movement trajectories by distorting original trajectories in different scales (i.e., additive noise) to account for different levels of similarities, (2) enriching limited-size datasets by adding augmented trajectories to the original dataset used purely for training. Thus enabling a self-supervised training process for the neural network model, and (3) simulating the practical imperfections found in location data acquisition systems in the form of large-scale noise (i.e., large spatio-temporal displacements or jumps), aiming to train a model that is robust to issues such as noise and other imperfections in the data. To achieve these goals, the augmented trajectory $\bar{T}_i$ is created by applying pre-defined transformations to each original trajectory T_i of a moving object i, as formulated in Eq. 1:

$$\bar{T}_i(\bar{\mathbf{x}}, \bar{\mathbf{y}}) = (\mathbf{x} + \mathbf{x}_i', \mathbf{y} + \mathbf{y}_i'), \qquad \begin{cases} \mathbf{x}_i' = \mathcal{U}(l_x, h_x).d_s, \\ \mathbf{y}_i' = \mathcal{U}(l_y, h_y).d_s, \end{cases} \tag{1}$$

where $\bar{T}_i(\bar{\mathbf{x}}, \bar{\mathbf{y}})$ is the augmented trajectory of moving object i. $\mathbf{x}_i'$ is the coordinate of the original trajectory T_i. $\mathbf{x}_i'$ is the distortion vector across $\mathbf{x}$. $\mathcal{U}(l_x, h_x)$ is the uniform distribution bounded between lower bound $l_x = -\sigma(\Delta\mathbf{x})$ and higher bound $h_x = \sigma(\Delta\mathbf{x})$, in which $\sigma(.)$ is the standard deviation over $\Delta\mathbf{x}$. and $\Delta\mathbf{x}$ is the vector of positional differences across $\mathbf{x}$ coordinates. d_s is a distortion scale randomly selected among the given scaling range. In all the mentioned annotations, replacing $\mathbf{x}$ with $\mathbf{y}$ gives the same definition across the $\mathbf{y}$ coordinate (instead of $\mathbf{x}$). In our experiments, we acquire a single augmented trajectory by applying two types of distortions namely *additive noise* and *jumps* (by setting two values for d_s). While additive noise applies to all coordinates of the original data, jumps only apply randomly to k number of samples. Another difference between additive noise and jumps is the scale of the distortion. In the additive noise, d_s includes a lower distortion scale to create similar movements, satisfying aims (1) and (2). In the jumps, d_s includes a larger distortion scale to create imperfections in trajectories, satisfying aims (2) and (3). Thus, the semi-synthesized trajectory, $\bar{T}_i(\bar{\mathbf{x}}, \bar{\mathbf{y}})$, creates one variation of the object i in the database per distortion scale. In Sect. 4.2, we will discuss our proposed neural network model.

4.2 Siamese Network Architecture

This section discusses the design of our proposed neural network model. The unique aspect of this design is the model's ability to be directly applied to raw 2-D trajectories without implementing any preliminary pre-processing steps. The rationale behind this choice is to include as much information as possible without abstracting spatial details, e.g., trajectory griding, or extracting spatio-temporal features, e.g., speed. To achieve this goal, we adopted a Triplet-based Siamese network, which typically consists of three identical networks with shared weights [2]. The primary purpose of Siamese networks is to learn the similarity between inputs by comparing their representations. In this design, the model takes an anchor trajectory together with a positive sample (i.e., a trajectory from the same class as the anchor) and a negative sample (i.e., a trajectory from a different class than the anchor) as a triplet input, pass each sample through its identical branch, and hand over the generated embeddings to a triplet-based loss function. Specifically, our model contains two sections:

2D-CNN/Average Pooling/Dropout. Inspired by the use of neural networks in time-series forecasting [25], this section employs a 2D convolutional layer (2D-CNN) as the initial feature extraction step. The 2D-CNN is particularly effective in capturing spatial-temporal dependencies within trajectory data, as it slides learnable filters over the input to identify local patterns.

Following the 2D-CNN, an average pooling layer is introduced to reduce dimensionality and filter out noise from the extracted feature maps. This pooling mechanism computes the average values within a small window, which helps the network generalize better by emphasizing essential trajectory trends while discarding less significant details.

To further enhance model robustness, a dropout layer is applied after pooling. Dropout randomly deactivates a subset of neurons during training, which prevents the network from overfitting. This regularization technique ensures that the model remains adaptable and performs well on unseen trajectory data.

In summary, this section of the framework transforms raw trajectory data into a compact yet informative representation by extracting relevant spatial-temporal patterns, filtering out noise, and preventing overfitting.

Fully-Connected Layer. The outputs from the previous layer representing high-level embedding features are fed to a fully connected (FC) layer. This allows the model to learn the non-linear combinations of these features. This FC layer serves as a crucial component in transforming the extracted spatial-temporal features into a more discriminative representation suitable for trajectory ranking and clustering tasks. The generated embeddings will be given to Circle Loss as detailed in Sect. 4.3.

4.3 Loss Function: Circle Loss

Most loss functions used for measuring similarity, including the triplet loss [20], directly incorporate the pairwise distances between trajectory pairs into a similarity score. This optimization approach is rigid as the similarity score linearly impacts the loss. Ideally, we would like to give greater emphasis (or penalty) when a similarity score significantly deviates from the optimum. To this end, Sun. et al. [23] proposed Circle Loss, which re-weights each similarity to highlight the less optimized similarity scores. The Circle Loss was initially proposed in computer vision with a unified formula for two elemental deep feature learning paradigms, i.e., learning with class-level and pair-wise labels, formulated as follows:

$$L = log \left[1 + \sum_{j=1}^{L} exp\left(\gamma \alpha_n^j (s_n^j - \Delta_n)\right) \sum_{i=1}^{K} exp\left(-\gamma \alpha_p^i (s_p^i - \Delta_p)\right) \right], \begin{cases} \alpha_p^i = [O_p - s_p^i]_+, \\ \alpha_n^j = [s_n^j - O_n]_+, \end{cases} \quad (2)$$

in which n and p annotations in a pairwise label paradigm refer to dissimilar and similar pairs. γ is the scale factor. L and K are the number of dissimilar and similar samples. Δ_n and Δ_p are dissimilar and similar margins, and α_n^j and α_p^i are non-negative weighting factors for pairwise distances between dissimilar samples (i.e., s_n) and similar samples (i.e., s_p), respectively. In their calculation, $[\cdot]_+$ is the "cut-off at zero" operation to ensure α_p^i and α_n^j are non-negative. O_p and O_n is the optimum similarity score for s_p and s_n respectively. To reduce the number of hyperparameters, $O_p = 1 + m$, $O_n = -m$, $\Delta_p = 1 - m$, and $\Delta_n = m$. Hence, there are only two hyper-parameters, i.e., the scale factor γ and the relaxation margin m. In a class-level paradigm, the dissimilar and similar terminologies are replaced with between-class and within-class terms using the same equation as Eq. 2. Analytically, Circle Loss offers a more flexible optimization approach towards a more definite convergence target.

In our proposed model, Circle Loss is adopted for the first time in TRL via a triplet learning paradigm using a Siamese network. We collect an equal number of dissimilar and similar pairs in each batch (i.e., $L = K$, equal to a pre-defined batch size) to create a triplet of anchor, positive, and negative samples. This also creates a balanced batch (with an equal number of positive and negative samples) in each epoch to prevent overfitting and bias towards a particular class. Moreover, we adopted the "Hard" triplet strategy [20]. In comparison with the random triplet strategy, where triplets are selected randomly, applying this strategy disregards triplets that are too easy, thereby reducing the risk of overfitting.

5 Experiments

We evaluate SiamCircle on five real trajectory datasets using seven evaluation metrics in trajectory ranking and clustering tasks. The following sections describe the experimental settings, datasets, baselines, and evaluation metrics.

Experimental Settings. SiamCircle is implemented in Tensorflow. The details of the implementation is provided in the Github repository.[1] We report the average and standard deviation of results across ten runs for each experiment. The Wilcoxon signed rank test [24] has been applied to investigate the significant differences between the top two performing models and to rank algorithms based on their performances in different metrics. In the following sections, we present the details of datasets, baselines, and evaluation metrics used in our study.

[1] The code is available at https://anonymous.4open.science/r/SiamCircle-Trajectory-Representation-Learning-17C3/.

Table 1. The characteristics of Opentraj datasets. The columns indicate the name of the dataset, the captured area, and the data size for Train, augmented train (i.e., Train($\overline{T}$)) and Test splits in the form of *(Sample Size, Number of groups).

Datasets	Area	Train (T)	Train($\overline{T}$)	Test
Eth	224.704	(40, 15)	(5098, 15)	(96, 37)
Hotel	67.210	(12, 6)	(5027, 6)	(31, 15)
Zara01	171.504	(34, 16)	(5050, 16)	(80, 35)
Zara02	158.063	(71, 35)	(5164, 35)	(167, 82)
Student03	242.884	(185, 61)	(5242, 61)	(433, 194)

Datasets. Five pedestrian datasets -*eth*, *hotel* [18], *zara01*, *zara02*, and *students03* [14]-all captured in a free setting-are utilized in the experiments. These are widely recognized benchmarks for group detection tasks using spatio-temporal data [1]. They contain the location and velocity of movements across multiple timeframes, including ground truth information on group membership. Table 1 shows the main characteristics of these datasets. Since the number of samples per dataset is limited, we adopted our novel data augmentation process described in Sect. 4.1 to generate augmented trajectories per class (or group).

Baseline. We compared SiamCircle with three other deep learning models, namely S-BiLSTM [7], TrajCL [4], and Trajectory2vec [28]. We use the released code and default parameters for all baseline methods except S-BiLSTM, which has no released code. We implement this method following their original study [7]. In TrajCL, the cell size is set to 5 as it performed better than the default value, i.e., cell size = 100, as it was designed for city-scale datasets.

Evaluation Metrics. We have adopted seven evaluation metrics in trajectory ranking and clustering tasks. The trajectory ranking task is evaluated by the top-k hitting ratio, while the clustering task is evaluated by two supervised metrics and two unsupervised metrics as follows:

Top-K Hitting Ratio. This metric examines the overlap of the *top-k*, for $k = 1, 5$, and 10, k, sorted distances between embeddings and the ground truth. Specifically, we use the Euclidean measure to calculate the distances between the original trajectory T and other trajectories in the test set. Analogously, we calculate the Euclidean distances between $E(T)$ and the embeddings of other trajectories in the test set. We then sort the obtained distances in both sets and find the number of common indexes across the window size of k. The higher overlap ratio shows better effectiveness of the measure.

Clustering Performance. Retrieving the embedding of raw trajectories allows a clustering algorithm to better identify groups of similar trajectories. To test this, we selected two supervised evaluation metrics, i.e., Normalized Mutual Information (NMI) [22] and Fowlkes Mallows (FM) score [8], and two unsupervised metrics, i.e., Davies Bouldin (DB) score [5] and Silhouette (Si) [19], to evaluate the performance of a K-mean Clustering Algorithm [16]. The number of ground truth groups, as indicated in Table 1, defines the number of clusters in the K-mean Algorithm. In summary, the NMI score measures the agreement of the two assignments, with the highest score being 1. The FM score is the geometric mean of the pairwise precision and recall, rating from 0 to 1. The Si score indicates how well clusters are separated from each other, ranging between -1 and 1. DB score is the average similarity measure of each cluster with its most similar cluster, where the minimum score is zero, with lower values indicating better clustering. In all other metrics, the higher score indicates better-defined clusters.

6 Results

In this section, the experiment's results are presented in two studies: (i) an ablation study to explore different design options and (ii) a comparative study to compare the performance of our proposed model against three baselines.

Ablation Study. Our ablation study includes loss function analysis to investigate the impact of different loss functions in our framework. Thus, we trained our proposed neural network design, i.e., $1 \times Conv2D$, with Triplet loss [20], Contrastive loss [12], USR loss [9], and Circle Loss. As shown in Table 2, the model $1 \times Conv2D$ with Circle Loss, on average, ranked higher than all the other models in almost all metrics across all datasets. There is only one exception on the unsupervised clustering task where $1 \times Conv2D$ with Circle Loss ranked last in average rank (M_R) and the rank of each unsupervised metric (i.e., DB and Si). A deeper analysis of the data shows that the worst performance of these two measures is in the *Hotel* dataset, which includes the smallest area with the smallest sample size in the original datasets (See Table 1). This makes the clusters less distinguishable, which might explain the poor performance of these two metrics, as both indicate how well clusters are separated. Yet, the other loss functions did not consistently outperform in either of these measures. Hence, $1 \times Conv2D$ with Circle Loss, i.e., SiamCircle, is selected and used in the following sections to conduct comparative experiments.

Comparative Study. In this section, we compare the performance of our proposed model against three baselines, namely S-BiLSTM, TrajCL, and Trajectory2vec, in trajectory ranking and clustering tasks using the seven evaluation metrics. The results are depicted in Table 3.

Trajectory Ranking. SiamCircle performed significantly higher than the baselines in the top Hit-k measure in almost all cases except in the Hit-1 metric of the

Table 2. The result of the ablation study, in which the highest ranks in each evaluation metric are shown in Boldface. R and M_R denote the performance ranking and the average rank per category. Horizontal lines separate categories.

Loss Function	Metric	Datasets					R	M_R
		ETH	Hotel	Student03	Zara01	Zara02		
Triplet loss	Hit-1	0.76	0.677	0.479	0.6	0.669	1.8	1.9
	Hit-5	0.79	0.852	0.67	0.758	0.801	2.0	
	Hit-10	0.857	0.935	0.69	0.814	0.815	2.0	
	NMI	0.869	0.934	0.864	0.774	0.836	2.4	2.3
	FM	0.406	0.765	0.174	0.123	0.063	2.2	
	DB	0.669	0.588	0.504	0.521	0.288	2.7	2.7
	Si	0.345	0.434	0.322	0.286	0.334	2.8	
Contrastive loss	Hit-1	0.49	0.613	0.407	0.312	0.585	2.8	2.9
	Hit-5	0.623	0.761	0.544	0.503	0.754	3.0	
	Hit-10	0.749	0.848	0.587	0.585	0.77	3.0	
	NMI	0.79	0.845	0.87	0.788	0.844	2.6	2.6
	FM	0.227	0.436	0.212	0.164	0.07	2.6	
	DB	0.606	0.336	0.454	0.643	0.239	2.0	**2.0**
	Si	0.395	0.38	0.34	0.253	0.394	**2.0**	
USR loss	Hit-1	0.302	0.452	0.172	0.162	0.296	4.0	4.0
	Hit-5	0.331	0.742	0.193	0.268	0.337	4.0	
	Hit-10	0.424	0.806	0.229	0.35	0.352	4.0	
	NMI	0.801	0.82	0.858	0.751	0.838	3.6	3.7
	FM	0.248	0.34	0.152	0.056	0.049	3.8	
	DB	0.548	0.22	0.504	0.305	0.293	**1.7**	**2.0**
	Si	0.392	0.595	0.301	0.479	0.285	2.4	
Circle Loss	Hit-1	0.771	0.523	0.686	0.724	0.732	**1.4**	**1.1**
	Hit-5	0.84	0.88	0.738	0.791	0.909	**1.0**	
	Hit-10	0.892	0.953	0.744	0.843	0.877	**1.0**	
	NMI	0.858	0.878	0.911	0.811	0.849	**1.4**	**1.4**
	FM	0.39	0.517	0.399	0.223	0.102	**1.4**	
	DB	0.627	0.671	0.526	0.589	0.324	3.6	3.2
	Si	0.328	0.218	0.331	0.287	0.366	2.8	

Hotel dataset. However, the highest performance (obtained by S-BiLSTM) is not statistically significant. Moreover, the SiamCircle outperformed with an average performance gap of 19% compared with the second best-performing model, i.e., BiLSTM. This shows that the embedding vectors retrieved by our model are helpful for trajectory ranking tasks to estimate the top similar trajectories.

Table 3. The comparative study result (mean±standard division). The statistically significant results compared to the second best results are shown by *. The highest ranks per evaluation metric are shown in Boldface. R and M_R denote the performance ranking and the average rank per category.

Model	Metric	Datasets					R	M_R
		ETH	Hotel	Student03	Zara01	Zara02		
S-BiLSTM	Hit-1	0.449±0.497	0.561±0.496	0.42±0.494	0.34±0.474	0.554±0.497	1.8	1.9
	Hit-5	0.651±0.209	0.78±0.187	0.582±0.236	0.488±0.215	0.661±0.239	2.0	
	Hit-10	0.699±0.176	0.854±0.15	0.614±0.204	0.587±0.145	0.687±0.228	2.0	
	NMI	0.799±0.01	0.859±0.036	0.864±0.006	0.763±0.014	0.851±0.009	**1.8**	2.1
	FM	0.245±0.025	0.455±0.103	0.188±0.021	0.124±0.034	0.1±0.029	2.4	
	DB	0.474±0.023	0.391±0.041	0.466±0.023	0.39±0.048	0.281±0.021	1.6	1.9
	Si	0.415±0.014	0.4±0.05	0.31±0.012	0.323±0.032	0.388±0.019	2.2	
TrajCL	Hit-1	0.097±0.296	0.106±0.308	0.071±0.257	0.094±0.291	0.079±0.27	3.8	3.5
	Hit-5	0.194±0.188	0.476±0.323	0.182±0.209	0.204±0.179	0.192±0.216	3.4	
	Hit-10	0.288±0.19	0.566±0.167	0.234±0.172	0.32±0.167	0.276±0.181	3.2	
	NMI	0.731±0.01	0.754±0.015	0.836±0.005	0.756±0.008	0.782±0.002	4.0	3.8
	FM	0.158±0.024	0.171±0.055	0.136±0.014	0.126±0.024	0.068±0.002	3.6	
	DB	0.486±0.069	0.422±0.102	**0.405 ± 0.024***	0.335±0.108	**0.073 ± 0.022***	**1.4**	**1.2**
	Si	0.435±0.056	0.412±0.082	**0.436 ± 0.048***	**0.438 ± 0.066***	**0.518 ± 0.007***	**1.0**	
Trajectory2vec	Hit-1	0.198±0.398	0.319±0.466	0.063±0.243	0.154±0.361	0.275±0.446	3.2	3.5
	Hit-5	0.134±0.142	0.401±0.272	0.096±0.131	0.23±0.237	0.376±0.304	3.6	
	Hit-10	0.16±0.107	0.455±0.141	0.117±0.11	0.268±0.135	0.385±0.291	3.8	
	NMI	0.775±0.008	0.843±0.018	0.841±0.003	0.807±0.008	0.85±0.003	2.4	2.4
	FM	0.189±0.017	0.416±0.064	0.08±0.009	0.237±0.034	**0.149 ± 0.01***	2.4	
	DB	0.638±0.028	0.472±0.028	0.611±0.013	0.564±0.035	0.31±0.014	3.2	3.5
	Si	0.277±0.011	0.388±0.028	0.221±0.006	0.262±0.014	0.324±0.009	3.8	
SiamCircle	Hit-1	**0.771 ± 0.420***	0.523±0.449	**0.686 ± 0.464***	**0.724 ± 0.447***	**0.732 ± 0.443***	**1.2**	**1.0**
	Hit-5	**0.840 ± 0.144***	**0.880 ± 0.118***	**0.738 ± 0.195***	**0.791 ± 0.146***	**0.909 ± 0.116***	**1.0**	
	Hit-10	**0.892 ± 0.086***	**0.953 ± 0.055***	**0.744 ± 0.155***	**0.843 ± 0.1***	**0.877 ± 0.146***	**1.0**	
	NMI	**0.849 ± 0.009***	0.858±0.026	**0.921 ± 0.004***	**0.805 ± 0.007***	0.838±0.003	**1.8**	**1.7**
	FM	**0.366 ± 0.025***	0.458±0.083	**0.451 ± 0.019***	0.213±0.018	0.07±0.007	**1.6**	
	DB	0.668±0.025	0.618±0.046	0.529±0.015	0.584±0.023	0.328±0.012	3.8	3.4
	Si	0.329±0.012	0.257±0.024	0.331±0.007	0.289±0.011	0.379±0.006	3.0	

Clustering. SiamCircle performed higher than the other baselines in supervised clustering measures, i.e., NMI and FM scores (with S-BiLSTM performing similarly to SiamCircle in the NMI metric), which was in contrast with the result of unsupervised clustering measures, i.e., DB and Si. This could be due to providing explicit positive and negative samples for training in these two models, which allows the models to learn more specific patterns and relationships. Additionally, the dynamic penalty strategy by Circle Loss likely helped SiamCircle achieve a higher rank than S-BiLSTM. On the other hand, TrajCL has ranked best in unsupervised clustering metrics, likely because TrajCL trains without using group membership. Thus, the obtained embeddings via TrajCL form clusters in the embedding space that do not necessarily reflect individual group memberships. This makes TrajCL particularly useful when ground truth is not available. Yet, a deep understanding of data is required to translate the findings into an implication.

Overall, SiamCircle has ranked highest compared to the baselines in Trajectory Ranking. In Trajectory Clustering, the highest performance is obtained only in supervised metrics. This demonstrates the strong capability of representations obtained by SiamCircle to be used in various domains and applications.

7 Conclusion

This study revisits the problem of TRL, specifically when trajectories are collected from free movements in small areas. We introduced a novel framework, SiamCircle, which includes a Siamese framework with Circle Loss. We conducted experiments with five benchmark datasets, using seven evaluation metrics in trajectory ranking and clustering tasks. In the ablation study, we demonstrated that one 2-dimensional convolutional layer together with Circle Loss, on average, outperformed other candidates. In a comparative study, SiamCircle consistently outperformed other models in trajectory ranking with an average performance gap of 19% compared with the second best-performing model, i.e., BiLSTM, and in unsupervised clustering measures. In future research, automatically setting the parameter of Circle loss based on the characteristics of the given datasets can be explored to make it more adaptable to diverse datasets without requiring manual hyperparameter adjustments which improves its usability in real-world applications. SiamCircle can be used in various tasks, such as analyzing social interactions in free settings, e.g., schoolyards and sports clubs. This opens up possibilities for behavioral studies, crowd analysis, and human-computer interaction research.

References

1. Amirian, J., Zhang, B., Castro, F.V., Baldelomar, J.J., Hayet, J.B., Pettré, J.: Opentraj: assessing prediction complexity in human trajectories datasets. In: Proceedings of the Asian Conference on Computer Vision (2020)
2. Bromley, J., et al.: Signature verification using a "siamese" time delay neural network. Int. J. Pattern Recogn. Artif. Intell. **7**(04) (1993)
3. Cardoso-Pereira, I., Borges, J.B., Viana, A.C., Loureiro, A.A., Ramos, H.S.: Popayi: muscling ordinal patterns for low-complex and usability-aware transportation mode detection. IEEE Internet Things J. (2024)
4. Chang, Y., Qi, J., Liang, Y., Tanin, E.: Contrastive trajectory similarity learning with dual-feature attention. In: 2023 IEEE 39th International Conference on Data Engineering (ICDE), pp. 2933–2945. IEEE (2023)
5. Davies, D.L., Bouldin, D.W.: A cluster separation measure. IEEE Trans. Pattern Anal. Mach. Intell. **2**, 224–227 (1979)
6. Ermagun, A., Levinson, D.: Spatiotemporal traffic forecasting: review and proposed directions. Transp. Rev. **38**(6), 786–814 (2018)
7. Fan, Z., Song, X., Chen, Q., Jiang, R., Shibasaki, R., Tsubouchi, K.: Trajectory fingerprint: one-shot human trajectory identification using siamese network. CCF Trans. Perv. Comput. Interact. **2**(2), 113–125 (2020)

8. Fowlkes, E.B., Mallows, C.L.: A method for comparing two hierarchical clusterings. J. Am. Stat. Assoc. **78**(383), 553–569 (1983)
9. Franceschi, J.Y., Dieuleveut, A., Jaggi, M.: Unsupervised scalable representation learning for multivariate time series. Adv. Neural Inf. Process. Syst. **32** (2019)
10. Jiang, C., et al.: Hallucination augmented contrastive learning for multimodal large language model. In: Proceedings of the IEEE/CVF Conference on Computer Vision and Pattern Recognition, pp. 27036–27046 (2024)
11. Ju, W., et al.: Cool: a conjoint perspective on spatio-temporal graph neural network for traffic forecasting. Inf. Fusion **107**, 102341 (2024)
12. Khosla, P., et al.: Supervised contrastive learning. Adv. Neural. Inf. Process. Syst. **33**, 18661–18673 (2020)
13. Kudari, M., Nandeppanavar, A.S., Thotad, P., Jatti, S., Koti, S.: A machine learning based approach for bird migration detection and feature analysis with shap. In: 2024 IEEE North Karnataka Subsection Flagship International Conference (NKCon), pp. 1–7. IEEE (2024)
14. Lerner, A., Chrysanthou, Y., Lischinski, D.: Crowds by example. In: Computer Graphics Forum, vol. 26, pp. 655–664. Wiley Online Library (2007)
15. Li, X., Zhao, K., Cong, G., Jensen, C.S., Wei, W.: Deep representation learning for trajectory similarity computation. In: 2018 IEEE 34th International Conference on Data Engineering (ICDE), pp. 617–628. IEEE (2018)
16. MacQueen, J., et al.: Some methods for classification and analysis of multivariate observations. In: Proceedings of the Fifth Berkeley Symposium on Mathematical Statistics and Probability, Oakland, CA, USA, vol. 1, pp. 281–297 (1967)
17. Oord, A.V.D., Li, Y., Vinyals, O.: Representation learning with contrastive predictive coding. arXiv preprint arXiv:1807.03748 (2018)
18. Pellegrini, S., Ess, A., Schindler, K., Van Gool, L.: You'll never walk alone: modeling social behavior for multi-target tracking. In: 2009 IEEE 12th International Conference on Computer Vision, pp. 261–268. IEEE (2009)
19. Rousseeuw, P.J.: Silhouettes: a graphical aid to the interpretation and validation of cluster analysis. J. Comput. Appl. Math. **20** (1987)
20. Schroff, F., Kalenichenko, D., Philbin, J.: Facenet: a unified embedding for face recognition and clustering. In: Proceedings of the IEEE Conference on Computer Vision and Pattern Recognition, pp. 815–823 (2015)
21. Shi, J., Gao, P., Qin, J.: Transformer-based no-reference image quality assessment via supervised contrastive learning. In: Proceedings of the AAAI Conference on Artificial Intelligence, vol. 38, pp. 4829–4837 (2024)
22. Strehl, A., Ghosh, J.: Cluster ensembles—a knowledge reuse framework for combining multiple partitions. J. Mach. Learn. Res. **3**(Dec) (2002)
23. Sun, Y., et al.: Circle loss: a unified perspective of pair similarity optimization. In: Proceedings of the IEEE/CVF Conference on Computer Vision and Pattern Recognition (2020)
24. Wilcoxon, F.: Individual comparisons by ranking methods. Biometr. Bull. **1**(6), 80–83 (1945)
25. Yan, A., Cheng, S., Kang, W.C., Wan, M., McAuley, J.: Cosrec: 2d convolutional neural networks for sequential recommendation. In: Proceedings of the 28th ACM International Conference on Information and Knowledge Management (2019)
26. Yang, B., Fan, F., Ni, R., Wang, H., Jafaripournimchahi, A., Hu, H.: A multi-task learning network with a collision-aware graph transformer for traffic-agents trajectory prediction. IEEE Trans. Intell. Transp. Syst. **25**(7), 6677–6690 (2024)

27. Yang, P., Wang, H., Yang, J., Qian, Z., Zhang, Y., Lin, X.: Deep learning approaches for similarity computation: a survey. IEEE Trans. Knowl. Data Eng. (2024)
28. Yao, D., Zhang, C., Zhu, Z., Huang, J., Bi, J.: Trajectory clustering via deep representation learning. In: 2017 International Joint Conference on Neural Networks (IJCNN), pp. 3880–3887. IEEE (2017)

Synthetic Tabular Data Detection in the Wild

G. Charbel N. Kindji[1,2(✉)], Elisa Fromont[2], Lina M. Rojas-Barahona[1], and Tanguy Urvoy[1]

[1] Orange Innovation, Lannion, France
{charbel.kindji,lina.rojas,tanguy.urvoy}@orange.com
[2] Université de Rennes, CNRS, Inria, IRISA UMR 6074, Rennes, France
elisa.fromont@irisa.fr

Abstract. Detecting synthetic tabular data is essential to prevent the distribution of false or manipulated datasets that could compromise data-driven decision-making. This study explores whether synthetic tabular data can be reliably identified across different tables. This challenge is unique to tabular data, where structures (such as number of columns, data types, and formats) can vary widely from one table to another. We propose four table-agnostic detectors combined with simple preprocessing schemes that we evaluate on six evaluation protocols, with different levels of "wildness". Our results show that cross-table learning on a restricted set of tables is possible even with naive preprocessing schemes. They confirm however that cross-table transfer (*i.e.* deployment on a table that has not been seen before) is challenging. This suggests that sophisticated encoding schemes are required to handle this problem.

Keywords: Tabular Data · Tabular Generative Models · Fake Data Detection · Classification · Transfer learning

1 Introduction

Most studies on synthetic data detection focus on images [2,4,7,27], texts [13, 21,22,28,38], audio files [26], videos (face-swap) [31], or their combination [34]. Nevertheless, a growing number of generative models for tabular data generation has emerged recently; some are general-purpose [19,44], while others are tailored to specific domains like finance [33] or healthcare [14]. These advancements will make it easier to create realistically manipulated datasets, enabling the falsification of scientific results or the concealment of fraud and accounting loopholes. Consequently, it is crucial to prioritize research on identifying synthetic tabular data and to develop detection methods that match the remarkable capabilities of the generative models.

Detecting synthetic content issued from a known generative model on a restricted domain is a relatively manageable task. The performance of such a predictor is indeed commonly used for adversarial training [10] and as a metric to assess generation performance [26,43]. However, the challenge intensifies

G. Krempl et al. (Eds.): IDA 2025, LNCS 15669, pp. 81–96, 2025.
https://doi.org/10.1007/978-3-031-91398-3_7

when attempting to detect synthetic data “in the wild” [36], namely, when the deployed system has to face modalities and content generators it has never seen during training. It is well-established that, even for homogeneous formats such as images or text, synthetic content detection systems lack robustness against *cross-generator* and *cross-domain* distribution shifts [20].

In the context of tabular data, we encounter a more pronounced form of domain shift, that we refer to as *cross-table* shift. The term “distribution shift” refers to a change in distribution within the same table as in [9]. For instance, by training on “Adult” table for a specific gender and deploying on another one. On the other hand, “cross-table shift” is a new setting involving distinct tables. For instance, by training on “Adult” and deploying on “Insurance”. We focus here on the cross-table shift.

A first requirement for a synthetic table detection system to be effective “in the wild”, is to be “table-agnostic” (*i.e.* able to take diverse table formats with varying numbers of columns and data types as input). A stronger requirement is for the model to generalize across different tables (*i.e.* to be able to work on a table structure it has never seen before). We call this “cross-table shift” adaptation.

We propose and study four “table-agnostic” baselines for synthetic tabular data detection “in the wild”. Two are based on standard models (logistic regression and gradient boosting) and two, potentially stronger, are based on Transformers. These baselines rely on different preprocessing schemes: three textual linearizations of the rows and one specific *column-based* encoding. We focus our evaluation on cross-table robustness.

The remainder of this paper is structured as follows. The related work is introduced in Sect. 2. We present datasets, models and encodings in Sect. 3. We further describe the detection in the wild in Sect. 4. Finally, we discuss the results and conclude in Sects. 5 and 6 respectively.

2 Related Work

In this section, we present prior research related to synthetic data detection as well as tabular data generation and pretrained models for representation learning on tables.

2.1 Synthetic Data Detection

Although a significant amount of work focus on image [2,27], text [28,38] and audio synthetic data detection [23], relatively few studies have focused on tabular data. Most synthetic tabular data detectors assess the quality of the generated data through frameworks like the Synthetic Data Vault (SDV) [30], which generally rely on comparing the distributions of two samples from the same dataset. The primary metric used to measure data realism is the Classifier two-Sample

Test (C2ST) [26], which evaluates how effectively a detector can distinguish real from synthetic data. Unlike previous work, we explore a broader approach that can detect synthetic data across diverse tables and in a variety of settings. Another key difference in our work is the training set used for the generators. In C2ST, the generator has only access to the train subset. Since we want to evaluate the detector, we opted for a less favorable yet more realistic scenario where the generator has access to the test subset and the detector has to be table-agnostic.

2.2 Tabular Data Generation

Several tabular data generation models have been developed in the last few years. The state of the art comprises iterative models such as diffusion models and non iterative (or "push-forward") ones. Among diffusion models we consider TabDDPM [19] and TabSyn [44]. These models aim to gradually change random Gaussian noise into samples from the target data distribution. They involve two phases: a training phase where a forward diffusion process adds Gaussian noise to the samples, and a generation phase where a backward diffusion process removes progressively the noise to generate samples from the target distribution. Following the idea in [32,37], TabSyn embeds the diffusion in a latent space by using a Variational Autoencoder (VAE) [18]. Among the *push-forward* generative models, we evaluate TVAE and CTGAN [40]. TVAE is built upon a Variational Autoencoder (VAE), while CTGAN is based on a Wasserstein GAN [10]. Both models utilize one-hot encoding for categorical features and a Gaussian Mixture Model (GMM) normalization approach for numerical features. Additionally, CTGAN incorporates specialized techniques, such as a conditional vector and oversampling, to address imbalances in categorical values.

2.3 Table-Agnostic Embeddings and Encodings

Research on tabular data has also explored representation learning for tables. Most off-the-shelf table representation models are inspired by BERT-like architectures and are designed for small tables, such as those found in popular datasets like WikiTableQuestions [29], WikiSQL [45], and TabFact [6]. These models include TaBERT [42], which is pretrained on 26 million tables, jointly learning representations for both tables and their accompanying descriptions. Other models, such as TAPAS [12] and TAPEX [25], focus on question answering over tables. TAPAS is a BERT-based model trained using a Masked Language Modeling (MLM) objective, where table cells or text segments are masked during training. TAPEX, on the other hand, is a BART-based model pretrained to emulate the functionality of an SQL execution engine when interacting with tables. All these encoders rely on a text linearization of the tables or rows which is fed to a standard subword tokenizer. Some recently proposed architectures like CARTE [16] and PORTAL [35] replace the text linearization by a table-specific encoding scheme.

We focus here on large tables commonly found in databases, which are often too extensive to be processed directly by BERT-like models due to their context-token limits (512 for BERT, 1024 for BART). For both representation learning and classification tasks, it is crucial to decide on an appropriate level of *granularity*, meaning the number of rows to be included as input. In our approach, we utilize classifiers that process a single row at a time, while recognizing that certain table properties may only become apparent when considering multiple rows together.

Concerns such as the lack of control, potential data leakage between test sets and pretraining tables, discrepancies between pretraining and fine-tuning objectives, model size, and environmental impact (e.g., carbon footprint) led us to develop and train our own lightweight transformers rather than relying on a pretrained model.

3 Datasets, Encodings and Detectors

In this section, we present the real and synthetic datasets in Sect. 3.1, the text and *column-based* encodings in Sect. 3.2 as well as the detection models in Sect. 3.3.

3.1 Datasets

Real Data: We use 14 common public tabular datasets from the UCI[1] repository with different sizes, dimensions and domains. These datasets are described in Table 1.

Synthetic Data: The *data generators* are heavily tuned versions of TabDDPM [19], TabSyn [44], TVAE, and CTGAN [40] provided by [17].

To generate a synthetic version of a specific table using a particular generator, we first train the generator on the entire table using the hyperparameters derived from the extensive tuning in [17]. After training, the model generates synthetic rows. These are then mixed with real and other synthetic data to form the detectors' training and testing sets. In our setup, each generator is used to create a synthetic version of the 14 real tables (Table 1) resulting in $4*14 = 56$ synthetic tables.

[1] https://archive.ics.uci.edu/.

Table 1. Description of the datasets. "#Num" refers to the number of numerical attributes and "#Cat" the number of categorical ones for each dataset.

Name	Size	#Num	#Cat
Abalone[a]	4177	7	2
Adult[a]	48842	6	9
Bank Marketing[a]	45211	7	10
Black Friday[a]	166821	6	4
Bike Sharing[a]	17379	9	4
Cardio[b]	70000	11	1
Churn Modelling[b]	4999	8	4
Diamonds[a]	26970	7	3
HELOC[b]	5229	23	1
Higgs[a]	98050	28	1
House 16H[a]	22784	17	0
Insurance[b]	1338	4	3
King[b]	21613	19	1
MiniBooNE[a]	130064	50	1

[a] https://www.openml.org
[b] https://www.kaggle.com/datasets

3.2 Table-Agnostic Data Encodings

To be practical "in the wild", a detection model must be "table-agnostic," meaning it should be capable of processing inputs from various table formats.

We explore four simple table-agnostic preprocessing schemes: three *text-based* data encodings and one *column-based* encoding. For the *text-based* approaches, the table is first linearized into text. From this, we derive three representations: (i) character trigrams (referred to as "*3gram-char*"), (ii) word trigrams (referred to as "*3gram-word*"), and (iii) a general linearized representation (referred to as "*Flat Text*"). For the *column-based* encoding (referred to as "*Column*" in the results), we use a coarse-grained column-level encoding of tables.

Text-Based Encodings: A natural solution to build a table-agnostic model is to consider the tables as raw text. This approach is used in pretrained models such as TaBERT [42], TAPAS [12], or TAPEX [25]. These models are designed to encode small tables like the ones found on Wikipedia. They are derived from BERT and rely on a text encoding of the whole table. In order to work with larger tables we opted, as in [3], to work at the row level. The left-hand side of Fig. 1 illustrates the row linearization process.

We converted each table row into a shuffled sequence of `<column>:<value>` patterns. For example, the first row of Table 1 can be encoded as the string: `"Name: Abalone,Size:4177,#Num:7,#Cat:2"` or any permutation of its columns. This random column permutation is designed to enhance generalization

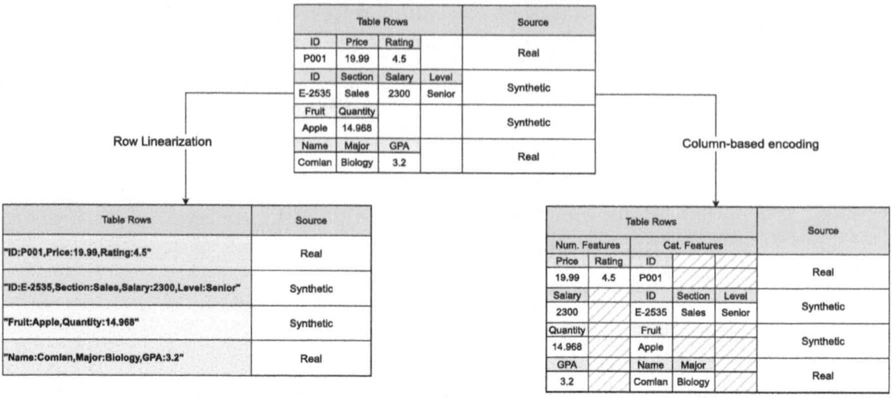

Fig. 1. Table-agnostic data encodings used in this study. **At the top**: Training data with rows from multiple table structures. **On the left**: Flat-text encoding, where each row is converted to text. **On the right**: Column-based encoding with padding (represented by hatched cells). "Num. Features" stands for "Numerical Features" and "Cat. Features" stands for "Categorical Features".

across different tables. Then three options are considered: (i) For *3gram-char*, the string is simply split into a bag of character-level trigrams such as `"Nam"`, `"e:A"`, `":41"` or ,`"t:2"` (ii) For *3gram-word*, it is split into bags of word-level trigrams such as `"Name Abalone Size"` or `"4177 #Num 7"`; (iii) For the *Flat Text* encoding, the string is tokenized into a sequence of characters indexes.

These basic text-based encodings have the advantage that they can be applied to any table, even in multitable setups with varying numbers of columns. They can also help to highlight obvious patterns in the data. We also employ traditional binary text classification methods for establishing baselines for our problem, particularly the Bag-Of-Words (BOW) approach [1], and explore both word-level and character-level trigrams.

Column-Based Encoding: In this approach, all datasets are encoded using the same procedure: numerical features are normalized using the *QuantileTransformer*, and categorical features are encoded with the *OrdinalEncoder*, both from scikit-learn[2]. Notably, each dataset is processed independently, ensuring that the methods for encoding numerical and categorical features are applied separately to avoid any unintended interactions between the cross-table features. This approach preserves the natural relationships between features, helping detectors learn meaningful interactions. It also emphasises unique feature characteristics, which may improve pattern recognition.

A key challenge when working with tabular data is its heterogeneous nature. For instance, tabular datasets often vary in the number of features they contain. Unlike image data, which can typically be normalized to a consistent range, standardizing tabular data is less straightforward. To address this characteristic, we

[2] https://scikit-learn.org/stable/.

first split the columns according to their types: the numerical and the categorical features. For models with a fixed input size like logistic regression or decision trees, we then apply a padding scheme or a crop to match their input dimensions. The procedure is illustrated on the right-hand side of Fig. 1. A notable advantage of this – very basic – approach is that it is universally applicable to any new table encountered by the detector. It also effectively mitigates data type conflicts by treating numeric and categorical features separately.

3.3 Detectors

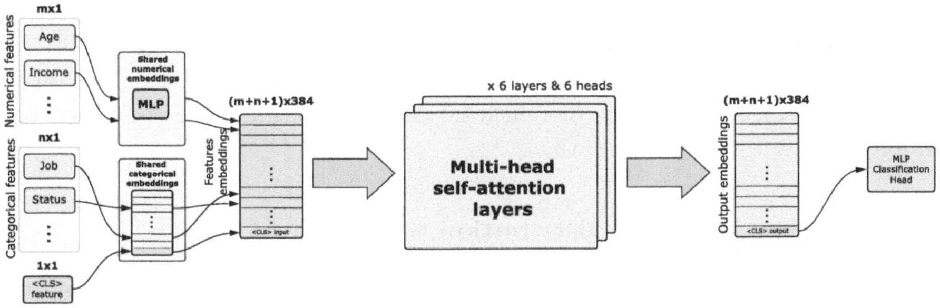

Fig. 2. Column-based Transformer model architecture for m numerical features and n categorical features.

We deployed two Transformer-based detectors: one utilizing *column-based* encoding and the other employing *Flat Text* encoding. They share the following core components: (i) an embedding block, (ii) a Transformer encoder block, and (iii) a classification head. For the *column-based* Transformer, as shown in Fig. 2, the feature embedding module uses a shared feed-forward layer for numerical features and a shared embedding layer for categorical features. Additionally, we have incorporated a positional encoding layer into the detector. While this baseline is relatively simple, more advanced strategies are discussed in [39] and [35]. For the *Flat Text* Transformer, the tokenized input is mapped, as usual for Transformers, into a sequence of embedding vectors that are combined with a positional encoding.

Similar to BERT-like [8] Transformer architectures, our Transformer-based detectors utilize a *CLS* embedding, which is appended to the input and whose representation is extracted from the output of the Transformer encoder block. This *CLS* representation is then passed to the classification head to predict the binary target class (real or synthetic data).

Transformers were chosen primarily because they can be designed to be permutation invariant and are capable of handling inputs with varying lengths. This flexibility is particularly advantageous for a table-agnostic detector, which must operate on sets of features with variable sizes. Furthermore, while tree-based models tend to outperform neural networks on medium-sized tabular data [11], transformer-based models offer an advantage over tree-based competitors when handling large tables [15].

We trained *XGBoost* [5] and *Logistic Regression* detectors on all four preprocessing schemes. All detectors are trained for binary classification.

4 Detection Setups

All dataset rows are combined into a list with two additional labels: the dataset name (to simulate the *cross-table shift*) and the origin, which can be either "real" or "synthetic". In the following sections, we outline two setups: training detectors without distribution shift (Sect. 4.1) and with a cross-table shift (Sect. 4.2). For each setup, we apply the preprocessing schemes described in Sect. 3.3 to evaluate the detectors. We use a 3-fold cross-validation protocol which evaluates discrepancies across three runs of synthetic data generation, each combined with a different subset of the real data.

4.1 Detection Without Distribution Shift

First, we train detectors on the detection of synthetic data generated by each generator, resulting in four setups, namely: (i) *TVAE vs Real*, (ii) *CTGAN vs Real*, (iii) *TabSyn vs Real*, and (iv) *TabDDPM vs Real*. In each one of these setups, we mix all 14 synthetic tables produced by each generator along with real tables. Rows are labelled accordingly. We include a fifth setup where each synthetic dataset is generated using a combination of all generators, mixed along with real data, and labelled accordingly. This setup is referred to as *All Models vs Real*.

With these setups, we aim to evaluate the effectiveness of the detectors in accurately identifying data generated by each generator independently, using a restricted set of datasets. While our primary focus is on the *cross-table shift*, we consider this an essential step for both validating the detectors and ensuring their effectiveness in realistic use cases that practitioners may encounter (especially the *All Models vs Real*). Moreover, these setups can also be viewed as an advanced version of the C2ST performed on multiple tables instead of table-specific like the approach in [17]. They are also more realistic and challenging for the detectors as the generators are exposed to the full available real data and not only a train subset.

4.2 Detection Under Cross-Table Shift

As explained in Sect. 1, the cross-table shift is a setup that simulates the realistic situation where the model is deployed on table structures it has never seen before. This setup can be implemented with GroupKFold from scikit-learn[3]. For

[3] https://scikit-learn.org/stable/.

Table 2. Example of a *cross-table shift* split. The white cells indicate the training elements, while the grey cells represent the test set.

		Tables		
		Adult	Cardio	King
Real Data				
Generators	TVAE			
	TabSyn			
	CTGAN			
	TabDDPM			

example, as shown in Table 2, if Tables *Adult* and *Cardio* are used in training, they cannot be used in testing. Similarly, if Table *King* is used in the test, it cannot be used in the training phase.

5 Results

We present the results of our baselines across different setups, both without (Sect. 5.1) and with (Sect. 5.2) *cross-table shift*. These results are summarized in Table 3 and Table 4 using standard metrics including *ROC-AUC*, *Accuracy*, and *F1* score.

Table 3. Performance report for various table-agnostic setups without distribution shift under 3-Fold cross-validation. "3gram-char" stands for "trigrams of characters", "3gram-word" stands for "trigrams of words", and "Transf." stands for "Transformer". "Column" refers to *column-based* encoding. "no shift" refers to the absence of cross-table shift. Best performance are highlighted in **bold**.

Setup	Model	Encoding	Metrics		
			AUC	Accuracy	F1
TVAE vs Real (All tables no shift)	LReg.	3gram-char	0.72 ± 0.00	0.65 ± 0.00	0.66 ± 0.00
		3gram-word	0.57 ± 0.00	0.53 ± 0.00	0.54 ± 0.00
		Column	0.59 ± 0.00	0.56 ± 0.00	0.57 ± 0.00
		Flat Text	0.63 ± 0.00	0.59 ± 0.00	0.60 ± 0.00
	XGBoost	3gram-char	0.51 ± 0.02	0.51 ± 0.01	0.54 ± 0.09
		3gram-word	0.51 ± 0.00	0.51 ± 0.00	0.67 ± 0.00
		Column	0.84 ± 0.00	0.75 ± 0.00	0.76 ± 0.00
		Flat Text	0.77 ± 0.00	0.69 ± 0.00	0.70 ± 0.00
	Transf.	Column	$\mathbf{0.92 \pm 0.00}$	$\mathbf{0.83 \pm 0.00}$	$\mathbf{0.83 \pm 0.00}$
		Flat Text	0.76 ± 0.01	0.67 ± 0.01	0.67 ± 0.03
CTGAN vs Real (All tables no shift)	LReg.	3gram-char	0.61 ± 0.00	0.57 ± 0.00	0.56 ± 0.00
		Column	0.53 ± 0.00	0.52 ± 0.00	0.53 ± 0.00
		Flat Text	0.56 ± 0.00	0.55 ± 0.00	0.53 ± 0.00

(*continued*)

Table 3. (*continued*)

Setup	Model	Encoding	Metrics		
			AUC	Accuracy	F1
	XGBoost	3gram-char	0.51 ± 0.00	0.50 ± 0.00	0.33 ± 0.02
		3gram-word	0.50 ± 0.00	0.50 ± 0.00	0.00 ± 0.00
		Column	0.70 ± 0.00	0.63 ± 0.00	0.60 ± 0.00
		Flat Text	0.64 ± 0.00	0.60 ± 0.00	0.56 ± 0.00
	Transf.	Column	**0.86 ± 0.00**	**0.77 ± 0.00**	**0.76 ± 0.01**
		Flat Text	0.62 ± 0.02	0.58 ± 0.01	0.53 ± 0.04
TabSyn vs Real (All tables no shift)	LReg.	3gram-char	0.78 ± 0.00	0.68 ± 0.00	0.68 ± 0.00
		3gram-word	0.84 ± 0.00	0.75 ± 0.00	**0.75 ± 0.00**
		Column	0.52 ± 0.00	0.51 ± 0.00	0.51 ± 0.00
		Flat Text	0.79 ± 0.00	0.68 ± 0.00	0.67 ± 0.00
	XGBoost	3gram-char	0.51 ± 0.01	0.50 ± 0.00	0.43 ± 0.16
		3gram-word	0.53 ± 0.00	0.53 ± 0.00	0.12 ± 0.00
		Column	0.72 ± 0.00	0.64 ± 0.00	0.64 ± 0.00
		Flat Text	**0.87 ± 0.00**	**0.76 ± 0.00**	**0.75 ± 0.00**
	Transf.	Column	0.82 ± 0.00	0.71 ± 0.00	0.71 ± 0.00
		Flat Text	0.86 ± 0.01	0.73 ± 0.01	0.72 ± 0.06
TabDDPM vs Real (All tables no shift)	LReg.	3gram-char	0.75 ± 0.00	0.65 ± 0.00	0.65 ± 0.00
		3gram-word	0.83 ± 0.00	**0.74 ± 0.00**	**0.75 ± 0.00**
		Column	0.52 ± 0.00	0.51 ± 0.00	0.50 ± 0.00
		Flat Text	0.70 ± 0.00	0.61 ± 0.00	0.61 ± 0.00
	XGBoost	3gram-char	0.51 ± 0.00	0.51 ± 0.00	0.03 ± 0.00
		3gram-word	0.51 ± 0.00	0.51 ± 0.00	0.03 ± 0.00
		Column	0.66 ± 0.00	0.60 ± 0.00	0.60 ± 0.00
		Flat Text	0.81 ± 0.00	0.70 ± 0.00	0.69 ± 0.00
	Transf.	Column	0.74 ± 0.00	0.65 ± 0.00	0.65 ± 0.00
		Flat Text	**0.86 ± 0.00**	**0.74 ± 0.00**	**0.75 ± 0.04**
All Models vs Real (All tables no shift)	LReg.	3gram-char	0.64 ± 0.00	0.59 ± 0.00	0.58 ± 0.00
		3gram-word	0.57 ± 0.00	0.55 ± 0.00	0.56 ± 0.00
		Column	0.52 ± 0.00	0.52 ± 0.00	0.53 ± 0.00
		Flat Text	0.63 ± 0.00	0.58 ± 0.00	0.54 ± 0.00
	XGBoost	3gram-char	0.50 ± 0.00	0.50 ± 0.00	0.39 ± 0.27
		3gram-word	0.51 ± 0.00	0.51 ± 0.00	0.03 ± 0.00
		Column	0.66 ± 0.01	0.61 ± 0.00	0.59 ± 0.00
		Flat Text	0.73 ± 0.00	0.66 ± 0.00	0.60 ± 0.00
	Transf.	Column	**0.77 ± 0.00**	**0.69 ± 0.00**	**0.68 ± 0.01**
		Flat Text	0.73 ± 0.04	0.66 ± 0.05	0.65 ± 0.03

5.1 Without Distribution Shift

The first observation from Table 3 is that Transformer-based detectors perform well across all metrics and setups with an AUC above 0.70 across all setups, except for *CTGAN vs Real* where the AUC is 0.62 for the *Flat Text* encoding.

This suggests that these table-agnostic detectors have good generalization capabilities when there is no distribution shift. Note however that *XGBoost* slightly outperforms the Transformers on the *TabSyn vs Real* setup (0.86 ± 0.01 for the best Transformer versus 0.87 ± 0.00 for *XGBoost*).

The encoding scheme has also a strong impact on the performance for all detectors. *XGBoost* performs better with *column-based* or *Flat Text* encodings in all setups. However, its performance is close to random guessing with an *AUC* around 0.50 for the trigram encodings, and it shows poor performance on other metrics as well. As for *Logistic Regression*, performance varies across setups, with either encoding yielding better results depending on the specific setup. Nonetheless, it delivers its best performance with the three textual encodings, achieving an average *AUC* of 0.70 under the *3gram-char* encoding.

Regarding the generators, we note that TVAE is easier to detect than the other generators. Indeed, despite its simple design, the table-agnostic *column-based* Transformer achieves an *AUC* of 0.92 for detecting TVAE rows against real rows. This is an improvement from the results reported in [17] where the median *AUC* for *XGBoost* was 0.81 with the less challenging C2ST setup (see Sect. 2.1). Similarly, the median *AUC* for detecting TabSyn samples is 0.63 in [17], whereas we achieve an *AUC* above 0.85 across multiple detectors (Logistic Regression with *3gram-word*, *XGBoost*, and *Flat Text* Transformer) when combining datasets generated by TabSyn.

The task becomes more challenging under the *All models vs Real* setup, but overall performance remains stable across all detectors. As a secondary observation, we note that there is still considerable room for improvement in achieving realistic tabular data generation. The synthetic tabular data generators appear to exhibit patterns that even a simple table-agnostic classifier can detect.

Table 4. Same performance report as Table 3 but with cross-table shift.

Setup	Model	Encoding	Metrics		
			AUC	Accuracy	F1
Cross-table shift (All tables all models)	LReg.	3gram-char	**0.60 ± 0.05**	**0.52 ± 0.03**	0.45 ± 0.17
		3gram-word	0.50 ± 0.00	0.50 ± 0.00	0.00 ± 0.00
		Column	0.50 ± 0.01	0.50 ± 0.00	0.45 ± 0.12
		Flat Text	0.52 ± 0.06	0.50 ± 0.00	0.30 ± 0.27
	XGBoost	3gram-char	0.49 ± 0.01	0.49 ± 0.01	0.06 ± 0.06
		3gram-word	0.50 ± 0.00	0.50 ± 0.00	**0.67 ± 0.00**
		Column	0.51 ± 0.01	0.50 ± 0.00	0.26 ± 0.12
		Flat Text	0.49 ± 0.03	0.49 ± 0.01	0.05 ± 0.04
	Transf.	Column	0.51 ± 0.00	0.50 ± 0.00	0.32 ± 0.03
		Flat Text	**0.60 ± 0.07**	**0.52 ± 0.01**	0.40 ± 0.14

5.2 Cross-Table Shift

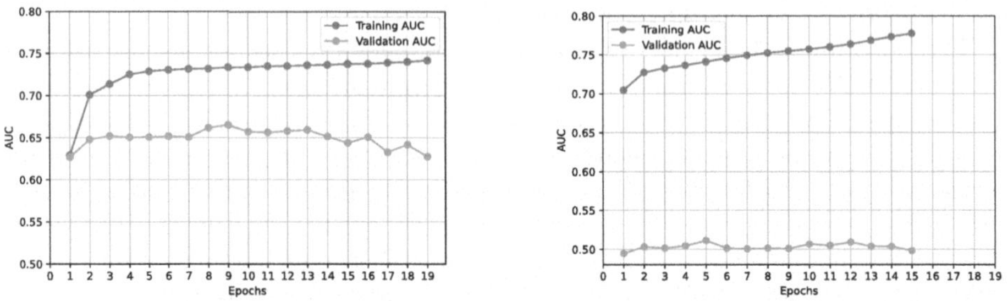

(a) *Flat Text* Transformer detector (b) *Column-based* Transformer detector

Fig. 3. Training and validation AUC performance of the Transformer-based detectors trained under *cross-table shift* setup.

The *cross-table shift* results (Table 4) highlight the difficulty of this setup, as all detectors struggle to achieve correct performance. The *column-based* Transformer experiences a significant drop in performance, with an AUC of 0.51. The detector fails to identify meaningful patterns and is unable to generalize to unseen datasets, essentially making random guesses on the test set.

It is interesting to note that the *Flat Text* Transformer outperforms the *column-based* one with an AUC of 0.60. The training curves shown in Fig. 3 confirm that, with a *cross-table shift* across all training, validation, and test sets, the *Flat Text* Transformer (on the left-hand side) is more robust than the *column-based* Transformer (on the right-hand side). The *table-agnostic* encoding used in the *column-based* method shows its limitations when evaluated on unseen tables. Being closely tied to the specific characteristics of the datasets, this encoding does not generalize well to datasets with different features (e.g., the number of features, range of numerical values, categories in categorical features, and sample size). In contrast, the *Flat Text* representation yields more promising results in presence of a *cross-table shift.*

The *Logistic Regression* baseline shows poor performance with most encodings, particularly with *3gram-word*, which scores 0.0 for *F1*. Nethertheless, it achieves decent performance with the *3gram-char* encoding ($AUC = 0.60$), but, unlike Transformers [24,41,46], its potential for improvement is limited. As for *XGBoost*, it now produces random guesses (AUC around 0.50) and performs poorly on the accuracy as well. Our analysis also shows that the F1 score of 0.67 results from predicting all samples as the same class, leading to a recall of 1.0 and a precision of 0.5.

Taken together, these results suggest that the textual preprocessing schemes in the *cross-table* setup lead to more promising results, especially the character-based encodings. These preliminary findings indicate the need for further investigation into Transformer-based detectors with both *Flat Text* and *column-based* encodings. Additionally, the potential for transfer learning from pre-trained models could further enhance performance, making Transformer-based approaches a valuable asset in the *cross-table* setup.

5.3 Limitations

The results demonstrate that our *column-based* Transformer, combined with its simple table-agnostic preprocessing and feature embedding scheme, performs well when there is no distribution shift. However, it struggles to generalize effectively under *cross-table shift*. A better encoding scheme is hence required.

In contrast, the *text-based* Transformer is simpler and more flexible. However, it results in longer row-encoding sequences and it lacks tabular-specific inductive bias, potentially limiting its ability to capture patterns that are unique to tabular data.

A limitation of this study is that although we have deployed light transformers which are much smaller than most *on-the-shelf* table embedding models, we did not analyze their cost and scalability for large tables. Another limitation of this study is that we only considered row-by-row detection. But we acknowledge that some statistical properties, especially for time-series or sequential data, may become apparent only when considering multiple rows together.

6 Conclusion

We investigated the detection of synthetic tabular data "in the wild" using 14 datasets and 4 state-of-the-art, highly-tuned tabular data generation models. By evaluating various detectors with different tabular data representations as inputs, we demonstrated that synthetic data detection is both feasible and shows promising performance. Our analysis spanned different levels of "wildness", considering scenarios both with and without *cross-table shift*, which was a central focus of our study. While our preliminary findings are encouraging, they underscore the challenges of adapting to *cross-table* shifts. Moving forward, we aim to refine these results by exploring more advanced encodings, addressing other types of distribution shifts, and developing improved adaptation strategies. We also plan to evaluate recent approaches, such as [35], while also developing a benchmark platform for cross-table synthetic tabular data detection, inspired by [9]. The benchmark will provide a detailed evaluation of the costs and scalability of the detectors for large tables.

References

1. Alsammak, I., Itwee, W.H., Mahmoud, M.A., Jasim, N.I.: Text classification: a comprehensive survey from traditional approaches to deep learning methods. In: Al-Sharafi, M.A., Al-Emran, M., Mahmoud, M.A., Arpaci, I. (eds.) Current and Future Trends on AI Applications. Studies in Computational Intelligence, vol. 1178, pp. 247–267. Springer, Cham (2025). https://doi.org/10.1007/978-3-031-75091-5_14

2. Bammey, Q.: Synthbuster: towards detection of diffusion model generated images. IEEE Open J. Signal Process. **5**, 1–9 (2024). https://doi.org/10.1109/OJSP.2023.3337714
3. Borisov, V., Sessler, K., Leemann, T., Pawelczyk, M., Kasneci, G.: Language models are realistic tabular data generators. In: The Eleventh International Conference on Learning Representations (2023)
4. Chai, L., Bau, D., Lim, S.-N., Isola, P.: What makes fake images detectable? Understanding properties that generalize. In: Vedaldi, A., Bischof, H., Brox, T., Frahm, J.-M. (eds.) ECCV 2020. LNCS, vol. 12371, pp. 103–120. Springer, Cham (2020). https://doi.org/10.1007/978-3-030-58574-7_7
5. Chen, T., Guestrin, C.: XGBoost: a scalable tree boosting system. In: Proceedings of the 22nd ACM SIGKDD International Conference on Knowledge Discovery and Data Mining, pp. 785–794 (2016)
6. Chen, W., et al.: Tabfact: a large-scale dataset for table-based fact verification. In: International Conference on Learning Representations (2020). https://openreview.net/forum?id=rkeJRhNYDH
7. Corvi, R., Cozzolino, D., Zingarini, G., Poggi, G., Nagano, K., Verdoliva, L.: On the detection of synthetic images generated by diffusion models, pp. 1–5 (2023). https://doi.org/10.1109/ICASSP49357.2023.10095167
8. Devlin, J., Chang, M.W., Lee, K., Toutanova, K.: BERT: pre-training of deep bidirectional transformers for language understanding. In: North American Chapter of the Association for Computational Linguistics (2019). https://api.semanticscholar.org/CorpusID:52967399
9. Gardner, J., Popovic, Z., Schmidt, L.: Benchmarking distribution shift in tabular data with tableshift. In: Advances in Neural Information Processing Systems, vol. 36 (2024)
10. Goodfellow, I., et al.: Generative adversarial networks. Commun. ACM **63**(11), 139–144 (2020)
11. Grinsztajn, L., Oyallon, E., Varoquaux, G.: Why do tree-based models still outperform deep learning on typical tabular data? In: Thirty-sixth Conference on Neural Information Processing Systems Datasets and Benchmarks Track (2022). https://openreview.net/forum?id=Fp7__phQszn
12. Herzig, J., Nowak, P.K., Müller, T., Piccinno, F., Eisenschlos, J.: TaPas: weakly supervised table parsing via pre-training. In: Jurafsky, D., Chai, J., Schluter, N., Tetreault, J. (eds.) Proceedings of the 58th Annual Meeting of the Association for Computational Linguistics, pp. 4320–4333. Association for Computational Linguistics, Online (2020)
13. Hu, X., Chen, P.Y., Ho, T.Y.: Radar: robust AI-text detection via adversarial learning. Adv. Neural. Inf. Process. Syst. **36**, 15077–15095 (2023)
14. Hyun, J., Lee, S.H., Son, H., Park, J.U., Chung, T.M.: A synthetic data generation model for diabetic foot treatment. In: International Conference on Future Data and Security Engineering (2020). https://api.semanticscholar.org/CorpusID:228085848
15. Karlsson, A., Wang, T., Nowaczyk, S., Pashami, S., Asadi, S.: Mind the data, measuring the performance gap between tree ensembles and deep learning on tabular data. In: Miliou, I., Piatkowski, N., Papapetrou, P. (eds.) Advances in Intelligent Data Analysis XXII, pp. 65–76. Springer, Cham (2024)
16. Kim, M.J., Grinsztajn, L., Varoquaux, G.: Carte: pretraining and transfer for tabular learning. In: Forty-first International Conference on Machine Learning (2024)
17. Kindji, G.C.N., Rojas-Barahona, L.M., Fromont, E., Urvoy, T.: Under the hood of tabular data generation models: the strong impact of hyperparameter tuning (2024). https://arxiv.org/abs/2406.12945

18. Kingma, D.P., Welling, M.: Auto-encoding variational bayes. Stat **1050**, 1 (2014)
19. Kotelnikov, A., Baranchuk, D., Rubachev, I., Babenko, A.: Tabddpm: modelling tabular data with diffusion models. In: International Conference on Machine Learning, pp. 17564–17579. PMLR (2023)
20. Kuznetsov, K., et al.: Robust AI-generated text detection by restricted embeddings. In: Findings of the Association for Computational Linguistics: EMNLP 2024, pp. 17036–17055 (2024)
21. Lahby, M., Aqil, S., Yafooz, W.M., Abakarim, Y.: Online fake news detection using machine learning techniques: a systematic mapping study. In: Combating Fake News with Computational Intelligence Techniques, pp. 3–37 (2022)
22. Lavergne, T., Urvoy, T., Yvon, F.: Filtering artificial texts with statistical machine learning techniques. Lang. Resour. Eval. **45**, 25–43 (2011)
23. Li, X., Chen, P.Y., Wei, W.: Sonar: a synthetic AI-audio detection framework and benchmark (2024). https://arxiv.org/abs/2410.04324
24. Li, Y., McClelland, J.: Systematic generalization and emergent structures in transformers trained on structured tasks (2023). https://openreview.net/forum?id=pXDmbfVL_SB
25. Liu, Q., et al.: TAPEX: table pre-training via learning a neural SQL executor. In: International Conference on Learning Representations (2022). https://openreview.net/forum?id=O50443AsCP
26. Lopez-Paz, D., Oquab, M.: Revisiting classifier two-sample tests. In: International Conference on Learning Representations (2016)
27. Marra, F., Gragnaniello, D., Verdoliva, L., Poggi, G.: Do GANs leave artificial fingerprints? . In: 2019 IEEE Conference on Multimedia Information Processing and Retrieval (MIPR), pp. 506–511. IEEE Computer Society (2019)
28. Mitchell, E., Lee, Y., Khazatsky, A., Manning, C.D., Finn, C.: DetectGPT: zero-shot machine-generated text detection using probability curvature. In: Proceedings of the 40th International Conference on Machine Learning, ICML 2023. JMLR.org (2023)
29. Pasupat, P., Liang, P.: Compositional semantic parsing on semi-structured tables. In: Zong, C., Strube, M. (eds.) Proceedings of the 53rd Annual Meeting of the Association for Computational Linguistics and the 7th International Joint Conference on Natural Language Processing (Volume 1: Long Papers), pp. 1470–1480. Association for Computational Linguistics, Beijing, China (2015). https://doi.org/10.3115/v1/P15-1142
30. Patki, N., Wedge, R., Veeramachaneni, K.: The synthetic data vault. In: IEEE International Conference on Data Science and Advanced Analytics (DSAA), pp. 399–410 (2016). https://doi.org/10.1109/DSAA.2016.49
31. Pu, J., et al.: Deepfake videos in the wild: analysis and detection. In: Proceedings of the Web Conference 2021, pp. 981–992 (2021)
32. Rombach, R., Blattmann, A., Lorenz, D., Esser, P., Ommer, B.: High-resolution image synthesis with latent diffusion models. In: Proceedings of the IEEE/CVF Conference on Computer Vision and Pattern Recognition, pp. 10684–10695 (2022)
33. Sattarov, T., Schreyer, M., Borth, D.: Findiff: diffusion models for financial tabular data generation. In: Proceedings of the Fourth ACM International Conference on AI in Finance, ICAIF 2023, pp. 64–72. Association for Computing Machinery (2023)
34. Singhal, S., Kabra, A., Sharma, M., Shah, R.R., Chakraborty, T., Kumaraguru, P.: Spotfake+: a multimodal framework for fake news detection via transfer learning (student abstract). In: Proceedings of the AAAI Conference on Artificial Intelligence, vol. 34, pp. 13915–13916 (2020)

35. Spinaci, M., Polewczyk, M., Klein, T., Thelin, S.: Portal: scalable tabular foundation models via content-specific tokenization. In: NeurIPS 2024 Third Table Representation Learning Workshop (2024)
36. Stadelmann, T., Amirian, M., Arabaci, I., Arnold, M., Duivesteijn, G.F., Elezi, I., Geiger, M., Lörwald, S., Meier, B.B., Rombach, K., Tuggener, L.: Deep learning in the wild. In: Pancioni, L., Schwenker, F., Trentin, E. (eds.) ANNPR 2018. LNCS (LNAI), vol. 11081, pp. 17–38. Springer, Cham (2018). https://doi.org/10.1007/978-3-319-99978-4_2
37. Vahdat, A., Kreis, K., Kautz, J.: Score-based generative modeling in latent space. Adv. Neural. Inf. Process. Syst. **34**, 11287–11302 (2021)
38. Wang, Q., Zhang, L., Guo, Z., Mao, Z.: IDEATE: detecting AI-generated text using internal and external factual structures. In: Proceedings of the 2024 Joint International Conference on Computational Linguistics, Language Resources and Evaluation (LREC-COLING 2024), pp. 8556–8568 (2024)
39. Wang, Z., Sun, J.: Transtab: learning transferable tabular transformers across tables. Adv. Neural. Inf. Process. Syst. **35**, 2902–2915 (2022)
40. Xu, L., Skoularidou, M., Cuesta-Infante, A., Veeramachaneni, K.: Modeling tabular data using conditional GAN. In: Advances in Neural Information Processing Systems (NeurIPS), vol. 32 (2019)
41. Yadlowsky, S., Doshi, L., Tripuraneni, N.: Can transformer models generalize via in-context learning beyond pretraining data? In: NeurIPS 2023 Workshop on Distribution Shifts: New Frontiers with Foundation Models (2024)
42. Yin, P., Neubig, G., Yih, W., Riedel, S.: Tabert: pretraining for joint understanding of textual and tabular data. In: Proceedings of the 58th Annual Meeting of the Association for Computational Linguistics, ACL, pp. 8413–8426 (2020)
43. Zein, E.H., Urvoy, T.: Tabular data generation: can we fool XGBoost ? In: NeurIPS 2022 First Table Representation Workshop (2022)
44. Zhang, H., et al.: Mixed-type tabular data synthesis with score-based diffusion in latent space. In: The Twelfth International Conference on Learning Representations (2024)
45. Zhong, V., Xiong, C., Socher, R.: Seq2SQL: generating structured queries from natural language using reinforcement learning. CoRR abs/1709.00103 (2017)
46. Zhou, H., et al.: What algorithms can transformers learn? A study in length generalization. In: The Twelfth International Conference on Learning Representations (2024)

Assessing the Graph Structure Learning in Graph Deviation Networks

Canberk Ozen[1(✉)], Sławomir Nowaczyk[1], Prayag Tiwari[1], and Sepideh Pashami[1,2]

[1] Halmstad University, Halmstad, Sweden
{canberk.ozen,slawomir.nowaczyk,prayag.tiwari,sepideh.pashami}@hh.se
[2] RISE Research Institute of Sweden, Gothenburg, Sweden

Abstract. Statistical modeling of multivariate time-series data poses significant challenges due to their high dimensionality and complex inter-variable relationships. Reliable forecasts or anomaly detection on these datasets require capturing such relationships within and between the features. While traditional deep learning architectures are good at capturing temporal non-linear patterns within features, they are less efficient at modeling inter-variable relationships explicitly structured as graphs-a capability where Graph Neural Networks (GNNs) excel. Inspired by the success of GNNs, Graph Deviation Network (GDN) was originally proposed for anomaly detection on industrial multivariate time-series data. After proving its merits through experiments with real-world data, GDN gained significant popularity in the research community, claiming to learn the hidden graph structure in any multivariate time-series data. Various modifications to GDN were proposed over the years, but essentially all of them kept its Graph Structure Learning (GSL) module intact. However, until now, this module has never been rigorously evaluated. This work scrutinizes the contribution of the GSL module. Our experiments reveal that the graph learned by GSL is relatively ineffective, and the key to the overall performance achieved by GDN lies almost entirely in the downstream Graph Attention Network (GAT) module. We hope our findings will garner attention for further development of the GSL module of GDN, whose fidelity can improve the performance of GDN variants.

Keywords: Graph Neural Networks · GNNs for Time-Series Anomaly Detection · Graph Structure Learning · Graph Deviation Network

1 Introduction

Multivariate time-series data often shows complex dependencies between and within features. These dependencies are expected to capture important information for understanding the current state and future evolution of the underlying processes generating the observed time-series data, regardless of where they may originate from: industrial plants, network traffic, power grids, or vehicle components. Correctly learning these pairwise relationships between features paves the

G. Krempl et al. (Eds.): IDA 2025, LNCS 15669, pp. 97–109, 2025.
https://doi.org/10.1007/978-3-031-91398-3_8

way for better modeling performance, which in turn makes tasks such as time series forecasting or time series anomaly detection easier.

Before the advent of deep learning, time-series anomaly detection methods were based on variants of autoregressive-moving average models or machine learning-based approaches that rely on less complex models than Neural Networks. While being easy to run on sensor hardware, such methods may fail to capture the non-linear relationships that exist between the features, limiting their performance. As the data collection increased and computing power developed further, deep learning models became more and more popular in statistical modeling. Initially, new methods based on LSTMs, GANs, and Autoencoder variants [8,9,21,22] were proposed to detect anomalies in multivariate time-series datasets. While such methods are certainly powerful for capturing non-linear relationships in time-series datasets, they may still not fully capture the dependencies between features due to their relational inductive bias not being optimally suited for learning arbitrary relations between features [28].

The introduction of graph neural networks has shown promise to address this gap. Models such as Graph Convolutional Networks [27] and Graph Attention Networks [23] explicitly model the relationships between data features by representing every feature as a node in a graph. This made them to utilize the presence or absence of any arbitrary dependency between features, which helped the network to learn more efficiently.

Graph Deviation Network (GDN) [12] is one of the seminal methods utilizing GNNs for time-series anomaly detection. Besides being one of the first GNN-based methods, the importance of GDN also lies in its ability to take any multivariate time-series dataset and learn its underlying graph during training. At every training run, GDN refined both the learned graph and the parameters of its graph attention network based on the forecasting loss. As a result, it is assumed that the graph learned by GDN converged to a topology that is faithful to the unknown ground truth graph of the multivariate dataset. In light of this, later works on GNN-based time series anomaly detection either copied GDN's graph structure learning method or did minor modifications to it [3,4,13–16,25].

However, this assumption has never been tested rigorously. Several methods build on top of GDN either by modifying its anomaly scoring module or the model it uses for forecasting. To our knowledge, only [5] compared GDN's learned graph with a given ground truth graph, in addition to testing the usefulness of the learned graph for forecasting. While we consider [5] to be valuable, it is not definitive enough to understand the learning process of the GSL module. Therefore, a comprehensive test of the effectiveness of GDN's graph learning module is lacking. Following that observation, the main contribution of this work is a rigorous evaluation of GDN's graph structure learning (GSL) module, which reveals its limited effectiveness in learning to capture meaningful relationships between features.

The rest of the paper is organized as follows. Section 2 gives an overview of work related to our research. We further explain our problem statement in Sect. 3 and describe the GDN framework in Sect. 4. Section 5 outlines our experimental setup for evaluating the GSL module. Section 6 presents and analyzes the results of the experiments. Section 7 summarizes our findings and suggests further research topics.

2 Related Work

2.1 Time Series Anomaly Detection

Anomaly detection is the process of identifying patterns that significantly deviate from the expected behavior in a dataset. This task can be challenging in practice since some data points that are part of the healthy data may appear very similar to genuine anomalies. This similarity creates problems not only in accurately labeling the data but also in modeling it effectively. Since anomalous events are rare, supervised learning approaches are prone to overfitting when used for anomaly detection. As a result, primarily unsupervised approaches are used for this task.

Several approaches have been used for time-series anomaly detection before deep learning. Forecasting-based methods using simple models like ARIMA [11] predict the expected future values and then identify deviations by discrepancy between actual values and model predictions. Density-based methods [24] detect anomalous data comparing the local densities of the data points; anomalies are assumed to be found in low-density regions. Advances in deep learning have led to neural network-based anomaly detection methods [8,21,22], with better performance on some complex datasets. However, a recent benchmark article testing the performance of state-of-the-art networks demonstrated their failures, even against weak baselines [20]. Only one Graph Neural Network (GNN) based approach, GDN, showed reasonable robustness in said benchmark. This prompted us to perform a deeper analysis of the method to understand its success.

2.2 Anomaly Detection with Graph Time Series

GNNs' ability to effectively capture the pairwise relationships between features and the ease of interpretability of their results made them a promising candidate for anomaly detection on complex data [7,26].

Numerous methods have been proposed for solving time-series anomaly detection tasks using graphs, with most of these methods not necessarily requiring the ground truth graph beforehand [7,26]. Among these approaches, Graph Deviation Network(GDN) managed to achieve a high amount of interest; but subsequent works such as [4,13–16,25] all made significant contributions on top of the achievements of GDN by improving its original results. For example, [13] used a reconstruction loss in parallel with forecasting loss to detect anomalies. [14] used a variational inference together with a graph neural network, while [4] changed the anomaly scoring module of GDN. Surprisingly, none of these methods addressed the graph structure learning approach used by GDN.

In addition to GDN and the architectures based on it, other methods using graph neural networks for anomaly detection have been proposed, such as [6,17–19]. These methods do not rely on using node embedding similarity; instead, they typically exploit a simple metric like mutual information or Gumbel-softmax trick for creating the underlying graph.

2.3 Graph Structure Learning

Graph structure learning (GSL) is crucial because the correctness of the underlying graph is often unknown in real-world tasks *a priori*. GSL aims to produce a high-fidelity output graph from datasets of various modalities (e.g., text, image, time series). An alternative to learning the graph is to use simple edge-discovery approaches such as measuring the cosine similarity, mutual information, or Pearson correlation directly on the features.

DGM proposed by [1] focused on learning the graph together with the downstream task and tested their method on tasks from computer vision and healthcare domains. [2] proposed two different metrics for judging the quality of the learned graph for the downstream node classification task. [6] used mutual information measure between the dataset features to construct the graph for detecting vehicle failures.

GDN, which this article focuses on, uses cosine similarity between node embeddings to capture the dependencies between sensors. These node embeddings are randomly initialized and then trained together with the parameters of the graph attention network. This approach, at least in principle, refines the graph based on the MSE loss for the downstream forecasting task. Many other methods followed a similar logic with small changes [13,14], if not the same as the original GDN [4,16,25]. For example, [13] used a recurrent neural network for learning the embeddings, which are then optimized in the same way as in GDN.

Other approaches, like [18,19], utilize different methods for GSL. A preliminary analysis of these methods, including GDN, is provided in [5]. For further information on graph time-series tasks and GSL, refer to surveys [7,26]

3 Problem Statement

Given the simplicity of GDN's graph structure learning approach, the node embeddings used by GDN may not accurately capture the underlying relationships between variables in real-world datasets. The most likely reasons are that node embeddings may not be able to capture the characteristics of the time series features well enough or that they do not receive a sufficiently effective learning signal during training.

To the best of our knowledge, the fidelity of GDN's learned graph against a given ground truth graph and the benefits of using its GSL module for forecasting have only been tested in [5]. For the latter, the authors changed the learned graph to a random Erdös-Renyi graph and checked the differences in forecasting performance against vanilla GDN. For the former, they measured the correlation of edge scores between GDN's learned graph and ground truth graphs coming from two different traffic datasets. The original GDN article [12] also included an ablation study in which the learned graph was replaced with a static complete graph. Yet this experiment is less rigorous than those of [5], since it seems to be done only once.

In this work, we instead focus on the analysis of the *graph learning process*. We hypothesize that the algorithm employed by GDN is of limited effectiveness, suggesting it as a potential bottleneck worth exploring further.

4 Background

In this paper, we address the unsupervised anomaly detection problem. Consider a multivariate time-series dataset of N features and T samples. The training set is then constructed by taking the first T_{train} number of samples, which consists only of healthy values. The remaining samples, which we denote as $T_{\text{test}} = T - T_{\text{train}}$, constitute the test set, and it consists of both healthy and anomalous values. Both datasets can be seen as ordered collections of vectors $s_{\text{train}} = \left[s^1, \cdots, s^{T_{\text{train}}}\right]$ and $s_{\text{test}} = \left[s^1, \cdots, s^{T_{\text{test}}}\right]$. where $s^t \in \mathbb{R}^N$ is an N-dimensional vector representing the feature values at timestep t.

4.1 Graph Deviation Network (GDN)

GDN is a popular GNN-based anomaly detection algorithm. The output of GDN is a set of T_{test} binary labels, $a(t) \in \{0,1\}$, where $a(t) = 1$ indicates an anomaly at timestep t. GDN consists of three primary components: Graph Structure Learning, Graph Attention-Based Forecasting, and Anomaly Scoring. We discuss the first two in the following subsections and refer the reader to the original paper [12] concerning the third since that module is less relevant to our work.

Graph Structure Learning. GDN learns embeddings $v_i \in \mathbb{R}^d$, one for each time series feature, $i \in \{1, 2, \cdots, N\}$. These embeddings are initialized randomly and are then fine-tuned based on the training signal.

The graph representing relations between the features is obtained via the cosine similarity of these embeddings. For any feature i, there exists a set of candidate features $C_i \subseteq \{1, 2, \cdots, N\} \setminus \{i\}$, i.e., all other features except for i itself. In order to select among these candidates, the cosine similarity between embedding vectors v_i and v_j are calculated:

$$e_{ji} = \frac{v_i^\top v_j}{\|v_i\| \cdot \|v_j\|}, \text{ where } j \in C_i. \tag{1}$$

The neighbors for feature i are defined by selecting the features j whose cosine similarity score is among the *top-k* scores, with k being a hyperparameter:

$$A_{ji} = \mathbb{1}\left\{ j \in \textit{top-k}(\{e_{ki} : k \in C_i\}) \right\} \tag{2}$$

Performing these operations for all the features will produce a directed graph with an adjacency matrix A, where A_{ij} represents the existence or absence of a directed edge from node i to node j.

Graph Attention-Based Forecasting. GDN's forecasting module takes as an input a sliding window of size w, $x^{(t)} := \left[s^{(t-w)}, s^{(t-w+1)}, \cdots, s^{(t-1)}\right]$, and the learned graph A to predict the feature values at the current timestep t, i.e., s^t.

Consider $x_i^{(t)} \in \mathbb{R}^w$ which represents the sliding window for feature i as a node in the graph, and a shared weight matrix $W \in \mathbb{R}^{d \times w}$. The new representation for node i is then given by:

$$z_i^{(t)} = \max\left(0, \sum_{j:A_{ji}>0} \alpha_{ij} W x_j^{(t)}\right), \tag{3}$$

where α_{ij} is the attention coefficient of node i to its neighbor j, calculated by:

$$\pi_{ij} = \text{LeakyReLU}\left[a^\top \left(v_i \oplus W x_i^{(t)} + v_j \oplus W x_j^{(t)}\right)\right], \tag{4}$$

$$\alpha_{ij} = \frac{\exp(\pi_{ij})}{\sum_{k:A_{ki}>0} \exp(\pi_{ik})}, \tag{5}$$

where $\oplus$ denotes concatenation, and a is a vector of learned coefficients for the attention mechanism. After obtaining $z_i^{(t)}$ for every node, the predictions $\hat{s}^{(t)}$ for time t are given by a fully-connected neural network f_θ, with θ representing fully-connected neural network's parameters:

$$\hat{s}^{(t)} = f_\theta\left(v_1 \circ z_1^{(t)}, \ldots, v_N \circ z_N^{(t)}\right), \tag{6}$$

where $\circ$ is element-wise multiplication.

5 Experimental Setup

For all the experiments, we set the training epochs to a maximum of 50 with an early stopping criterion of "no improvement for 10 epochs". The sliding window size was 15 with a stride of 1. We set the top-k parameter to 15 and the dimension of node embeddings to 64. Lastly, we trained the models with Adam optimizer and set its learning rate to 0.001.

5.1 Experiment 0: Baseline

We first conducted an experiment to measure the average performance of vanilla GDN, to serve as a baseline for comparisons in the following experiments. To this end, we ran the original GDN 10 times on the SWaT dataset [10] and calculated the mean square error(MSE) loss on the validation set at every epoch. If a particular run was finished before reaching 50 epochs, we used that epoch's latest recorded loss value for future epochs.

5.2 Experiment I: Embedding Freeze

We hypothesize that *if* GDN learns a graph useful for forecasting, turning off the graph learning module will cause a drop in forecasting performance in the GAT module. In light of this hypothesis, we designed Experiment I by freezing the embedding parameters at the early stages of training, effectively fixing the graph structure. This allowed us to train only the GAT module. By freezing these parameters very early in the process, when the learned graph might still be uninformative or nearly random, we aimed to isolate the impact of the GSL module on the final forecasting performance.

We created three different setups for Experiment I. Each setup was evaluated over 10 different GDN runs, with the embedding parameters frozen at different epochs: 1, 3, and 5. Note that the parameters associated with embeddings include not only the embedding vectors but also the attention weights for the embeddings themselves.

After completing the runs for each setup, we calculated the mean and standard deviation of the loss values, following the same procedure as in Experiment 0. We then compared the final epochs' mean loss values from the three different setups with the mean loss value observed in Experiment 0 using Welch's t-test.

5.3 Experiment II: Embedding Replacements

We, moreover, test the usefulness of the GSL module by comparing the learned graph at the final epochs against one learned at the initial epochs. Naturally, we expect the graph learned at later epochs to be more refined, leading to improved forecasting results – in comparison to the one learned at initial epochs.

We picked a reference run from Experiment 0, which achieved the lowest MSE loss on the validation dataset. We then copied the embedding parameters obtained at the end of epochs 1 and 21 (which is when the learning in this particular case converged) from this reference run. Those two embeddings were used as in two variants of Experiment II as drop-in replacements for the embedding parameters at the end of epoch 1 for 10 new runs in Experiment II. The average loss calculation procedure followed the same steps as in Experiment 0. For comparing the mean losses, we again resorted to Welch's t-test.

6 Results and Discussion

6.1 Results

The results of Experiments 0, I, and II are illustrated in Figs. 1, 2, and 3, respectively. Interestingly, in the majority of our experiments, the validation loss was lower than the training loss. This phenomenon could be attributed to various factors, such as the validation set containing more easily learnable samples, the negative impact of dropout layers on training accuracy, or a combination of both. While this observation is noteworthy, we will not delve deeper into its causes for the purposes of our current analysis.

In Fig. 1, we observe that the average performance of vanilla GDN runs resulted in a mean F1 score of 0.74, which is considerably lower than 0.81, the score mentioned in the original GDN paper [12]. Moreover, two GDN runs barely exceeded an F1 score of 0.5, which is equivalent to random guess in binary classification. This discrepancy might be attributed to differences in hyperparameter selection, even if our changes were minimal.

Another important observation is the differences in F1 scores for the two variants of Experiment II. Standard deviation and mean of F1 scores change significantly when we exchange the current run's embeddings with those of the reference run's epoch 21 embeddings – an effect that is not visible for epoch 1 embeddings. A similar effect occurs for the validation loss means, while the validation loss standard deviations increase at more or less in the same amount. We suspect that this may be due to the GSL module not being able to update the current run's graph well enough after replacing it with a graph fine-tuned for another run, thus resulting in a drastic decrease in F1 scores obtained on the test set.

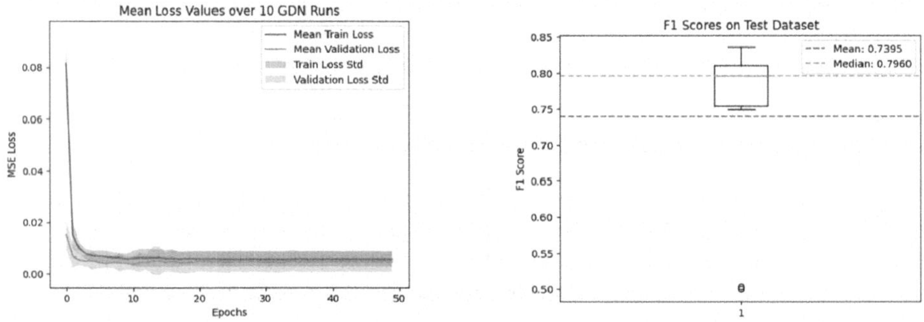

Fig. 1. Left: Mean MSE losses with their confidence intervals over 10 vanilla GDN runs. **Right:** Boxplot of F1 scores on the test set.

6.2 Analysis

Table 1 summarizes the means and their standard deviations at epoch 50 for all our experiments – by comparing them, we can identify any statistically significant differences.

Table 1. Average MSE values and the Standard Deviation of all Experiments

	train_loss_mean	train_loss_std	val_loss_mean	val_loss_std
Vanilla_GDN	0.005780	0.002870	0.004779	0.004065
Freeze_Epoch1	0.016959	0.002131	0.010798	0.001798
Freeze_Epoch3	0.007123	0.000625	0.003941	0.000709
Freeze_Epoch5	0.008895	0.004160	0.006388	0.004930
ChangeEp1withEp21	0.008427	0.003611	0.012621	0.008361
ChangeEp1withEp1	0.007308	0.003568	0.005414	0.007267

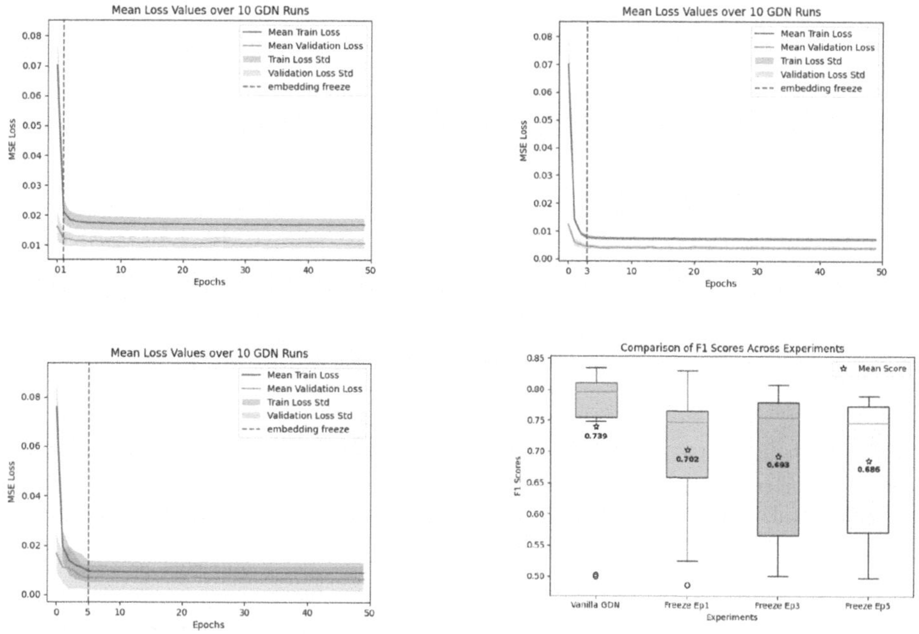

Fig. 2. Loss curves on train&validation data and boxplots of F1 Scores on test data from 10 runs. **Top Left:** Fixing the GSL module at epoch 1.**Top Right:** Fixing the GSL module at epoch 3. **Bottom Left:** Fixing the GSL module at epoch 5. **Bottom Right:** Boxplot of F1 Scores

Analysis of Experiment I. Although the differences between vanilla GDN runs and those where the graph was fixed at epoch 1 appear notable, applying Welch's t-test with the **null hypothesis** "the MSE of vanilla GDN runs is bigger than or equal to the MSE of runs with frozen embeddings" shows that they are statistically significant only in the case of epoch 1 freeze. The results of this test can be seen in Table 2.

Table 2. p-values obtained after Welch's t-test for comparing Vanilla GDN to settings of Experiment I. $\alpha = (0.05/3) \approx 0.0167$, after applying Bonferroni correction for multiple comparisons. Bold numbers mark the significant differences.

	Vanilla vs Freeze1	Vanilla vs Freeze3	Vanilla vs Freeze5
p-value	**0.000496**	0.731996	0.218353

Except for the setup where embeddings were frozen at epoch 1, the p-values for the other two setups exceed the significance level. While we reject the null hypothesis for the first setup, we cannot reject the null hypothesis in the cases where we fixed the graph at epoch 3 and epoch 5. This indicates that fixing

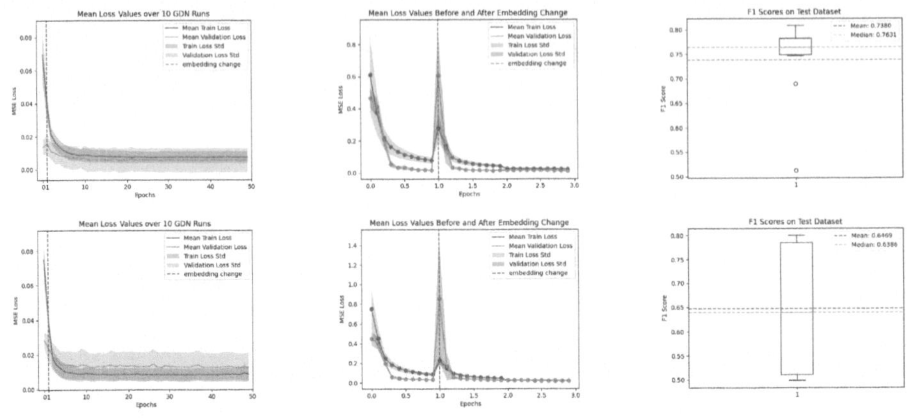

Fig. 3. Results of replacing node embeddings of current run's epoch 1 with embeddings from reference run, **Top:** epoch 1 and **Bottom:** epoch 21. **Left:** MSE losses with confidence intervals. **Middle:** MSE right before and right after changing the embeddings. **Right:** Boxplot of F1 scores on test dataset.

the graph at epochs 3 or 5 does not lead to a statistically significant increase in MSE, suggesting that the graph learning process is only helpful for a very limited part of the overall GDN learning process. To further investigate this, we conducted another test comparing loss values across the three setups, with their results presented in Table 3. The null hypothesis for this second test states that freezing the GSL module at earlier epochs results in MSE values less than or equal to those where GSL was frozen at later epochs.

Table 3. Welch's t-test p-values comparing loss metrics across three different setups of Experiment I. Bold: significant at $\alpha = (0.05/3) \approx 0.0167$, after applying Bonferroni correction for multiple comparisons.

	Freeze1vFreeze3	Freeze1vFreeze5	Freeze3vFreeze5
p-value	**6.32e–08**	**0.010881**	0.923319

The p-values presented in Table 3 show that while training the GSL module has some benefits, it does not have a significantly positive effect on final forecasting performance after epoch 3.

Analysis of Experiment II. Here we hypothesize that using the graph learned at epoch 21 of the reference run will not increase the MSE values compared to using the graph learned at epoch 1, when these graphs are copied to the current run. So, the null hypothesis states that the MSE loss right after copying embeddings from epoch 21 of the reference GDN run is bigger than or equal to the MSE loss after copying embeddings from epoch 1.

Table 4. Mean MSE and Standard Deviation right after Embedding Replacement

	mean	std
Epoch1withEpoch21	0.849996	0.638599
Epoch1withEpoch1	0.604779	0.304731

Applying Welch's t-test on the mean and std values in Table 4 returned us a p-value of **0.85**, which is much higher than 0.05, the significance level. As a consequence, we cannot reject the null hypothesis. This result suggests that we cannot conclude with statistical confidence that the reference GDN run's learned graph at epoch 21 produces better forecasts than its learned graph at epoch 1.

7 Conclusion

Our experiments demonstrate that GDN's GSL module stops improving the graph structure after the first epoch. We hope this finding encourages further development of the GSL module, potentially enhancing GDN's forecasting capabilities. Analyzing graph learning modules of other GNN-based frameworks for time-series data can be another avenue for future research.

Our analysis highlights the GSL module's shortcomings, but further examination of GDN's forecasts together with its anomaly scoring module is necessary to fully understand its performance, as we hinted in Sect. 6.1. Secondly, investigating the GSL module's graph updates after embedding replacement could explain the significant differences between F1 scores in Experiment II.

Acknowledgments. The work was carried out with support from The Knowledge Foundation and from Vinnova (Sweden's innovation agency) through the Vehicle Strategic Research and Innovation Programme, FFI.

Disclosure of Interests. The authors have no competing interests to declare that are relevant to the content of this article.

References

1. Kazi, A., Cosmo, L., Ahmadi, S.A., Navab, N., Bronstein, M.M.: Differentiable graph module (DGM) for graph convolutional networks. IEEE Trans. Pattern Anal. Mach. Intell. **45**(2), 1606–1617 (2022)
2. Hou, Y., et al.: Measuring and improving the use of graph information in graph neural networks [Conference paper]. In: International Conference on Learning Representations, Addis Ababa, Ethiopia (2020)
3. Wu, Z., Pan, S., Long, G., Jiang, J., Chang, X., Zhang, C.: Connecting the dots: multivariate time series forecasting with graph neural networks. In: Proceedings of the 26th ACM SIGKDD International Conference on Knowledge Discovery and Data Mining, pp. 753–763 (2020)

4. Buchhorn, K., Santos-Fernandez, E., Mengersen, K., Salomone, R.: Graph neural network-based anomaly detection for river network systems. F1000Research 12 (2023)
5. Zügner, D., Aubet, F.X., Satorras, V.G., Januschowski, T., Günnemann, S., Gasthaus, J.: A study of joint graph inference and forecasting. arXiv preprint arXiv:2109.04979 (2021)
6. Hojjati, H., Sadeghi, M., Armanfard, N.: Multivariate time-series anomaly detection with temporal self-supervision and graphs: application to vehicle failure prediction. In: Joint European Conference on Machine Learning and Knowledge Discovery in Databases. Springer, Cham (2023)
7. Jin, M., et al.: A survey on graph neural networks for time series: forecasting, classification, imputation, and anomaly detection. IEEE Trans. Pattern Anal. Mach. Intell. (2024)
8. Li, D., et al.: MAD-GAN: multivariate anomaly detection for time series data with generative adversarial networks. In: International Conference on Artificial Neural Networks. Springer, Cham (2019)
9. Park, D., Hoshi, Y., Kemp, C.C.: A multimodal anomaly detector for robot-assisted feeding using an LSTM-based variational autoencoder. IEEE Robot. Autom. Lett. **3**(3), 1544–1551 (2018)
10. Goh, J., et al.: A dataset to support research in the design of secure water treatment systems. In: Critical Information Infrastructures Security: 11th International Conference, CRITIS 2016, Paris, France, 10–12 October 2016, Revised Selected Papers 11. Springer, Cham (2017)
11. Pena, E.H.M., de Assis, M.V.O., Proença, M.L.: Anomaly detection using forecasting methods Arima and HWDS. In: 2013 32nd International Conference of the Chilean Computer Science Society (SCCC). IEEE (2013)
12. Deng, A., Hooi, B.: Graph neural network-based anomaly detection in multivariate time series. In: Proceedings of the AAAI Conference on Artificial Intelligence, vol. 35, no. 5, pp. 4027–4035 (2021)
13. Han, S., Woo, S.S.: Learning sparse latent graph representations for anomaly detection in multivariate time series. In: Proceedings of the 28th ACM SIGKDD Conference on Knowledge Discovery and Data Mining (2022)
14. Chen, W., et al.: Deep variational graph convolutional recurrent network for multivariate time series anomaly detection. In: International Conference on Machine Learning. PMLR (2022)
15. Dai, E., Chen, J.: Graph-augmented normalizing flows for anomaly detection of multiple time series. In: International Conference on Learning Representations
16. Liu, Z., et al.: Multivariate time-series anomaly detection based on enhancing graph attention networks with topological analysis. In: Proceedings of the 33rd ACM International Conference on Information and Knowledge Management (2024)
17. Zhang, W., Zhang, C., Tsung, F.: GRELEN: multivariate time series anomaly detection from the perspective of graph relational learning. IJCAI (2022)
18. Shang, C., Chen, J.: Discrete graph structure learning for forecasting multiple time series. In: Proceedings of International Conference on Learning Representations (2021)
19. Kipf, T., et al.: Neural relational inference for interacting systems. In: International Conference on Machine Learning. PMLR (2018)
20. Kim, S., et al.: Towards a rigorous evaluation of time-series anomaly detection. In: Proceedings of the AAAI Conference on Artificial Intelligence, vol. 36, no. 7 (2022)

21. Hundman, K., et al.: Detecting spacecraft anomalies using LSTMs and nonparametric dynamic thresholding. In: Proceedings of the 24th ACM SIGKDD International Conference on Knowledge Discovery & Data Mining (2018)
22. Su, Y., et al.: Robust anomaly detection for multivariate time series through stochastic recurrent neural network. In: Proceedings of the 25th ACM SIGKDD International Conference on Knowledge Discovery & Data Mining (2019)
23. Veličković, P., Cucurull, G., Casanova, A., Romero, A., Lio, P., Bengio, Y.: Graph attention networks. In: International Conference on Learning Representations (ICLR) (2018)
24. Breunig, M.M., et al.: LOF: identifying density-based local outliers. In: Proceedings of the 2000 ACM SIGMOD International Conference on Management of Data (2000)
25. Ray, S., et al.: Learning graph neural networks for multivariate time series anomaly detection. arXiv preprint arXiv:2111.08082 (2021)
26. Cini, A., et al.: Graph deep learning for time series forecasting. arXiv preprint arXiv:2310.15978 (2023)
27. Kipf, T.N., Welling, M.: Semi-supervised classification with graph convolutional networks. arXiv preprint arXiv:1609.02907 (2016)
28. Battaglia, P.W., et al.: Relational inductive biases, deep learning, and graph networks. arXiv preprint arXiv:1806.01261 (2018)

Foundations of Data Science

The When and How of Target Variable Transformations

Loren Nuyts(✉) and Jesse Davis

Department of Computer Science, Research Unit DTAI, KU Leuven, 3000 Leuven, Belgium
{loren.nuyts,jesse.davis}@kuleuven.be

Abstract. The machine learning pipeline typically involves the iterative process of (1) collecting the data, (2) preparing the data, (3) learning a model, and (4) evaluating a model. Practitioners recognize the importance of the data preparation phase in terms of its impact on the ability to learn accurate models. In this regard, significant attention is often paid to manipulating the feature set (e.g., selection, transformations, dimensionality reduction). A point that is less well appreciated is that transformations on the target variable can also have a large impact on whether it is possible to learn a suitable model. These transformations may include accounting for subject-specific biases (e.g., in how someone uses a rating scale), contexts (e.g., population size effects), and general trends (e.g., inflation). However, this point has received a much more cursory treatment in the existing literature. The goal of this paper is three-fold. First, we aim to highlight the importance of this problem by showing when transforming the target variable has been useful in practice. Second, we will provide a set of generic "rules of thumb" that indicate situations when transforming the target variable may be needed. Third, we will discuss which transformations should be considered in a given situation.

Keywords: Data Preparation · Target Transformation · Target Manipulation

1 Introduction

The machine learning pipeline typically involves the iterative process of (1) collecting the data, (2) preparing the data, (3) learning a model, and (4) evaluating a model. While the machine learning literature primarily focuses on learning algorithms, there is still substantial interest in the steps to prepare or transform the data prior to applying a learning algorithm. Practitioners also recognize the importance of this step in terms of both the time it requires [1] and its impact on the quality of the learned model [2,3].

During the data preparation phase, significant attention is often paid to manipulating the feature set. There are three broad categories of operations that are typically considered. First, a variety of techniques exist for transforming an

G. Krempl et al. (Eds.): IDA 2025, LNCS 15669, pp. 113–126, 2025.
https://doi.org/10.1007/978-3-031-91398-3_9

individual feature's values. These approaches range from simple ideas such as standardization or min-max normalization to more complex domain-dependent operations such as removing unwanted biases [4,5]. Second, there has been significant work in areas such as feature selection [6] and dimensionality reduction [7]. Third, there is work on automatically transforming the given input space to a more optimal representation by applying techniques such as automated feature construction [8,9], representation learning [10,11], or manifold learning [12,13]. The relevance of the data preparation is also reinforced by the fact that there have been significant efforts made to *automate* this part of the machine learning pipeline: many automated machine learning toolkits explicitly consider including these operations when trying to search for the best configuration of a learning algorithm [10,14]. For example, *Auto-sklearn* [14,15] automatically decides whether to include scaling, imputation, balancing, or PCA factorization.

Interestingly, *transforming the target variable* can also play a crucial role in facilitating learning a model. As an illustrative example, consider the task of predicting a ship's arrival time at a lock based on historical arrival time data. Using the historical arrival time as the target variable is bad because it strongly depends on the departure time, which in turn is influenced by factors like the crews' shifts and the availability of equipment. Instead, one should *transform the target variable by computing the ship's travel time* and predict that instead [16]. Similarly, one of the most practically impactful insights from the Netflix challenge was the importance of isolating the user-movie interaction component of a rating by deriving a new variable that controlled for factors related to the user, the movie, and time [17,18]. The necessity of deriving a new target variable has also arisen in Kaggle competitions [19] and problems like predicting running fatigue [20].

Despite the importance of selecting the appropriate form of the target variable, it has received a much more cursory treatment in existing literature. With a few exceptions (e.g., [21]), existing work typically mentions manipulations to the target variable briefly (if at all) without an attempt to understand the general task properties that lead to the need for the transformation [17,20,22,23]. Consequently, we lack a systematic description of the problem characteristics that indicate when a transformation may be useful and what transformations should be considered. The goal of this paper is to fill this gap. Specifically, our contributions are: (1) providing a set of "rules of thumb" for when a given target variable might be unsuitable, (2) describing possible mathematical transformations that can be useful in each situation, (3) conducting a set of experiments that demonstrate the utility of the transformation, and (4) providing open-source code for performing them.[1]

2 Preliminaries and Related Work

We assume an d-dimensional input space $\mathcal{X} \subset \mathbb{R}^d$ and an output space $\mathcal{Y} \subset \mathbb{R}$. In this work, we tackle the problem of transforming the target variable. Intuitively,

[1] https://github.com/ML-KULeuven/target_transformations.

the idea is that applying a function f to observed the target values and training a model to predict $f(\mathcal{Y})$ may be easier than learning a model to directly predict $\mathcal{Y}$. This problem setting can be formalized as follows:

Target Transformation Problem. *Given a dataset* $\mathcal{D}_=\{(\mathbf{x}_1, y_1)\dots(\mathbf{x}_n, y_n)\}$, *a model class* $\mathcal{H} : \mathcal{X} \mapsto \mathcal{Y}$, *and a loss function* $\mathcal{L} : \mathcal{Y} \times \mathcal{Y} \mapsto \mathbb{R}$, *the target transformation problem aims to find a bijective transformation* $f : \mathcal{Y} \to \mathcal{Y}$, *s.t.*

$$\frac{1}{n}\sum_{i=1}^{n}\mathcal{L}(f^{-1}(\hat{h}_*(\mathbf{x}_i)), y_i) < \frac{1}{n}\sum_{i=1}^{n}\mathcal{L}(\hat{h}(\mathbf{x}_i), y_i) \tag{1}$$

where we use the inverse transformation because some losses are scale-dependent (e.g., mean absolute error, mean squared error), and $\hat{h}_*$ *and* $\hat{h}$ *are trained to predict respectively the transformed target variable and the given target variable:*

$$\hat{h}_* = \arg\min_{h\in\mathcal{H}} \frac{1}{n}\sum_{i=1}^{n}\mathcal{L}(h(\mathbf{x}_i)), f(y_i))$$

$$\hat{h} = \arg\min_{h\in\mathcal{H}} \frac{1}{n}\sum_{i=1}^{n}\mathcal{L}(h(\mathbf{x}_i)), y_i))$$

While prior work [17,20,22] briefly mentions target transformations, Hyndman and Athanasopoulos [21] offer a more detailed discussion of common transformations in forecasting: calendar adjustments, population adjustments, inflation adjustments, and mathematical transformations. Building on this, we (1) synthesize these transformations into three high-level categories, (2) generalize them beyond forecasting, (3) propose additional approaches for the listed dependencies, (4) identify and address additional dependencies that were not previously discussed, and (5) provide a publicly available package for applying these transformations.

Sklearn also provides a *TransformedTargetRegressor*[2] that trains a given regression model on a transformed target by applying a user-defined transformation. However, the choice of whether and which transformation to apply is left to the user. Commonly, a logarithmic transformation or a *QuantileTransformer*[3] is used [23], but, to our knowledge, no study has systematically examined the impact of these transformations on the predictive performance. Notably, this has been neglected by the AutoML literature: *Auto-sklearn* [14,15] does not support performing target variable transformations and the book on Automated Machine Learning [24] does not discuss this problem.

3 A Guide to Target Transformations

In this section, we will provide motivating examples of when target transformations have been useful in practice. Specifically, we have synthesized the transfor-

[2] https://scikit-learn.org/stable/modules/generated/sklearn.compose.TransformedTargetRegressor.html.

[3] https://scikit-learn.org/stable/modules/generated/sklearn.preprocessing.QuantileTransformer.html.

mations from the literature into three high-level categories and formulated five heuristic "rules of thumb" to help practitioners identify characteristics that suggest when a transformation might be beneficial. Finally, within each category, we will suggest several possible transformations.

3.1 Subjective Target Variable

Often the target variable is a subjective rating such as how much a user likes a movie (e.g., number of stars given) [17], an athlete's rating of perceived exertion (which can be a proxy for fatigue) [20], or a person's perceived stress level [25]. Comparing values across subjects can be difficult because they may inherently use the rating scale differently or unmeasured external factors (e.g., mood, baseline fatigue level) can affect a person's rating. Therefore, performing subject-level transformations of the target variable can enable learning a more accurate model.

Heuristic 1 (Subjective Target). *The target variable is subjective if it depends on someone's judgment (e.g., rating or questionnaire-derived label) meaning that for subject i,* $P(y) \neq P(y|i)$*. Additionally, the target should be ordinal to allow for transforming it.*

The most straightforward way to normalize a subjective target variable is to center it w.r.t. the mean target of that person (e.g., as was done for Netflix ratings [17]). This changes the task from predicting the absolute rating to predicting the subject's deviation from their average:

$$f(y_{i,j}) = y_{i,j} - \overline{y_i} \quad \text{with} \quad \overline{y_i} = \frac{\sum_{j=0}^{S_i} y_{i,j}}{S_i} \tag{2}$$

where $y_{i,j}$ is the jth sample of subject i and S_i total number of samples for subject i.

The previous transformation accounts for differences between subjects, but there can also be within-subject variation. This can be particularly acute when subjects give a sequence of responses in close succession. For example, Koren [18] discusses how a user's movie rating may be given in relation to other recently rated films. Similarly, the evolution of a runner's fatigue during a workout will depend on how tired they are at the start [20]. In these cases, a local within-subject transformation may be needed. However, determining the appropriate transformation is inherently more challenging because it will rely on the nature of the ratings. If a subject performs several trials where they provide multiple ratings within a trial, it is possible to independently min-max normalize all target variables per trial in order to eliminate any trial-specific effects (e.g., the athlete's starting fatigue state).

3.2 Contextual Dependencies

Often, there are contextual factors that affect a target's value. To give an illustrative example, consider the production of solar energy. This depends on the number of daylight hours which is determined by two important contextual variables:

the season and the latitude of the location. Consequently, production numbers across time and place are not directly comparable. Therefore, it may be simpler to transform the target to solar energy production per daylight hour to remove these effects from the data. We have identified three high-level categories of contexts that may arise: frame of reference, trends, and domain-dependent contexts.

Frame Dependency. A target variable is often constructed by *aggregating events within a semantically meaningful frame of reference (i.e., context).* This may be a time frame such as a month or a sports game. It could also be a geographic reference such as a country or region. However, these aggregates are often not directly comparable [21,26,27]. In the case of time, months have a varying number of days. Even if the time frame is of a fixed duration, issues can arise: matches within a competition for a sport have a standardized time[4] but the number of points scored depends on the pace (i.e., how quickly teams play) [27,28]. Geographic areas vary in terms of their population and size (i.e., land area). An increase in the number of hospital beds for example may reflect population growth rather than a true increase in capacity. To correctly evaluate changes in hospital bed availability, the number of beds per capita should be considered [21,29]. Similarly, crop yield depends on both the yield per plant and the number of plants per unit area [30].

Heuristic 2 (Frame Dependency). *A frame dependency arises if (1) the target variable is aggregated over a frame of reference and (2) its value depends on the choice of this frame of reference.*

In these cases, it is natural to normalize the target variable w.r.t. the frame of reference R (e.g., time, area, number of possessions, number of sunlight hours):

$$f(y_i) = \frac{y_i}{R} \tag{3}$$

Practically, this transformation normalizes the target to, e.g., sales per day, points per fixed number of possessions in sports [28], or solar energy production per daylight hour.

Trend Dependency. A common form of historical data involves examples that are collected over an extended period of time. For example, this could be house pricing data, transfers of soccer players, or measurements of stellar noise. In these cases, part of the value of a target variable is explained by a trend that arises due to the passage of time such as inflation [21], seasonal fluctuations [31], or stellar trends [32].

[4] Some obvious caveats exist such as extra time, and soccer's subjective time added.

Heuristic 3 (Trend Dependency). *A trend dependency arises if (1) the target is (partially) explained by a monotonically changing variable, and (2) it is tracked over a sufficiently long time such that the explanatory variable has a significant influence on the target.*

What trends may be present is highly domain-dependent. Consequently, picking the best way to transform the target to remove or mitigate the effect of the trend depends on domain knowledge. For monetary values, a common approach is to transform the price to be expressed as a specific year's (t_0) cost:

$$f(y_{i,t}) = \frac{y_{i,t}}{z_t} * z_{t_0} \tag{4}$$

where $y_{i,t}$ represents the target of subject i at time t, and z_t is the price index at time t.

Several trend estimation and detrending techniques have been proposed [33–35]. The most appropriate technique often depends on the exact application. However, a detailed discussion is beyond the scope of this paper.

Specific Domain Knowledge. A target variable is often influenced by several domain-specific external factors. For example, a person's stride length is proportional to their height [4]. Similarly, how much a machine vibrates also depends on the amount of vibrations in nearby machines [5]. While some of these causal relationships are known, they are often left unmodeled.

Heuristic 4 (General Contextual Dependency). *A general contextual dependency arises if (1) the target depends on external factors that are known in advance, and (2) the relationship between the target and the external factors is not modeled explicitly.*

Transforming the target to account for these factors helps eliminate part of its variability. Consequently, it allows the learned model to focus on capturing more complex patterns. We provide two methods to transform the target y_i w.r.t. k contextual variables $\varphi_{i,k}$. The first, proposed by Turney and Halasz [5] for predicting faulty gas turbines, removes the context of a healthy turbine from all dataset features. While originally applied to features, this method can also be extended to the target variable by subtracting its expected value given the contextual variables and dividing by the expected variation σ_i:

$$f(y_i) = \frac{y_i - \mathbb{E}(y_i|\varphi_{i,1} \cdots \varphi_{i,k})}{\mathbb{E}(\sigma_i|\varphi_{i,1} \cdots \varphi_{i,k})} \tag{5}$$

The second method, proposed by Wahid et al. [4], first fits a multiple regression model to the labels:

$$y_i = \beta_0 + \sum_{j=1}^{k} \beta_j \varphi_{i,j} + \epsilon_i \tag{6}$$

where β_0 represents the linear regression line intercept term, β_j represents the coefficient for the jth contextual feature, and $\epsilon_i \sim N(0, \sigma^2)$ represents the independent and identically distributed residual error for the ith instance. Then, they normalize the gait features by dividing the value of the original feature by the value predicted by the multiple regression model. When applied to a target variable, this becomes:

$$f(y_i) = \frac{y_i}{\beta_0 + \sum_{j=1}^{k} \beta_j \varphi_{i,j}} \tag{7}$$

where the β_js are the fitted coefficients from the multiple regression model.

3.3 Unsuitable Target Distribution

The distributional characteristics of the target values can also pose problems for learning. On the one hand, models make assumptions about the target variable, such as being homoscedastic, which are often violated in practice [36,37]. On the other hand, regression approaches struggle with imbalance and yield less accurate predictions for rare values or in situations where there are large gaps between target values [38,39].

Heuristic 5 (Unsuitable Target Distribution). *An unsuitable target distribution is either heteroscedastic or imbalanced, meaning that specific ranges of target values have significantly fewer observations than others.*

To mitigate heteroscedasticity, variance-stabilizing transformations such as the logarithm, square root, or Box-Cox transformations can be applied to normalize the variance [37,40].

To address an imbalanced target distribution, one can either (1) incorporate the imbalance into the model or loss function [39], or (2) transform the target to achieve a more balanced distribution. As the first approach does not involve direct target transformation, it lies outside the scope of this paper. For the second approach, transformations such as the logarithm [22] are widely used to address skewness. Transforming the target value's distribution to approximate a normal distribution can also mitigate imbalance, as shown in the experiments.

4 Experiments

The previously referenced literature contains numerous examples that empirically demonstrate the benefits of both the subjective [17,20] and contextual transformations [26–28]. Instead of replicating those experiments, we will focus on transformations related to unsuitable target distributions as to our knowledge there currently is not a systematic empirical study of these. All code and results are available at https://github.com/ML-KULeuven/target_transformations.

Table 1. For each dataset, the used acronym, number of instances n, the number of features d, and the Fisher-Pearson coefficient of skewness γ are given. It also indicates the reason for imbalance in the dataset (skewness or gap of missing values).

Dataset full name	Acronym	n	d	Skewed/Gap	γ
Auto MPG	AMPG	398	7	Skewed	0.455
Bike Sharing	BS	17389	13	Skewed	1.277
Combined Cycle Power Plant	CCPP	9568	4	Gap	0.306
Concrete Compressive Strength	CCS	1030	8	Skewed	0.416
Energy Efficiency: Heating Load	EF1	768	8	Gap	0.360
Energy Efficiency: Cooling Load	EF2	768	8	Gap	0.360
Liver Disorders	LD	345	5	Skewed	1.537
Online News Popularity	ONP	39797	58	Skewed	33.963
Real Estate Valuation	REV	414	6	Skewed	0.598
Servo	SRV	167	4	Skewed + Gap	1.775

Setup. We evaluate the performance of a Lasso (*sklearn's Lasso*), Ridge (*sklearn's Ridge*), Gradient Boosted Trees (*sklearn's GradientBoostingRegressor*), and Support Vector (*sklearn's SVR*) regression models on 10 public datasets with an imbalanced distribution. Table 1 shows (1) the general characteristics of the datasets, (2) whether it is skewed, has a gap of missing values, or both, and (3) the Fisher-Pearson coefficient of skewness.

We compare the baseline regression model, without transformations, with the following 4 transformations:

1. **Quantile Normal (QN)** transforms the target with *sklearn's QuantileTransformer* to an approximately normal distribution.
2. **Quantile Uniform (QU)** transforms the target with *sklearn's QuantileTransformer* to an approximately uniform distribution.
3. **Yeo-Johnson (YJ)** transforms the target with *sklearn's PowerTransformer* to also approximate a normal distribution.
4. The **natural logarithm (Ln)** transformation applies $ln(y_i + \mathit{offset})$ to the target, where $\mathit{offset} = \max_i(\lceil -y_i \rceil, 1)$ ensures that all $y_i > 0$.

Each experiment ran on a single thread on an x86_64 Intel(R) Xeon(R) CPU @3.20 GHz with 32 Gb RAM. To mitigate the randomness, we repeat all steps using a five-times repeated two-fold-cross-validation (5x2CV) approach [41] and report the average and standard deviation across all folds.

Evaluation Metrics. We employ two evaluation metrics: the relative squared error (RSE) and the symmetric mean absolute percentage error (SMAPE), which are defined as follows:

$$\text{RSE} = \frac{\sum_{i=1}^{n}(y_i - \hat{y}_i)^2}{\sum_{i=1}^{n}(y_i - \bar{y})^2} \tag{8}$$

$$\text{SMAPE} = \frac{1}{n}\sum_{i=1}^{n}\frac{|y_i - \hat{y}_i|}{\frac{|y_i|+|\hat{y}_i|}{2}} \times 100\% \tag{9}$$

where y_i is the actual value, $\bar{y}$ the mean of the actual values, n the number of instances, and $\hat{y}_i$ corresponds to either $\hat{h}(\mathbf{x}_i)$ or $\hat{h}_*(\mathbf{x}_i)$ if the model was trained on respectively the original targets or the transformed targets.

Results. Tables 2 and 3 provide the RSE and SMAPE results for all four included regression models. The QN transformation significantly improves performance for Lasso, Ridge, and SVR, yielding median improvements of between 11% and 63% for RSE and between 10% and 19% for SMAPE. In contrast, the GBTR seems to be adversely affected by the normal distribution with a median increase in RSE of 19%, and SMAPE of 4%.

Both the YJ and Ln transformations substantially improve over the baseline for Lasso, Ridge, and SVR, with median improvements for YJ (Ln) in RSE ranging from 5% to 40% (5% to 50%), and in SMAPE from 8% to 16% (9% to 11%). The advantages for the GBTR model are less clear: while the median RSE increases with 6% for Yeo-Johnson and 1% for the logarithm, the median SMAPE decreases with 2% for both transformations.

In contrast to the previous transformations, the QU does not convincingly outperform the baseline for Lasso, Ridge, and GBTR. Although the median SMAPE improves by between 0% and 5%, the median RSE drops by between -2% and -12%, which makes it unclear whether these regression models benefit from the uniform distribution. Only SVR consistently benefits from the QU transformation, with a median decrease in RSE of 48% and in SMAPE of 7%.

To conclude, applying the QN, YJ, and Ln transformations to an imbalanced target distribution significantly improves the predictions for Lasso, Ridge, and SVR. In contrast, GBTR shows a general robustness to the distribution of the target variable. The linearity of the model likely plays a role in how susceptible the model is to an imbalanced target distribution. Lasso and Ridge are both linear regression models and SVR fits a linear model after applying a kernel function. GBTR on the other hand does not require the target to be linear in any hyperplane of the data, making it more suitable for an imbalanced target distribution. Finally, the QU transformer only shows considerable improvements for the SVR model and mixed results for the other approaches, indicating that a normal distribution is preferred over a uniform one for these algorithms.

Table 2. Avg. ± std. RSE (↓) across all folds for the Lasso, Ridge, Gradient Boosted Trees (GBTR), and Support Vector Regressor (SVR) regression models, comparing the baseline without any target transformations (Base), and after applying the Quantile Normal (QN), Quantile Uniform (QU), Yeo-Johnson (YJ), and logarithmic (Ln) transformations.

		Base	QN	QU	YJ	Ln
Lasso	AMPG	0.241 ± 0.02	**0.207** ± 0.03	0.279 ± 0.03	0.219 ± 0.02	0.213 ± 0.02
	BS	1.601 ± 0.07	1.393 ± 0.06	1.91 ± 0.1	1.494 ± 0.05	**1.238** ± 0.05
	CCPP	0.077 ± <0.01	**0.073** ± <0.01	0.08 ± <0.01	50.964 ± 52.31	0.075 ± <0.01
	CCS	0.682 ± 0.05	0.625 ± 0.04	0.906 ± 0.07	0.666 ± 0.04	**0.621** ± 0.04
	EF1	0.095 ± 0.01	0.088 ± 0.01	**0.087** ± 0.01	0.093 ± 0.01	0.094 ± 0.01
	EF2	0.131 ± 0.01	**0.121** ± 0.01	0.131 ± 0.01	0.129 ± 0.01	0.129 ± 0.01
	LD	17.172 ± 21.18	**3.935** ± 2.19	7.48 ± 8.7	4.776 ± 2.09	5.303 ± 2.41
	ONP	21.59 ± 21.52	3.618 ± 2.67	**2.45** ± 1.4	11.554 ± 18.14	20.701 ± 18.54
	REV	0.867 ± 0.26	**0.781** ± 0.31	0.978 ± 0.31	0.847 ± 0.24	0.823 ± 0.21
	SRV	2.666 ± 3.71	1.666 ± 2.49	**1.357** ± 0.88	2.3 ± 1.21	2.795 ± 2.43
Ridge	AMPG	0.242 ± 0.03	**0.197** ± 0.03	0.283 ± 0.04	0.219 ± 0.02	0.213 ± 0.02
	BS	1.582 ± 0.06	1.359 ± 0.04	1.877 ± 0.09	1.471 ± 0.04	**1.206** ± 0.04
	CCPP	0.077 ± <0.01	**0.072** ± <0.01	0.08 ± <0.01	**0.072** ± <0.01	0.075 ± <0.01
	CCS	0.683 ± 0.06	0.623 ± 0.05	0.9 ± 0.09	0.665 ± 0.04	**0.62** ± 0.03
	EF1	0.095 ± 0.01	0.091 ± 0.01	**0.086** ± 0.01	0.09 ± 0.01	0.091 ± 0.01
	EF2	0.131 ± 0.01	**0.12** ± 0.01	0.131 ± 0.01	0.127 ± 0.01	0.127 ± 0.01
	LD	3.723 ± 0.75	**1.827** ± 0.41	3.124 ± 1.22	2.889 ± 0.82	3.119 ± 0.8
	ONP	21.256 ± 21.38	3.503 ± 2.69	**2.311** ± 1.32	11.573 ± 18.18	21.693 ± 17.22
	REV	0.826 ± 0.24	**0.721** ± 0.3	0.946 ± 0.32	0.808 ± 0.23	0.766 ± 0.18
	SRV	1.255 ± 0.34	**0.851** ± 0.42	1.296 ± 0.78	2.246 ± 1.09	2.056 ± 0.72
GBTR	AMPG	0.183 ± 0.03	0.216 ± 0.06	0.201 ± 0.03	0.182 ± 0.03	**0.181** ± 0.03
	BS	**0.206** ± 0.01	0.286 ± 0.02	0.342 ± 0.02	0.251 ± 0.02	0.282 ± 0.04
	CCPP	**0.057** ± <0.01	0.058 ± <0.01	0.062 ± <0.01	50.958 ± 52.31	**0.057** ± <0.01
	CCS	0.156 ± 0.02	0.153 ± 0.01	0.191 ± 0.02	**0.152** ± 0.02	0.155 ± 0.02
	EF1	**0.003** ± <0.01	0.011 ± 0.01	0.005 ± <0.01	**0.003** ± <0.01	**0.003** ± <0.01
	EF2	**0.032** ± <0.01	0.041 ± 0.01	0.035 ± 0.01	**0.032** ± <0.01	**0.032** ± <0.01
	LD	1.933 ± 0.4	**1.417** ± 0.18	1.938 ± 0.38	2.056 ± 0.34	2.088 ± 0.33
	ONP	**11.281** ± 8.56	25.422 ± 8.83	29.567 ± 5.62	42.679 ± 5.72	47.789 ± 6.25
	REV	0.46 ± 0.07	0.442 ± 0.11	**0.412** ± 0.09	0.463 ± 0.07	0.462 ± 0.08
	SRV	**0.197** ± 0.1	0.238 ± 0.1	0.226 ± 0.08	0.391 ± 0.27	0.199 ± 0.11
SVR	AMPG	0.393 ± 0.05	0.165 ± 0.02	0.211 ± 0.02	0.165 ± 0.02	**0.155** ± 0.02
	BS	3.16 ± 0.07	0.826 ± 0.02	0.812 ± 0.04	0.802 ± 0.02	**0.754** ± 0.02
	CCPP	0.066 ± <0.01	**0.059** ± <0.01	0.098 ± 0.01	58.833 ± 99.49	40.14 ± 29.91
	CCS	3.428 ± 0.47	0.351 ± 0.03	0.41 ± 0.07	0.367 ± 0.03	**0.323** ± 0.04
	EF1	0.126 ± 0.02	**0.073** ± 0.01	0.097 ± 0.01	0.074 ± 0.01	0.08 ± 0.01
	EF2	0.177 ± 0.02	**0.115** ± 0.01	0.125 ± 0.01	0.118 ± 0.01	0.116 ± 0.01
	LD	3.578 ± 1.06	2.0 ± 0.34	**1.912** ± 0.48	2.233 ± 0.57	2.163 ± 0.52
	ONP	35.717 ± 4.79	8.021 ± 4.59	**1.874** ± 0.53	37.266 ± 5.01	37.637 ± 5.01
	REV	2.657 ± 0.77	**0.585** ± 0.09	0.742 ± 0.12	0.689 ± 0.14	0.594 ± 0.12
	SRV	1.931 ± 0.62	0.635 ± 0.45	0.98 ± 0.45	1.378 ± 0.8	**0.529** ± 0.23

Table 3. Avg. ± std. SMAPE (↓) across all folds for the Lasso, Ridge, Gradient Boosted Trees (GBTR), and Support Vector Regressor (SVR) regression models, comparing the baseline without any target transformations (Base), and after applying the Quantile Normal (QN), Quantile Uniform (QU), Yeo-Johnson (YJ), and logarithmic (Ln) transformations.

		Base	QN	QU	YJ	Ln
Lasso	AMPG	11.6 ± 0.48	10.033 ± 0.77	11.062 ± 0.75	9.651 ± 0.46	**9.44** ± 0.34
	BS	75.372 ± 0.61	**68.625** ± 0.42	68.8 ± 0.4	68.756 ± 0.3	71.82 ± 0.31
	CCPP	0.8 ± <0.01	**0.773** ± <0.01	0.819 ± 0.01	2.253 ± 1.21	0.789 ± <0.01
	CCS	**26.183** ± 0.48	26.401 ± 0.53	27.743 ± 0.6	26.917 ± 0.46	28.641 ± 0.56
	EF1	9.989 ± 0.31	9.1 ± 0.79	8.627 ± 0.13	8.546 ± 0.27	**8.533** ± 0.27
	EF2	9.073 ± 0.31	8.682 ± 0.33	8.599 ± 0.35	8.409 ± 0.21	**8.401** ± 0.23
	LD	**83.908** ± 3.92	84.781 ± 2.78	85.488 ± 3.21	87.843 ± 3.25	87.034 ± 3.68
	ONP	77.221 ± 1.1	**54.842** ± 0.14	55.166 ± 0.14	55.699 ± 0.1	57.154 ± 0.15
	REV	19.372 ± 2.65	17.634 ± 1.93	19.886 ± 2.36	18.162 ± 3.73	**15.871** ± 0.58
	SRV	77.731 ± 7.02	44.934 ± 4.16	**41.114** ± 2.89	44.545 ± 2.42	59.915 ± 5.2
Ridge	AMPG	11.691 ± 0.41	10.164 ± 0.81	11.109 ± 0.73	9.668 ± 0.51	**9.454** ± 0.38
	BS	75.488 ± 0.52	**68.674** ± 0.41	68.783 ± 0.4	68.732 ± 0.29	71.773 ± 0.32
	CCPP	0.8 ± <0.01	0.775 ± <0.01	0.821 ± 0.01	**0.773** ± <0.01	0.79 ± <0.01
	CCS	**26.22** ± 0.47	26.41 ± 0.63	27.743 ± 0.65	26.942 ± 0.48	28.678 ± 0.56
	EF1	9.988 ± 0.31	9.798 ± 0.39	**8.626** ± 0.12	8.867 ± 0.27	8.94 ± 0.28
	EF2	9.071 ± 0.3	8.768 ± 0.25	8.584 ± 0.34	**8.4** ± 0.22	8.401 ± 0.22
	LD	83.031 ± 3.33	**82.812** ± 3.12	84.483 ± 2.32	86.545 ± 2.48	85.675 ± 2.78
	ONP	77.306 ± 1.15	**54.841** ± 0.14	55.161 ± 0.14	55.703 ± 0.1	57.146 ± 0.15
	REV	19.359 ± 2.6	17.619 ± 1.92	19.79 ± 2.37	18.262 ± 3.7	**15.928** ± 0.67
	SRV	78.886 ± 5.49	44.006 ± 4.02	**40.979** ± 2.72	44.42 ± 2.35	59.591 ± 4.87
GBTR	AMPG	9.223 ± 0.43	10.06 ± 0.78	9.361 ± 0.44	9.209 ± 0.41	**9.191** ± 0.45
	BS	45.62 ± 0.77	32.653 ± 0.91	34.414 ± 0.61	31.519 ± 0.97	**31.347** ± 1.24
	CCPP	**0.66** ± 0.01	0.671 ± <0.01	0.677 ± 0.01	2.209 ± 1.26	0.661 ± 0.01
	CCS	13.545 ± 0.74	13.757 ± 0.82	15.519 ± 1.05	**13.155** ± 0.52	13.353 ± 0.67
	EF1	1.805 ± 0.11	2.992 ± 0.45	2.319 ± 0.33	1.776 ± 0.11	**1.77** ± 0.13
	EF2	4.007 ± 0.24	4.537 ± 0.41	4.046 ± 0.26	**3.803** ± 0.22	3.815 ± 0.21
	LD	87.184 ± 3.71	90.263 ± 3.0	**85.091** ± 3.75	86.575 ± 3.41	86.818 ± 3.43
	ONP	74.417 ± 0.61	**53.787** ± 0.13	54.131 ± 0.15	54.809 ± 0.12	56.213 ± 0.13
	REV	14.54 ± 0.95	15.175 ± 1.15	**13.86** ± 0.63	14.368 ± 0.94	14.475 ± 1.07
	SRV	24.719 ± 3.63	27.944 ± 4.97	24.467 ± 2.63	24.077 ± 1.32	**22.017** ± 2.0
SVR	AMPG	10.605 ± 0.4	**8.507** ± 0.52	9.606 ± 0.45	8.511 ± 0.52	8.736 ± 0.45
	BS	68.705 ± 0.46	54.697 ± 0.24	55.962 ± 0.45	55.428 ± 0.26	**54.447** ± 0.19
	CCPP	0.716 ± 0.01	**0.686** ± 0.01	0.807 ± 0.02	2.221 ± 1.26	3.356 ± 0.04
	CCS	30.865 ± 0.46	**19.823** ± 0.53	20.602 ± 0.58	20.389 ± 0.58	20.323 ± 0.76
	EF1	9.542 ± 0.4	**7.912** ± 0.31	10.406 ± 0.41	8.105 ± 0.28	9.047 ± 0.2
	EF2	9.326 ± 0.32	**7.882** ± 0.39	8.925 ± 0.23	8.031 ± 0.39	9.075 ± 0.32
	LD	84.227 ± 5.28	**82.418** ± 2.79	82.923 ± 2.43	83.98 ± 3.43	84.141 ± 3.72
	ONP	57.568 ± 0.17	54.582 ± 0.17	54.747 ± 0.16	**54.528** ± 0.18	54.546 ± 0.18
	REV	19.748 ± 1.82	15.07 ± 0.98	16.434 ± 1.75	15.245 ± 1.0	**14.638** ± 0.82
	SRV	41.102 ± 3.45	**31.626** ± 4.04	32.313 ± 3.27	34.047 ± 3.18	34.16 ± 2.68

5 Conclusion

This paper highlights that transforming the target variable can also be crucial for obtaining accurate models. It provides a systematic description of the characteristics that indicate when a transformation may be beneficial and what transformations should be considered. Specifically, we (1) propose a set of heuristic "rules of thumb" that indicate when a target might be unsuitable, (2) describe several mathematical transformations for each situation, (3) empirically evaluate the effect of transforming the target variable's distribution on the predictive performance of four well-known regression approaches, and (4) offer an open-source repository to facilitate applying these transformations. Together, these contributions aim to emphasize the often-overlooked dependencies within target variables and lower the barrier to use target transformations.

Acknowledgments. This work received funding from the Interuniversity Special Research Fund (IBOF/21/075) and the Flemish Government under the "Onderzoeksprogramma Artifciële Intelligentie (AI) Vlaanderen" program. We thank Wannes Meert and Lorenzo Perini for their feedback.

Disclosure of Interests. The authors have no competing interests to declare that are relevant to the content of this article.

References

1. Khan, S., Alam, M.: Preprocessing framework for scholarly big data management. Multimed. Tools Appl. **82**(25), 39719–39743 (2023)
2. Luengo, J., García-Gil, D., Ramírez-Gallego, S., García, S., Herrera, F.: Big Data Preprocessing. Springer, Cham (2020)
3. Burdack, J., Horst, F., Giesselbach, S., Hassan, I., Daffner, S., Schöllhorn, W.I.: Systematic comparison of the influence of different data preprocessing methods on the performance of gait classifications using machine learning. Front. Bioeng. Biotechnol. **8**, 260 (2020)
4. Wahid, F., Begg, R., Lythgo, N., Hass, C.J., Halgamuge, S., Ackland, D.C.: A multiple regression approach to normalization of spatiotemporal gait features. J. Appl. Biomech. **32**(2), 128–139 (2016)
5. Turney, P., Halasz, M.: Contextual normalization applied to aircraft gas turbine engine diagnosis. Appl. Intell. **3**(2), 109–129 (1993)
6. Dhal, P., Azad, C.: A comprehensive survey on feature selection in the various fields of machine learning. Appl. Intell. **52**(4), 4543–4581 (2022)
7. Zebari, R., Abdulazeez, A., Zeebaree, D., Zebari, D., Saeed, J.: A comprehensive review of dimensionality reduction techniques for feature selection and feature extraction. J. Appl. Sci. Technol. Trends **1**(1), 56–70 (2020)
8. Davis, J., et al.: View learning for statistical relational learning with an application to mammography. In: Proceedings of Nineteenth International Joint Conference on Artificial Intelligence, pp. 677–683 (2005)
9. De Brabandere, A., Op De Beéck, T., Hendrickx, K., Meert, W., Davis, J.: TSFuse: automated feature construction for multiple time series data. Mach. Learn. 113(8), 5001–5056 (2024)

10. Yang, Z., Xu, B., Luo, W., Chen, F.: Autoencoder-based representation learning and its application in intelligent fault diagnosis: a review. Measurement **189**, 110460 (2022)
11. Yi, H.-C., You, Z.-H., Huang, D.-S., Kwoh, C.K.: Graph representation learning in bioinformatics: trends, methods and applications. Briefings Bioinform. **23**(1), bbab340 (2021)
12. Meilă, M., Zhang, H.: Manifold learning: what, how, and why. Ann. Rev. Stat. Appl. **11**(1), 393–417 (2024)
13. Izenman, A.J.: Introduction to manifold learning. Wiley Interdisc. Rev. Comput. Stat. **4**(5), 439–446 (2012)
14. Feurer, M., Klein, A., Eggensperger, K., Springenberg, J., Blum, M., Hutter, F.: Efficient and robust automated machine learning. In: Advances in Neural Information Processing Systems, pp. 2962–2970 (2015)
15. Feurer, M., Eggensperger, K., Falkner, S., Lindauer, M., Hutter, F.: Auto-Sklearn 2.0: hands-free AutoML via meta-learning. J. Mach. Learn. Res. **23**(261), 1–61 (2022)
16. Janssen, V., De Raedt, L.: Predicting the travel time of vessels to multiple locks using support vector regression. MA thesis, KU Leuven (2019)
17. Koren, Y.: Factorization meets the neighborhood: a multifaceted collaborative filtering model. In: Proceedings of the 14th ACM SIGKDD International Conference on Knowledge Discovery & Data Mining, pp. 426–434 (2008)
18. Koren, Y.: Collaborative filtering with temporal dynamics. Commun. ACM **53**(4), 89–97 (2010)
19. Wind, D.K.: Concepts in predictive machine learning: a conceptual framework for approaching predictive modelling problems and case studies of competitions on Kaggle. MA thesis, Technical University of Denmark (2014)
20. Op De Beéck, T., Meert, W., Schütte, K., Vanwanseele, B., Davis, J.: Fatigue prediction in outdoor runners via machine learning and sensor fusion. In: Proceedings of the 24th ACM SIGKDD International Conference on Knowledge Discovery & Data Mining, pp. 606–615 (2018)
21. Hyndman, R.J., Athanasopoulos, G.: Forecasting: Principles and Practice. OTexts (2018)
22. Devos, L., Meert, W., Davis, J.: Fast gradient boosting decision trees with bit-level data structures. In: Proceedings of the ECML PKDD European Conference on Machine Learning and Principles and Practice of Knowledge Discovery in Databases, pp. 590–606 (2020)
23. Patel, R.S., Akolekar, H.D.: Machine-learning based optimization of a biomimiced herringbone microstructure for superior aerodynamic performance. Eng. Res. Express **5**(4) (2023)
24. Hutter, F., Kotthoff, L., Vanschoren, J.: Automated Machine Learning: Methods, Systems, Challenges. Springer, Cham (2019)
25. Flesia, L., et al.: Predicting perceived stress related to the COVID-19 outbreak through stable psychological traits and machine learning models. J. Clin. Med. **9**(10), 3350 (2020)
26. Decroos, T., Bransen, L., Van Haaren, J., and Davis, J.: Actions speak louder than goals: valuing player actions in soccer. In: Proceedings of the 25th ACM SIGKDD International Conference on Knowledge Discovery & Data Mining, pp. 1851–1861 (2019)
27. Davis, J., et al.: Methodology and evaluation in sports analytics: challenges, approaches, and lessons learned. Mach. Learn. **113**(9), 6977–7010 (2024)

28. Kubatko, J., Oliver, D., Pelton, K., Rosenbaum, D.T.: A starting point for analyzing basketball statistics. J. Quant. Anal. Sport **3**(3), 1–24 (2007)
29. Nagpal, S., Athavale, V.A., Saini, A.K., Sharma, R.: Indian health care system is ready to fight against COVID-19: a machine learning tool for forecast the number of beds. In: Proceedings of the 6th International Conference on Parallel, Distributed and Grid Computing (PDGC), pp. 61–65 (2020)
30. Bleasdale, J., Nelder, J.A.: Plant population and crop yield. Nature **188**(4747), 342 (1960)
31. He, R., Zhang, L., Chew, A.: Modeling and predicting rainfall time series using seasonal-trend decomposition and machine learning. Knowl- Based Syst. **251**, 109125 (2022)
32. Hippke, M., David, T.J., Mulders, G.D., Heller, R.: Wōtan: comprehensive time-series detrending in Python. Astron. J. **158**(4), 143–158 (2019)
33. Alexandrov, T., Bianconcini, S., Dagum, E.B., Maass, P., McElroy, T.S.: A review of some modern approaches to the problem of trend extraction. Economet. Rev. **31**(6), 593–624 (2012)
34. Seip, K.L., Zhang, D.: Scoring six detrending methods on timing, leadlag relations, and cycle periods: an empirical study of US and UK recessions 1977–2020. Comput. Econ. **64**(5), 3087–3116 (2024)
35. Kopel, R., et al.: No time for drifting: comparing performance and applicability of signal detrending algorithms for real-time fMRI. Neuroimage **191**, 421–429 (2019)
36. Yang, K., Tu, J., Chen, T.: Homoscedasticity: an overlooked critical assumption for linear regression. Gen. Psychiatry **32**(5) (2019)
37. Payne, N.: Evaluating the impact of heteroscedasticity on the predictive ability of modern regression techniques. MA thesis, Simon Fraser University (2014)
38. Krawczyk, B.: Learning from imbalanced data: open challenges and future directions. Progress Artif. Intell. **5**(4), 221–232 (2016)
39. Yang, Y., Zha, K., Chen, Y., Wang, H., Katabi, D.: Delving into deep imbalanced regression. In: Proceedings of the 38th International Conference on Machine Learning, pp. 11842–11851 (2021)
40. Box, G.E., Cox, D.R.: An analysis of transformations. J. R. Stat. Soc. Ser. B Stat Methodol. **26**(2), 211–243 (1964)
41. Dietterich, T.G.: Approximate statistical tests for comparing supervised classification learning algorithms. Neural Comput. **10**(7), 1895–1923 (1998)

Balancing Performance and Scalability of Demand Forecasting ML Models

Mateusz Żarski[1,2](✉) and Sławomir Nowaczyk[2]

[1] Institute of Theoretical and Applied Informatics, Polish Academy of Sciences, Bałtycka 5, 44-100 Gliwice, Poland
mzarski@iitis.pl

[2] Halmstad University, Kristian IV:s väg 3, 301 18 Halmstad, Sweden
slawomir.nowaczyk@hh.se

Abstract. Balancing performance and scalability is a major concern when developing robust ML models for diverse, big-data scenarios, such as predicting demand for a number of products across multiple locations. The two mutually opposite approaches are to use a single ML model for maximizing scalability, often at the expense of performance, or to use a specialized model for each specific use case, which is often prohibitive in terms of computational costs. In this paper, we propose to balance those two approaches using our methods of model clustering and grouping. We achieve the performance level of a single use-case model while preserving the global scalability of the solution. In our experiments, we use a publicly available demand forecasting dataset as a use case. We develop and train baseline shallow ML models and DL models for both maximizing performance and scalability. Then, we showcase a desirable balance that can be achieved using our proposed methods, one that outperforms both shallow ML and specific use-case models.

Keywords: Demand forecasting · Deep Learning · Hybrid models · Transfer learning · Model fine-tuning

1 Introduction

With the increase of applicability of Machine Learning (ML) in new areas, the importance of accurately modeling complex real-world relationships increases significantly. In the past, ML was employed to solve narrowly defined problems with hand-tailored data. Today, with the emergence of Deep Learning (DL), it is applied to a broad range of challenges under the expectation of working out-of-the-box. In particular, a single ML model typically cannot cover the full breadth of the problem, while creating many specialized ML models leads to inefficiencies (both in data utilization and computational resources).

Performance-scalability trade-off is not a new concept in the field of ML, as it was tackled multiple times in the past using various techniques, such as statistical methods, neural networks, and genetic algorithms [10, 23]. Today, the

G. Krempl et al. (Eds.): IDA 2025, LNCS 15669, pp. 127–140, 2025.
https://doi.org/10.1007/978-3-031-91398-3_10

areas where this problem is most pronounced are issues related to processing large amounts of data, such as social media [31], IoT [19], and training models on large sets of imagery data [30]. In the most recent applications, there are two most prominent approaches covered by researchers. The former emphasizes building data models [29], while the latter focuses on creating scalable ML architectures [21]. In this paper, we tackle the problem of demand forecasting over big data scenarios using hybrid approaches – by developing transfer-learned models and training them on structured data, thus adapting them to specific sub-cases as needed.

Forecasting the demand using numerical and statistical approaches, along with the performance-scalability trade-off, has been present in business management for at least half a century [2]. Moreover, with the emergence of Big Data methods [18], and data aggregation [20,25], companies are able to collect large amounts of data on the past performance of demand for their products, which can significantly help to accurately predict the future. However, traditionally used data-driven [9,17] or linear regression-based methods (*e.g.* with Support Vector Machines [12]) proved to be highly unreliable in accurately rendering other-than-linear relationships in big-data setups.

The most recent, highly successful approaches tend to use ML and, most notably, DL methods to increase future demand accuracy by capturing nonlinearities in the data patterns for time-series analysis. The task is mostly performed by employing Artificial Neural Networks (ANNs), Recurrent Neural Networks (RNNs), and Transformer-type models. Currently, ML methods are used in various forecasting tasks, including specific product demand [11], passenger demand in scheduled transportation [1], energy consumption forecasting [27] and others.

The main contribution of our paper is proposing a method to train and manage a family of MLP models, striking a desired balance between a single global model covering all use cases at one end and a multitude of highly specialized models, each covering as narrow as possible set of circumstances, at the other. We demonstrate the usefulness of the proposed approach on a time series regression problem, specifically demand forecasting of many products over many locations. We showcase that classic fine-tuning and transfer learning methods allow models trained for a specific product-location pair to significantly outperform a global model. In order to handle the scalability challenge, we then concentrate on employing model grouping and clustering to reduce the number of models necessary from over one thousand to twenty or fifteen, which actually happens while slightly increasing the overall performance.

The rest of the paper is organized as follows: in Sect. 2, we describe previous works that are related to our paper. In Sect. 3, we define the problem at hand and the goal of the experiments performed. Next, in Sect. 4, we describe the dataset used (4.1) and our experiments starting with the baselines (4.2), and then our implementation of ANNs models (4.3), as well as compare the results obtained (4.4). Lastly, in Sect. 5, we provide our conclusions on the experiments performed and describe our plans for future work.

2 Related Works

The current trends of engaging with the problem of demand forecasting on vast datasets lean to focus on the use of ML methods, with most emphasis on Deep Learning models. There are comparative studies in which researchers confirmed that in diverse demand forecasting scenarios, ML models tend to outperform traditional statistical models [26]. However, methods like ARIMA are still well in use in many use cases as a robust baseline or target solution [6].

In the scope of employing ML models, distinctive approaches can be separated. Researchers use Multilayer Perceptron (MLP), RNN, and Transformer models for the task.

MLPs have successfully been applied for the tasks of forecasting transportation needs [7] and customer information, in comparison to Shallow Learning methods [33]. Other researchers explored the influence of various model architectures and evaluation metrics on the model training process [14] or used Dynamic Artificial Neural Networks (DANs) [8].

In the scope of other architectures used for demand forecasting on time-series data, a distinctive use of Recurrent and Transformer-type models can be seen. Researchers explored the use of LSTM [3] and GRU-based [28] models, as well as a hybrid approach with both of the aforementioned RNN cells [13]. Also, since their emergence, Transformers were used for the task [32].

Lastly, there are also works devoted to using hybrids of ANN models with other techniques of ML or general numerical methods. Approaches employing hybrids of MLP and Auto-regressive models were used [4] as well as MLP-Fourier [22] or MLP-SVM [18] models. Another interesting branch of such research is models employing trainable categorical embeddings within MLP architecture in tasks where products are highly differentiated or when additional product categorization is needed [5].

In our approach, we focus on using the latter approach, with hybrid models and algorithmic-supported data structuring techniques along with trainable categorical embeddings. We use lightweight ANN models to perform the task of demand forecasting on an extensive dataset with multiple products and vendors. We use methods of fine-tuning, clustering, and model grouping in order to maintain a high level of prediction accuracy while preserving the models' versatility and low computational costs. We compare our approach with baseline models, and provide insight into the dataset preparation and management, as well as the ANNs training process.

3 Problem and Goal Definition

The problem we are tackling is how to efficiently apply ML to a large number of related but diverse prediction problems. This requires evaluating the trade-off between building a single, global, general model versus building a large number of possibly independent models, one per use case – or anything in between these two ends of the spectrum.

In the use case scenario covered by our experiments, we focused on the issue of forecasting future product demands in the data consisting of multiple items and final point-of-sale IDs. As we demonstrate, the data is sufficiently diverse that it cannot be accurately solved with a single forecasting model.

For our goal, we have chosen to lower the forecasting error of the accuracy metric $\mathcal{E}_{ev}$ on the evaluation dataset. As a metric – root mean square error (RMSE) was chosen, calculated between the ground truth y and predicted $\hat{y}$ values for n samples. RMSE metric was also chosen because it provides easily interpretable measurements in the same unit as the forecasted value (in contrast to *e.g.* mean squared error), and we confirmed that our dataset is free of outliers, which RMSE is known to be sensitive to.

Using the RMSE metric for obtaining the error of our trained models did not, however, mean that it was necessarily used as our loss function – throughout the experiments we performed, multiple loss functions were tested, but in order to retain consistency in test-to-test comparison, RMSE was always used to measure the forecasting error.

4 Experiments

In our experiments, we aimed to minimize two criteria: the RMSE metric and model count. We divided our experiments into **Baseline** (4.2) with ARIMA and traditional ML methods, and **MLP experiments** (4.3) with our MLP architecture training along with our methods of increasing its performance.

All experiments were performed on a Linux-based operating system and Python 3.11 environment with a Ryzen 7 5800x CPU, 32 GB RAM, and Nvidia RTX 3060 with 12 GB VRAM. For baseline experiments, we used *Scikit-Learn* and *statsmodels* libraries, while our MLP models were trained using *PyTorch* and *PyTorch-Lightning* libraries.

4.1 Main Dataset

In our experiments, we used a public demand forecasting dataset freely available at Kaggle [24]. It contains weekly sales data from the period of three years for multiple vendors and products, and additional information regarding each sale, with a full feature vector containing 8 elements of varied data types. The whole dataset consists of over 150k data points. A full description of the data included in the dataset is presented in Table 1. Additionally, in Appendix A we also present results on two additional, popular demand forecasting datasets.

In our main dataset, in addition to continuous variables (*e.g.* `base_price`), there are also categorical variables. Although the `week` variable may be excluded from the dataset, as we treated the data as a continuous time series, two variables – `store_id` and `sku_id` form a total of 1155 valid *SKU-Store* pairs (out of possible 2128 pairs, but not all locations sell all products). Encoding those variables efficiently requires creating a meaningful representation that helps effective training process and data generalization, while allowing for the lower model count.

Table 1. Demand forecasting dataset summary.

Feature name	Description	Data type	Data count/ Data range
`week`	Week-time of the performed sale	*calendar-type format*	130
`store_id`	ID number of the vendor	integer (categorical)	76
`sku_id`	ID (Stock Keeping Unit) of the product	integer (categorical)	28
`total_price`	Price for a single unit sold in current week including discounts	decimal	41.325 - 562.1625
`base_price`	Price for a single unit excluding discounts	decimal	61.275 -562.1625
`is_featured_sku`	Is the product on discount in current week	integer (categorical)	2
`is_display_sku`	Is the product promoted in current week	integer (categorical)	2
`units_sold`	Total number of units sold in week	integer	1 - 2876

We also made sure that the data remained consistent and without outliers throughout the dataset. In Fig. 1, we show monthly-aggregated sales of all product SKUs in the dataset. It can be seen that the sales of each product remain in the same magnitude for three years and there is no missing data or visible outliers. We also checked the week-wise standard deviation of the data and confirmed that it is consistent for all SKUs with population standard deviation $\sigma = 50.22$.

The results shown in the remained of this article consist of the mean, minimal, and maximal RMSE score computed for every *SKU-Store* pair within the evaluation dataset. Where applicable, feature vectors were normalized (MLP experiments), and the results for nondeterministic training processes are presented as the mean out of five random seed trainings.

4.2 Baseline

ARIMA. Experiments on the Autoregressive Integrated Moving Average model were performed considering the data as non-seasonal, with parameter grid search for all components of the model (AR (p), MA (q) and I (d)). Due to its limitations, the feature vector consisted only of the total sales in a given lag period.

For autoregressive model – AR, order parameter p was selected from range $p = 1, \ldots, 15$ with step 1, and for both moving average (MA) parameter q and degree of differencing (I) parameter d – from range $q, d = 1, \ldots, 3$, also with step 1. This resulted in a total of 135 ARIMA models for the experiment.

In Table 2, we present the results of a set of the best-performing ARIMA models for *SKU-Store* pairs in the dataset. Note that due to training every *SKU-Store* pair model separately, each of them could be prepared with a different set of parameters.

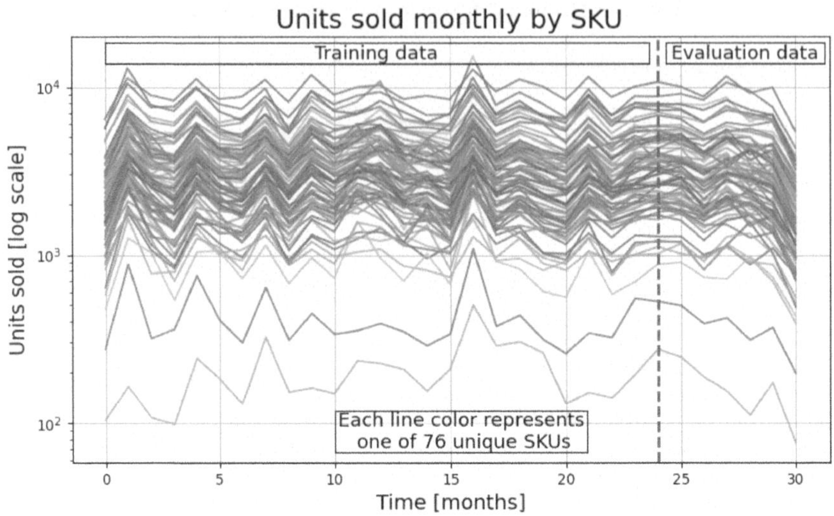

Fig. 1. A chart of the course of product sales during the dataset period. In the graph, a line marks the division between the training and evaluation sets.

Decision Trees and Random Forests. Experiments were also performed on *SKU-Store* pairs for consistency, with a set of grid-searched parameters for each algorithm run on a given pair. Unlike in our ARIMA experiments, for tree-based models, we used the whole feature vector from the dataset. In grid-search, parameters commonly used for both Decision Trees and Random Forests included a number of lags, $lags = 1, \ldots, 15$ with step 1, maximum depth of the regressor $depth_{max} = 4, \ldots, 12$ with step 2, and maximum count of leaf nodes $nodes_{max} = \{10, 15, 20\}$. For Random Forests, we additionally performed a grid search over the maximum number of estimators in the ensemble $est_{max} = \{100, 150, 200, 250\}$.

We also performed additional experiments regarding post-pruning of the trained Decision Trees estimators using various levels of leaves' total impurity for every Decision Tree trained. However, in those experiments, the RMSE metric tended to be much higher than both DT and RF estimators.

In Table 2, we present the results of the best tree-based estimators trained. Like previously with ARIMA experiments, each model could represent a different set of parameters used for training.

4.3 MLP Experiments

We performed four MLP experiments in total, delineating the spectrum between the two extreme approaches to scalability-performance trade-off. The first one, Base MLP, consists of training a single model to cover all *SKU-Store* pairs at once. The other end of the spectrum is Fine-tuned MLP models, focusing on building the best possible model for each *SKU-Store* pair individually. Finally,

Clustered MLP and Grouped MLP use the algorithms we have developed to balance the scalability-performance characteristics.

To ensure consistency in our MLP experiments, for all experiments within this section, we used the same model architecture, optimizer, loss function, and learning rate management and scheduling techniques. The selection of model architecture and tuning of hyperparameters was performed on a subset of *SKU-Store* pairs with a grid-search over learning rate, weight decay, and scheduler parameters. The selected hyperparameters remained consistent for all experiments.

The architecture of our models assumes an input vector with *lag=15* and consists of two embedding layers for encoding *store_id* and *sku_id* into 5-element vectors, as well as the main body of the MLP. It was built from four sequentially connected blocks: $\{Linear, ReLU, Dropout\}$ with hidden dimensions $linear_{dim} = \{512, 512, 256, 64\}$ and consistent dropout of 0.5. The last layer was a linear, single-neuron regressor.

As a weight optimizer, we used *AdamW* with an initial learning rate of 0.001 and additional weight decay. For the loss function, multiple versions were tested, including RMSE, L1 Loss, and root mean squared log error (RMSLE), but in the end, we found out that mean squared log error (MSLE) provides the best results. Lastly, for learning rate management, we used *Reduce on Plateau* algorithm with 0.8 learning rate reduction. Early stopping was not used due to the relatively small amount of datapoints per *SKU-Store* pair.

Base MLP. In our initial MLP experiments, we tested the approach of extreme scalability of the model by training only one MLP model for all *SKU-Store* pairs together. This approach allows for major resource savings, however, forces retraining or fine-tuning of the whole model in the case when a new data pair is added to the dataset. The model was trained for 200 epochs, with *Reduce on Plateau* scheduler set to decrease the learning rate after the 20-epoch plateau. The results of the experiment are included in Table 2 and indicate clearly that a single model cannot solve the task with sufficient accuracy.

This initially trained model was later used in the rest of our experiments.

Fine-Tuned MLP. Our fine-tuning experiments were performed with *Base MLP* as an initially pre-trained model. For the extreme performance approach, each of the *SKU-Store* pairs was selected for training separately, and each of the models was trained for an additional 20 epochs with a 5-epoch plateau. The results of this approach are included in Table 2, showing significant performance improvements.

Clustered MLP. The first of our proposed methods for balancing scalability-performance was based on the use of the two embedding layers trained broadly on the whole dataset along *Base MLP* model. For the clustering, we initially used those layers for encoding every *SKU-Store* ID and concatenate them into

10-element representation vectors. Next, we used K-means clustering in order to divide the set of vectors into clusters of *SKU-Store* pairs that are the most similar to each other. The pairs were then used for the training of new models, the same way as *Fine-tuned MLP* models.

The total number of clusters was selected with a grid search over a set of numbers: $cluster_{num} = \{10, 15, 20, 50, 100\}$. During experiments, clustering with 15 models performed the best, and the results from this model are included in Table 2.

Other methods of data clustering can also be considered, *e.g.* clustering on the whole or a part of input data vectors themselves. However, some datasets, such as presented in Appendix A do not present data outside categorical labels, and such clustering would not apply to them.

Grouped MLP. Our second method of improving the scalability-performance trade-off employed an algorithmic approach during the training of the models. The full algorithm is described in detail in Algorithm 1.

During the training, in the first step, a random *SKU-Store* pair is selected, and the model is trained on its data. Next, the trained model is tested on the rest of the pairs, and only those obtaining $test\_metric$ value lower than $threshold$ are grouped together and removed from pairs available in the $Pair\_bank$. This training-testing loop is executed for a total assumed number of models $Models_{max}$ or until there are no pairs left in $Pair\_bank$. After grouping, a model is trained separately for each group in the same way as *Fine-tuned MLP* model.

In experiments, we grid-searched over various values for algorithm parameters, however $Models_{max} = 35$ and $threshold = 25$ performed the best and are included in Table 2.

4.4 Results Comparison

The results obtained from all our experiments performed on the evaluation dataset, divided into groups of algorithms used, are shown in Table 2. Additonal experiments are available in Appendix A, and sample results in Appendix B.

As can be seen in the results, the ARIMA model provided a robust baseline for our experiments, achieving low mean RMSE metric, however, allowing for no scalability, as it was trained independently for every *SKU-Store* pair. Similarly, the tree-based models do not allow for scalability, and despite using extended feature vectors, performed poorly in both mean and max RMSE.

On the other hand, our plain MLP-based models offer a distinctive trade-off between scalability and performance. Two extreme approaches to this matter are presented with *Base MLP* and *Fine-tuned MLP* models. The highly scalable *Base MLP* model, while handling all *SKU-Store* pairs at once, offers significantly worse performance than ARIMA in both mean and maximum RMSE terms. The *Fine-tuned MLP* model tuned for each pair separately, manages to notably outperform ARIMA in mean and minimum RMSE; however, it still performs poorly in maximum RMSE, indicating overfitting for specific *SKU-Store* pairs.

Algorithm 1. Grouping algorithm

Require:
Model: M
Data bank of all pairs: $Pair\_bank$
Max number of models: $Models_{max}$
Min number of data pairs for model: $Pairs_{min}$
Metric threshold value: $threshold$
Ensure:
$n \leftarrow 0$
$matches\_array \leftarrow \{\}$
while $n \leq Models_{max}$ **and** $|Pair\_bank| \geq Pairs_{min}$ **do**
  $initialize(M)$ ⊳ initialize model randomly for selected pair
  $random\_pair \leftarrow random(Pair\_bank)$
  $train(M, random\_pair)$
  $matches\_per\_model \leftarrow \{\}$ ⊳ this set will store matching pairs for given model
  for $pair \in Pair\_bank$ **do**
    $test\_metric \leftarrow test(M, pair)$
    if $test\_metric \leq threshold$ **then**
      $matches\_per\_model \xleftarrow{+} pair$
    end if
  end for
  if $|matches\_per\_model| \geq Pairs_{min}$ **then**
    $matches\_array \xleftarrow{+} matches\_per\_model$
    $Pair\_bank \xleftarrow{-} matches\_per\_model$
  end if
  $matches\_per\_model \leftarrow \{\}$ ⊳ the set is emptied for the next loop
  $n \leftarrow n + 1$
end while
$matches\_array \xleftarrow{+} Pair\_bank$ ⊳ leftover pairs are added as the last group

A significantly improved balance between scalability and performance is achieved with our proposed hybrid approaches. Both *Clustered MLP* and *Grouped MLP* models significantly outperform ARIMA (and all other tested models) in terms of mean RMSE and, most strikingly, in maximum RMSE, more than three times improving the result of ARIMA. Notably, they also perform better than models fine-tuned for every *SKU-Store* pair, despite being much less numerous than them, indicating that our grouping methods can also help with overfitting.

Table 2. Experimental results with standard deviation measurements of successive models

Model name	Model count	RMSE mean (σ)	RMSE min (σ)	RMSE max (σ)
ARIMA	1155	25.34 (4.35)	3.43 (3.15)	196.90 (7.22)
Decision Trees	1155	28.98 (6.43)	4.09 (3.11)	360.56 (9.95)
Random Forests	1155	26.93 (5.55)	3.32 (1.17)	308.62 (9.21)
Pruned DTs	1155	37.40 (6.32)	4.26 (2.76)	383.92 (11.70)
Base MLP	**1**	40.71 (8.31)	4.90 (2.13)	361.92 (12.15)
Fine-tuned MLP	1155	19.97 (3.41)	**2.91** (1.77)	224.18 (9.57)
Clustered MLP	15	**18.48** (3.81)	8.72 (2.11)	69.19 (7.92)
Grouped MLP	20	19.12 (3.33)	8.09 (2.06)	**53.35** (6.39)

5 Conclusions and Future Work

In this paper, we covered ML-based and algorithmic methods for tackling the problem of scalability-performance when working on large-scale, diverse datasets. We have also shown extreme cases of approaches to this issue and our own methods of balancing it, along with their description and the results achieved. In our experiments, we proved that efficient model training and data division in diverse scenarios is possible with little additional effort by combining ML hybrid models and algorithmic data grouping methods, creating a usable and implementable solution.

However, our work does not exhaust the subject, as it is only an introduction to the vast topic of developing effective methods for balancing performance and scalability in issues where a single model can not solve the task efficiently. Our next steps will include employing other model architectures to verify how they will behave in the presented scenario.

Acknowledgments. The work was carried out with support from The Knowledge Foundation and from Vinnova (Sweden's innovation agency) through the Vehicle Strategic Research and Innovation Programme, FFI.

Disclosure of Interests. The authors have no competing interests to declare that are relevant to the content of this article

A Experiments on Additional Datasets

To better illustrate the performance of the methods we developed, we performed experiments on two additional datasets available at Kaggle – *Store Item Demand Forecasting Challenge (**SIDF**)* [16] and *Product Demand Forecast (**PDF**)* [15]. Datasets contain only categorical variables, making it more challenging to accurately predict the demand, and also preventing clustering of the similar entries by using feature vectors. Both datasets use a different range of demand value, making them not feasible to compare results in dataset-to-dataset manner. The experimental results from both dataset are shown in Table 3 and Table 4. Note that the lack of statistical data in tables is due to space requirements, and will be provided upon a request.

Table 3. Experimental results – **SIDF**

Model name	Model count	RMSE mean	RMSE min	RMSE max
ARIMA	500	12.43	4.82	23.61
Decision Trees	500	17.42	5.45	35.29
Random Forests	500	17.40	5.45	35.30
Pruned DTs	500	17.44	5.62	33.93
Base MLP	**1**	7.46	3.44	12.31
Fine-tuned MLP	500	8.23	**3.23**	**11.65**
Clustered MLP	15	11.51	6.48	16.50
Grouped MLP	4	**6.64**	3.33	14.24

B Sample Results

In Fig. 2 we provide a comparison of model inferences. Ground truth (GT) series is marked with a widened, black line for better visibility.

Table 4. Experimental results – **PDF**

Model name	Model count	RMSE mean	RMSE min	RMSE max
ARIMA	93	6232	3.36	8217
Decision Trees	93	7294	2.71	10095
Random Forests	93	7288	2.71	10095
Pruned DTs	93	7294	2.71	10095
Base MLP	**1**	2679	**0.98**	4624
Fine-tuned MLP	93	2535	3.24	3168
Clustered MLP	20	1653	7.78	**1992**
Grouped MLP	5	**858.2**	26.76	2253

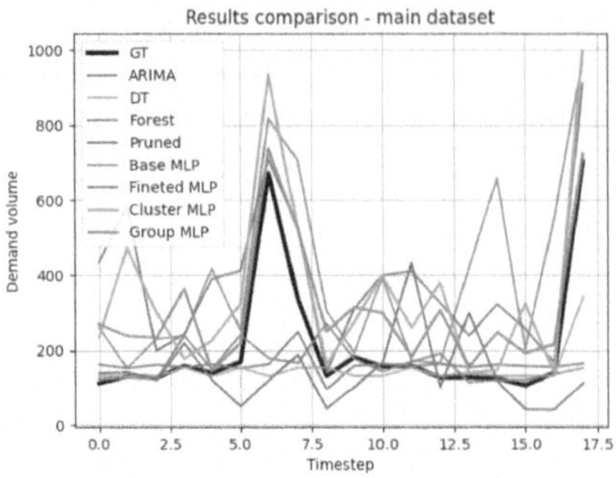

(a) Main dataset

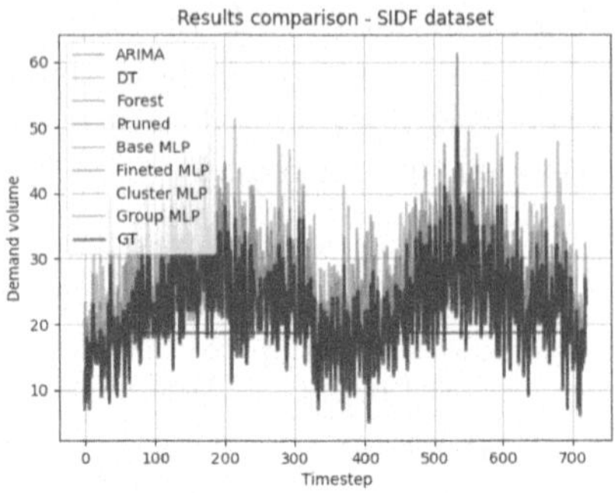

(b) SIDF dataset

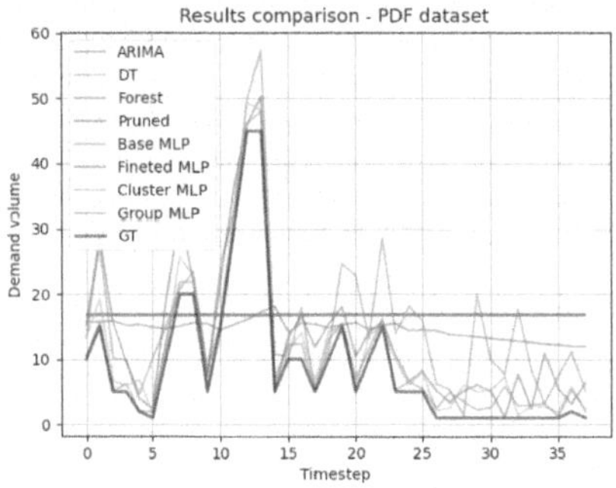

(c) PDF dataset

Fig. 2. Inference with models on the datasets used

References

1. Banerjee, N., Morton, A., Akartunalı, K.: Passenger demand forecasting in scheduled transportation. Eur. J. Oper. Res. **286**(3), 797–810 (2020). https://doi.org/10.1016/j.ejor.2019.10.032
2. Box, G.E., Jenkins, G.M.: Some recent advances in forecasting and control. J. R. Stat. Soc. Ser. C (Appl. Stat.) **17**(2), 91–109 (1968)
3. Chandriah, K.K., Naraganahalli, R.: RNN/LSTM with modified adam optimizer in deep learning approach for automobile spare parts demand forecasting. Multimed. Tools Appl. (2021). https://doi.org/10.1007/s11042-021-10913-0
4. Choi, G.S., Yu, C., Jin, R.M., Yu, S.K., Chun, M.G.: Short-term water demand forecasting algorithm using AR model and MLP. J. Korean Inst. Intell. Syst. **19**(5), 713–719 (2009). https://doi.org/10.5391/jkiis.2009.19.5.713
5. Dholakia, R., Randeria, R., Dholakia, R., Ashar, H., Rana, D.: Cognitive demand forecasting with novel features using word2vec and session of the day. In: Gunjan, V.K., Zurada, J.M., Raman, B., Gangadharan, G.R. (eds.) Modern Approaches in Machine Learning and Cognitive Science: A Walkthrough. SCI, vol. 885, pp. 59–72. Springer, Cham (2020). https://doi.org/10.1007/978-3-030-38445-6_5
6. Fattah, J., Ezzine, L., Aman, Z., El Moussami, H., Lachhab, A.: Forecasting of demand using arima model. Int. J. Eng. Bus. Manag. **10**, 1847979018808673 (2018)

7. Ferreira, R., Martiniano, A., Ferreira, A., Ferreira, A., Sassi, R.: Study on daily demand forecasting orders using artificial neural network. IEEE Latin Am. Trans. **14**(3), 1519–1525 (2016). https://doi.org/10.1109/TLA.2016.7459644
8. Ghiassi, M., Zimbra, D.K., Saidane, H.: Urban water demand forecasting with a dynamic artificial neural network model. J. Water Resour. Plan. Manag. **134**(2), 138–146 (2008). https://doi.org/10.1061/(ASCE)0733-9496(2008)134:2(138)
9. Guo, F., Diao, J., Zhao, Q., Wang, D., Sun, Q.: A double-level combination approach for demand forecasting of repairable airplane spare parts based on turnover data. Comput. Ind. Eng. **110**, 92–108 (2017). https://doi.org/10.1016/j.cie.2017.05.002
10. Gupta, P., Sharma, A., Jindal, R.: Scalable machine-learning algorithms for big data analytics: a comprehensive review. WIREs Data Min. Knowl. Discov. **6**(6), 194–214 (2016). https://doi.org/10.1002/widm.1194
11. Ali, Ö.G., Sayın, S., van Woensel, T., Fransoo, J.: SKU demand forecasting in the presence of promotions. Expert Syst. Appl. **36**(10), 12340–12348 (2009). https://doi.org/10.1016/j.eswa.2009.04.052
12. Hansen, J.V., McDonald, J.B., Nelson, R.D.: Some evidence on forecasting time-series with support vector machines. J. Oper. Res. Soc. **57**(9), 1053–1063 (2006)
13. Jadli, A., Lahmer, E.H.B., et al.: A novel LSTM-GRU-based hybrid approach for electrical products demand forecasting. Int. J. Intell. Eng. Syst. **15**(3) (2022)
14. Jeyaraman, J., Krishnamoorthy, G., Konidena, B., Sistla, S.: Machine learning for demand forecasting in manufacturing. Int. J. Multidisc. Res. (2024). https://doi.org/10.17613/ct0h-bn97
15. Kaggle: Product demand forecast (2017). https://www.kaggle.com/code/flavioakplaka/product-demand-forecast
16. Kaggle: Store item demand forecasting challenge (2018). https://www.kaggle.com/competitions/demand-forecasting-kernels-only
17. Khan, N., et al.: Big data: survey, technologies, opportunities, and challenges. Sci. World J. **2014**(1), 712826 (2014). https://doi.org/10.1155/2014/712826
18. Kilimci, Z.H., et al.: An improved demand forecasting model using deep learning approach and proposed decision integration strategy for supply chain. Complexity **2019**(1), 9067367 (2019). https://doi.org/10.1155/2019/9067367
19. Luntovskyy, A., Globa, L.: Performance, reliability and scalability for IoT. In: 2019 International Conference on Information and Digital Technologies (IDT), pp. 316–321 (2019). https://doi.org/10.1109/DT.2019.8813679
20. Babai, M.Z., Boylan, J.E., Rostami-Tabar, B.: Demand forecasting in supply chains: a review of aggregation and hierarchical approaches. Int. J. Prod. Res. **60**(1), 324–348 (2022). https://doi.org/10.1080/00207543.2021.2005268
21. Mehendale, P.: Scalable architecture for machine learning applications. J. Sci. Eng. Res. **11**, 111–117 (2024)
22. Odan, F.K., Reis, L.: Hybrid water demand forecasting model associating artificial neural network with Fourier series. J. Water Resour. Plan. Manag. **138**(3), 245–256 (2012). https://doi.org/10.1061/(ASCE)WR.1943-5452.0000177
23. Paliouras, G.: Scalability of machine learning algorithms. Ph.D. thesis, University of Manchester (1993)
24. Rao, A.V.S.: Demand forecasting dataset (2020). https://www.kaggle.com/datasets/aswathrao/demand-forecasting
25. Rostami-Tabar, B., Babai, M.Z., Syntetos, A., Ducq, Y.: Demand forecasting by temporal aggregation. Nav. Res. Logist. (NRL) **60**(6), 479–498 (2013). https://doi.org/10.1002/nav.21546

26. Spiliotis, E., Makridakis, S., Semenoglou, A.A., Assimakopoulos, V.: Comparison of statistical and machine learning methods for daily SKU demand forecasting. Oper. Res. Int. J. (2022). https://doi.org/10.1007/s12351-020-00605-2
27. Suganthi, L., Samuel, A.A.: Energy models for demand forecasting-a review. Renew. Sustain. Energy Rev. **16**(2), 1223–1240 (2012). https://doi.org/10.1016/j.rser.2011.08.014
28. Tian, R., Wang, B., Wang, C.: MAGRU: multi-layer attention with GRU for logistics warehousing demand prediction. KSII Trans. Internet Inf. Syst. **18**(3) (2024)
29. Vajpayee, A., Mohan, R., Chilukoori, V.V.R.: Building scalable data architectures for machine learning. Int. J. Comput. Eng. Technol. **14**, 308–320 (2024). https://doi.org/10.5281/zenodo.13234031
30. Yan, F., Ruwase, O., He, Y., Chilimbi, T.: Performance modeling and scalability optimization of distributed deep learning systems. In: Proceedings of the 21th ACM SIGKDD International Conference on Knowledge Discovery and Data Mining, KDD 2015, pp. 1355–1364. Association for Computing Machinery, New York (2015). https://doi.org/10.1145/2783258.2783270
31. Yogi, K.S., et al.: Scalability and performance evaluation of machine learning techniques in high-volume social media data analysis. In: 2024 11th International Conference on Reliability, Infocom Technologies and Optimization (Trends and Future Directions) (ICRITO), pp. 1–6 (2024). https://doi.org/10.1109/ICRITO61523.2024.10522361
32. Zhang, G.P., Xia, Y., Xie, M.: Intermittent demand forecasting with transformer neural networks. Ann. Oper. Res. **339**(1), 1051–1072 (2024)
33. Zohdi, M., Rafiee, M., Kayvanfar, V., Salamiraad, A.: Demand forecasting based machine learning algorithms on customer information: an applied approach. Int. J. Inf. Technol. (2022). https://doi.org/10.1007/s41870-022-00875-3

Balancing Global Importance and Source Proximity for Personalized Recommendations Using Random Walk Length

Tsuyoshi Yamashita[1(✉)] and Kunitake Kaneko[2,3]

[1] Graduate School of Science and Technology, Keio University, Kanagawa, Japan
tullys@inl.ics.keio.ac.jp

[2] Research Institute for Digital Media and Content, Keio University, Kanagawa, Japan

[3] Faculty of Science and Technology, Keio University, Kanagawa, Japan

Abstract. For personalized recommendation in graphs, random walks starting from the user's interest node are known as a general-purpose and fast analysis method. Specifically, Personalized PageRank (PPR), which quantifies the importance of each node by the distribution of visited nodes in random walks, determines nodes with high global importance and source proximity. However, it is difficult to balance both influences monotonically. This paper clarifies that the random walks length is effective to monotonically control the balance of both influences on PPR vectors. In particular, we exploit the fact that correlation between PPR and PageRank values monotonically changes depending on PPR parameter that controls the average random walk length. Here, PageRank is a metric that quantifies the global importance by the probability of random walks from all nodes visiting each node. A case study using the movie rating dataset showed that nodes that are considered to be directly related to the source node get high PPR value by shortening the average random walk length. Moreover, statistical evaluation on nine real-world datasets revealed that changing the average random walk length from 1.01 to 100 resulted in a monotonic increase of the cosine similarity between PPR and PageRank vectors from 0.002 to 0.76 at the maximum.

Keywords: Graph · Personalization · Random Walks · Global Importance · Source Proximity

1 Introduction

Graphs are widely used to represent relationships between digital data, and random walks on graphs provide fast and general-purpose personalized recommendations [4,7,8,20–22,27,29]. Random walks are computations taking a source node s and a parameter α to control its length as inputs and output a list of

G. Krempl et al. (Eds.): IDA 2025, LNCS 15669, pp. 141–153, 2025.
https://doi.org/10.1007/978-3-031-91398-3_11

paths starting from s. By setting s to the user's interest node and analyzing paths starting from s, personalized recommendations are realized. The path analysis methods include centrality computations [20,27], embeddings [21,22], and graph neural networks [7,29]. In particular, Personalized PageRank (PPR) [20], one of the centrality computation methods, is a general-purpose, fast, independent of external information analysis method [4,8,20]. Here, PPR quantifies the importance of each node with respect to s by the probability of visiting each node in random walks.

Although PPR determines nodes with high global importance and source proximity, it is necessary to monotonically balance the influences of global importance and source proximity on the recommendation results according to the user's requirements. When the influence of global importance is dominant, nodes that are important from the perspective of the whole graph tend to be recommended, regardless of the source node. While it is likely to get nodes that are considered important for many users, the drawback is that the source node has little impact and the results are obvious. In contrast, when the influence of source proximity is dominant, nodes that are close to the source node tend to be recommended even if their global importance is not necessarily high. While it is likely to get nodes that are strongly affected by the source node, the drawback is that the results tend to be minor from a global perspective. Thus, there is a trade-off between global importance and source proximity, and in real-world applications, users need to adjust the balance between them depending on the source node or their preferences. It is also required that users observe the recommendation results while gradually changing the parameters. In such a case, it is important to monotonically balance the influences of them according to the parameters.

However, existing work on PPR-based recommendations has discussed the way of setting the source node [18,19], and the characterization of the recommendation results for a fixed source node has not been adequately discussed. Although some methods utilize external information such as node labels to improve the recommendation quality [18,25,28], these methods are not general-purpose because they require to determine the necessary external information and collect them for each graph. In addition, various recommendation criteria including diversity [11,25,28,30], novelty [11,18], serendipity [11,19], and fairness [24,30] have been proposed. To the best of our knowledge, no paper focuses on balancing the influences between global importance and source proximity.

This paper discusses a method to monotonically balance the influences of global importance and source proximity on PPR results. In particular, we change the characteristics of the nodes that random walks tend to visit in terms of global importance and source proximity by α, which is a parameter of PPR that controls the random walk length. As the random walks get longer, nodes with high PageRank values, where PageRank indicates the global importance, are more likely to be visited. On the other hand, as the random walks get shorter, nodes that are close to the source node are more likely to be visited, regardless of their global importance. A case study using a movie rating dataset confirmed that movies that were considered directly relevant to the source node got a higher

PPR value by shortening the random walks. Furthermore, statistical evaluation using nine real-world datasets confirmed that the similarity between PPR and PageRank vectors weakens by shortening the random walks. In particular, changing the average random walk length from 1.01 to 100 resulted in a monotonic change in the cosine similarity of PPR and PageRank vectors from 0.002 to 0.76 at the maximum.

The main contributions of this paper are summarized as follows.

- We proposed a method to monotonically balance the influences of global importance and source proximity on Personalized PageRank-based recommendation results by the random walk length.
- Using a movie rating dataset, we show that the random walk length can monotonically control the ranking of movies that are considered to be directly related to the source movie.
- Using nine real-world datasets, we find that the similarity between PPR vector and PageRank vector, which represents global importance, changes monotonically with the random walk length.

Finally, the remainder of this paper is structured as follows: Sect. 2 discusses related work; Sect. 3 describes the existing personalized recommendation method; Sect. 4 presents our approach; Sect. 5 provides the case study using the movie rating dataset; Sect. 6 shows the statistical experimental results; and Sect. 7 concludes the paper.

2 Related Work

Personalized recommendation to the source node is realized by quantifying how much each node is worthy of recommendation. This section discusses the existing quantification methods from three perspectives: the recommendation criteria, the information used for recommendations, and the controllability of the recommendation results. We also show the characteristics of our method.

Firstly, regarding the recommendation criteria, while general personalized recommendations focus only on the proximity to the source node [31], various criteria including diversity [11,25,28,30], novelty [11,18], serendipity [11,19], and fairness [24,30] have been proposed. Each recommendation criteria focuses on different aspects in addition to a certain level of proximity. In particular, diversity emphasizes that the similarity between nodes in the recommendation list is small, novelty emphasizes that the user has not checked the recommended nodes before, serendipity emphasizes that the recommended nodes belong to a field that the user has not checked before, and fairness emphasizes that every node will have the opportunity to be recommended. On the other hand, this paper focuses on the global importance and source proximity.

Secondly, regarding the information used for recommendations, personalized recommendation methods are categorized into two types: those using external information such as node labels [18,25,28] and those using only graph topology [19,22,31]. For example, the movie recommendations using external information recommends movies that have the same director, cast, or genre as the user's favorite one [18,25,28]. However, these methods rely on external information such as node labels and require to select and collect the necessary external information for each graph, making them less general-purpose. On the other hand, the methods using only graph topology utilize only the adjacent information between nodes. For example, these methods exploit the random walk paths starting from the source node [19,31] and the similarity between the embedding vectors [22]. These methods are general-purpose because the same analysis method can be applied to any graph. Furthermore, comparing the methods focusing on the random walk paths and embedding vectors, the former requires computation only on the source node, while the latter requires computation on all nodes for the similarity computation, resulting in a lack of scalability. From the above, we focus on the general-purpose and lightweight method exploiting random walks.

Thirdly, regarding the controllability of the recommendation results, the existing methods are categorized into two types: those that can control how much the recommendation criteria are incorporated into the results by parameters [18,25,28], and those that cannot [19]. Furthermore, the controllable methods are divided into two types: those that can control monotonically [28] and those that cannot [18,25]. The controllability is an important element for flexible recommendation according to the users' preferences and interest nodes. Moreover, the monotonic control is also an important element in finding the desired parameter settings with a small amount of computation, e.g., by binary search. Therefore, this paper aims to monotonically control the influence of global importance on the recommendation results by the random walk length. Here, a random walk has two parameters: the source node s and the termination probability α that controls the random walk length. However, existing work has mainly discussed the effect of s on the recommendation results [18,19]. An existing random walk-based work, where novelty is the recommendation criteria and external information is used, reports that α affects the novelty of the recommendation results [18]. However, the influence is non-monotonic, highly dependent on the dataset, and small range. Therefore, the influence of α needs more discussion.

3 Existing Recommendation Method Using Random Walks: Personalized PageRank

In this paper, random walks in a graph $G = (V, E)$ are the computations taking the source node s, the termination probability α and the number of random walks ω as inputs. In a random walk, s is the initial visited node and continues to move to a random adjacent node until termination. At each visited node, the random walk terminates with probability α and moves to a next node with probability $1 - \alpha$. This operation is repeated until ω walks terminate. Note that the random walk length is defined by the number of visited nodes until the termination. Moreover, the way of selecting the next node depends on whether the edge is weighted or not. If the edges do not have weights, the next node is randomly and equally selected from all adjacent nodes. Otherwise, the next node is randomly selected with a probability proportional to the weight of the corresponding edge.

Personalized PageRank (PPR) is a source-dependent centrality metric defined with random walks [20,27]. PPR quantifies the importance of each node v with respect to the source node s by the probability that a random walk starting from s visits v. If the number of visits to node v in ω random walks is N_v and the sum of the ω random walk lengths is N_ω, PPR value of v with respect to s, denoted as $\pi_s(v)$, is N_v/N_ω. A n-dimensional vector that stores PPR values of all nodes with respect to s is called a PPR vector and is denoted as $\boldsymbol{\pi}_s$. In addition, PageRank is a metric that quantifies the global importance in a graph by the probability that a random walk equally starting from all nodes visits each node [20]. A n-dimensional vector that stores PageRank values of all nodes is called a PageRank vector and is denoted as $\boldsymbol{\pi}$. Note that PPR is the scalable analysis method for large graphs. The time complexity of PPR is determined by the total length of the random walks N_ω. To guarantee the accuracy of the approximated PPR vector, N_ω must be set to $O(n \log n)$ [5,23]. However, if a high enough accuracy is accepted for practical use, it is sufficient to set N_ω to 10^7–10^8 [26]. Moreover, previous work has shown that the accuracy of the PPR vector depends on N_ω [16]. As a result, it is possible to decrease ω for small α because $N_\omega = \omega/\alpha$. Thus, α does not affect the time complexity.

PPR-based recommendation is realized by effectively setting the source node s and the termination probability α. Existing PPR-based methods have mainly discussed a way of selecting the source node and combining multiple PPR vectors [4,8,18,19]. However, α is fixed as 0.15 or 0.2 in most methods. Therefore, we focus on α as a parameter for personalizations. In the following sections, we will show that α can balance the influences of global importance and source proximity on PPR vectors.

4 Approach

This section describes an approach to balance the influences of global importance and source proximity on PPR vector. First, we show that the random walk length is controlled by the termination probability α. Next, we describe how the visited nodes of the random walks changes by the random walk length. We then present the hypothesis that this change can be used to balance the influences of global importance and source proximity on PPR vector.

First, the expected random walk length $\mathbb{E}[l_\alpha]$ is $1/\alpha$. This is because the probability that the random walk length is $k (k \geq 1)$ is expressed by the Eq. (1), and $\mathbb{E}[l_\alpha]$ is derived as shown in the Eq. (2).

$$P(l_\alpha = k) = \alpha(1-\alpha)^{k-1} \tag{1}$$

$$\mathbb{E}[l_\alpha] = \Sigma_{k=1}^{\infty} k \cdot P(l_\alpha = k) = \frac{1}{\alpha} \tag{2}$$

Thus, we see that α controls the random walk length.

Next, when the random walk length is changed by α, the visited nodes of the random walks are expected to change in terms of how close they are to the source node or how high their PageRank values are. In particular, we expect that a smaller α and longer random walks decrease the influence of the source node on the visited node. As a result, nodes with high global importance, i.e., nodes with high PageRank values, are more likely to be visited even if they are distant from the source node. In other words, as α approaches 0, the visited nodes are no longer affected by the source node, and PPR vector approaches PageRank vector, which is independent of the source node. For example, when the graph is undirected and connected, $\lim_{\alpha \to 0} \boldsymbol{\pi}_s$ is known to converge to a vector proportional to the degree distribution [3]. Here, some work report that PageRank value of each node is strongly correlated with its degree [3,6].

On the other hand, when the random walks are shortened by increasing α, the source node becomes to have larger influence on the visited node. As a result, it is expected that nodes with high source proximity regardless of their global importance, i.e., nodes that are close to the source node even if they do not necessarily have a high PageRank value, are more likely to be visited. This means that as α approaches 1, the visited nodes of the random walks are no longer affected by PageRank value, and the PPR vector will approach the one-hot vector where only the element corresponding to the source node is 1. Moreover, some existing work shows that the short random walks are effective to collect source-dependent community information [1,9,12].

Although the behavior of PPR vector in the limit of α approaching 0 and 1 is obvious, the behavior when α changes between 0 and 1 is unpredictable. Therefore, experimental observations about the monotonicity or speed of the changes are required. In the following sections, we present the case study and statistical evaluation using real-world datasets.

5 Case Study Using Movie Rating Dataset

In this section, we use MovieLens dataset [10] and observe the top-20 nodes of PPR while changing the termination probability α. MovieLens dataset contains a total of 34 million ratings from 330,000 users for about 87,000 movies. At each rating, a user u gives a score s of [1.0, 5.0] to a movie m. This dataset is handled as a graph whose nodes are users and movies by putting an undirected edge weighted by s between u, m [18,19,28]. Personalized recommendation is realized by computing PPR vector whose source node is the user's interest movie. Although MovieLens dataset includes metadata such as the genre of each movie, we do not use information other than graph topology because this paper focuses on general-purpose personalized recommendation that does not depend on external information.

Table 1 shows the top-20 PPR movies for $\alpha = 0.2, 0.4, 0.6, 0.8$, and "Avengers: Infinity War Part I" is set to the source node. Note that the movies shown in bold font are produced by Marvel Studios. Table 1a–1d confirms that Marvel movies, which are directly related to the source node, monotonically rise to the upper rankings as α is increased. In particular, when $\alpha = 0.6, 0.8$, seven Marvel movies are ranked in the top-20, indicating that source proximity had a significant influence on the results. Specifically, the Marvel movies ranked in the top-20 are nodes with PageRank values between 91st and 706th place. This indicates that as α increases, nodes that do not necessarily have high global importance but have high source proximity are more likely to be visited by random walks. On the other hand, when $\alpha = 0.2$, only three Marvel movies including the source node are ranked in the top-20, and non-Marvel movies are nodes with high PageRank values. In particular, the all non-Marvel movies ranked in the top-20 are nodes with PageRank values in the top-46. This shows that when α is small, nodes with higher global importance are more likely to be visited by random walks.

The similar behaviour is also observed when other movies are taken as the source node. For example, when "Monsters, Inc." was taken as the source node, we found that Disney and Pixar movies monotonically rose to the higher ranks as α increased. In other case, when "Shoplifters" was taken as the source node, we confirmed that "Parasite" related in terms of Asian movies that won the Palme d'Or at the Cannes Film Festival rose to the top of the list by increasing α.

6 Statistical Evaluation

In this section, we quantitatively evaluate the relationships between the termination probability α and the characteristics of PPR vectors using nine real-world datasets. The datasets are listed in Table 2 [10,13,14,17]. In Sect. 6.1, we measure the relationships between α and the influence of PageRank vector on PPR vector, where PageRank represents the global importance. The results show that an increase of α causes a monotonic decrease of the influence of PageRank vector on PPR vectors, indicating that the influences of global importance and source proximity can be monotonically balanced by α. In addition, Sect. 6.2 analyzes how PPR values of the high-ranking nodes change depending on α. As a result, the decrease of α causes PPR values of the top-ranking nodes to be concentrated at large and close values, indicating that random walks tend to visit many nodes in a relatively uniform distribution.

6.1 α vs. the Influences of the Global Importance on PPR Vectors

This section evaluates the influence of global importance on PPR vector by measuring the similarity between PPR and PageRank vector while changing the termination probability α. In particular, we compute PPR vectors for randomly selected 100 nodes by changing α in 21 ways: [0.01, 0.05, 0.1, 0.15, 0.2, ..., 0.85, 0.9, 0.95, 0.99]. We then measure the similarity between the computed PPR vectors and PageRank vector. Here, α used to compute PageRank vector is fixed at 0.2. This is because when comparing PPR vectors for different α with PageRank vector as a baseline, the baseline vector needs to be fixed. The setting of $\alpha = 0.2$ in PageRank computation follows many existing work [4,8,20,27].

Table 1. Top-20 PPR nodes when "Avengers: Infinity War Part I" is the source node.

$\alpha = 0.2$

Rank	Title
1	**Avengers: Infinity War I**
2	The Shawshank Redemption
3	The Matrix
4	The Dark Knight
5	Inception
6	Star Wars: Episode IV
7	The Lord of the Rings III
8	The Lord of the Rings I
9	The Lord of the Rings II
10	Star Wars: Episode V
11	Fight Club
12	Forrest Gump
13	Raiders of the Lost Ark
14	**Thor: Ragnarok**
15	Star Wars: Episode VI
16	Pulp Fiction
17	Interstellar
18	**Avengers: Infinity War II**
19	Silence of the Lambs
20	Schindler's List

$\alpha = 0.4$

Rank	Title
1	**Avengers: Infinity War I**
2	The Matrix
3	The Shawshank Redemption
4	The Dark Knight
5	Inception
6	Star Wars: Episode IV
7	The Lord of the Rings III
8	**Thor: Ragnarok**
9	Star Wars: Episode V
10	The Lord of the Rings II
11	The Lord of the Rings I
12	**Avengers: Infinity War II**
13	Fight Club
14	Raiders of the Lost Ark
15	Interstellar
16	Star Wars: Episode VI
17	**Guardians of the Galaxy**
18	Forrest Gump
19	**Iron Man**
20	**Logan**

$\alpha = 0.6$

Rank	Title
1	**Avengers: Infinity War I**
2	The Matrix
3	The Shawshank Redemption
4	The Dark Knight
5	Inception
6	**Thor: Ragnarok**
7	Star Wars: Episode IV
8	The Lord of the Rings III
9	Star Wars: Episode V
10	**Avengers: Infinity War II**
11	The Lord of the Rings II
12	The Lord of the Rings I
13	Fight Club
14	Interstellar
15	**Guardians of the Galaxy**
16	Raiders of the Lost Ark
17	Star Wars: Episode VI
18	**Iron Man**
19	**Logan**
20	**Deadpool 2**

$\alpha = 0.8$

Rank	Title
1	**Avengers: Infinity War I**
2	The Matrix
3	The Shawshank Redemption
4	The Dark Knight
5	**Thor: Ragnarok**
6	Inception
7	Star Wars: Episode IV
8	The Lord of the Rings III
9	**Avengers: Infinity War II**
10	Star Wars: Episode V
11	The Lord of the Rings II
12	The Lord of the Rings I
13	**Guardians of the Galaxy**
14	Interstellar
15	Raiders of the Lost Ark
16	Fight Club
17	Star Wars: Episode VI
18	**Iron Man**
19	**Logan**
20	**Deadpool 2**

Table 2. Datasets (K = 10^3, M = 10^6, B = 10^9).

Name	Directed	Weighted	#Nodes	#Edges
DBLP [14]	No	No	317K	1.1M
Web-BerkStan [14]	Yes	No	685K	7.6M
MovieLens [10]	No	Yes	418K	34M
YouTube [17]	No	No	3.2M	9.4M
Flickr [17]	Yes	No	2.3M	33M
LiveJournal [14]	No	No	4.0M	35M
Orkut [14]	No	No	3.1M	117M
Twitter [13]	Yes	No	41.7M	1.5B
Friendster [14]	No	No	65.6M	1.8B

Similarity is quantified by Normalized Discounted Cumulative Gain (NDCG) focusing on the high-ranking nodes [2] and cosine similarity focusing on the whole vectors. NDCG quantifies the similarity of the top-k nodes of both vectors as a [0, 1] value, where $k = 128$ in this paper; the more nodes with high PageRank values are included in the top nodes of PPR vector, the larger NDCG is. NDCG is defined by the Eq. (3), where t_i, t'_i are the nodes with the ith highest PageRank and PPR values, respectively.

$$ndcg(\boldsymbol{\pi}_s, \boldsymbol{\pi}, k) = \frac{\sum_{i=1}^{k} \frac{2^{\pi(t'_i)}-1}{\log_2(i+1)}}{\sum_{i=1}^{k} \frac{2^{\pi(t_i)}-1}{\log_2(i+1)}} \quad (3)$$

Cosine similarity quantifies the correlation of the values in both vectors as a [0, 1] value; the more nodes with high PageRank values have high PPR values, the larger the cosine similarity is.

Figure 1 shows the similarity between PageRank and PPR vectors for each α. Figures 1a and 1b are the results with the similarity metrics as NDCG and cosine similarity, respectively. Note that in both figures, the y-axis represents the average of the 100 calculated similarities, and the scale of the y-axis is lenear in Fig. 1a and logarithmic in Fig. 1b.

Figure 1a indicates that in NDCG, a similarity metric focusing only on the top nodes, an increase of α leads to a monotonic decrease of the similarity between PageRank and PPR vectors, resulting in an increase of the source proximity in PPR vectors. In particular, by setting $\alpha = 0.01$, NDCG increases to [0.1, 0.94] for each dataset. Here, when NDCG is 0.94, the top nodes of PPR and PageRank vectors are almost identical, differing only in that their rankings are swapped. On the other hand, by setting $\alpha = 0.99$, NDCG is decreased to [0.05, 0.38] for each dataset. Here, when NDCG is 0.05, the top nodes in PPR and PageRank vectors hardly overlap. Moreover, some datasets, such as Web-BerkStan and Friendster, showed a monotonous decrease of NDCG over α, but the decrease ranges were [0.18, 0.28] and [0.05, 0.1], and the decrease rates were about 36 and 50%, respectively. Note that, as described later, a large decrease rates were observed in Web-BerkStan and Friendster when the similarity metric was the cosine similarity.

As shown in Fig. 1b, we also confirmed a similar trend in the cosine similarity. In particular, by setting $\alpha = 0.01$, the cosine similarity increases to [0.03, 0.76] for each dataset, and by setting $\alpha = 0.99$, it decreases to [0.0001, 0.002]. In Web-BerkStan and Friendster, where the decrease rates were small in the case of NDCG, the decrease ranges of the cosine similarity were [0.0017, 0.23], [0.0001, 0.03], and the decrease rates were both over 99%. Note that the cosine similarity is strongly affected by the dimension of the vectors (in this paper, the number of nodes), thus we should be careful about simply comparing the results across datasets. For example, in Fig. 1b, we observe that datasets with a larger number of nodes tend to have lower cosine similarity. On the other hand, it is worth noting that the shape of the decrease among datasets are similar.

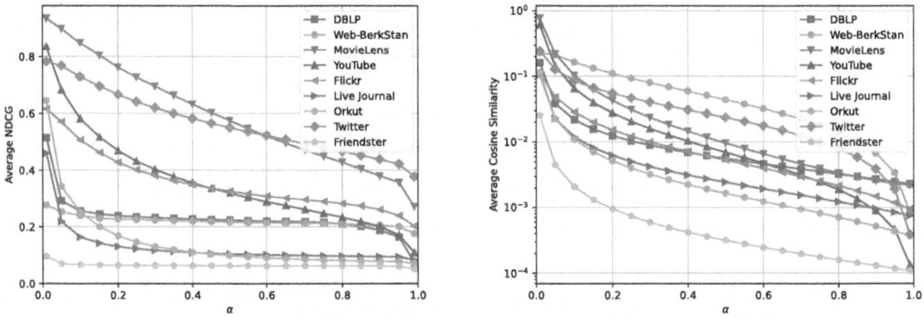

(a) Similarity metric: NDCG (b) Similarity metric: Cosine similarity

Fig. 1. α vs. the similarity between PageRank and PPR vectors.

6.2 α vs. the Distribution of PPR Values of High-Ranking Nodes

This section evaluates the distribution of PPR values of top-ranking nodes while changing α. In particular, we compute PPR vectors for 100 randomly selected source nodes while changing α to [0.01, 0.05, 0.1, 0.15, 0.2, ..., 0.85, 0.9, 0.95, 0.99] and measure the average PPR value for each ranking from 1st to 128th. Here, based on the report that the average PPR value at each ranking is distributed to the power [15], we fit the average PPR value y for the ranking x to the distribution of $y = ax^b$ and observe the scaling exponent b ($b < 0$). The reason for focusing on the scaling exponent b is that b can quantify the degree of concentration of PPR values in the high-ranking nodes. The larger b means that PPR values are concentrated in close values, and the smaller b means that PPR values largely differs among rankings.

Figure 2 shows the relationships between α and the scaling exponents. Figure 2 indicates that the scaling exponent decreases monotonically against

α. Therefore, by decreasing α, PPR values of top nodes are likely to be concentrated, indicating that random walks visit a large number of nodes in a relatively uniform distribution. The ratio of scaling exponents for $\alpha = 0.01, 0.99$ was [1.2, 3.3] for each dataset. Note that the coefficient of determination when calculating the scaling exponent was always greater than 0.9.

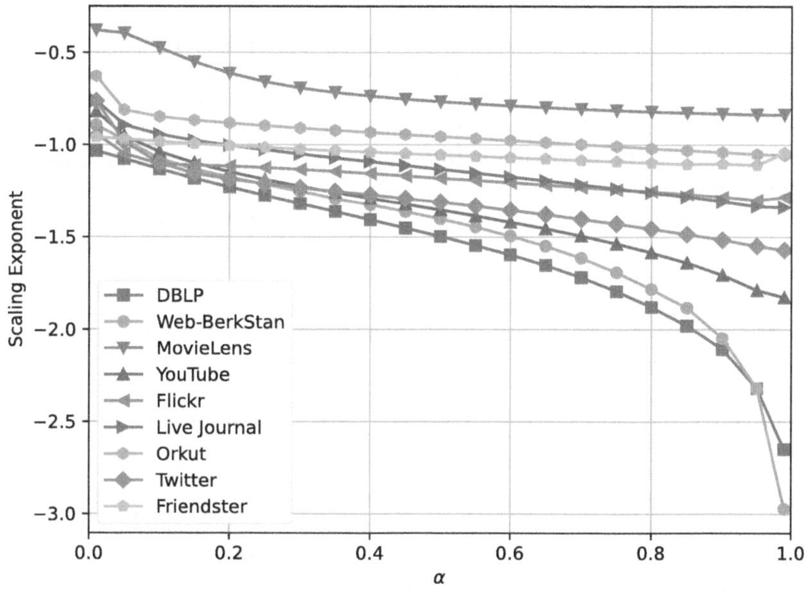

Fig. 2. α vs. scaling exponent of the distribution of PPR values.

7 Conclusion

It is important to balance the influences of global importance and source proximity on Personalized PageRank-based recommendations monotonically. This paper proposes the method to balance them by the termination probability α, which is a parameter to control the random walk length. In particular, a case study using MovieLens dataset showed that when we increased α, nodes with high source proximity rose to the higher rankings. Moreover, statistical evaluation on nine real-world datasets revealed that changing α from 0.01 to 0.99 resulted in a monotonic decrease of the cosine similarity between PPR and PageRank vectors from 0.76 to 0.002 at the maximum.

References

1. Berberidis, D., Giannakis, G.B.: Node embedding with adaptive similarities for scalable learning over graphs. IEEE Trans. Knowl. Data Eng. **33**(2), 637–650 (2021)

2. Burges, C., et al.: Learning to rank using gradient descent. In: Proceedings of International Conference on Machine Learning, pp. 89–96 (2005)
3. Chierichetti, F., Dasgupta, A., Kumar, R., Lattanzi, S., Sarlós, T.: On sampling nodes in a network. In: Proceedings of International Conference on World Wide Web, pp. 471–481 (2016)
4. Eksombatchai, C., et al.: Pixie: a system for recommending 3+ billion items to 200+ million users in real-time. In: Proceedings of International Conference on World Wide Web, pp. 1775–1784 (2018)
5. Fogaras, D., Rácz, B., Csalogány, K., Sarlós, T.: Towards scaling fully personalized PageRank: algorithms, lower bounds, and experiments. Internet Math. **2**(3), 333–358 (2005)
6. Fujii, K., Yamashita, T., Shin, A., Kaneko, K.: A flexible weighting framework for converting relational database to hypergraphs. In: Proceedings of International Conference on Information Networking (2025, to appear)
7. Gasteiger, J., Bojchevski, A., Günnemann, S.: Predict then propagate: graph neural networks meet personalized PageRank. In: Proceedings of International Conference on Learning Representations, pp. 1–15 (2019)
8. Gupta, P., Goel, A., Lin, J., Sharma, A., Wang, D., Zadeh, R.: WTF: the who to follow service at Twitter. In: Proceedings of International Conference on World Wide Web, pp. 505–514 (2013)
9. Hamilton, W.L., Ying, R., Leskovec, J.: Inductive representation learning on large graphs. In: Proceedings of International Conference on Neural Information Processing Systems, Long Beach, CA, USA, pp. 1025–1035 (2017)
10. Harper, F.M., Konstan, J.A.: The MovieLens datasets: history and context. ACM Trans. Interact. Intell. Syst. **5**(4), 1–19 (2015)
11. Kaminskas, M., Bridge, D.: Diversity, serendipity, novelty, and coverage: a survey and empirical analysis of beyond-accuracy objectives in recommender systems. ACM Trans. Interact. Intell. Syst. **7**(1), 1–42 (2016)
12. Kloumann, I.M., Kleinberg, J.M.: Community membership identification from small seed sets. In: Proceedings of ACM International Conference on Knowledge Discovery and Data Mining, New York, NY, USA, pp. 1366–1375 (2014)
13. Kunegis, J.: KONECT: the Koblenz network collection. In: Proceedings of International Conference on World Wide Web, pp. 1343–1350 (2013)
14. Leskovec, J., Krevl, A.: Stanford large network dataset collection. http://snap.stanford.edu/data
15. Litvak, N., Scheinhardt, W., Volkovich, Y.: In-degree and PageRank: why do they follow similar power laws? Internet Math. **4**(2–3), 175–198 (2007)
16. Liu, Q., Li, Z., Lui, J.C., Cheng, J.: PowerWalk: scalable personalized PageRank via random walks with vertex-centric decomposition. In: Proceedings of ACM International Conference on Information and Knowledge Management, Indianapolis, IN, USA, pp. 195–204 (2016)
17. Mislove, A.: Online social networks: measurement, analysis, and applications to distributed information systems. Ph.D. thesis, Rice University, Department of Computer Science (2009)
18. Nakatsuji, M., Fujiwara, Y., Tanaka, A., Uchiyama, T., Fujimura, K., Ishida, T.: Classical music for rock fans?: novel recommendations for expanding user interests. In: Proceedings of ACM International Conference on Information and Knowledge Management, pp. 949–958 (2010)
19. Onuma, K., Tong, H., Faloutsos, C.: Tangent: a novel, "Surprise-me", recommendation algorithm. In: Proceedings of ACM International Conference on Knowledge Discovery and Data Mining, pp. 657–665 (2009)

20. Page, L., Brin, S., Motwani, R., Winograd, T.: The PageRank citation ranking: bringing order to the web. Technical report, Stanford InfoLab (1998)
21. Perozzi, B., Al-Rfou, R., Skiena, S.: DeepWalk: online learning of social representations. In: Proceedings of ACM International Conference on Knowledge Discovery and Data Mining, pp. 701–710 (2014)
22. Tsitsulin, A., Mottin, D., Karras, P., Müller, E.: VERSE: versatile graph embeddings from similarity measures. In: Proceedings of World Wide Web, pp. 539–548 (2018)
23. Wang, S., Yang, R., Xiao, X., Wei, Z., Yang, Y.: FORA: simple and effective approximate single-source personalized PageRank. In: Proceedings of ACM International Conference on Knowledge Discovery and Data Mining, Halifax, NS, Canada, pp. 505–514 (2017)
24. Wang, Y., Ma, W., Zhang, M., Liu, Y., Ma, S.: A survey on the fairness of recommender systems. ACM Trans. Inf. Syst. **41**(3), 1–43 (2023)
25. Wang, Z., et al.: Diversity-aware deep ranking network for recommendation. In: Proceedings of ACM International Conference on Information and Knowledge Management, pp. 2564–2573 (2023)
26. Yamashita, T., Matsumoto, N., Kaneko, K.: Reducing re-indexing for top-k personalized PageRank computation on dynamic graphs. IEEE Trans. Big Data 1–13 (2025)
27. Yang, M., Wang, H., Wei, Z., Wang, S., Wen, J.R.: Efficient algorithms for personalized PageRank computation: a survey. IEEE Trans. Knowl. Data Eng. 1–20 (2024)
28. Zanon, A.L., da Rocha, L., Manzato, M.G.: Balancing the trade-off between accuracy and diversity in recommender systems with personalized explanations based on linked open data. Knowl.-Based Syst. **252**, 109333 (2022)
29. Zhang, Q., Ma, J., Lou, J., Yang, C., Xiong, L.: Towards training graph neural networks with node-level differential privacy. In: Proceedings of International Conference on World Wide Web, Singapore, Singapore (2024)
30. Zhao, Y., Wang, Y., Liu, Y., Cheng, X., Aggarwal, C.C., Derr, T.: Fairness and diversity in recommender systems: a survey. ACM Trans. Intell. Syst. Technol. (2024)
31. Zhou, Y., Hayes, C.: Graph-based diffusion method for top-N recommendation. In: Artificial Intelligence and Cognitive Science, pp. 292–304 (2023)

Counterintuitive Behavior of Clustering Quality: Findings for K-Means on Synthetic and Real Data

Marco Loog[1(✉)], Jesse H. Krijthe[2], and Manuele Bicego[3]

[1] Radboud University, Nijmegen, The Netherlands
marco.loog@ru.nl
[2] Delft University of Technology, Delft, The Netherlands
j.h.krijthe@tudelft.nl
[3] University of Verona, Verona, Italy
manuele.bicego@univr.it

Abstract. Little is known about how the quality of a clustering changes when changing the size of the set used to determine the clustering model. We show that, for K-means clustering, the relationship between dataset size and clustering quality can display counterintuitive behavior. Notably, the quality can significantly *deteriorate* with *more data* to build the model. More generally, using artificial datasets and data from bioinformatics, we uncover a variety of learning curve behaviors for K-means. Our results clearly illustrate that the training sample size can have a nontrivial influence on the clustering performance. Our findings should appeal to both the clustering practitioner and the clustering researcher concerned with developing basic insights.

Keywords: Clustering quality · K-means · Counterintuitive behavior · Monotonicity · Gene ontology enrichment analysis

1 Introduction

Clustering [14,17] is a classic problem studied in machine learning, data mining, and abutting fields. One of its goals is to aggregate objects into meaningful groups so that objects in the same group are, in some sense, similar whereas objects from different groups are dissimilar. This is sometimes referred to as clustering's realist aim [15]. The goal can also be "constructive," aiming to split up data for more pragmatic reasons [15].

Clustering is ubiquitous in science and extensively applied for unsupervised exploratory data analysis in many scenarios. To roughly quantify how pervasive it is, we searched Scopus[1] and found around 39,000 papers published since 2020 with the word "clustering" in the title. Putting this in perspective: a similar

[1] Search performed in `www.scopus.com` in November 2024 with the search string `TITLE (''Clustering'') AND PUBYEAR > 2019 AND PUBYEAR < 2025`.

G. Krempl et al. (Eds.): IDA 2025, LNCS 15669, pp. 154–166, 2025.
https://doi.org/10.1007/978-3-031-91398-3_12

search for "CNN"[2], a popular machine learning method, returned about 25,000 hits. These papers come from more than 25 subject areas, ranging from computer to material science, medicine to the social sciences, and geology via psychology to the arts and humanities.

The foundational and theoretical aspects of clustering have generally received relatively little attention, compared to its widespread application. Existing work in this area investigates axiomatic approaches to clustering and issues of model selection and validation, but also addresses the question of clusterability and the aim and usage of clustering as such are considered (e.g., [1,15,20,46]).

This paper contributes to this foundational part of the field as well. However, it focuses on a matter entirely different from the above and asks the basic question: "Does clustering become better with more data?" This issue has received virtually no attention in the community, for none of the clustering techniques. Possibly this is due to the common belief that, if we avoid adversarial choices [9,49], *more data* should indeed imply *better cluster models*. We show that this belief is incorrect.

Until now, the only work to study this question is our 2023 contribution to ECML [24]. It shows that, when considering K-means clustering in the classical context of empirical risk minimization (ERM), performance can actually get worse with more data. That is, in terms of the minimum sum-of-squares (MSS or the within-group squared loss [24]) that K-means minimizes at training time, the average test loss can increase with more data. While this is an interesting theoretical finding, it merely shows that this behavior can occur in a specifically constructed probability distribution, and only in terms of the MSS, rather than more commonly accepted clustering quality measures. Here, we show such counterintuitive behaviors also occur for measures of quality that are broadly accepted within the clustering community. Moreover, we demonstrate this on both simulated and real-world bioinformatics data.

Our work identifies unknown and surprising behavior and, as such, furthers our fundamental understanding of K-means. However, our work is not theoretical. Due to the use of more commonly accepted performance measures and the employment of real data especially, we have to forego mathematical proofs and rely on empirical investigations. Our findings offer basic insight into K-means that should be of interest to both practitioners and researchers. To start with, Sect. 2 covers some additional related work and initial concepts used in the experiments. Section 3 then presents the datasets, experiments, and results. Finally, Sect. 4 offers a discussion and conclusion.

2 Further Background and Experimental Preliminaries

We cover directly relevant technical works in Subsect. 2.1. Subsection 2.2 then covers performance and quality measures and Subsect. 2.3 explains learning curves. These last two subsections cover some essentials for the experiments. First, however, we will specify the K-means method in more detail.

[2] Search string: `TITLE (''CNN'') AND PUBYEAR > 2019 AND PUBYEAR < 2025`.

Many methods have been proposed since the advent of clustering [16,18]. The K-means algorithm [22,37] is probably the most famous clustering approach. It belongs to the family of the so-called minimum sum-of-squares (MSS) clustering methods, where the clustering is found by minimizing the sum of the squared Euclidean distance of each object from the mean of the cluster to which it belongs. That is, given a set of K means $M = \{m_1, m_2, \ldots, m_K\}$ and N objects x_i, K-means aims to minimize

$$R_{\text{K-means}} = \frac{1}{N} \sum_{i=1}^{N} \min_{m \in M} \|x_i - m\|^2. \tag{1}$$

To find the best K means, the basic iterative procedure often used is to assign every data point to its nearest mean and then update the K means to the average of all points assigned to it. Typically, these two steps are repeated until convergence. To overcome the sensitivity of K-means to the initialization, this procedure is then run multiple times from a new random set of K means, and the solution with the minimum MSS is kept.

While this algorithm has known drawbacks [16,38] and many variations and alternatives have been presented [18,48], K-means is still widely applied for clustering.[3] Three factors contribute to its popularity: it is simple to implement, it is effective in many different contexts, and it is widely known—especially researchers from life sciences often prefer a thoroughly studied algorithm, with known limitations, over less studied alternatives.

2.1 Earlier Work Using Empirical Risk Minimization

Our earlier-mentioned work [24] shows that if one measures the clustering quality on a test set using the same measure as the one that is used during training, i.e., the minimum sum-of-squares (MSS), one can find unexpected behavior in which the expected loss can actually go up rather than down when the training set grows. This behavior is demonstrated to be structural, i.e., this deterioration of performance is in expectation over all possible training sets of a particular size, and therefore not due to an unlucky draw or other coincidence.

What is important in [24], and also in this work, is that training and test sets do not necessarily coincide. This may not be the most common setting in clustering, but it has its precedents. For instance, [7,35,45] consider K-means clustering in the context of the statistical learning theory and approach it in terms of classical empirical risk minimization (ERM) [42]. The result derived in [24] hinges on ERM as well. Where our work differs essentially is that we evaluate the performance of K-means in terms of commonly accepted and practically employed clustering quality measures. This situation is not unlike the classification setting where the error rate or accuracy is the de facto evaluation measure, but hardly any classifier optimizes these criteria directly [25].

[3] Just to get an idea of the numbers: Google Scholar finds about 37,000 articles with "K-means" in 2024 only.

We finally note that in recent years some similarly surprising findings have been reported for other learners, where these can start to perform worse with increasing numbers of training data [5,8,23,26,27,29]. Examples can be found for classification, regression, and density estimation. In the clustering case, however, a comprehensive analysis is still missing.

2.2 Performance and Quality of Clustering

Our study is concerned with the effect of dataset size on the quality of clustering [2,36,44,47]. Within clustering, as opposed to the supervised context, the definition of "the quality of a result" can be rather contentious. This is mainly due to the ground truth not being available and the initial problem being posed in a qualitative way.

In a seminal paper offering a related result pertaining to the definition of the quality of clustering, Kleinberg [20] showed that it is impossible to design a general-purpose clustering function obeying three simple and reasonable properties: scale-invariance, richness, and consistency. Subsequently, interesting alternatives for theoretical cluster validation have been proposed, such as in [4]. Generally, the problem of measuring the quality of a clustering result is still open, and, as suggested by [46], should probably be made dependent on the application under consideration.

This work presents results that go beyond the MSS loss in Eq. (1) that K-means intrinsically minimizes. The three widely accepted and commonly used clustering quality measures we consider are the adjusted Rand index (ARI), the normalized mutual information (NMI, type 1), and the purity. A range of variations on these measures exist [36], but our specific selection of these three indices can be considered representative.

Before providing their precise definitions, it should be noted that these measures rely on the availability of a true underlying clustering. This follows the well-known external validation strategy for clustering [17], which is widely employed, especially when developing novel clustering methods. For our artificial datasets, we will have to define ground truth labels ourselves. The ground truth used for our first real-world bioinformatics dataset is externally provided. Next to these, we also offer a completely different and, in a way, even more realistic evaluation through a gene ontology enrichment analysis for which the details follow in Subsect. 3.3.

Given N objects for which a clustering in K clusters is obtained, while the true labels have K' possible values. Let n_{ij} be the number of instances in cluster i that have true label j. In addition, let c_i be the size of cluster i and t_j be the number of points with true label j. The adjusted Rand index (ARI) then equals

$$\frac{\sum_{i=1}^{K}\sum_{j=1}^{K'}\binom{n_{ij}}{2} - \frac{\sum_i \binom{c_i}{2}\sum_j \binom{t_j}{2}}{\binom{N}{2}}}{\frac{1}{2}\left(\sum_{i=1}^{K}\binom{c_i}{2} + \sum_{j=1}^{K'}\binom{t_j}{2}\right) - \frac{\sum_i \binom{c_i}{2}\sum_j \binom{t_j}{2}}{\binom{N}{2}}}, \quad (2)$$

the (type 1) normalized mutual information (NMI) calculates

$$\frac{\sum_{i=1}^{K}\sum_{j=1}^{K'}\frac{n_{ij}}{N}\log\frac{Nn_{ij}}{c_i t_j}}{-\frac{1}{2}\left(\sum_{i=1}^{K}\frac{c_i}{N}\log\frac{c_i}{N}+\sum_{j=1}^{K'}\frac{t_j}{N}\log\frac{t_j}{N}\right)}, \tag{3}$$

and the third index we use is the purity, which is given by $\frac{1}{N}\sum_{i=1}^{K}\max_j n_{ij}$.

Note that ARI, NMI, and purity are quality measures, so larger is better, whereas for MSS, smaller is.

2.3 Learning Curves

Our analysis is based on learning curves. Such curves are widely used in supervised learning scenarios [32,43] and provide insight into the behavior of a given learner when varying the size of the set used to learn from. Within the clustering domain, such curves have hardly been utilized. A notable exception is [28] where both generalization and computational efficiency are plotted against the training set size for various clustering methods. We will use learning curves in combination with the clustering quality measures.

To generate our learning curves, we follow two different regimes. For the synthetic data, we know the underlying distribution, so we can sample training sets of a given size at random, train K-means on every draw, and then test its performance on a potentially large test set sampled from the same distribution. For the real data, performance is measured using the full datasets. Training sets of different sizes are drawn, without replacement, from this full dataset, so the training set size cannot exceed the number of instances in the full dataset.

It should be noted that, at test time, we measure the performance of the means obtained during training. First, all points in the test set are clustered using the training means, after which the quality of this clustering is determined using the measures introduced in the previous subsection.

3 Datasets, Experiments, Results

First, on the real-world data, we highlight some interesting learning curve behavior using the quality measures from Subsect. 2.2. Subsequently, we provide two synthetic datasets, for which we can understand, at least in part, why the learning curve behaves unexpectedly. We present the experiments and results in parts: every subsection introduces the specific data, presents the relevant learning curves, and makes provisional remarks on the results. Before that, let us present the last experimental details.

3.1 Experimental Parameter Settings

On the datasets in Subsects. 3.2 and 3.4, we repeat the procedure described in the previous section 1,000 times and report the mean and its standard deviation. We consider 29 training set sizes for every dataset. The smallest size equals K, while the largest size S depends on the dataset: for the synthetic data, we consider 1,000; for the real-world data, we take the total set size as a maximum or, if the full size leads to excessive computation, we also settle for 1,000 as the maximum training set size. The other 27 sizes are taken linearly in between the extrema on a logarithmic scale and rounded to the nearest integer, i.e., the ith training set size is given by $\lfloor \log(e^K + \frac{i-1}{28}(e^S - e^K)) \rceil$. In Subsect. 3.3, we limit ourselves to 100 repetitions and a smaller number of training set sizes. Additional specifics are provided there.

Finally, where the theory in [24] assumes that we always find the globally optimal K-means for the particular training set at hand, we typically cannot guarantee this on the datasets we consider here. To get to acceptably good solutions, we use the widely used K-means++ algorithm [3] and take the best-performing solution over 100 random initializations from the training set. In the experiments of the first two subsections to follow, we consider two-class data for illustrative purposes and take $K = 2$.

3.2 Cancer Genome Atlas Data

We present results based on a dataset from The Cancer Genome Atlas (TCGA) with 4,434 specimens. The dataset has 19 cancer classes, but we consider cluster performance on subproblems comprising only two cancer types.[4] Originally, this dataset has 54,416 features made up of gene expressions, DNA-methylations, copy-number variations, and microRNA expressions. As suggested in [39], the dimensionality is reduced to 50 for every of these four feature subsets using PCA and then concatenated into a 200-dimensional final feature space. In bioinformatics, clustering is typically applied to datasets such as this. Reference [39] analyses this particular set with various techniques, among which K-means.

Figure 1 provides some interesting learning curves that show the surprising behavior these can display for K-means clustering. We picked these four to point out some particulars. Let us first note, however, that not every TCGA problem with two cancer subtypes leads to such nonmonotonic behavior. On the contrary, most of these learning curves show clustering improvement with more data to train on. However, the important point is that such non-monotonic learning curve behavior can at all happen and the community has no explanation for it.

First, on all four problems from Fig. 1, MSS behaves similarly monotonic, but the quality measures show quite some different behavior. The first row shows curves decreasing twice with increasing training set sizes. The second shows non-monotonic behavior for a clustering problem that K-means can solve perfectly, given enough data. The third row shows a drop in performance early on in the learning curve, after which its behavior becomes regular. This may not pose a problem practically, but it does challenge our understanding of K-means at a more fundamental level. The bottom row shows that, at least in for ARI and

[4] TCGA is kindly acknowledged for making this data available. The original data was generated by the TCGA Research Network: https://www.cancer.gov/tcga.

NMI, the performance loss can be rather dramatic. Granted, even their best performance may not be good, but K-means being a bad fit for the problem is not an a priori reason for it to lead to nonmonotonic learning.

3.3 Clustering Gene Expression Data

We now turn to another important clustering challenge involving gene expressions [11,19,33]. K-means is a typical strategy [13,40] and also industrially, it is routinely applied (for instance in the Affymetrix technology from Thermo Fisher Scientific, a prominent biotechnology company[5]). Where for TCGA the gene expressions are used as features to describe cancer types, in the current application every gene is described by its expression at different time points. Importantly, this clustering problem is one of the few real-world clustering problems where the quality can be assessed using a large collection of external information. This offers well-established procedures that we can use and complements quality measures like ARI, NMI, and purity. It offers an essentially different illustration of the counterintuitive clustering behavior this work exposes.

The specific procedure we employ is a gene ontology (GO) enrichment analysis [6]. The GO provides a computational representation of all available information on gene functions at molecular, cellular, and organism levels. This can be used to assess the coherence of a cluster of genes by checking if certain biological processes, molecular functions, or cellular components, as characterized by GO terms, appear more often than expected by chance in that cluster. The more clusters contain such an overrepresentation, the more successful the clustering.

We use the well-known `yeast cycle` dataset [10] describing the expression levels of 6,400 genes at seven different time points, i.e., the number of instances is 6,400 with a dimensionality of 7. As typically done in gene expression analysis [41], missing values were imputed using nearest neighbor imputation.

Now, determining the learning curve is somewhat different from the previous experiments. For a random draw of genes, we find the K-means solution and use the means to assign all 6,400 data points to the K clusters. The clustering quality is then measured by a GO enrichment analysis, following standard pipelines such as those from biclustering [12,30,34], in which only the largest gene clusters are evaluated, considering the smaller ones noise. In our case, we kept half of the clusters. A cluster is considered relevant if at least one GO term is overrepresented at a given significance level. The quality of the clustering is the fraction of evaluated clusters that are considered relevant.

The five significance levels, i.e., p-values, at which we perform the analysis are 0.05, 0.01, 0.005, 0.001, and 0.0001. We consider K-means with $K = 20$ and $K = 40$. Since the validation is very time-consuming,[6] we evaluated the learning curve for $K = 20$ over training set sizes 100, 200, 300, 500, 750, 1,000, 2,000,

[5] https://www.thermofisher.com/us/en/home/references/ambion-tech-support/transcriptome-analysis/tech-notes/analysis-of-microarray-data.html.

[6] The enrichment analysis is performed via the FuncAssociate web tool, available at http://llama.mshri.on.ca/funcassociate/, entailing a web call for every cluster at every level of significance.

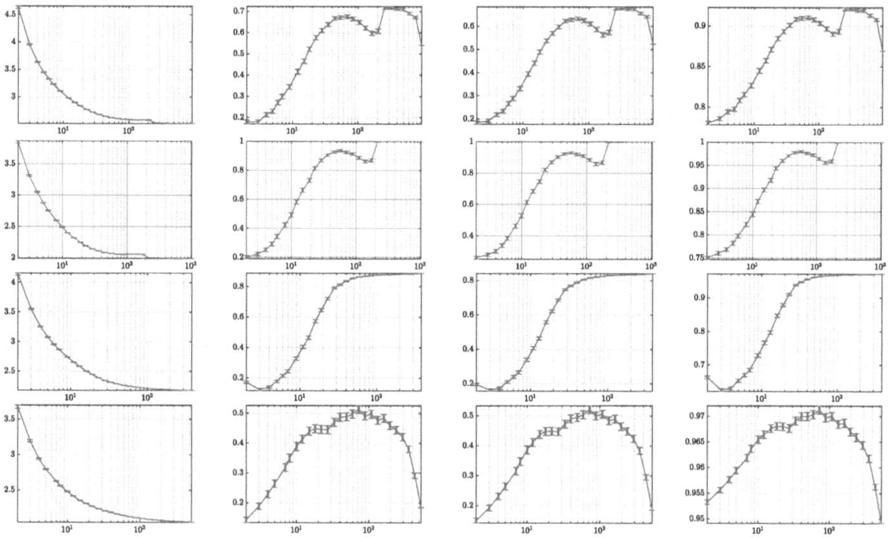

Fig. 1. Top to bottom, learning curves for a selection of two-class subproblems (4-letter abbreviations for the cancer types are from TCGA): `blca` vs `brca`, `brca` vs `prad`, `cesc` vs `lusc`, and `dlbc` vs `lusc`. Vertical axes: performance in terms of MSS, ARI, NMI, and purity, from left to right respectively. Horizontal axes are logarithmic and show the training set size used. Axes scalings may differ. The figure is best viewed electronically.

3,000, 4,000, and 6,400. For $K = 40$ we took 250, 500, 1,000, 2,000, 4,000, and 6,400 genes. We repeated the experiments 100 times.

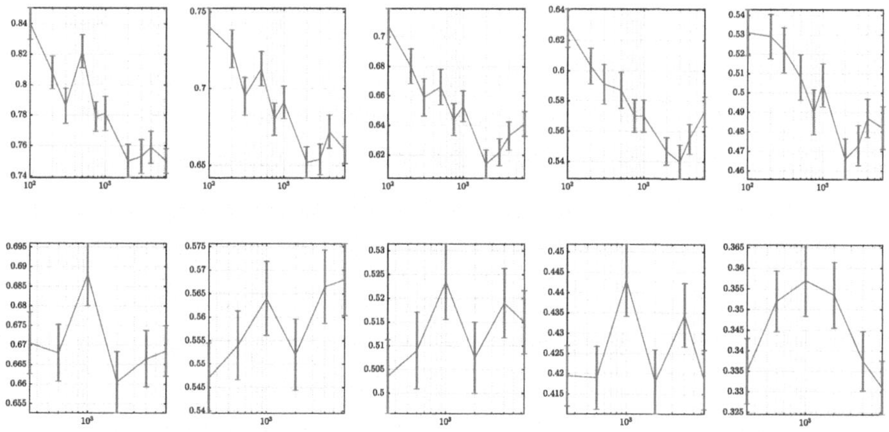

Fig. 2. Learning curves on `yeast cycle`. The top row is for $K = 20$, the bottom for $K = 40$. The columns contain the results for different p-values in the GO validation from left to right: 0.05, 0.01, 0.005, 0.001, and 0.0001. Vertical axes: performance in terms of the fraction of relevant clusters. Horizontal axes are logarithmic and show the training set size used. The figure is best viewed electronically.

The results in Fig. 2 further corroborate the general occurrence of counter-intuitive behavior. Especially when using $K = 20$ (top row), the best results (all statistically significant) are obtained with sample sizes considerably smaller than the complete dataset. For $K = 40$ (bottom row), only for the right two, the best performing training size is statistically significantly better (< 0.05) than using the full set; for the plot on the left we find $p = 0.059$.

3.4 Two Synthetic Datasets

In research, one way to try and understand more complex behavior is by designing toy models/examples that display some of the behavior we aim to understand [31]. In our final experiments, we consider two artificial datasets, where the true labeling follows directly from the generative process that defines the data.

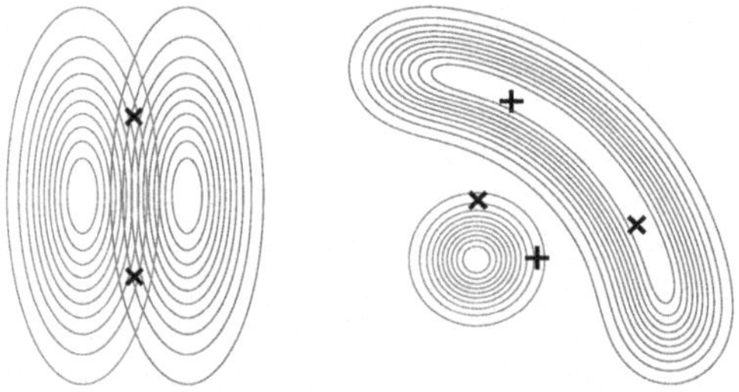

Fig. 3. Contour plots of the densities of the two artificial datasets leading to worsening learning behavior. The pairs of crosses indicate the globally optimal cluster centers. `Twogauss`, on the left, has one such optimum. For `oneeye`, on the right, there are two global optima: the two +s or the two ×s.

The first, referred to as `twogauss` and displayed on the left in Fig. 3, consists of two elongated Gaussian classes in two dimensions with means at $(-0.6, 0)$ and $(+0.6, 0)$ and equal covariance matrices $\begin{pmatrix} 0.16 & 0 \\ 0 & 1 \end{pmatrix}$. The class overlap is modest: the Bayes error is smaller than 0.07, and two well-placed means can attain this error. However, K-means will not find them because the variability in the vertical direction is too large. In the figure, the two black crosses indicate where the optimal means, in terms of the MSS objective, are located. These provide a solution that is completely at odds with the true classes. With more and more data, the K-means' solution gets closer to this suboptimum. As a result, the clustering quality becomes *worse* with increasing training set size (while the intrinsic MSS loss consistently improves). The top row in Fig. 4 provides the accompanying plots. This example shows an extreme case: the more data we train on, the worse the quality of clustering becomes.

The other dataset, referred to as `oneeye` and depicted on the right of Fig. 3, is a variation to mimic the possibility of, first, an improving quality measures,

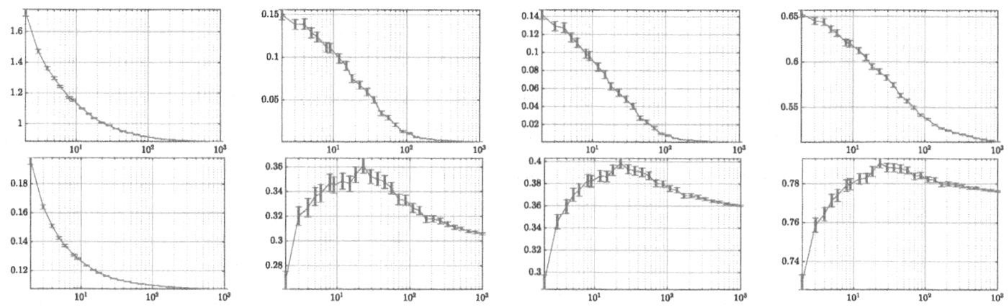

Fig. 4. Results on the `twogauss` (top) and `oneeye` (bottom) datasets. On the vertical axes is the performance in terms of MSS, ARI, NMI, and purity, from left to right respectively. The horizontal axes are logarithmic and show the training set size used, ranging from 2 to 1000.

followed by a deterioration. The one class is a Gaussian with mean $(0.3, 0, 3)$ and covariance $\frac{1}{81}I$. The other class is generated by first sampling a location uniformly at random from the arc in the top-right quadrant from the unit circle and then adding zero-mean Gaussian noise with the same covariance $\frac{1}{81}I$. The two global optima are given by the pair of pluses and the pair of crosses, respectively. With moderate amounts of training data, the two means we find are more often in between these two extremes, which leads the one mean to capture most of the red class and the other to capture most of the blue and therefore is better than the solutions that are globally optimal in terms of the MSS. The three bottom plots on the right in Fig. 4 show how the learning regime changes over increasing training set sizes. As for all other examples, the MSS shows regular monotonic behavior.

These two-dimensional examples, enable us to visualize how (initially) counterintuitive behavior may occur. However, this also makes it easy to see how K-means fails. One might then object that K-means should not be used in these settings. However, the point we want to make in the first place, is that it at all *can* happen that the performance is worse for a large training set than for a small one and that we understand how this could happen. Clearly, in real-world settings, one often cannot check any of this due to the absence of labels, or because of the high data dimensionality of the problem, etc. In those cases, one can only rely on the kind of insight such as the current work offers.

4 Discussion and Conclusion

The experiments provide the valuable insight that K-means clustering can show nontrivial behavior with, at times, deteriorating performance with an increasing amount of data used to build the clustering model. Qualitatively, the behavior seems rather similar between the different quality measures.

This information is relevant to the clustering practitioner because it is valuable to know that such behavior can at all happen and that the best performance is not necessarily attained by using the complete dataset. The insight in itself should be of interest to the clustering researcher. Furthermore, our findings suggest new ways to broaden the applicability of K-means as such. For example, one of the major surveys on gene expression clustering, [19], claims that research has moved away from K-means as this method has relevant drawbacks. One of those drawbacks is the computational burden large dataset sizes cause. However, our work suggests that in certain settings one can have both less data and better performance.

By constructing examples using synthetic data, we have shed some light on the potential origins of this behavior. Of course, we do not think this is where it should stop. In fact, the real-world data seems to display nonmonotonic behavior that can be quite a bit more dramatic. Are we able to construct synthetic data that leads to curves similar to those in Fig. 1? Can we, more generally, identify every underlying mechanism that can lead to such behavior? Can we characterize it for some quality measures? Such a deeper understanding would also be of much use to the practitioner as it provides some control over the influence of the training set size.

A major open question is of course whether similar observations can be made for other clustering methods. It is possible to set up similar studies if the clustering method has a mechanism that permits assigning new points to the clusters identified in the training set. Even if, for many clustering approaches, there may be no natural way of accomplishing this, the class of methods for which it is possible is extensive: all K-means variants (including the so-called Generalized K-means models, in which models replace means), mixture models like Gaussian mixture models, and exemplar-based methods [21], to name a few. While we do not offer solutions, identifying nonmonotonic behavior is an important step toward a better understanding of clustering, both in theory and in practice.

Disclosure of Interests. The authors have no competing interests relevant to the content of this article.

References

1. Adolfsson, A., Ackerman, M., Brownstein, N.C.: To cluster, or not to cluster: an analysis of clusterability methods. Pattern Recogn. **88**, 13–26 (2019)
2. Arbelaitz, O., Gurrutxaga, I., Muguerza, J., Pérez, J.M., Perona, I.: An extensive comparative study of cluster validity indices. Patt. Recog. **46**(1), 243–256 (2013)
3. Arthur, D., Vassilvitskii, S.: k-means++ the advantages of careful seeding. In: Proceedings ACM-SIAM Symposium on Discrete Algorithms, pp. 1027–1035 (2007)
4. Ben-David, S., Ackerman, M.: Measures of clustering quality: a working set of axioms for clustering. NeurIPS **21** (2008)
5. Ben-David, S., Loker, D., Srebro, N., Sridharan, K.: Minimizing the misclassification error rate using a surrogate convex loss. In: ICML, pp. 83–90 (2012)
6. Berriz, G.F., King, O.D., Bryant, B., Sander, C., Roth, F.P.: Characterizing gene sets with funcassociate. Bioinformatics **19**(18), 2502–2504 (2003)
7. Buhmann, J.: Empirical risk approximation: an induction principle for unsupervised learning. Technical report AIA-TR-98-3, 3 (1998)

8. Chen, Z., Loog, M., Krijthe, J.H.: Explaining two strange learning curves. In: Benelux Conference on Artificial Intelligence, pp. 16–30. Springer (2022)
9. Crussell, J., Kegelmeyer, P.: Attacking dbscan for fun and profit. In: Proceedings SIAM International Conference on Data Mining (SDM), pp. 235–243 (2015)
10. DeRisi, J.L., Iyer, V.R., Brown, P.O.: Exploring the metabolic and genetic control of gene expression on a genomic scale. Science **278**(5338), 680–686 (1997)
11. D'haeseleer, P.: How does gene expression clustering work? Nat. Biotechnol. **23**(12), 1499–1501 (2005)
12. Eren, K., Deveci, M., Küçüktunç, O., et al.: A comparative analysis of biclustering algorithms for gene expression data. Brief. Bioinf. **14**(3), 279–292 (2013)
13. Gasch, A.P., Eisen, M.B.: Exploring the conditional coregulation of yeast gene expression through fuzzy k-means clustering. Genome Biol. **3**(11), 1–22 (2002)
14. Hartigan, J.: Clustering Algorithms. Wiley (1975)
15. Hennig, C.: What are the true clusters? PRL **64**, 53–62 (2015)
16. Jain, A.: Data clustering: 50 years beyond k-means. PRL **31**(8), 651–666 (2010)
17. Jain, A., Dubes, R.: Algorithms for clustering data. Prentice Hall (1988)
18. Jain, A., Murty, M., Flynn, P.: Data clustering: a review. ACM Comput. Surv. **31**(3), 264–323 (1999)
19. Jiang, D., Tang, C., Zhang, A.: Cluster analysis for gene expression data: a survey. IEEE Trans. Knowl. Data Eng. **16**(11), 1370–1386 (2004)
20. Kleinberg, J.: An impossibility theorem for clustering. NeurIPS **15** (2002)
21. Lashkari, D., Golland, P.: Convex clustering with exemplar-based models. NeuRIPS **20** (2007)
22. Lloyd, S.: Least squares quantization in PCM. Bell Laboratories (1957)
23. Loog, M., Duin, R.P.: The dipping phenomenon. In: Proceedings of S+SSPR, pp. 310–317. Springer (2012)
24. Loog, M., Krijthe, J.H., Bicego, M.: Also for k-means: more data does not imply better performance. Mach. Learn. **112**(8), 3033–3050 (2023)
25. Loog, M., Krijthe, J.H., Jensen, A.C.: On measuring and quantifying performance. In: Handbook of PRCV, pp. 53–68. World Scientific (2016)
26. Loog, M., Viering, T.: A survey of learning curves with bad behavior: or how more data need not lead to better performance. arXiv preprint arXiv:2211.14061 (2022)
27. Loog, M., Viering, T., Mey, A.: Minimizers of the empirical risk and risk monotonicity. NeurIPS **32**, 7478–7487 (2019)
28. Meek, C., Thiesson, B., Heckerman, D.: The learning-curve sampling method applied to model-based clustering. J. Mach. Learn. Res. **2**(Feb), 397–418 (2002)
29. Nakkiran, P., Kaplun, G., Bansal, Y., Yang, T., Barak, B., Sutskever, I.: Deep double descent: where bigger models and more data hurt. In: ICLR (2020)
30. Padilha, V.A., Campello, R.J.: A systematic comparative evaluation of biclustering techniques. BMC Bioinform. **18**(1), 1–25 (2017)
31. Páez, A.: Understanding with toy surrogate models in machine learning. Mind. Mach. **34**(4), 45 (2024)
32. Perlich, C.: Learning curves in machine learning. In: Sammut, C., Webb, G.I. (eds.) Encyclopedia of Machine Learning, pp. 577–488. Springer (2010)
33. Petegrosso, R., Li, Z., Kuang, R.: Machine learning and statistical methods for clustering single-cell RNA-sequencing data. Brief. Bioinf. **21**(4), 1209–1223 (2020)
34. Prelić, A., Bleuler, S., et al.: A systematic comparison and evaluation of biclustering methods for gene expression data. Bioinformatics **22**(9), 1122–1129 (2006)
35. Rakhlin, A., Caponnetto, A.: Stability of k-means clustering. NeurIPS **19** (2006)
36. Rezaei, M., Fränti, P.: Set matching measures for external cluster validity. IEEE Trans. Knowl. Data Eng. **28**(8), 2173–2186 (2016)

37. Steinhaus, H.: Sur la division des corps matériels en parties. Bulletin de l'Académie Polonaise des Sciences IV **12**, 801–804 (1956)
38. Steinley, D.: K-means clustering: a half-century synthesis. Br. J. Math. Stat. Psychol. **59**(1), 1–34 (2006)
39. Taskesen, E., Huisman, S.M., Mahfouz, A., et al.: Pan-cancer subtyping in a 2D-map shows substructures that are driven by specific combinations of molecular characteristics. Sci. Rep. **6**(1), 24949 (2016)
40. Tavazoie, S., Hughes, J.D., Campbell, M.J., Cho, R.J., Church, G.M.: Systematic determination of genetic network architecture. Nature Gen. **22**(3), 281–285 (1999)
41. Troyanskaya, O., Cantor, M., Sherlock, G., et al.: Missing value estimation methods for DNA microarrays. Bioinformatics **17**(6), 520–525 (2001)
42. Vapnik, V.: Estimation of Dependences Based on Empirical Data. Springer Series in Statistics. Springer (1982)
43. Viering, T., Loog, M.: The shape of learning curves: a review. TPAMI (2022)
44. Vinh, N.X., Epps, J., Bailey, J.: Information theoretic measures for clusterings comparison: is a correction for chance necessary? In: ICML, pp. 1073–1080 (2009)
45. Von Luxburg, U., Ben-David, S.: Towards a statistical theory of clustering. In: Pascal Workshop on Statistics and Optimization of Clustering, pp. 20–26 (2005)
46. Von Luxburg, U., Williamson, R.C., Guyon, I.: Clustering: science or art? In: Workshop on Unsupervised and Transfer Learning, pp. 65–79 (2012)
47. Wu, J., Xiong, H., Chen, J.: Adapting the right measures for k-means clustering. In: Proceedings of the ACM SIGKDD, pp. 877–886 (2009)
48. Xu, R., Wunsch, D.: Survey of clustering algorithms. TNN **16**(3), 645–678 (2005)
49. Yang, X., Deng, C., Wei, K., Yan, J., Liu, W.: Adversarial learning for robust deep clustering. In: NeurIPS, vol. 33, pp. 9098–9108 (2020)

BOWSA: A Contribution of Sensitivity Analysis to Improve Bayesian Optimization for Parameter Tuning

Lise Kastner(✉), Bertrand Cuissart, and Jean-Luc Lamotte

Université Caen Normandie, ENSICAEN, CNRS, Normandie Univ, GREYC UMR 6072, 14000 Caen, France
{lise.kastner,bertrand.cuissart,jean-luc.lamotte}@unicaen.fr

Abstract. In applied sciences, an experiment is defined by a parameter setting, and it leads to an associated outcome. For example, when carried out under different conditions, a chemical reaction leads to different yields. Finding a satisfactory outcome is a difficult task when the number of possible settings is too large: it becomes unfeasible to explore all the possibilities due to material and time constraints. It is therefore important to quickly identify a setting that leads to a sufficient enough outcome. Bayesian optimization offers a strategy to efficiently explore parameter settings. However, the latter needs to be improved to identify interesting settings earlier. Our contribution, *BOWSA*, analyzes the outputs of Bayesian optimization through sensitivity analysis: it quantifies the association between each parameter and the outcome. Then, *BOWSA* uses this information to generate a score that refines the outputs of the Bayesian optimization.

This paper provides a methodological description of *BOWSA*. It proposes an implementation that uses the existing *EDBO, Experimental Design via Bayesian Optimization* tool, a practical implementation of Bayesian optimization designed to handle both discrete and categorical inputs, making it particularly well-suited for experimental science applications. An evaluation has been conducted on several datasets derived from well-known functions and two chemical synthesis reactions from the *EDBO* repository. Comparative results demonstrate that on average, *BOWSA* identifies productive settings earlier than Bayesian optimization alone.

Keywords: Bayesian optimization · Sensitivity analysis · Parameter tuning

1 Introduction

In experimental sciences, an outcome refers to the result of an experiment, the latter being determined by a specific parameter setting. An experiment may be conducted under numerous different settings: finding a parameter setting that

G. Krempl et al. (Eds.): IDA 2025, LNCS 15669, pp. 167–180, 2025.
https://doi.org/10.1007/978-3-031-91398-3_13

leads to a productive outcome can be very challenging, especially with only a few observations available. For example, in chemistry, a reaction can be carried out under different conditions and it is often difficult to find a productive enough yield.

It requires an expert knowledge and many trials to determine which settings will lead to a productive outcome. Active learning methods can be useful to quickly identify parameter settings to test in priority.

This work focuses on *Bayesian optimization (BO)*, an iterative process that suggests new parameter settings to test in order to quickly find a setting that leads to a productive outcome. This BO process is repeated until a satisfactory outcome is reached. Our goal is to improve the efficiency of the BO process by getting a productive outcome earlier. *Sensitivity analysis (SA)* handles this kind of issue, as it quantifies the influence of each parameter on the outcome.

Our contribution *Bayesian Optimization With Sensitivity Analysis (BOWSA)* proposes a new method to use SA in order to prioritize between the different settings suggested by a BO process. It operates in two steps. Firstly, BO returns a large number of potential settings to score. The latter are then ranked thanks to a sensitivity score. Secondly, the first ranked setting is finally proposed to be tested. To the best of our knowledge, *BOWSA* is the first method which uses SA to directly score settings proposed by BO. The paper also presents an experimental study which indicates that the reclassification carried out by *BOWSA* has a positive effect: it finds a productive outcome faster than BO alone.

This paper is organized in six sections. In Sect. 2, our work is motivated with a state of the art. BO and SA are detailed in Sect. 3, which constitute the fundamental concepts of our contribution. Then in Sect. 4, our contribution *BOWSA* is presented, combining BO with SA in order to improve the research of a productive set of parameters. Section 5 defines the experimental protocol conducted to evaluate the performance of this new method. Section 6 reports the evaluation results on several synthetic datasets, indicating that *BOWSA* improves BO.

2 Related Works

Different types of methods are used to guide the search for productive settings. These methods are commonly divided in two categories: model free and model based approaches. Among the model free approaches, genetic algorithms or simulated annealing have often been employed to address these issues, but they require many tests to gather information about the model: testing a new experimental setting can require significant time and resources. Model-based approaches are more adapted to deal with these kind of issues, like random forests [5], BO [19] or reinforcement learning. Over the past few years, several tools have been introduced to address the parametrization problem in experimental sciences, particularly *OED* [15] which focuses on complex dynamical systems, as well as specialized frameworks like *Nextorch* [23] and *EDBO* [19]. *EDBO,*

Experimental Design via Bayesian Optimization, is particularly useful to handle discrete and categorical data, which is essential when dealing with non-numerical parameters. Categorical parameters are independent, with no ordinal relationship between their values. To address this issue, categorical parameters can be encoded using one-hot encoding (OHE) [25], which converts each possible value of the variable into a unique binary vector: a categorical variable that accepts k different values is encoded using k binary variables, one per value. Advanced encodings, also available in *EDBO*, can better capture dependencies between values. Techniques like DFT [12] or Mordred [13] introduce such dependencies, often involving parameters specific to a category. For instance, in chemistry, a reactant can be characterized by attributes such as its molar mass or the number of atoms. *EDBO* is also able to propose several settings simultaneously. It is useful in experimental science if the experts need to conduct several experiments in parallel.

As the number of parameters increases, finding productive settings becomes more challenging. Several approaches explore the different possible settings and efficiently analyze parameter interactions. Some functional decomposition methods, such as functional ANOVA [9], decompose the variance of a function into contributions of individual input parameters. It provides a valuable insight of the importance of the different parameters. In addition to functional decomposition, ablation analysis [3] focuses on the most relevant dimensions. These techniques help to refine the search space and to improve optimization efficiency. However, when too few observations are available, these techniques may become overly limiting.

A solution is offered by SA [16], which may assist BO by quantifying each parameter's contribution on the output, using sensitivity indices. Selecting an appropriate type of index requires prior knowledge of the outcome's complexity and regularity [7]. In this context, outcome information is still limited, it is not possible to state that the outcome is either monotonic or continuous. Therefore, our contribution focuses on Sobol and HSIC indices, both adapted to this type of outcome. The Sobol indices [8] are variance-based measures. When outcome information is limited, kernel-based methods, such as HSIC indices [20], often provide a more suitable alternative. HSIC indices have already been applied to a BO context [20]: at each iteration, the goal is to select the most influential parameters and to run the BO on them only, the values of the less relevant parameter being determined independently.

Our contribution makes use of SA to guide the BO differently: rather than using sensitivity indices to divide the search space, they are used to evaluate the settings suggested by BO, in two steps.

3 Fundamental Concepts

3.1 Notations

Given m parameters X_i whose values lie in the domain $\mathcal{X}_i$, a setting e corresponds to a vector of m values $(e_1, ..., e_m)$, $e_i \in \mathcal{X}_i$, $\forall 1 \leq i \leq m$. The domain of all possible settings is denoted as $\mathcal{X} = \mathcal{X}_1 \times \ldots \times \mathcal{X}_m$. In experimental sciences, the parameters are often discrete, taking their values from a finite set: $|\mathcal{X}| = \prod_{i=0}^{m} |\mathcal{X}_i|$. The *outcome* associated to a setting is considered as an unknown application Y whose domain is $\mathcal{Y}$. The set of all the possible settings is partitioned into two parts. Firstly, the set of all the observations is denoted as $\mathcal{O} = \{o_1, ..., o_n\}$ with corresponding outcomes $\{y_1, ..., y_n\}$, considering that an *observation* is a setting for which the resulting output is known. Secondly, the part of the settings yet to explore is denoted as $\mathcal{V} = \mathcal{X} \setminus \mathcal{O}$.

3.2 Bayesian Optimization

Our objective is to find a setting that maximizes the outcome function Y. At the beginning, there are few or no observations available at all.

At each iteration, eventually using the observations $\mathcal{O}$, BO selects a new setting e from $\mathcal{V}$ and queries its corresponding outcome. Consequently, $\mathcal{O}$ is updated each time a new setting outcome is determined. A predictive mean μ and a predictive variance σ^2 are calculated from the observations, which together define a *Gaussian process* (GP), as displayed on Fig. 1. To choose the next setting e to be conducted, an *acquisition function* is calculated using the GP: the setting e that maximises this function is chosen. e is finally conducted to find its outcome $Y(e)$ and it becomes a new observation of $\mathcal{O}$. The Gaussian process and the acquisition function are also updated if another iteration is required.

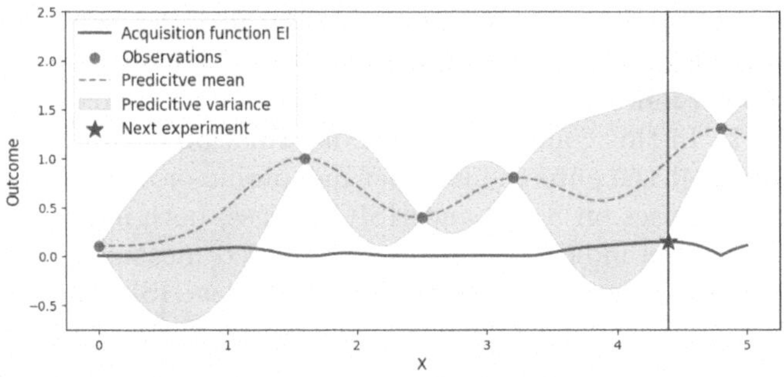

Fig. 1. A Gaussian process $GP(\mu, k)$ in a 1 dimensional space with 5 observations, with its predictive mean μ and variance σ^2. The acquisition function, represented in blue, uses the GP to select the next setting to explore, displayed with the ★ (Color figure online).

The Gaussian Process. $GP(\mu, k)$ is a probabilistic model defined by a prior *mean* μ and *covariance* k, used to interpolate the *prior distribution* of n observations $\{o_1, \ldots, o_n\}$ and their outcomes $\{y_1, \ldots, y_n\}$.

Let e be the next setting to explore.

The predictive mean μ and variance σ^2 of the *posterior distribution* are given by Eqs. (1) and (2) [18], where $\mu_0(e)$ is the prior mean, m the prior mean vector, y the observed outcomes vector, $k(e)$ the covariance vector between e and the observations, $k(e,e)$ the prior distribution variance, K the covariance matrix of $\mathcal{O}$, I the identity matrix, and σ_n^2 the noise variance of $\mathcal{O}$.

$$\mu(e) = \mu_0(e) + k(e)^T(K + \sigma_n^2 I)^{-1}(y - m) \tag{1}$$

$$\sigma^2(e) = k(e,e) - k(e)^T(K + \sigma_n^2 I)^{-1}k(e) \tag{2}$$

Hereafter, k will be referred to as a kernel. Matérn kernels are commonly used in BO [22], and this work specifically uses the k_{MATERN52} [19].

The Acquisition Function is used to propose the next setting to test by identifying where its maximum value is reached. It combines two strategies: *exploration* and *exploitation.* The *exploration* aims at minimizing uncertainty, measured by the predictive variance σ^2. The *exploitation* seeks high performance by maximizing the predictive mean μ. The *Expected Improvement* (EI) is a commonly used acquisition function [11]. For a setting e, it considers the expected improvement between the predictive mean $\mu(e)$ and the best observed value y^* in $\mathcal{O}$, together with its predictive standard deviation $\sigma(e)$, where Φ is a cumulative distribution function of a normal law.

$$EI(e) = (\mu(e) - y^*) \times \Phi\left(\frac{\mu(e) - y^*}{\sigma(e)}\right) + \sigma(e) \times \Phi\left(\frac{\mu(e) - y^*}{\sigma(e)}\right) \tag{3}$$

3.3 Sensitivity Analysis

Our goal is to improve *BO* to obtain a productive outcome earlier in the process. Any insight about each parameter influence on the outcome can help BO to identify productive settings more rapidly. SA provides this information by identifying the most influential parameters on the outcome [16]. This work focuses on quantitative methods, as they provide precise sensitivity indices. As precised in Sect. 2, this paper focuses on both Sobol and HSIC indices.

The Sobol Index is a variance-based method, involving variance decomposition [1,17]. Let X_i be a parameter, $V(Y)$ be the variance of the outcome Y, and $V[E(Y|X_i)]$ be the variance of the expectations of Y when X_i is fixed.

The first order Sobol index of X_i is:

$$S_i = \frac{V[E(Y|X_i)]}{V(Y)},\ 1 \leq i \leq m \tag{4}$$

The Hilbert-Schmidt Independence Criterion (HSIC) is a *kernel-based* method introduced to measure dependency between parameters. To compute the HSIC index of X_i on Y, let $\mathcal{F}_i$ and $\mathcal{G}$ be two reproducing kernel Hilbert spaces (RKHS) applied to X_i and Y distributions. Denote the marginal and joint probability distributions of X_i and Y as $P_Y P_{X_i}$ and P_{Y,X_i} respectively. μ denote the application used to embed these distributions into the considered Hilbert space [6]. The HSIC index of X_i is given by:

$$HSIC(X_i, Y)_{\mathcal{F}_i,\mathcal{G}} = \|\mu(P_Y P_{X_i}) - \mu(P_{Y,X_i})\| \tag{5}$$

This dependence measure transcribes the following hypothesis: a dissimilarity between the joint and marginal distributions indicates that X_i influences Y. The definition of embedding μ is not provided here, as it is clearly detailed in [6].

4 BOWSA: Bayesian Optimization With Sensitivity Analysis

4.1 Algorithm

Our method is detailed in Figs. 2, 3 and 4. The general procedure displayed in Fig. 2 computes the next settings to test, based on the current observations. The process is repeated until a productive outcome is achieved. To select the next settings to test, the two considered procedures are explained in the dashed boxes of Figs. 3 and 4. Figure 3 corresponds to the standard BO approach while Fig. 4 details our contribution, *BOWSA*: using the observations, sensitivity indices are calculated for the m parameters while BO proposes N candidate settings. Note that N must always be greater than the number of settings finally proposed by *BOWSA* (later defined as a *batch*). For each candidate setting, a sensitivity score S is calculated with Eq. 6. The method finally proposes the settings with the highest scores.

4.2 The Sensitivity Score

To integrate SA to BO, our algorithm assigns a *sensitivity score* for each setting generated by BO, thereby enabling a selection process. This score is calculated using the *sensitivity indices* of the parameters, along with prior knowledge on values and results: an *occurrence score* and a *mean outcome* defined below. Each ind_i sensitivity indices of each parameter is calculated on prior observations with first order Sobol indices or HSIC indices, both defined in Sect. 3.3.

Let $e = (e_1, ..., e_m)$ be a candidate setting returned by the BO. To evaluate the relevance of e, the algorithm calculates a sensitivity score $S(e)$ using the observations space $\mathcal{O}$ and their outcomes $\mathcal{Y}$ defined in Sect. 3. The Eq. 6 calculates the score S of the setting e, where ind_i is the *sensitivity index* of X_i, p_i is the *occurrence score* of the e_i value calculated with Eq. 7, and μ_i is the *mean outcome* associated to the observations containing e_i:

$$S(e) = \sum_{i=1}^{m} ind_i \times (1 - p_i) \times \mu_i \tag{6}$$

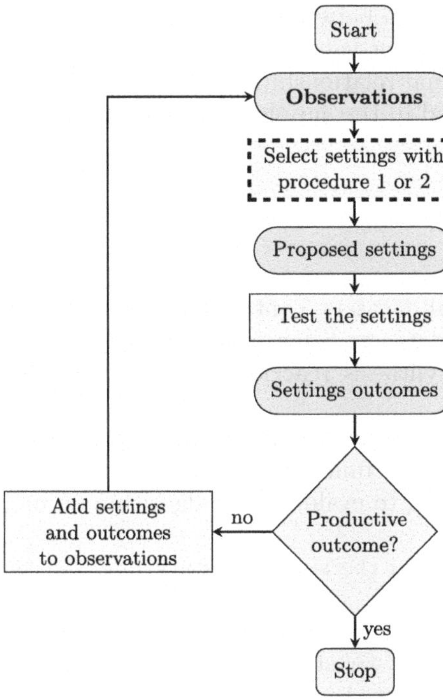

Fig. 2. General procedure

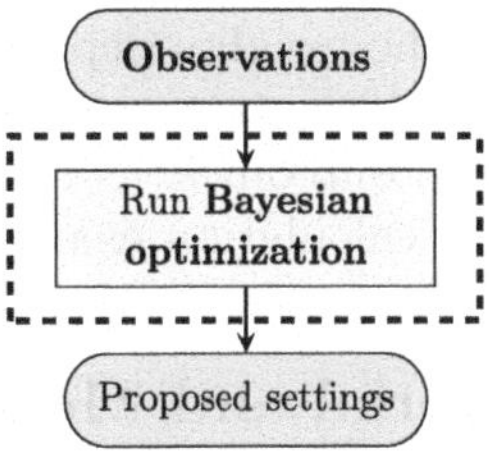

Fig. 3. Procedure 1 - BO

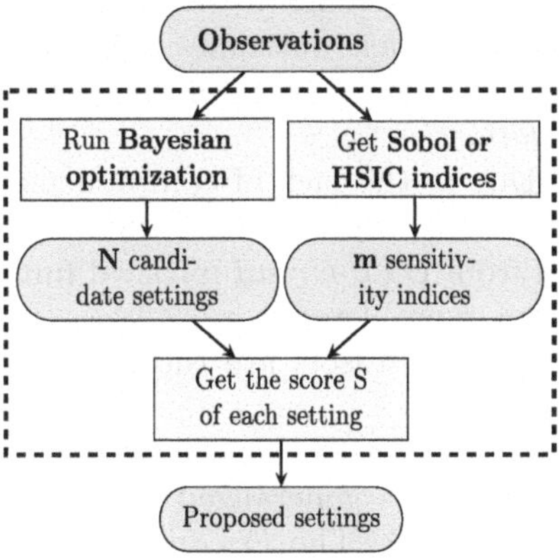

Fig. 4. Procedure 2 - BOWSA

The Occurrence Score p_i of the current value e_i is calculated for each parameter X_i based on prior observations $\mathcal{O}$. The goal is to prioritize unexplored values. When a parameter has significant influence on the outcome, it is interesting to know how the outcome varies depending on X_i value.

Let $occ(e_i)$ be the number of occurrences of e_i in the observations $\mathcal{O}$. To calculate p_i, it is important to consider the number of values a parameter can take: let $|\mathcal{X}_i|$ be the cardinal value of the current X_i parameter, and $|\mathcal{X}_{min}|$ the minimum cardinal among all parameters: $|\mathcal{X}_{min}| = MIN(card(\mathcal{X}_{1..m}))$. The number of observations is denoted as $|\mathcal{O}|$. The occurrence score p_i is defined as:

$$p_i = \frac{occ(e_i) \times |\mathcal{X}_{min}|}{|\mathcal{O}| \times |\mathcal{X}_i|} \tag{7}$$

If p_i were defined as $\frac{occ(e_i)}{|\mathcal{O}|}$, it would not account for differences in parameter ranges. For example, if a parameter value with a cardinality of 4 had been explored, its impact should be greater than a parameter with 20 possible values.

The Mean Outcome μ_i is the mean calculated on the observed outcomes containing the e_i value. If e_i is not in $\mathcal{O}$, two options are possible: if X_i takes discrete dependent values, μ_i is a weighted arithmetic mean of all Y outcomes in $\mathcal{O}$, and the weights depend on the distance between e_i and the values taken by X_i in $\mathcal{O}$; if X_i takes categorical independent values, μ_i becomes the mean of all observed outcomes, as there is no ordinal relationship between categories.

A score calculation example is provided in the supplementary material[1].

5 Material and Methods

5.1 Datasets

This new approach is evaluated on several datasets, containing all possible settings and their outcomes. The two first datasets come from the *EDBO* repository [19], and represents two chemical synthesis reactions: *Suzuki* and *Direct arylation*.

Due to the lack of comprehensive large dataset, four synthetic datasets are generated from *Friedman* function [24], *Park* function [4], *Shekel* function [4] and from *OTL circuit* physical function [21], to evaluate our algorithm on more complex problems.

These datasets are thoroughly detailed in the supplementary material (see footnote 1).

These synthetic functions have been specifically chosen because they are not structured as generalized sums; each of their variables has a distinct influence and behavior. The *Friedman* dataset is generated using the Friedman function, which is defined as $f(x) = 10sin(\pi x_1 x_2) + 20(x_3 - 0.5)^2 + 10x_4 + 5x_5$. The *Park* dataset is created using *Park* function $f(x) = \frac{2}{3}exp(x_1 + x_2) - x_4 sin(x_3) + x_3$ where each variable takes discrete values within the range $[0, 1]$. As for the *Shekel* dataset, it is constructed using the *Shekel* function [10], which has multiple local maxima, making it difficult to optimize. The *OTL circuit* function simulates the behavior of an electrical circuit: a dataset is created based on its modeling function and the variable domains provided in the following reference [21].

For the categorical parameters of *Suzuki* and *Direct arylation*, various encoding methods can be applied. In the most general case, where no additional information is available, these parameters are encoded using OHE, as explained in Sect. 2.

5.2 Experimental Protocol

The following protocol has been conducted on the 6 datasets described in Table 1, to evaluate the new *BOWSA* algorithm performance, compared to BO alone.

For each dataset, 30 runs were executed, each starting with one randomly selected setting. These 30 initial settings are the same for both standard BO and *BOWSA*. For each run, this setting constitutes the initial set of observations. Considering this set, BO and *BOWSA* iteratively propose one new setting. The number of settings proposed each time is called a batch.

[1] https://git.unicaen.fr/greyc/codag/bowsa.

Table 1. Summary of variable types and domain sizes of the datasets

Dataset	Parameter types	Number of parameters	Domain Size
Direct arylation	Numerical + Categorical	5	1 728
Suzuki	Categorical	5	3 696
Friedman	Numerical	5	100 000
Park	Numerical	4	160 000
OTL circuit	Numerical	6	254 100
Shekel	Numerical	4	194 481

These settings together with their outcomes are included to the set of observations, the latter being the base of the next BO iteration. Note that in our study, the outcomes are available in the related datasets: in a real experimental process, the outcomes would typically be provided by an expert.

5.3 Parameters

Many parameters have been tested in advance to determine the best way to use *BOWSA*: the acquisition function and the batch, which are used in both standard BO and *BOWSA* procedure, and two parameters specific to *BOWSA*: the sensitivity indices and the number of candidate settings N generated by BO.

The *Expected Improvement acquisition function* is used to conduct our protocol because it combines both exploration and exploitation strategy, and it has proven to be the best choice in the context of experimental sciences [19]. For the *batch* b, the evaluation tests a BO process (standard BO or *BOWSA*) returning one single setting per iteration, representing the most general case: $b = 1$.

As for the *number of candidate settings*, the BO step of *BOWSA* must propose N candidates in order to select the b best(s) among them with the sensitivity score: $N >> b$. To fix N, the procedure was tested with 10, 30, 50 and 70 candidate settings, in order to choose one of them to evaluate the method: it is provided in the supplementary material (see footnote 1). Finally our study uses $N = 50$, as it seems the most efficient to get a suitable setting early: increasing N further would unnecessarily prolong the evaluation and lead to an excessive dispersion of results.

For the *sensitivity indices*, it is possible to use Sobol or HSIC indices. On each graph of Fig. 5, the x-axis denotes the X parameters while the y-axis pictures the mean sensitivity indices calculated for each parameter, on the 100 sets of observations: HSIC indices demonstrate superior performance when working with sparse observations. Indeed, the graphs show a clearer distinction of the sensitivity indices of each parameter using HSIC compared to those obtained with Sobol. Sobol and HSIC indices are mainly applied to numerical parameters. However, they can be used with categorical parameters, by computing a smoothing target encoding [14], designed for sparse datasets. Let's consider a

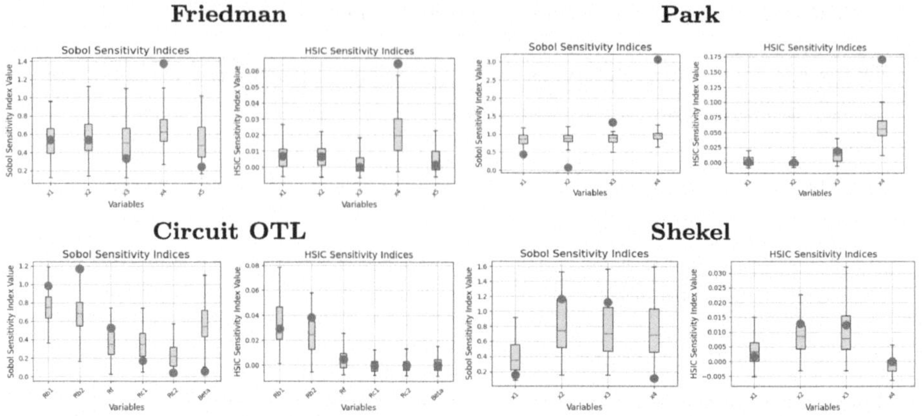

Fig. 5. Sobol and HSIC sensitivity indices on the whole datasets: the boxplots illustrate the indices distribution on 100 subsets of 25 observations. Red points are the true indices, calculated on the complete dataset. (Color figure online)

variable with different categories: each category is replaced by the mean outcome from existing observations including that category. For unseen categories, a global mean is used, with a smoothing adjustment applied if there are many.

Our study is implemented in *Python* language, using the *EDBO* implementation of BO [19], detailed in Sect. 2. As for SA, the indices have been calculated with the libraries *OpenTurns* and *statistics*.

6 Results and Discussion

6.1 Results

The main results are displayed on Figs. 6 and 7. On each graphic of Fig. 6, the x-axis denotes the iteration while the y-axis pictures outcome values. Note that each point corresponds to the mean value of maximum outcomes surrounded by the standard deviation, resulting from the 30 corresponding runs. The results achieved by standard BO are displayed with a black line, while those of our contribution *BOWSA* are pictured in red. Figure 7 illustrates the standardized improvement of *BOWSA* method over *BO*. Positive values indicate that *BOWSA* outperforms *BO*, while negative values suggest the opposite.

More details about the results are available in the supplementary material (see footnote 1).

6.2 Discussion

This study compares *BOWSA* with standard BO on six different datasets, and evaluates their performance in achieving productive outcomes.

For the *OTL circuit*, *Park*, *Shekel* and *Friedman* datasets, *BOWSA* achieves productive outcomes earlier than the standard BO method. This is especially

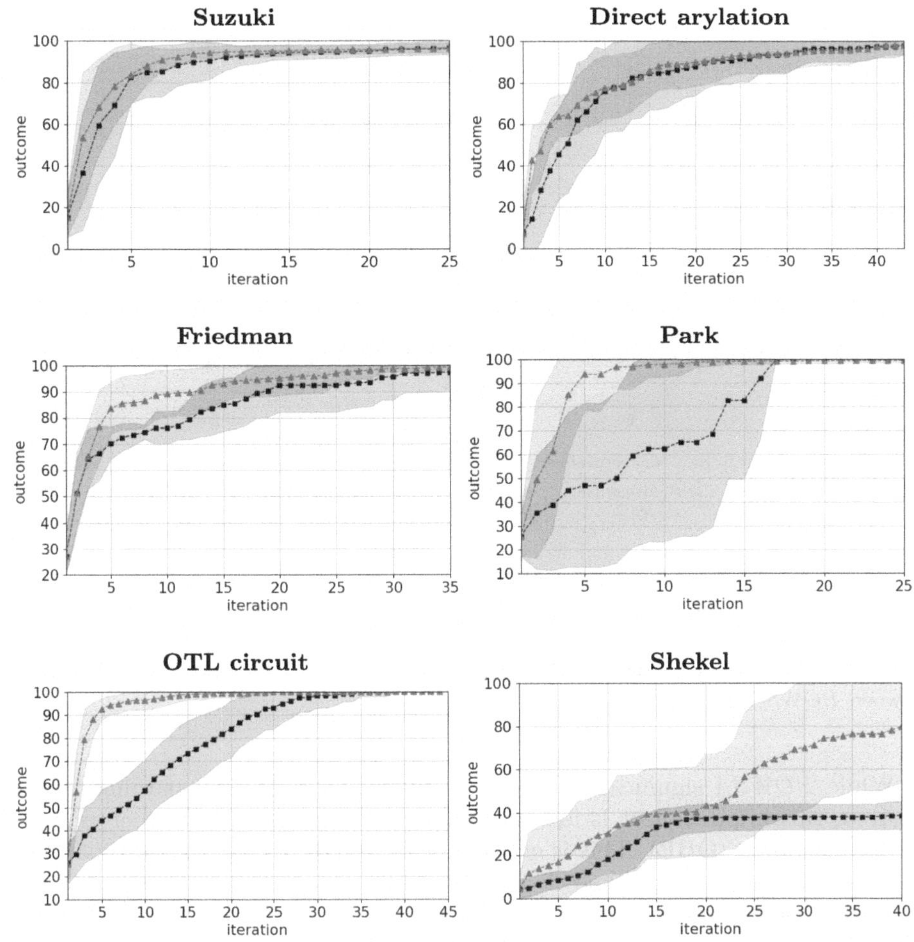

Fig. 6. Global evaluation on 6 datasets, with BO in black squares ■ and *BOWSA* in red triangles ▲: each point represents the mean of the maximum outcome values resulting from the 30 corresponding runs. at a given iteration, surrounded by the standard deviation

noticeable for the *OTL circuit* and *Park* datasets, for which it is more difficult to obtain a productive outcome using *BO* alone, making *BOWSA*'s improvement more significant. The *Shekel* dataset features multiple local maxima around 40%, making it challenging to reach the global maximum. The BO tends to remain trapped in local maxima for several iterations, while *BOWSA* explores alternative settings that lead to better outcomes. For the *Suzuki* dataset, *BOWSA*'s average curve is slightly above *BO*. The main improvement lies in the faster reduction of standard deviation, indicating more consistency between the different subsets.

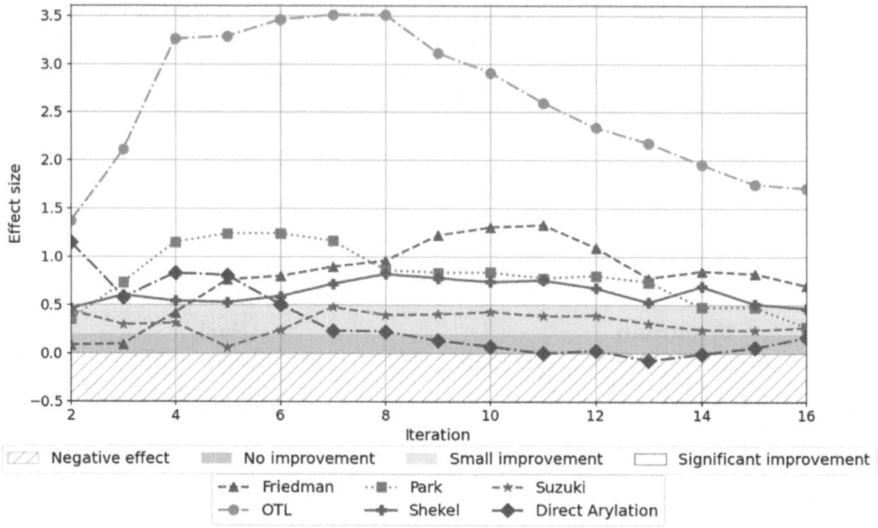

Fig. 7. Effect size: improvement of *BOWSA* over *BO*: each point corresponds to the effect size [2] for a given iteration, on the 30 subsets of a given dataset. The standardized effect size is $\frac{\mu_{BOWSA-BO}}{\sigma_{BOWSA-BO}}$, where $\mu_{BOWSA-BO}$ and $\sigma_{BOWSA-BO}$ are respectively the mean and standard deviation of the maximum outcome differences on each subset, between *BOWSA* and *BO*.

While *BOWSA* significantly improves BO when numerical parameters are involved, this improvement is less noticeable with categorical parameters using one-hot encoding (OHE). This is expected since OHE doesn't give ordinal relationships. If one has more knowledge about the parameters involved, different types of encoding can be used to introduce dependencies between the parameter values.

Figure 7 illustrates that *BOWSA* generally yields better outcomes than BO: its effect size is always positive, with only one exception at iteration 13 for *Direct arylation*. The improvement is significant when the effect size is superior to 0.5, small if the effect size is between 0.2 and 0.5. The graph highlights the specific iterations where *BOWSA* demonstrates an improvement over BO, on the six datasets. For the datasets with numerical parameters, *circuitOTL*, *Park*, *Shekel* and *Friedman*, *BOWSA* shows a significant improvement over standard *BO* for the first twelve iterations. A significant improvement is also noticeable on *Direct arylation* in the beginning of the process, while for *Suzuki*, the improvement is less significant, though still positive, likely due to the low standard deviation obtained with *BOWSA*.

BOWSA is particularly effective when the parameters have a limited range of possible values. When applied to parameters with broader ranges, sensitivity scores may become less informative at the beginning of the optimization, because many values remain unrepresented in the observations. This method is especially suited for experimental designs, where each parameter has a manageable number of possible values, and *BOWSA* can provide meaningful insights and guide the optimization process efficiently, right from the start.

BOWSA proves to be a valuable method, offering productive outcomes earlier than standard BO. Additionally, it is more reliable, as the standard deviation decreases earlier in the process: it provides a greater confidence in achieving good results, particularly during the initial stages of the active learning process.

7 Conclusion

This work introduces *BOWSA*, a method that improves Bayesian optimization by integrating sensitivity analysis to accelerate the search for productive outcomes in experimental sciences. By using sensitivity scores, *BOWSA* ranks the settings proposed by Bayesian optimization, and selects the most promising one for testing. Since *BOWSA* is based on the practical implementation of EDBO, it can handle both discrete and categorical inputs. The method has been evaluated on several synthetic datasets and real chemical synthesis reaction datasets. The results demonstrate that *BOWSA* outperforms standard Bayesian optimization, especially when the optimization task is complex. It identifies productive outcomes earlier and with more consistency. *BOWSA* is more effective when the number of possible values a parameter can take is limited, which is common in experimental designs.

Overall, *BOWSA* offers an improvement over standard Bayesian optimization, particularly in experimental sciences, where finding a productive setting early is crucial due to time and material constraints. It is important to precisely determine the goal, whether the priority is to achieve a satisfactory outcome in fewer iterations or achieve the best possible outcome with more experiments. *BOWSA* is especially useful for the first case, as it reduces the number of experiments needed to get a productive outcome early.

Currently, *BOWSA* is being tested on a real electrochemical reaction in collaboration with a chemistry lab. These reactions involve many chemical and electrical parameters, discrete and categorical, making the search for a productive setting more complex.

Disclosure of Interests. The authors have no competing interests to declare that are relevant to the content of this article.

References

1. Arnst, M., Soize, C., Bulthuis, K.: Computation of sobol indices in global sensitivity analysis from small data sets by probabilistic learning on manifolds. Int. J. Uncertain. Quantif. **11**(2), 1–23 (2021)
2. Cohen, J.: Statistical Power Analysis for the Behavioral Sciences. Routledge, New York (1988)
3. Eggensperger, K., Lindauer, M., Hoos, H.H., Hutter, F., Leyton-Brown, K.: Efficient benchmarking of algorithm configurators via model-based surrogates. Mach. Learn. **107**(1), 15–41 (2017). https://doi.org/10.1007/s10994-017-5683-z

4. Finck, S., Hansen, N., Ros, R., Auger, A.: Real-parameter black-box optimization benchmarking 2010: Presentation of the noiseless functions. [research report] (2010)
5. Greenhill, S., Rana, S., Gupta, S., Vellanki, P., Venkatesh, S.: Bayesian optimization for adaptive experimental design: a review. IEEE Access **8** (2020)
6. Gretton, A., Borgwardt, K., Rasch, M., Schölkopf, B., Smola, A.: A kernel method for the two-sample-problem (2006)
7. Iooss, B., Lemaître, P.: A review on global sensitivity analysis methods (2014)
8. Issan, O., Riley, P., Camporeale, E., Kramer, B.: Bayesian inference and global sensitivity analysis for ambient solar wind prediction. Space Weather Int. J. Res. Appl. **21**(9) (2023)
9. Kaufman, C.G., Sain, S.R.: Bayesian functional ANOVA modeling using Gaussian process prior distributions. Bayesian Anal. **5**(1), 123–149 (2010). https://doi.org/10.1214/10-BA505
10. Loeppky, J., Williams, B., Moore, L.: Global sensitivity analysis for mixture experiments. Technometrics **55** (2013). https://doi.org/10.1080/00401706.2012.727751
11. Mockus, J.: On Bayesian methods for seeking the extremum. In: Marchuk, G.I. (ed.) Optimization Techniques, vol. 27, pp. 400–404. Springer (1974)
12. Morishita, T., Kaneko, H.: Enhancing the search performance of Bayesian optimization by creating different descriptor datasets using density functional theory. ACS Omega **8**(36), 33032–33038 (2023). https://doi.org/10.1021/acsomega.3c04891
13. Moriwaki, H., Tian, Y.S., Kawashita, N., Takagi, T.: Mordred: a molecular descriptor calculator. J. Cheminform. **10**(1), 4 (2018). https://doi.org/10.1186/s13321-018-0258-y
14. Nugroho, H., Utama, N., Surendro, K.: Smoothing target encoding and class center-based firefly algorithm for handling missing values in categorical variable (2022)
15. Papadimitriou, C., Argyris, C.: Bayesian optimal experimental design for parameter estimation and response predictions in complex dynamical systems. Procedia Eng. **199**, 972–977 (2017). eURODYN 2017
16. Plischke, E., Borgonovo, E., Smith, C.L.: Global sensitivity measures from given data. Eur. J. Oper. Res. **226**(3), 536–550 (2013)
17. Saltelli, A., Tarantola, S., Campolongo, F., Ratto, M.: Sensitivity Analysis in Practice: A Guide to Assessing Scientific Models. Halsted Press, USA (2004)
18. Shahriari, B., Swersky, K., Wang, Z., Adams, R., De Freitas, N.: Taking the human out of the loop: a review of Bayesian optimization. Proc. Inst. Radio Eng. **104**(1), 148–175 (2016)
19. Shields, B., et al.: Bayesian reaction optimization as a tool for chemical synthesis. Nature **590**(7844), 89–96 (2021)
20. Spagnol, A., Le Riche, R., da Veiga, S.: Bayesian optimization in effective dimensions via kernel-based sensitivity indices. In: ICASP13. Séoul, South Korea (2019)
21. Surjanovic, S., Bingham, D.: Virtual library of simulation experiments: test functions and datasets (2013)
22. Swersky, K., Snoek, J., Adams, R.P.: Multi-task Bayesian optimization. In: Burges, C., Bottou, L., Welling, M., Ghahramani, Z., Weinberger, K. (eds.) Advances in Neural Information Processing Systems, vol. 26. Curran Associates, Inc. (2013)
23. Wang, Y., Chen, T.Y., Vlachos, D.G.: Nextorch: a design and Bayesian optimization toolkit for chemical sciences and engineering. J. Chem. Inf. Model. **61**(11), 5312–5319 (2021)
24. Yang, X.S.: Test problems in optimization. Wiley (2010)
25. Zhu, W., Qiu, R., Fu, Y.: Comparative study on the performance of categorical variable encoders in classification and regression tasks. CoRR (2024)

Overfitting in Combined Algorithm Selection and Hyperparameter Optimization

Sietse Schröder(✉), Mitra Baratchi, and Jan N. van Rijn

Leiden Institute of Advanced Computer Science (LIACS), Leiden University, Leiden, The Netherlands
s.schroder@liacs.leidenuniv.nl

Abstract. Hyperparameter optimization (HPO) aims to design machine learning algorithms that generalize well to unseen data by repeatedly evaluating hyperparameter configurations using a validation procedure. When the validation performance of these configurations is overly optimistic compared to the performance on an unseen test set, this is referred to as meta-overfitting. We decompose meta-overfitting into two types: (i) selection-based and (ii) adaptive overfitting. Selection-based overfitting occurs when testing many configurations, which increases the chance of finding a configuration that performs well on the validation set by chance but performs suboptimal on the test set. Adaptive overfitting arises from advanced HPO methods, such as Bayesian optimization, which iteratively utilize validation results of earlier configurations to propose new configurations increasingly tailored to the specific validation set. We provide one of the largest empirical studies of meta-overfitting in the context of HPO for the Combined Algorithm Selection and Hyperparameter Optimization (CASH) problem, analyzing random search and Bayesian optimization for 48 classification and 16 regression datasets using holdout validation. We show evidence of adaptive overfitting in Bayesian optimization for 41 classification datasets, and consistent with prior work, we show that multiclass datasets are less affected by this phenomenon. Additionally, we find that optimization procedures for regression datasets are surprisingly resilient to adaptive overfitting. Furthermore, we explore the effect of various design choices in the validation procedure (i.e., 10-fold cross-validation and varying holdout set sizes) on meta-overfitting.

Keywords: Overfitting · AutoML · Supervised Learning

1 Introduction

Designing a machine learning pipeline involves selecting various components and tuning their hyperparameters. Hyperparameter optimization (HPO) supports the human in the loop by automatically searching for an optimal pipeline that

G. Krempl et al. (Eds.): IDA 2025, LNCS 15669, pp. 181–194, 2025.
https://doi.org/10.1007/978-3-031-91398-3_14

best generalizes to unseen data, reducing manual effort and empowering non-experts to create complex machine learning workflows [3]. Simultaneously optimizing both the pipeline components as well as their hyperparameters is defined as the combined algorithm selection and hyperparameter optimization (CASH) problem [23].

The generalization performance of hyperparameter configurations is commonly estimated using nested resampling techniques such as k-fold cross-validation or a holdout set. To evaluate an HPO method on unseen data, the dataset is first split into a train and test set. Since the test set can only be used once, the train set is further divided into a train and validation set to evaluate the various configurations suggested throughout the HPO process. Typically, the best-performing configuration on the validation set is selected and evaluated on the test set. Instead of a single holdout set, cross-validation can be applied at either or both levels to improve robustness. These resampling splits are generally predetermined and remain fixed throughout the HPO process. In practical learning scenarios, limited data often restricts the size of resampling splits used for evaluation, leading to less statistically robust estimates of generalization performance [18]. When the performance estimates obtained during validation are overly optimistic compared to the actual performance the pipeline achieves on unseen test data, this is referred to as meta-overfitting, described by the meta-overfitting error (MOE) [1], which measures the difference between the validation estimation and the actual generalization performance. In the context of HPO, meta-overfitting arises because many hyperparameter configurations are evaluated, some naturally performing well on validation data due to chance alone [18]. Since HPO optimizes for the best validation performance, it tends to favor these overfitted configurations over others that might generalize better to unseen data. In this work, we refer to this type of meta-overfitting as *selection-based overfitting.*

Moreover, we hypothesize that meta-overfitting occurs due to the sequential nature of many hyperparameter optimization strategies, which leverage information from previous evaluations to propose configurations increasingly tailored to the specific validation procedure. While this approach may improve validation performance, it does not necessarily enhance the ability to generalize to unseen data. This phenomenon is in a more general context called adaptive data analysis [6], highlighting the risks of adaptively refining solutions based on limited validation data. When this adaptivity leads to overly optimistic performance estimates, it is termed *adaptive overfitting.* Given that state-of-the-art HPO techniques, such as Bayesian optimization, inherently rely on adaptive mechanisms to iteratively refine hyperparameter configurations, it is critical to assess whether the solutions they propose are overfitted due to either adaptive or selection-based effects.

In this study, we conduct a comprehensive empirical investigation into meta-overfitting arising from random search and Bayesian optimization, offering one of the first large-scale analyses of this phenomenon in HPO for the CASH problem. Our key contributions are as follows:

1. We advance the understanding of meta-overfitting in HPO by examining its effects across 64 diverse classification and regression datasets.
2. On four binary classification datasets, we explore the mitigating effect of 10-fold cross-validation (10CV).
3. We identify adaptive overfitting as a significant contributor to meta-overfitting in HPO and provide, to the best of our knowledge, the first evidence of its presence in Bayesian optimization across most of the classification datasets analyzed.
4. We examine the impact of varying validation set sizes on selection-based and adaptive overfitting, extrapolating the practical implications of meta-overfitting across scenarios with differing levels of data availability.

2 Related Work

In this section, we review the background of selection-based and adaptive overfitting and argue their inherent connection to HPO techniques. Furthermore, we review related work on meta-overfitting in HPO, specifically when using state-of-the-art Bayesian optimization.

Selection-Based Overfitting: Selecting the hypothesis with the highest cross-validation performance from multiple evaluations can lead to biased results and poor generalization [16,18], the phenomenon we refer to as selection-based overfitting. Subsequent research demonstrated that algorithms for variable selection, also referred to as feature selection, frequently overfit the cross-validation process [21]. Empirical studies reveal that manual model selection and hyperparameter tuning frequently result in models that perform better on validation data than they do on unseen test sets, regardless of whether this performance has been determined by a single nested holdout set or nested cross-validation procedures [5]. Circling back, since HPO algorithms repeatedly evaluate hyperparameter configurations [3], they are susceptible to selection-based overfitting.

Adaptive Data Analysis and Overfitting: Beyond the context of HPO, a famous example of adapting analyses based on validation results is Freedman's paradox [10], which demonstrates that selecting randomly sampled features based on their correlation with a target can yield statistically significant regression results on the validation set, ultimately leading to poor generalization. Similarly, adaptive data analysis, introduced by Dwork et al. [6], explores how adapting models or hypotheses based on intermediate evaluation undermines the statistical validity of the validation results. They show the risk of overfitting to the validation process, often referred to as adaptive overfitting, grows with the number of adaptively chosen evaluations but decreases as the size of the validation dataset increases. Theoretically, multiclass classification was shown to be more resilient to adaptive overfitting compared to binary classification [8].

This effect was empirically observed in HPO using Bayesian optimization in this work.

Despite theoretical concerns, a meta-analysis of Kaggle competitions reported minimal signs of adaptive overfitting [22]. Similarly, studies on widely-used benchmarks like MNIST, CIFAR-10, and ImageNet have found limited evidence of adaptive overfitting [19,20,24], suggesting that the practical impact of adaptive overfitting may be context-dependent. Investigating adaptive overfitting in the context of HPO is particularly interesting since Bayesian optimization, which is one of the most commonly used techniques for HPO, inherently selects new hyperparameter configurations based on the performance of previous configurations evaluated on the validation set [2].

Overfitting in HPO: Meta-overfitting in HPO, also known as overtuning or oversearching [16,17], remains a relatively underexplored topic in the literature [3,15]. Recent studies have begun addressing this gap. Lévesque [13] conducted a comprehensive evaluation, revealing significant meta-overfitting in Bayesian optimization when tuning SVM hyperparameters across 118 datasets. This study also demonstrated that reshuffling the validation split after each evaluation improved both stability and overall performance. Subsequent research confirmed these findings, showing that generalization performance with reshuffled holdout splits per iteration often matches that of 5-fold cross-validation [15]. Furthermore, early stopping was identified as a viable approach to mitigate meta-overfitting when optimizing hyperparameters for XGBoost and random forests [14]. Fabris and Freitas [7] identified evidence of meta-overfitting in the AutoML tool Auto-sklearn [9], offering a notable exploration of this phenomenon in the context of the CASH problem. However, the generalizability of their results is constrained by two factors: the reliance on a single run per dataset and the incorporation of post-hoc ensembles (i.e., instead of utilizing the best configuration based on the HPO procedure, several of the best configurations are combined into an ensemble). The use of ensembles is reported to improve generalization over individual pipelines, thus reducing meta-overfitting [9].

Our work extends prior research by analyzing 64 datasets (with 100 repetitions each), spanning classification and regression, and uniquely isolating adaptive overfitting as a key contributor to meta-overfitting in HPO for the CASH problem.

3 Problem Definition

We formalize the CASH problem and describe our method for quantifying meta-overfitting in HPO. The CASH problem involves finding the optimal combination of a learning algorithm $A \in \mathcal{A}$ and its hyperparameters $\lambda \in \Lambda$ to minimize a loss function $\mathcal{L}$ (e.g., error rate, RMSE) on validation data $\mathcal{D}_{val}$ after training on $\mathcal{D}_{train}$. As introduced by Thornton et al. [23], the CASH problem is defined as:

$$A^*, \lambda^* \in \underset{A \in \mathcal{A}, \lambda \in \Lambda}{\arg\min}\, \mathcal{L}(A_\lambda, \mathcal{D}_{train}, \mathcal{D}_{val}) \tag{1}$$

During the HPO process, many hyperparameter configurations are evaluated on $\mathcal{D}_{val}$, possibly with an overly optimistic estimation compared to evaluation on a completely unseen, ideally large, test set $\mathcal{D}_{test}$, which remains inaccessible to the HPO algorithm during optimization. We measure this effect with the *meta-overfitting error (MOE)*, which we formally define as:

$$MOE(A_\lambda, \mathcal{D}_{train}, \mathcal{D}_{val}, \mathcal{D}_{test}) := \mathcal{L}(A_\lambda, \mathcal{D}_{train}, \mathcal{D}_{test}) - \mathcal{L}(A_\lambda, \mathcal{D}_{train}, \mathcal{D}_{val}) \tag{2}$$

In this work, an *iteration* is defined as a single step in the HPO process where one configuration is proposed and evaluated. A *repetition* refers to a complete HPO procedure, consisting of multiple iterations. We assess meta-overfitting in HPO using two metrics: the average meta-overfitting error (MOE_{avg}) and the resulting meta-overfitting error (MOE_{res}). Let $[A_\lambda]_{i,j}$ denote the hyperparameter configuration from iteration i during repetition j. Additionally, let $[A_\lambda]^*_{i,j}$ be the best-found hyperparameter configuration in an HPO process for a given repetition j when completed until iteration i. The resulting meta-overfitting error is defined as:

$$MOE_{res}(\mathcal{D}_{train}, \mathcal{D}_{val}, \mathcal{D}_{test}, i) := \frac{1}{n}\sum_{j=1}^{n} MOE([A_\lambda]^*_{i,j}, \mathcal{D}_{train}, \mathcal{D}_{val}, \mathcal{D}_{test}) \tag{3}$$

In the context of HPO applications, this corresponds to the meta-overfitting error of the configuration that the algorithm would select if stopped at that iteration, averaged over various repetitions.

Furthermore, we define the average meta-overfitting error for iteration i as the mean meta-overfitting error of all configurations suggested in that iteration across n repetitions:

$$MOE_{avg}(\mathcal{D}_{train}, \mathcal{D}_{val}, \mathcal{D}_{test}, i) := \frac{1}{n}\sum_{j=1}^{n} MOE([A_\lambda]_{i,j}, \mathcal{D}_{train}, \mathcal{D}_{val}, \mathcal{D}_{test}) \tag{4}$$

This measure focuses on the meta-overfitting of configurations selected specifically at iteration i, rather than tracking the best configuration found up to that point (as in Eq. 3). If the MOE_{avg} increases across iterations of i, we found empirical evidence for adaptive overfitting, since the effect of selection-based meta-overfitting has been accounted for.

4 Experimental Setup

We aim to experimentally assess the impact of both selection-based overfitting and adaptive overfitting over a wide range of datasets. Additionally, we aim to assess which factors can mitigate this effect.

Datasets: We investigate meta-overfitting using holdout evaluation on 64 datasets: 48 classification datasets from the OpenML-CC18 benchmark [4] and 16 regression datasets from the AutoML benchmark [11]. Datasets with more than 50 features and two regression datasets (due to compatibility issues) were excluded. For experiments on 10-fold cross-validation and varying validation set sizes, we used the same four binary classification datasets, chosen for their large size (necessary for validation size experiments) and to manage computational resource constraints.

Search Space: The search space consists of all algorithms and hyperparameters that can be searched over by the hyperparameter optimization algorithm. The pipeline includes the following components: two categorical value encoders, three missing value imputers, four feature selectors, eight scalers, three dimensionality reduction techniques, 17 classifiers, and 15 regressors. Additionally, most of these components are configurable with hyperparameters.

Implementation: Bayesian optimization was implemented using SMAC3 [12]. For comprehensive details on the search space, datasets, and experimental implementation, we refer to our GitHub repository.[1]

Experiments: The following three experiments were conducted:

1. **Large-scale holdout validation**: We employ a holdout set for both validation, used by the HPO algorithm to evaluate and select configurations, and test, which remains fully inaccessible to the HPO algorithm. For datasets over 3 000 instances, splits were fixed at 1 000 (train), 500 (validation), and the remainder as test data. Smaller datasets used a 40/20/40% split. This division between training, validation and test data is broadly adopted in machine learning [3].
2. **10-fold cross-validation (10CV)**: Due to computational constraints, experiments were restricted to four datasets. We performed cross-validation, generating 10 different validation scores, which were averaged to estimate validation performance. For testing, we used a single large holdout set. Rather than retraining models on the full training set to obtain test performance, we evaluated the predictions from all 10 models on the test set and averaged their individual performances.
3. **Holdout with varying validation sizes**: Validation sizes ranged from 100 to 10 000 instances, with 1 000 train instances fixed for comparability. The experiment was conducted on the same four datasets used in the 10-fold cross-validation experiment.

Each dataset was evaluated using 100 repetitions, with 250 iterations of both random search and Bayesian optimization per repetition, to assess the various

[1] https://github.com/ADA-research/OverfittingCASH.

forms of meta-overfitting across two HPO methods with different search characteristics. This resulted in 6.6 million pipelines trained and tested on validation and test sets. The experiments spanned roughly 13 days, utilizing 20 Intel Xeon 2.20GHz machines with two cores each.

5 Results and Discussion

We present our results visually using the *adult* dataset, which consists of 15 features, 48,842 instances, and a binary target variable. We extend these findings to all 64 datasets analyzed. First, we examine the impact of meta-overfitting (Sect. 5.1). Next, we investigate adaptive overfitting and provide empirical evidence of its occurrence (Sect. 5.2). Finally, we analyze how the validation set size influences meta-overfitting (Sect. 5.3).

5.1 Analysis of Meta-Overfitting in HPO

As discussed previously, meta-overfitting can be categorized into two types: (i) selection-based overfitting and (ii) adaptive overfitting. In Fig. 1, we demonstrate this effect by comparing performance evaluations on both the validation and the test set, where each point represents a hyperparameter configuration suggested by an HPO algorithm. Note that generally configurations are evaluated only on the validation set, and the best-performing configuration is then assessed on the test set. However, for experimental purposes, we evaluate all configurations on both datasets. The validation set consists of 500 instances, while the independent test set contains approximately 47 000 instances. Configurations located to the right of the diagonal in the figure perform better on the validation set than on the test set and thus exhibit meta-overfitting. Because an HPO procedure typically selects the right-most configuration, which is generally (but not necessarily) on the right side of the diagonal, it is likely to produce a meta-overfitted configuration – the effect we refer to as selection-based meta-overfitting. This effect becomes apparent in the case of random search, illustrated in Fig. 1 (left). Additionally, Bayesian optimization tends to concentrate its search around more overfitted regions in later iterations, as it gains more information from previous evaluations on the validation set, as illustrated in Fig. 1 (right). The additional meta-overfitting introduced by this effect is referred to as adaptive overfitting. Both types of meta-overfitting are visually present across the datasets analyzed.

To better understand and quantify meta-overfitting in HPO, we analyze the hyperparameter configurations ultimately chosen during the optimization process. Specifically, at each iteration, we select the configuration that achieves the highest performance according to the validation procedure up to that point. The extent of meta-overfitting exhibited by these chosen configurations is quantified using the resulting meta-overfitting error, as defined in Eq. 3.

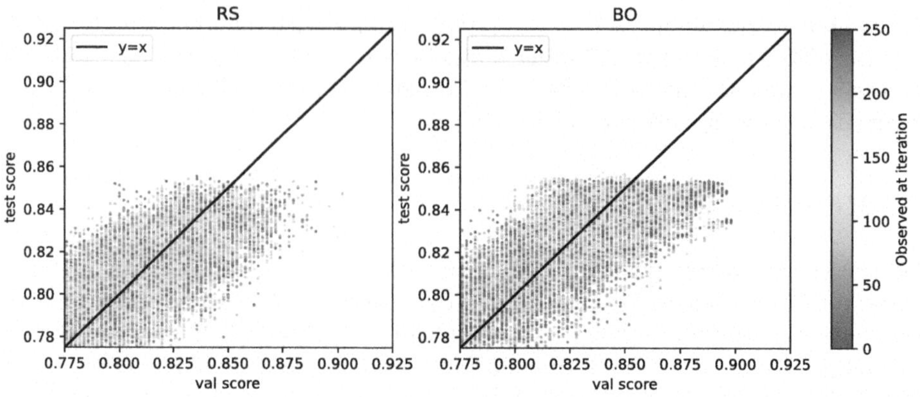

Fig. 1. Test and validation accuracy scores of highest performing configurations produced by random search (left) and Bayesian optimization (right) on the *adult* dataset. The color indicates in what iteration they were suggested.

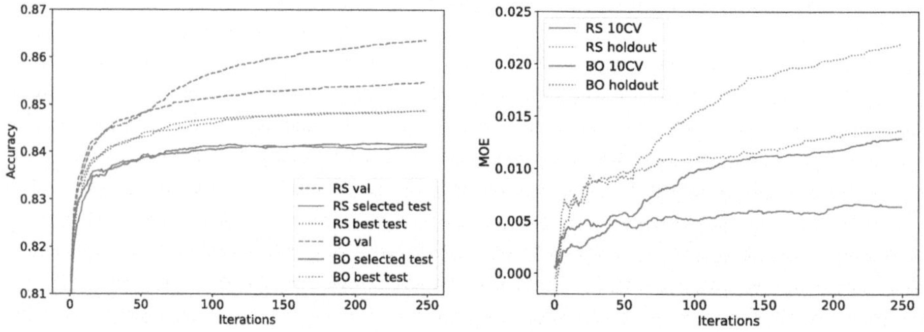

Fig. 2. Left: Average accuracy of the selected configuration from Bayesian optimization and random search on the holdout validation and test sets, along with the test set accuracy of the best configuration on the test set, using the *adult* dataset. **Right:** Resulting MOE for holdout validation and 10CV from Bayesian optimization and random search on the *adult* dataset.

Figure 2 (left) illustrates the average progression of accuracy scores throughout the HPO optimization process. Two key insights can be drawn from this figure. First, the difference between the best and selected configurations shows that there are configurations that outperform the selected ones but are not chosen, highlighting the potential for performance improvement if more reliable validation estimates were employed. Second, after a certain number of iterations, no significantly better configurations are identified for the test set, while the validation performance continues to improve, particularly in the case of Bayesian optimization. This suggests that the selected configurations increasingly overfit the validation set, with no corresponding improvement in generalization performance beyond a certain point in the optimization process. Figure 2 (right) illustrates the meta-overfitting error of the configurations selected during each iteration, averaged across 100 repetitions. These findings highlight the practical implications of meta-overfitting. Specifically, after 250 iterations of Bayesian optimization, the selected configuration is expected to exhibit an average overfitting error of 1.3% when using 10-fold cross-validation as the internal validation procedure. This error increases to 2.2% when holdout validation is used.

Across Datasets: These results are generalizable to other datasets. On the four binary classification datasets we used to investigate 10-fold cross-validation, the final configurations identified by Bayesian optimization exhibited meta-overfitting errors ranging from 1.0% to 3.7% in accuracy, with an average of 1.8%. These findings raise important questions about the use of 10-fold cross-validation as a benchmarking practice, especially when the same fixed cross-validation strategy is applied for both internal validation and final assessment. In such cases, a portion of the observed performance improvement may result from overfitting the validation procedure rather than genuinely enhancing the generalizability of the hyperparameter optimization process.

Furthermore, on 29 binary classification datasets using holdout validation, we observe an average meta-overfitting error of 3.3% in accuracy for the final selected configuration identified by Bayesian optimization, with a median of 2.3%. Similarly, for 19 multiclass classification datasets, the average meta-overfitting error is 2.9% accuracy, with a median of 2.3%. These findings are particularly relevant to common practices in widely used HPO frameworks, such as Auto-sklearn [9], which employs internal holdout validation by default. They highlight the importance of addressing meta-overfitting when designing robust and reliable machine learning pipelines with HPO.

5.2 Adaptive Overfitting Across Datasets

In this section, we present evidence of adaptive overfitting in Bayesian optimization applied to the CASH problem. Adaptive overfitting occurs when hyperparameter configurations become increasingly tailored to the validation procedure. This arises because Bayesian optimization incorporates growing amounts of information from prior evaluations into its surrogate model, leading to configurations that are progressively fine-tuned to the validation set. Therefore, adaptive overfitting can be observed as a relationship between the degree of meta-overfitting in a hyperparameter configuration and the iteration in which that configuration is suggested by Bayesian optimization. This means that hyperparameter configurations proposed in later iterations tend to exhibit greater levels of meta-overfitting, on average. In contrast, for random search, the iterations are independent of one another. As a result, there is no systematic pattern of meta-overfitting across iterations, and we would expect the average meta-overfitting for any given iteration to be approximately zero.

Figure 3 (left) illustrates the relationship between the average meta-overfitting error (defined in Eq. 4) of a hyperparameter configuration and the

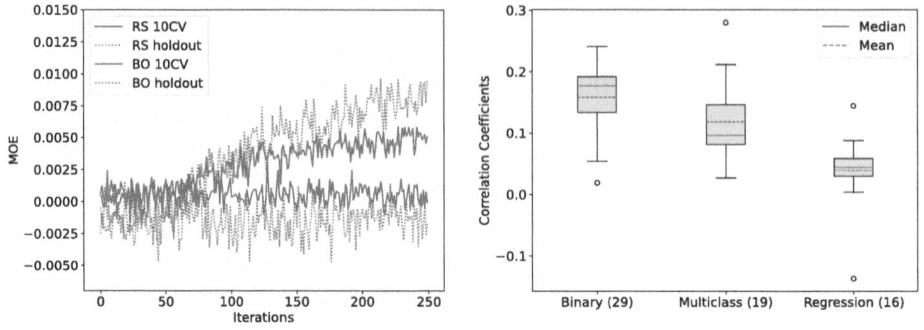

Fig. 3. Left: Average MOE of different HPO algorithms using holdout validation and 10CV on the *adult* dataset, showing a relationship for Bayesian optimization. **Right:** Box-plots of Spearman's coefficients of all 64 datasets considered using holdout validation, calculated using 100 repetitions per dataset.

iteration in which it was suggested on the *adult* dataset. This relationship is evident regardless of whether holdout evaluation or 10-fold cross-validation is employed, providing strong evidence of adaptive overfitting in both validation procedures. This implies that configurations suggested in the early iterations of Bayesian optimization are generally not overfitted, but configurations proposed in the final iterations tend to exhibit an average overfitting of approximately 0.5% accuracy for 10-fold cross-validation and 0.8% accuracy for holdout validation.

Extrapolating these observations to the full range of 64 datasets, we quantify adaptive overfitting using Spearman's rank-order correlation coefficient, which is suited for assessing relationships between variables that are not all normally distributed, such as the iterations uniformly spanning from 1 to 250. A higher coefficient indicates a stronger relationship between iterations and the average meta-overfitting error, signifying a more pronounced effect of adaptive overfitting. To account for testing multiple hypotheses, we apply the highly conservative Bonferroni correction within dataset types (binary, multiclass, and regression). This adjustment raises the threshold for statistical significance, ensuring that observed relationships are unlikely to be false positives. Using this approach, we assess the relationships between the average meta-overfitting error of a hyperparameter configuration and the iteration in which it was suggested. When using the holdout method, we find statistically significant results for 25 out of 29 binary classification datasets, 16 out of 19 multiclass classification datasets, and 3 out of 16 regression datasets. For 10-fold cross-validation, all 4 binary classification datasets show significant adaptive overfitting. Figure 3 (right) shows the distribution of Spearman's correlation coefficients of all datasets in each problem type for the holdout method. We statistically compare the coefficients of binary and multiclass datasets. Shapiro-Wilk tests confirm normality for both samples, and Levene's test verifies homogeneity of variances, allowing the use of an independent samples t-test. Supported by this test, we conclude that multiclass classifi-

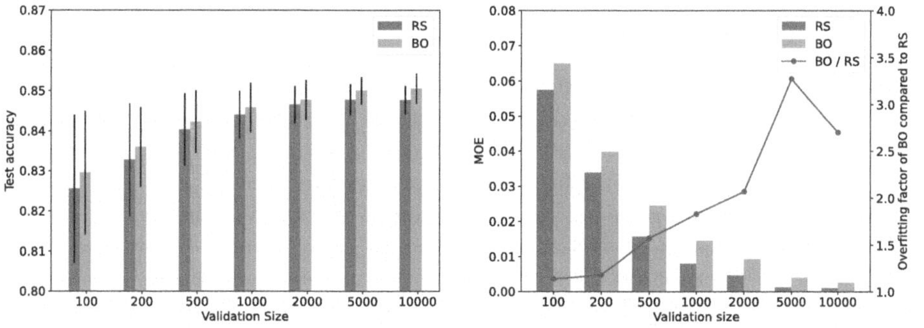

Fig. 4. Left: Effect of the validation set size on the stability and generalization of HPO for the *adult* dataset. **Right:** Meta-overfitting error for random search and Bayesian optimization across validation set sizes for the *adult* dataset. The BO/RS line illustrates the factor by which the selected configurations of Bayesian optimization exhibit greater resulting MOE compared to random search.

cation datasets exhibit significantly less adaptive overfitting compared to binary classification datasets ($t(46) = 2.3$, $p = .03$), confirming prior work [8]. However, further research is needed to substantiate this finding and explore the influence of the number of classes on adaptive overfitting. Additionally, our results indicate that regression datasets are highly resilient to adaptive overfitting.

5.3 Effects of Different Validation Set Sizes

Finally, we study the effect of validation set size on meta-overfitting and the generalizability of HPO algorithms using four binary classification datasets, as motivated in Sect. 4. Figure 4 (left) illustrates how larger validation sets improve stability and generalization performance. Figure 4 (right) highlights differences in meta-overfitting for random search and Bayesian optimization across validation sizes. For small validation sets, the selection-based overfitting in random search is comparable to the meta-overfitting in Bayesian optimization, suggesting a minimal role for adaptive overfitting. As validation sizes increase, the meta-overfitting in random search diminishes, as expected. However, for Bayesian optimization, meta-overfitting maintains a factor for increasing validation sizes. For instance, configurations selected by Bayesian optimization are 1.1 times more overfitted than those of random search at 100-instance validation sets, rising to 3.3 times for 5 000 instances. Similar results are observed across all four datasets. These findings suggest adaptive overfitting is still persistent with large validation sets, warranting caution even when validation sets contain thousands of instances. However, despite the effects of meta-overfitting, it remains clear that Bayesian optimization performs better than random search.

6 Conclusions

This study explores meta-overfitting in HPO for the CASH problem through an empirical analysis of 64 datasets covering classification and regression. We formalized meta-overfitting and decomposed it into two primary sources: (i) selection-based and (ii) adaptive overfitting. We show that both random search and Bayesian optimization exhibit meta-overfitting. Our results empirically confirm the logical fact that in random search, meta-overfitting arises solely from selection-based overfitting, whereas in Bayesian optimization, there is also evidence of adaptive overfitting. Using a single holdout validation set, Bayesian optimization produced configurations with an average meta-overfitting error of 3.3% accuracy on binary classification datasets and 2.9% accuracy on multiclass datasets. While 10-fold cross-validation reduced this error, configurations still showed an average meta-overfitting error of 1.8% accuracy on four binary classification datasets. Furthermore, we identified statistically significant adaptive overfitting in Bayesian optimization for HPO in the CASH problem, an effect not previously measured in this context to the best of our knowledge. Significant adaptive overfitting was observed in 25 of the 29 binary classification datasets and 16 of the 19 multiclass classification datasets, with multiclass datasets showing significantly less adaptive overfitting, consistent with prior theory [8]. Regression datasets, however, appeared notably resistant to this effect. Finally, we investigated how validation set size affects meta-overfitting. In random search, the meta-overfitting error of selected hyperparameter configurations diminished with larger validation sets. However, in Bayesian optimization, the meta-overfitting error remained more robust, suggesting that adaptive overfitting can be a concern even with validation sets containing thousands of instances.

Recommendations: We advise machine learning practitioners to be cautious about potentially overfitted results being produced by HPO algorithms, particularly when validation results are directly used to guide the suggestion of new hyperparameter configurations. Similarly, we recommend avoiding fixed 10-fold cross-validation when benchmarking HPO algorithms and instead resampling the training and validation splits, either at each iteration, following [15], or at least during the evaluation of the selected hyperparameter configuration.

Future Work: Future research could further explore meta-overfitting, especially in relation to HPO techniques like ensembling and meta-learning, commonly used in state-of-the-art HPO frameworks. Investigating adaptive overfitting in HPO and applying strategies from adaptive data analysis to mitigate its effects also presents a compelling avenue for future studies.

Disclosure of Interests. The authors have no competing interests to declare that are relevant to the content of this article.

References

1. Arora, S., Zhang, Y.: Rip van Winkle's razor: a simple estimate of overfit to test data. Computing Research Repository abs/2102.13189 (2021)
2. Baratchi, M., et al.: Automated machine learning: past, present and future. Artif. Intell. Rev. **57**(5), 122 (2024)
3. Bischl, B., et al.: Hyperparameter optimization: foundations, algorithms, best practices, and open challenges. Wiley Interdisc. Rev. Data Mining Knowl. Discov. **13**(2) (2023)
4. Bischl, B., et al.: OpenML benchmarking suites. In: Proceedings of the Neural Information Processing Systems Track on Datasets and Benchmarks (2021)
5. Cawley, G.C., Talbot, N.: On over-fitting in model selection and subsequent selection bias in performance evaluation. J. Mach. Learn. Res. **11**, 2079–2107 (2010)
6. Dwork, C., Feldman, V., Hardt, M., Pitassi, T., Reingold, O., Roth, A.L.: Preserving statistical validity in adaptive data analysis. In: Proceedings of the Forty-Seventh Annual ACM on Symposium on Theory of Computing, STOC 2015, pp. 117–126. ACM (2015)
7. Fabris, F., Freitas, A.A.: Analysing the overfit of the auto-sklearn automated machine learning tool. In: Nicosia, G., Pardalos, P., Umeton, R., Giuffrida, G., Sciacca, V. (eds.) LOD 2019. LNCS, vol. 11943, pp. 508–520. Springer, Cham (2019). https://doi.org/10.1007/978-3-030-37599-7_42
8. Feldman, V., Frostig, R., Hardt, M.: The advantages of multiple classes for reducing overfitting from test set reuse. In: Proceedings of the 36th International Conference on Machine Learning, ICML 2019. Proceedings of Machine Learning Research, vol. 97, pp. 1892–1900. PMLR (2019)
9. Feurer, M., Klein, A., Eggensperger, K., Springenberg, J.T., Blum, M., Hutter, F.: Efficient and robust automated machine learning. In: Advances in Neural Information Processing Systems 28, NeurIPS 2015, pp. 2962–2970 (2015)
10. Freedman, D.A.: A note on screening regression equations. Am. Stat. **37**(2), 152–155 (1983)
11. Gijsbers, P., et al.: AMLB: an AutoML benchmark. J. Mach. Learn. Res. **25**, 101:1–101:65 (2024)
12. Lindauer, M., et al.: SMAC3: a versatile Bayesian optimization package for hyperparameter optimization. J. Mach. Learn. Res. **23**, 54:1–54:9 (2022)
13. Lévesque, J.C.: Bayesian hyperparameter optimization: Overfitting, ensembles and conditional spaces. Ph.D. thesis, Université Laval (2018). http://hdl.handle.net/20.500.11794/28364
14. Makarova, A., et al.: Overfitting in Bayesian optimization: an empirical study and early-stopping solution. Computing Research Repository abs/2104.08166 (2021)
15. Nagler, T., Schneider, L., Bischl, B., Feurer, M.: Reshuffling resampling splits can improve generalization of hyperparameter optimization. In: Advances in Neural Information Processing Systems 38, NeurIPS 2024 (2024)
16. Ng, A.Y.: Preventing "overfitting" of cross-validation data. In: Proceedings of the Fourteenth International Conference on Machine Learning, ICML 1997, pp. 245–253. Morgan Kaufmann (1997)
17. Quinlan, J.R., Cameron-Jones, R.M.: Oversearching and layered search in empirical learning. In: Proceedings of the Fourteenth International Joint Conference on Artificial Intelligence, IJCAI 1995, pp. 1019–1024. Morgan Kaufmann (1995)
18. Rao, R.B., Fung, G.: On the dangers of cross-validation. An experimental evaluation. In: Proceedings of the SIAM International Conference on Data Mining, SDM 2008, pp. 588–596. SIAM (2008)

19. Recht, B., Roelofs, R., Schmidt, L., Shankar, V.: Do CIFAR-10 classifiers generalize to CIFAR-10? Computing Research Repository abs/1806.00451 (2018)
20. Recht, B., Roelofs, R., Schmidt, L., Shankar, V.: Do ImageNet classifiers generalize to ImageNet? In: Proceedings of the 36th International Conference on Machine Learning, ICML 2019. Proceedings of Machine Learning Research, vol. 97, pp. 5389–5400. PMLR (2019)
21. Reunanen, J.: Overfitting in making comparisons between variable selection methods. J. Mach. Learn. Res. **3**, 1371–1382 (2003)
22. Roelofs, R., et al.: A meta-analysis of overfitting in machine learning. In: Advances in Neural Information Processing Systems 32, NeurIPS 2019, pp. 9175–9185 (2019)
23. Thornton, C., Hutter, F., Hoos, H.H., Leyton-Brown, K.: Auto-WEKA: combined selection and hyperparameter optimization of classification algorithms. In: The 19th ACM SIGKDD International Conference on Knowledge Discovery and Data Mining, KDD 2013, pp. 847–855. ACM (2013)
24. Yadav, C., Bottou, L.: Cold case: the lost MNIST digits. In: Advances in Neural Information Processing Systems 32, NeurIPS 2019, pp. 13443–13452 (2019)

Local Subgroup Discovery on Attributed Network Graphs

Carl Vico Heinrich, Tommie Lombarts, Jules Mallens, Luc Tortike, David Wolf, and Wouter Duivesteijn(✉)

Technische Universiteit Eindhoven, Eindhoven, the Netherlands
{c.v.heinrich,t.lombarts,j.m.mallens,l.m.tortike,d.wolf}@student.tue.nl, w.duivesteijn@tue.nl

Abstract. We find locally exceptional subgroups of nodes in attributed graphs, combining both node attributes and structural information to assess subgroup exceptionality. Subgroups are locally exceptional if their behavior deviates from the behavior of a well-chosen local peer group, as opposed to the more common generic subgroup discovery approach where behavior is compared to the global behavior across the full dataset. This notion of Local Subgroup Discovery had been introduced for traditional flat-table data; to the best of our knowledge, we are the first to incorporate this notion explicitly in graph data. Our approach combines shortest-path distance with Gower's Distance, integrating both network structure and node attributes to rank nodes in relation to a prototype node. Combining this notion of exceptionality with existing LSD techniques, we discover local subgroups in three attributed graph datasets.

Keywords: Subgroup Discovery · Local Subgroup Discovery · Network Analysis · Attributed Graphs · Graph Distances · Pattern Mining

1 Introduction

The increasing availability of large-scale network data [35,41], combined with advances in analytical techniques [12], has enabled researchers to explore interesting patterns within various types of networks [11], such as biological [39], communication [2], or social networks [6]. Networks provide valuable insights not only into global graph structures but also into local phenomena and relations within the graph. Community structures, or clusters, are especially valuable in these analyses, as they reflect meaningful groupings [19].

One powerful method for uncovering these insights is Subgroup Discovery (SD) [29], which identifies subsets of nodes that deviate from expected norms. In traditional SD, methods typically compare characteristics of the subgroups to global characteristics [7,32]. This approach works well in environments where the global norm or tendency is clear. In many real-world networks, nodes represent entities with a variety of characteristics and behaviors [11]. In such heterogeneous

G. Krempl et al. (Eds.): IDA 2025, LNCS 15669, pp. 195–208, 2025.
https://doi.org/10.1007/978-3-031-91398-3_15

settings, comparing the subgroups to the entire network can obscure relevant patterns [13,19].

We propose a framework for applying Local Subgroup Discovery (LSD) [30] to general network structures. In LSD, the interestingness of a subgroup is not compared to the global population but rather to a reference group, in which people share similar characteristics. Thus, rather than finding exceptional nodes or behavior within a graph concerning the global setting, we define subgroups based on their relation to local reference groups. This localized approach allows the discovery of subgroups whose traits differ from expected behavior compared to their peers, capturing both structural relationships and node attributes.

We introduce a novel application of LSD to the domain of network analysis, combining both node attributes and structural information to identify exceptional local subgroups. We contribute a framework adaptable to diverse quality and distance measures, solving ties according to the node attributes.

2 Related Work

2.1 Subgroup Discovery, Local Patterns, Subgraphs

Subgroup Discovery (SD) seeks subgroups with respect to a variable of interest [7]. A quality measure evaluates the interestingness of a subgroup compared to the overall dataset [24]. In Local Subgroup Discovery (LSD) [30], the quality of a subgroup is instead compared to a subset of the data: the reference group. This method makes use of a ranking, often generated using a distance measure, to identify relevant reference groups. Different data structures will require different methods for creating a ranking and quality measure. Social network data is often structured as an attributed graph. For attributed network data, traditional SD techniques exist in literature [8], but LSD techniques do not.

Exception Rules [47,48] seek local patterns that deviate w.r.t. a local context, specifically: an association rule with a single item on the right-hand side, forming an exception to a more general common sense rule. Similarly, Anomalous Association Rules [10] seek association rules hidden by a dominant rule. None of these papers concern themselves with graph data. Conversely, the work on Exceptional Contextual Subgraph Mining [28] does discover interesting subgraphs defined in terms of graph attributes (the "context"), but subgraph behavior is compared to behavior on the entire graph as opposed to a local reference group.

2.2 Network Graphs

Network graphs are widely used in data mining to represent complex systems, such as social, biological, and communication networks, where entities are modeled as nodes and relationships as edges [40]. One key feature of network graphs is their ability to capture both local and global structural properties of the data, allowing researchers to study interactions at various scales [16]. The study of attributed networks is a specific extension of traditional graph theory, where nodes and/or edges have associated attributes (e.g., demographic information in social networks or molecular characteristics in biological networks) [52].

In recent years, significant attention has been given to mining interesting patterns from attributed networks, particularly in the context of subgroups with shared structural and attribute-based characteristics [8]. Such methods often focus on attributes of nodes and edges alone, such as an attributed 'to' and 'from' approach to edges, using these attributed edges as datapoints [26]. As data becomes more complex, integrating network structure into the analysis becomes increasingly important for finding novel patterns [1].

2.3 EgoNets

Ego Networks (EgoNets) represent a specific perspective in network analysis, focusing on a central node (the "ego") and its immediate neighbors (the "alters"). The structure of EgoNets is especially useful for studying local properties and immediate environments of nodes, allowing exploration of network dynamics at the micro-level [17]. EgoNets have been widely used in Social Network Analysis (SNA) [4,36,37], where relationships between individuals or entities are key areas of analysis. In social networks, EgoNets enable the examination of personal networks, including measures such as size, density, and centrality [20].

EgoNets provide insights into local connectivity and the specific role of the ego within its surrounding network, often enriched with attribute data to assess relationships at both the structural and attribute levels [37]. The relevance of EgoNets beyond SNA, for instance in LSD which considers both node attributes and graph structure, remains underexplored [8]. While EgoNets focus on local structures centered on the ego and its immediate connections [5], they do not encompass the broader pattern discovery that LSD entails, aimed at identifying recurring substructures across the network [49].

3 Methods

A *network* is represented as a graph $G = (V, E)$, where V is the set of nodes (vertices) and $E \subseteq V \times V$ is the set of edges between nodes. Each node $v \in V$ may have associated attributes $\boldsymbol{a}(v) = \{a_1(v), \ldots, a_h(v)\}$ and a binary target variable $t(v) \in \{0, 1\}$. A *subgroup* is defined as any subset of nodes $S \subset V$. Our methods require a distance measure $d : V \times V \to \mathbb{R}^+$ which quantifies the distance between any two nodes $u, v \in V$, to be instantiated in Sect. 3.1.

A *prototype* is any node $x \in V$ that serves as the center for defining a subgroup S. Given prototype x we define two parameters: $\sigma, \rho \in \mathbb{N}$, where $\sigma < \rho$. The *local subgroup* S_σ consists of the σ nearest neighbors to x, according to the distance measure d. The *reference group* R_ρ contains the ρ nearest neighbors to x. Therefore, by construction, $S_\sigma \subset R_\rho$. The goal is to find subgroups R_ρ and S_σ such that the distribution of the target variable t within these subgroups reveals distinct local patterns. Specifically, we seek to identify R_ρ and S_σ where:

1. the target distribution within R_ρ differs from the entire graph G;

2. the target distribution within S_σ differs from the reference group R_ρ.

To gauge the interestingness of local subgroups S_σ, a *quality measure* $q \in \mathbb{R}$ assigns a numerical value to any subset of nodes $S \subset V$. Given a *prototype* x and a distance measure d, we obtain a ranking of all nodes $v_i \in V$ based on their distance to x. This ranks the target values $t(v_i)$ for the nodes:

$$\boldsymbol{r}(x) = \{t(v_1),\, t(v_2),\, \ldots,\, t(v_\sigma),\, \ldots,\, t(v_\rho),\, \ldots\}$$

where the nodes v_i are ordered such that:

$$d(x, v_1) \leq d(x, v_2) \leq \ldots \leq d(x, v_\sigma) \leq \ldots \leq d(x, v_\rho) \leq \ldots$$

where $S_\sigma = \{x, v_1, v_2, \ldots, v_\sigma\}$ and $R_\rho = \{x, v_1, v_2, \ldots, v_\sigma, \ldots, v_\rho\}$.

The objective, in the end, is to identify pairs (R, S), such that q is maximized, indicating that subgroup S is exceptional compared to its reference group R.

3.1 Distance Measure Shortest Path Distance

We propose to use the shortest-path distance, denoted by d_{uv}, which quantifies the distance between any two nodes $u, v \in V$. The shortest-path distance between nodes u, v is defined as the minimum number of edges that need to be traversed from node u to v:

$$d_{uv} = \min_{P_{uv}} \left(\sum_{(u,v) \in P_{uv}} w(u, v) \right)$$

where P_{uv} represents the set of all possible paths between u and v, and $w(u, v)$ represents the weight of the edge between nodes u and v.

Several algorithms exist for finding the shortest path in a social network graph, such as the Bellman-Ford algorithm [9,18], Breadth-First Search [38], Dijkstra's algorithm [15], and the A* algorithm [22]. We choose to compute the shortest path distance through Dijkstra's algorithm.

3.2 Distance Tie-Breaker - Gower's Distance

When identifying reference groups, the shortest path distance frequently assigns ties to multiple nodes. We use node attributes to break these ties, applying Gower's Distance [21]. Since each node $v \in V$ not only has edges but also an associated set of attributes $\boldsymbol{a}(v)$, we can evaluate the similarity between the prototype x and any node v. Gower's Distance is particularly useful here, as it accommodates various types of attributes, including binary, categorical, ordinal, and continuous data. For any two nodes $u, v \in V$, Gower's Distance $d(u, v)$ is computed by considering the differences across all node attributes:

$$d(u, v) = \frac{1}{h} \sum_{k=1}^{h} \delta_k(u, v)$$

where h is the attribute dimensionality, and $\delta_k(u, v)$ is the partial distance between nodes u and v with respect to attribute k.

For *numerical* attributes, the partial distance $\delta_k(u, v)$ is computed as the scaled absolute difference:

$$\delta_k(u, v) = \frac{|a_k(u) - a_k(v)|}{\max_{x \in V} \big(a_k(x)\big) - \min_{x \in V} \big(a_k(x)\big)}$$

with the value $a_k(x)$ being replaced with the rank of the attribute for node x in case of *ordinal* attributes, as proposed by [27, pp. 35–36].

For *categorical* attributes, the distance is binary:

$$\delta_k(u, v) = \mathbb{1}_{\{a_k(u) \neq a_k(v)\}}$$

Gower's Distance allows us to integrate both node connectivity and node attributes into the subgroup detection process. Gower's Distance is only applied when nodes are equidistant according to the primary distance measure.

Whilst Gower's Distance is not guaranteed to break all ties, such as with 2 nodes with identical attributes at an identical distance from the prototype node, it does become significantly better at breaking ties for data with more features, specifically features that have a large domain size. This is because an increase in dimensions and feature domain size adds more unique sets of attributes, and thus more unique possibilities arise. In the case that Gower's Distance still results in a tie, the nodes will be added to the group in random order (based on their node names in the original graph).

3.3 Quality Measure

To quantify the exceptionality of the local subgroup S relative to its reference group R, we employ Weighted Kullback-Leibler divergence (WKL) [34] as our quality measure $q(S, R)$. WKL is an adaptation of Kullback-Leibler (KL) divergence [31], a fundamental concept in information theory [45]. It measures how one probability distribution diverges from a second, reference distribution, with the key enhancement of weighting by subgroup size. KL divergence is widely used in many fields, including Subgroup Discovery [3,33,51], as a tool for comparing distributions. In this context, KL divergence evaluates the difference between the distributions of the target variable $t(x)$ within the subgroup S and the distribution of the same target variable in the reference group R.

This measure captures how distinct the subgroup S is in comparison to its reference group R, a critical aspect of LSD. Specifically, it quantifies how much the presence of the target variable differs between the two subgroups:

$$q(S, R) = \frac{|S|}{|R|} \cdot \sum_{y \in \{0,1\}} P_S(y) \log\left(\frac{P_S(y)}{P_R(y)}\right)$$

where $P_A(y) = \mathbb{P}\big(t(x) = y \mid x \in A\big)$.

Algorithm 1. Generate Distance Ranking

```
procedure RANKING(x, S)
    D ← Dijkstra(x, S)
    R ← SortOn(R, D)
    for d in unique(R) do
        r, l ← idxmin(R, d), idxmax(R, d)
        gowerlist ← Gower(x, n_i) ∀ n ∈ R[r : l]
        SortOn(R[r : l], gowerlist)
    return R
```

Algorithm 2. Generate ρ and σ

```
procedure DISCOVERY(R ← ranking)
    ρ, σ, b ← 0, 0, 0
    for i in len(R) do
        q ← Quality(R[0 : i], S)
        if q ≥ b then
            b ← q
            ρ ← i
    b ← 0
    for i in [0 : ρ] do
        q ← Quality(R[0 : i], R[0 : ρ])
        if q ≥ b then
            b ← q
            σ ← i
    return ρ, σ
```

3.4 Local Subgroup Discovery

To discover our distance-based subgroups we build from a *prototype*, which can be any point x within our nodes V. The subgroup S_σ is then the σ closest neighbors to our prototype x and the reference group R_ρ the ρ closest neighbors. Firstly, we select a prototype x randomly from our total population V and store the attributes $[a_1(x), \ldots, a_h(x)]$. The shortest path and Gower's distance between nodes are used to create a ranking, such that the ranking can be used to easily compute quality features for different size references and local subgroups. To generate this ranking, data points are all sorted based on their distance to the prototype x using Dijkstra's algorithm [15], breaking ties with Gower's Distance.

Algorithm 1 describes how such a ranking is created using *Dijkstra*(x, S) to generate a list of size $|S|$ with shortest path distances starting from x to all nodes in S. Here, *SortOn*(A, B) sorts a list A based on the values in list B, and *idxmin*(L, x) and *idxmax*(L, x) find the lowest and highest index in list L where value x appears. We then traverse the ranked target variables and for every possible reference group size $i \in [0 : n]$ we find the reference group size ρ such that $q(R_\rho, G)$ is maximized. Then, we repeat the process of finding a local subgroup within said reference group for all subgroup sizes of $i \in [0 : \rho]$ to find the subgroup size σ such that *Quality*(S_σ, R_ρ) is maximized (cf. Algorithm 2).

This is done by first iteratively generating a quality measure for every possible size of the reference group and storing the best size, and then generating a quality measure for every possible size of the local subgroup and generating their quality measure with regards to the reference group, storing the most effective one.

3.5 Top-*k* Postprocessing

Selecting two prototypes u, v such that the d_{uv} is very low causes nearly identical rankings due to the properties of the shortest path distance. Hence, subgroups of neighboring nodes overlap substantially. We postprocess the set of discovered subgroups, based on the Dice-Sørensen coefficient [14,46] measuring the overlap between subgroups S_1 and S_2:

$$O(S_1, S_2) = \frac{2}{|S_1| + |S_2|} \sum_{n \in S_1} \mathbb{1}_{\{n \in S_2\}}$$

We set a threshold θ, as an upper bound of the maximum allowed overlap between subgroups. We sort all candidate subgroups in order of decreasing quality measure, considering them from the top down. While considering candidate subgroup S_i, we check whether $\exists_{j \in \{1,\ldots,i-1\}}$ s.t. $O(S_j, S_i) > \theta$; if so, subgroup S_i is discarded. We continue this process until k subgroups are retained, or until we run out of candidates.

4 Experimental Setup

We demonstrate our version of LSD on attributed graph data[1] by deploying it on three datasets: `OGBG-MolPCBA` [42], `Twitch PT` [44], and `WebKB Cornell` [43]. Each dataset offers other structural challenges and attributes, making them suitable for testing our method on various domains.

The datasets must meet several conditions to be considered. First, they need to be graph-based, with clearly defined nodes and edges. Second, each dataset must contain meaningful node attributes, ensuring the identification and interpretation of subgroups within the graph. Third, the datasets must be small enough to allow for computationally intensive calculations, as larger datasets can impose significant computational challenges. Thus, we aim to balance rich feature sets and complex structures with a manageable size. For the primary dataset, interpretability is essential. In the context of LSD, non-anonymized features are preferable to facilitate clear subgroup interpretation. However, finding datasets that do not contain sensitive information, and thus are not anonymized, can be challenging. We select the `OGBG-MolPCBA` dataset [42] as our primary dataset due to its non-anonymized features, ensuring interpretability. We augment these experiments with two further datasets where this interpretability is obscured by anonymization, `Twitch PT` [44] and `WebKB Cornell` [43], to demonstrate that our method generalizes beyond a single dataset.

[1] Source code at https://github.com/TUeEMM/LSD-ATNG and [23].

Table 1. Features and Possible Values in `OGBG-MolPCBA` [50]

Feature	Possible Values
Atomic Number	1 to 118, misc
Chirality	CHI_UNSPECIFIED, CHI_TETRAHEDRAL_CW, CHI_TETRAHEDRAL_CCW, CHI_OTHER, misc
Degree	0, 1, 2, 3, 4, 5, 6, 7, 8, 9, 10, misc
Formal Charge	−5, −4, −3, −2, −1, 0, 1, 2, 3, 4, 5, misc
No. of Hydrogen	0, 1, 2, 3, 4, 5, 6, 7, 8, misc
Radical Electrons	0, 1, 2, 3, 4, misc
Hybridization	SP, SP2, SP3, SP3D, SP3D2, misc
Aromaticity	False, True
In a Ring	False, True

4.1 OGBG-MolPCBA (Primary Dataset)

The `OGBG-MolPCBA` dataset from the Open Graph Benchmark [42], introduced as part of a benchmark study on molecular machine learning [50], is selected as the primary dataset for this study. Each graph in this dataset represents a molecule, with nodes corresponding to atoms and edges corresponding to chemical bonds. The node features are 9-dimensional [42], describing various atomic properties such as atomic number, chirality, and hybridization [25].

The detailed atomic features of the `OGBG-MolPCBA` dataset enable the discovery of subgroups of molecules that exhibit unique patterns in their chemical structure, potentially leading to insights that could guide future molecular studies. This is critical for the field of drug discovery and molecular biology, making the dataset relevant for scientific applications. The clear interpretability of the node features was a significant factor in selecting this dataset as our primary focus, as it allows for a thorough understanding of the subgroups identified by the LSD algorithm. Another key reason for choosing this dataset is its size. With 41 127 molecular graphs, each containing an average of 25.5 nodes and 27.5 edges [42], the dataset is large enough to test the scalability of the LSD algorithm while remaining computationally manageable.

Table 1 describes the various atomic properties that are present in the dataset. The *atomic number* represents the number of protons in the atom (with a *misc* category for undefined atoms). This feature acts as the binary target variable in our model, which evaluates to true if the atomic number is 6 or greater. *Chirality* refers to the geometric arrangement of atoms. The *degree* describes the number of bonds an atom forms and the *formal charge* indicates the hypothetical charge of an atom. The *number of hydrogen* atoms attached to an atom varies from 0 to 8, and the number of *radical electrons* (unpaired electrons) ranges from 0 to 4. *Hybridization* describes how atomic orbitals combine to form bonds with other atoms. *Aromaticity* tells whether an atom is part of an aromatic ring, and *in a ring* indicates if the atom is part of any ring system.

4.2 Twitch PT

The Twitch PT dataset, introduced as part of a study on multi-scale attributed node embedding [44], is a social network graph representing Twitch streamers based in Portugal. The 1 912 nodes represent individual streamers, and the 64 510 edges represent their followership relationships. The target represents explicit language usage. The 128-dimensional node attributes represent anonymized embeddings based on the games played by the streamers.

This dataset is chosen for its dense social network structure, allowing us to test the LSD algorithm in a context where connections between individuals form tight followership clusters. While the features are anonymized, the network itself is rich in structural information, enabling the discovery of subgroups of streamers who exhibit similar followership behaviors or content preferences. Thus, the Twitch PT experiment complements experiments on the molecular data in OGBG-MolPCBA by showing how LSD can identify communities of streamers based on behavioral patterns rather than explicit node attributes.

4.3 WebKB Cornell

The WebKB Cornell dataset, introduced in a study on geometric graph convolutional networks [43], is another graph-based dataset with anonymized features. It represents the network of web pages within Cornell University and consists of nodes representing individual web pages and edges representing hyperlinks between them. Each node has 1 703 features, which are bag-of-words representations of the web page's content.

The high-dimensional bag-of-words features provide a detailed description of each web page's content, making this dataset ideal for testing the LSD algorithm's ability to handle textual and content-based node attributes. The WebKB Cornell dataset serves as a test case for how well LSD performs on a relatively small network with large feature vectors, demonstrating the algorithm's versatility in handling different types of graphs.

5 Experimental Results

5.1 OGBG-MolPCBA

Figure 1 displays the top-two non-redundant local subgroups (blue nodes), along with their respective prototypes (black node) and reference groups (purple+blue nodes), found in three molecules (sampled to vary in size). A node $v \in V$ is depicted as a circle if $t(v) = 1$, and as a square if $t(v) = 0$. We mostly find subgroups in less dense parts of the network.

To further investigate the resulting subgroups, we run our discovery algorithm on the largest protein in the dataset. Table 2 lists the top-10 subgroups discovered, with values for ρ, σ, q, and the accuracy of the binary target variable within R_ρ and S_σ. The subgroups display relatively high q values, also reflected by target variable accuracy differences. Values are near identical for

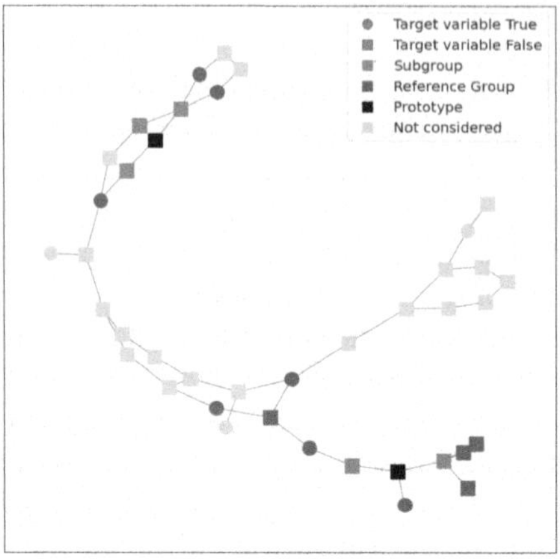

(a) Protein #200289

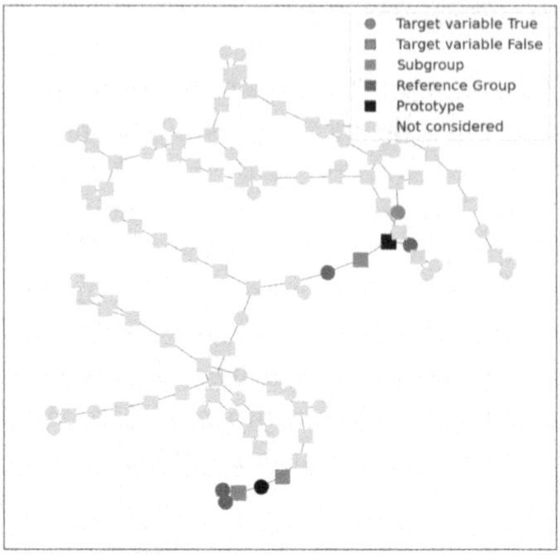

(b) Protein #135257

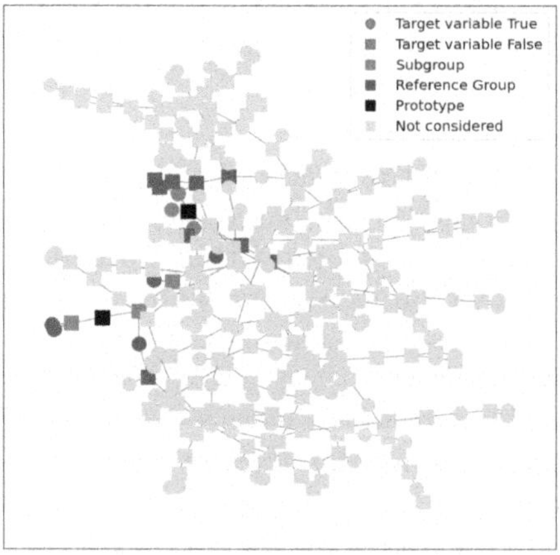

(c) Protein #182987

Fig. 1. Molecule graphs: each figure displays one protein, with the top-two non-redundant local subgroups and their respective reference groups.

the subgroups ranked 2 – 5, which stands to reason: molecular structures can recur within a protein. If such a structure is an interesting subgroup, its 'copies' should also be found. Since the copies concern graph isomorphisms across possibly distinct sets of nodes, these copies will slip through our redundancy filtering (cf. Sect. 3.5) which merely identifies identical node sets.

5.2 Twitch PT

The top-five subgroups discovered on the `Twitch PT` dataset (cf. Table 3) display more variation in size. We find mostly local subgroups with a very low target variable distribution compared to that of their reference group. This shows that the

Table 2. Protein 349519 subgroups

Prot.	ρ	σ	q	**Acc.** R_ρ	**Acc.** S_σ
46	9	3	0.321	0.56	0.00
28	14	5	0.231	0.21	0.60
253	13	5	0.229	0.23	0.60
35	13	5	0.229	0.23	0.60
147	13	5	0.229	0.23	0.60
11	11	7	0.207	0.55	0.29
167	5	3	0.207	0.60	0.33
268	5	3	0.207	0.60	0.33
45	9	4	0.204	0.56	0.25
272	16	5	0.196	0.25	0.60

Table 3. `Twitch PT` subgroups

Prot.	ρ	σ	q	**Acc.** R_ρ	**Acc.** S_σ
385	9	4	0.278	0.67	0.25
1378	26	13	0.218	0.62	0.31
97	12	4	0.192	0.83	0.50
403	103	14	0.173	0.61	0.14
1032	220	58	0.164	0.49	0.17

Table 4. `WebKB Cornell` subgroups

Prot.	ρ	σ	q	**Acc.** R_ρ	**Acc.** S_σ
78	21	6	0.255	0.19	0.67
135	36	10	0.252	0.22	0.70
30	39	6	0.246	0.21	0.83
60	22	4	0.242	0.18	0.75
16	33	3	0.228	0.24	1.00

local subgroups are mostly groups of Twitch users that use no explicit language, within a network/reference group of streamers that do use explicit language.

5.3 WebKB Cornell

On the `WebKB Cornell` dataset, we find more subgroups with a high q score (cf. Table 4). Contrary to the `Twitch PT` dataset, here most subgroups have a significantly higher target variable accuracy than their respective reference groups, indicating a subgroup of web pages classified as project/staff within a reference group of papers that are mostly student/course/faculty related.

5.4 Ablation Study: Full Graph as Reference Group

The motivation of LSD is that subgroups can show interesting exceptional behavior when compared to a local reference group. Here, we perform an ablation study, observing whether the preceding experimental results change when we take $R_\rho = V$: if not, LSD provides no benefits over standard SD.

Setting $R_\rho = V$ consistently produces subgroups markedly different from and less effective than those discovered by LSD. On `OGBG-MolPCBA`, the top-10 ablation subgroups are fully disjoint (in covered nodes) from the top-10 LSD subgroups; node overlaps are 0.62% on `Twitch PT` and 8.99% on `WebKB Cornell`. Measuring discriminative power through WRAcc [7, Equation (1)], the LSD subgroups outperform the ablation subgroups (`OGBG-MolPCBA`: 0.215 vs. 0.0533; `Twitch PT`: 0.2349 vs. 0.0798; `WebKB Cornell`: 0.2211 vs. 0.1103). LSD subgroups cover substantially higher proportions of nodes v with $t(v) = 1$ than ablation subgroups on `OGBG-MolPCBA` (31.56% vs. 0.00%) and `WebKB Cornell` (81% vs. 48.6%), but lower on `Twitch PT` (28.3% vs. 63.84%). We conclude that LSD provides substantial benefits on attributed network graphs over SD with

$R_\rho = V$: discovered subgroups are different, more discriminative, and ultimately more valuable.

6 Conclusions

We propose a method to apply Local Subgroup Discovery (LSD) on attributed network graphs. Our method employs a network-based distance between nodes to find interesting subgroups, which have a target variable distribution that differs from their local reference group. We expand on previously established LSD methods by using network-based distances and attribute-based similarity for the creation of a subgroup ranking. The algorithm finds reference groups and subgroups with high quality measures by iteratively exploring the dataset. A top-k post-processing procedure is used to prune highly similar subgroups. We analyze the results of running the LSD algorithm on three datasets, showing that the algorithm identifies pairs of groups (R_ρ, S_σ) such that S_σ is exceptional to R_ρ with regards to our proposed quality feature q.

The introduction of a new distance-based ranking method enables the application of existing Subgroup Discovery techniques to network-structured data, paving the way for future research into using Subgroup Discovery or Exceptional Model Mining methods on graph structures. Future work could involve adapting this approach to handle more complex target data, including numerical or multivariate targets, enabling richer subgroup discoveries.

Disclosure of Interests. The authors have no competing interests to declare that are relevant to the content of this article.

References

1. Aggarwal, C. (ed.): Social Network Data Analytics. Springer, New York (2011)
2. Ahmad, W., Hasan, O., Pervez, U., Qadir, J.: Reliability modeling and analysis of communication networks. J. Netw. Comput. Appl. **78**, 191–215 (2016)
3. Arab, A., Arora, D., Lu, J., Ester, M.: Subgroup discovery in unstructured data. CoRR abs/2207.07781 (2022)
4. Arnaboldi, V., Conti, M., Passarella, A., Pezzoni, F.: Analysis of ego network structure in online social networks. In: Proceedings of SocialCom-PASSAT, pp. 31–40 (2012)
5. Atzmueller, M.: Mining social media: key players, sentiments, and communities. WIREs Data Min. Knowl. Discovery **2**(5), 411–419 (2012)
6. Atzmueller, M.: Data mining on social interaction networks. J. Data Mining Digit. Humanit. **1** (2014)
7. Atzmueller, M.: Subgroup discovery. WIREs Data Mining Knowl. Discov. **5**(1), 35–49 (2015)
8. Atzmueller, M.: Compositional subgroup discovery on attributed social interaction networks. In: Proceedings of DS, pp. 259–275 (2018)
9. Bellman, R.: On a routing problem. Q. Appl. Math. **16**(1), 87–90 (1958)

10. Berzal, F., Cubero, J.C., Marín, N., Gámez, M.: Anomalous association rules. In: Proceedings of IEEE ICDM Workshop on Alternative Techniques for Data Mining and Knowledge Discovery (2004)
11. Boccaletti, S., et al.: The structure and dynamics of multilayer networks. Phys. Rep. **544**(1), 1–122 (2014)
12. Boccaletti, S., Latora, V., Moreno, Y., Chavez, M., Hwang, D.U.: Complex networks: structure and dynamics. Phys. Rep. **424**(4), 175–308 (2006)
13. Coscia, M., Giannotti, F., Pedreschi, D.: A classification for community discovery methods in complex networks. Stat. Anal. Data Mining ASA Data Sci. J. **4**(5), 512–546 (2011)
14. Dice, L.R.: Measures of the amount of ecologic association between species. Ecology **26**, 297–302 (1945)
15. Dijkstra, E.: A note on two problems in connexion with graphs. Numer. Math. **1**(1), 269–271 (1959)
16. Easley, D., Kleinberg, J.: Networks, Crowds, and Markets: Reasoning about a Highly Connected World. Cambridge University Press (2010)
17. Everett, M., Borgatti, S.: Ego network betweenness. Soc. Netw. **27**, 31–38 (2005)
18. Ford, L.: Network Flow Theory. RAND Corporation, Santa Monica, CA (1956)
19. Fortunato, S., Hric, D.: Community detection in networks: a user guide. Phys. Rep. **659**, 1–44 (2016)
20. Freeman, L.C.: Centered graphs and the structure of ego networks. Math. Soc. Sci. **3**(3), 291–304 (1982)
21. Gower, J.C.: A general coefficient of similarity and some of its properties. Biometrics **27**(4), 857–871 (1971)
22. Hart, P., Nilsson, N., Raphael, B.: A formal basis for the heuristic determination of minimum cost paths. IEEE Trans. Syst. Sci. Cybern. **4**(2), 100–107 (1968)
23. Heinrich, C.V., Lombarts, T., Mallens, J., Tortike, L., Wolf, D., Duivesteijn, W.: Local subgroup discovery on attributed network graphs (2025). https://doi.org/10.5281/zenodo.14919732
24. Herrera, F., Carmona, C., González, P., Del Jesus, M.: An overview on subgroup discovery: foundations and applications. Knowl. Inf. Syst. **29**(3), 495–525 (2010)
25. Hu, W., Milesi, A.: Ogb/ogb/utils/features.py at master · snap-stanford/ogb (2022). https://github.com/snap-stanford/ogb/blob/master/ogb/utils/features.py
26. Jorge, C.C., Atzmueller, M., Heravi, B.M., Gibson, J.L., de Sá, C.R., Rossetti, R.J.: Mining exceptional social behaviour. In: Proceedings of EPIA, pp. 460–472 (2019)
27. Kaufman, L., Rousseeuw, P.J.: Finding Groups in Data: An Introduction to Cluster Analysis. Wiley, New York (1990)
28. Kaytoue, M., Plantevit, M., Zimmermann, A., Bendimerad, A., Robardet, C.: Exceptional contextual subgraph mining. Mach. Learn. **106**(8), 1171–1211 (2017). https://doi.org/10.1007/s10994-016-5598-0
29. Klösgen, W.: Explora: A multipattern and multistrategy discovery assistant. In: Advances in Knowledge Discovery and Data Mining, pp. 249–271. AAAI/MIT Press (1996)
30. Konijn, R., Duivesteijn, W., Kowalczyk, W., Knobbe, A.: Discovering local subgroups, with an application to fraud detection. In: Proceedings of PAKDD, pp. 1–12 (2013)
31. Kullback, S., Leibler, R.A.: On information and sufficiency. Ann. Math. Stat. **22**(1), 79–86 (1951)

32. Lavrac, N.: Subgroup discovery techniques and applications. In: Proceedings of PKDD, pp. 2–14 (2005)
33. van Leeuwen, M.: Maximal exceptions with minimal descriptions. Data Mining Knowl. Discov. **21**(2), 259–276 (2010)
34. van Leeuwen, M., Knobbe, A.: Non-redundant subgroup discovery in large and complex data. In: Proceedings of ECML PKDD, pp. 459–474 (2011)
35. Leskovec, J., Lang, K., Dasgupta, A., Mahoney, M.: Statistical properties of community structure in large social and information networks. In: Proceedings of WWW, pp. 695–704 (2008)
36. Leskovec, J., Mcauley, J.: Learning to discover social circles in ego networks. Neural Inf. Process. Syst. **25**, 539–547 (2012)
37. Marsden, P.: Network data and measurement. Ann. Rev. Sociol. **16**, 435–463 (1990)
38. Moore, E.: The shortest path through a maze. In: Proceedings of an International Symposium on the Theory of Switching, Part II, pp. 285–292. Harvard University Press, Cambridge (1959)
39. Muzio, G., O'Bray, L., Borgwardt, K.: Biological network analysis with deep learning. Briefings Bioinform. **22**(2), 1515–1530 (2020)
40. Newman, M.E.J.: The structure and function of networks. Comput. Phys. Commun. **147**(1), 40–45 (2002)
41. Newman, M.E.J.: The structure and function of complex networks. SIAM Rev. **45**(2), 167–256 (2003)
42. Open Graph Benchmark: Graph property prediction (2024). https://ogb.stanford.edu/docs/graphprop/
43. Pei, H., Wei, B., Chang, K.C., Lei, Y., Yang, B.: Geom-GCN: geometric graph convolutional networks. In: Proceedings of ICLR (2020)
44. Rozemberczki, B., Allen, C., Sarkar, R.: Multi-scale attributed node embedding (2021). https://arxiv.org/abs/1909.13021
45. Shlens, J.: Notes on Kullback-Leibler divergence and likelihood. CoRR abs/1404.2000 (2014)
46. Sørensen, T.: A method of establishing groups of equal amplitude in plant sociology based on similarity of species and its application to analyses of the vegetation on Danish commons. Biol Skrifter/Kongelige Danske Videnskabernes Selskab **5**, 1 (1948)
47. Suzuki, E.: Autonomous discovery of reliable exception rules. In: Proceedings of KDD, pp. 259–262 (1997)
48. Suzuki, E., Kodratoff, Y.: Discovery of surprising exception rules based on intensity of implication. In: Proceedings of PKDD, pp. 10–18 (1998)
49. Wasserman, S., Faust, K.: Social Network Analysis: Methods and Applications. Structural Analysis in the Social Sciences, Cambridge University Press (1994)
50. Wu, Z., et al.: Moleculenet: a benchmark for molecular machine learning. Chem. Sci. **9**, 513–530 (2018)
51. Xu, S., Walter, N.P., Kalofolias, J., Vreeken, J.: Learning exceptional subgroups by end-to-end maximizing KL-divergence. CoRR abs/2402.12930 (2024)
52. Zhou, Y., Cheng, H., Yu, J.: Graph clustering based on structural/attribute similarities. In: Proceedings of VLDB, vol. 2, pp. 718–729 (2009)

Imposing Constraints in Probabilistic Circuits via Gradient Optimization

Soroush Ghandi[1(✉)], Benjamin Quost[1,2], and Cassio de Campos[1]

[1] TU Eindhoven, Eindhoven, The Netherlands
{s.ghandi,b.c.quost,c.decampos}@tue.nl, benjamin.quost@utc.fr
[2] Université de technologie de Compiègne, Compiègne, France

Abstract. Probabilistic Circuits (PCs) are a class of tractable models that allow a range of efficient and exact computations while achieving state-of-the-art performance in some domains. In this work, we propose a sample-based procedure to let the distribution encoded by a PC satisfy probabilistic propositional logic constraints. This sample-based method is proposed as a direct competitor to a mathematical approach previously introduced, which is based on optimizing a convex upper bound of the KL divergence. In our empirical study, we compare both methods in two different scenarios, where constraints are utilized to: i) apply fairness to a distribution; and ii) improve the performance of a PC model under scarce data. Our results indicate that although both methods are competitive to one another in the case of fairness, sample-based method has an advantage in scenarios with scarce data.

Keywords: Probabilistic Circuits · Constraints · Gradient

1 Introduction

Generative probabilistic graphical models are typically employed to learn the underlying distribution of data. Among those, *tractable* probabilistic graphical models provide access to the learned probability distribution and allow for a wider range of exact inferences. Probabilistic Circuits (PCs) are tractable models that use a graph-based representation to encode high-dimensional distributions and can process various inferences in polynomial time [17]. Learning a PC (graph structure and parameters) from data is an active line of research that has seen several proposals in the past few years [8,12,15,16,18].

In this paper, we propose an approach for combining a PC with constraints [2,5] using gradient optimization to re-tune the parameters of the model. Our strategy can be seen as a post-processing, which gives the advantages of versatility as well as modularity. We optimise the KL divergence to the given PC, while enforcing constraints via a penalization of the objective function. It is worth mentioning that a method to learn the structure and parameters of PCs with domain constraints is also proposed in [6], however with important differences: the learning method in this latter work is data-driven (sample-based), and hence

G. Krempl et al. (Eds.): IDA 2025, LNCS 15669, pp. 209–220, 2025.
https://doi.org/10.1007/978-3-031-91398-3_16

not applicable as a post-learning process; the constraints are limited to linear functions of conditional probabilities; and the proposed method is only tractable for deterministic PCs.

The benefits of taking user-specified constraints into account are multi-fold. In order to illustrate them, we employ Probabilistic propositional logic (PPL) constraints in a few scenarios: (1) we improve the quality of models by enforcing that the yielded model matches the empirical marginal distributions under situations of (a) scarce data or (b) missing data; (2) we enforce fairness constraints into the model while at the same time avoiding a decrease in fitness. Overall, the experiments indicate that using PPL constraints often yields a better model (without compromising efficiency or accuracy), which is likely possible because of typical over-parametrizations that current large machine learning models impose. We emphasize that these applications of constraints are only a few examples of possible use, as we believe there are many other possibilities ahead to be tried.

2 Probabilistic Circuits

Probabilistic circuits (PCs) are a family of distribution representations facilitating many exact and efficient inference routines (see [1] for a nice introduction). A PC encodes a probabilistic model over a collection of variables $\mathbf{X}$; it is structured as a rooted directed acyclic graph $\mathcal{G}$, containing three types of nodes: (i) distribution nodes, (ii) sum nodes, and (iii) product nodes. Distribution nodes are the leaves of the graph $\mathcal{G}$, while sum and product nodes are the internal nodes. Each distribution node (leaf) v computes a probability distribution over some subset $\mathbf{X}' \subseteq \mathbf{X}$, i.e. an integrable function $p_v(\mathbf{x}') \colon \boldsymbol{\mathcal{X}}' \to \mathbb{R}^+$ from the sample space of $\mathbf{X}'$ to the non-negative real numbers. The *scope* of v is the set of variables $\mathrm{sc}(v) := \mathbf{X}'$ over which the leaf computes a distribution. The scope of any internal node v (sum or product) is recursively defined as $\mathrm{sc}(v) = \cup_{u \in \mathrm{ch}(v)} \mathrm{sc}(u)$, where $\mathrm{ch}(v)$ is the set containing the children of v. Sum nodes compute convex combinations over their children, i.e. if v is a sum node, then v computes $v(\mathbf{x}) = \sum_{u \in \mathrm{ch}(v)} w_{v,u} u(\mathbf{x})$, where $w_{v,u} \geq 0$. In a normalized PC, we have $\sum_{u \in \mathrm{ch}(v)} w_{v,u} = 1$. Product nodes compute the product over their children, i.e. if v is a product node, then $v(\mathbf{x}) = \prod_{u \in \mathrm{ch}(v)} u(\mathbf{x})$. The support of a node is the region where its associated function is strictly positive.

The main feature of PCs is that they facilitate a wide range of *tractable* inference routines, which go hand in hand with certain structural properties [1,3]: (i) a sum node v is called *smooth* if its children have all the same scope: $\mathrm{sc}(u) = \mathrm{sc}(u')$, for any $u, u' \in \mathrm{ch}(v)$; (ii) a product node v is called *decomposable* if its children have non-overlapping scopes: $\mathrm{sc}(u) \cap \mathrm{sc}(u') = \emptyset$, for any $u, u' \in \mathrm{ch}(v)$, $u \neq u'$; (iii) a node is *consistent* if its support is non-empty. A PC is smooth if all its sum nodes are smooth, decomposable if all its product nodes are decomposable, and consistent if all its nodes are consistent. The distribution $p(\mathrm{sc}(v))$ represented by a node v in the PC is the function computed by the rules of the previous paragraph, and can be evaluated with a feed-forward pass.

In order to ensure the tractability of queries, we can rely on smoothness and decomposability, but we also need leaf distribution nodes to compute inferences efficiently. This is a reason for many proposed PCs in the literature to assume that leaf nodes are univariate with some known distribution, such as Bernoulli, categorical, Gaussian, etc. Now, assume that we wish to compute a *marginal query*, that is, to evaluate the probability value over $\mathbf{X}_o \subset \mathbf{X}$ for evidence $\mathbf{X}_o = \mathbf{x}_o$, while marginalizing $\mathbf{X}_{\neg o} = \mathbf{X} \setminus \mathbf{X}_o$. In smooth and decomposable PCs, this task reduces to performing marginalization at the leaves [13]: for each leaf v, one marginalizes $\mathrm{sc}(v) \cap \mathbf{X}_{\neg o}$, and evaluates it for the values corresponding to $\mathrm{sc}(v) \cap \mathbf{X}_o$. The desired marginal $p_{\mathbf{X}_o}(\mathbf{x}_o)$ results from evaluating internal nodes as in computing the complete distribution. Smoothness and consistency are sufficient to guarantee that the function of a PC represents a distribution. We also assume PCs are normalized.

An important feature of normalized valid PCs is their interpretation as hierarchical, discrete mixture models [11,19]:

$$p(\mathbf{x}) = \sum_z p(\mathbf{x}|z)p(z) = \sum_z p_z(\mathbf{x})p(z), \tag{1}$$

where Z is a discrete latent vector, which originates from the sum nodes of the structure. The number of states of Z, and thus of represented mixture components $p(\mathbf{x}|z)$, grows exponentially in the depth of the PC [10,19]. While we use this notation here and throughout the paper, we do not run computations directly in this formulation, but instead we make use of the graphical structure of the PC in order to perform efficient tractable inference, as usual for PCs. Figure 1 shows an example of the graphical structure.

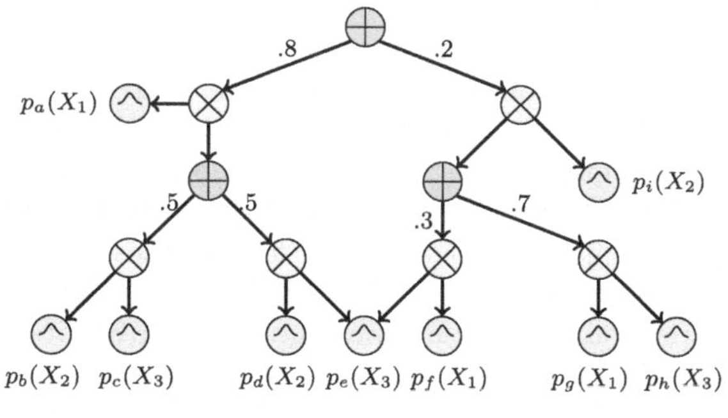

Fig. 1. Example of PC with variables $X_1, \ldots, X_3$. Sum nodes are in blue, product nodes in green, distribution leaf nodes in salmon. In this example, all leaf nodes are univariate. Subscriptions on each p in the figure are used to indicate that those are different leaf distributions (even if sometimes over the same variable). (Color figure online)

3 Probabilistic Circuits with Constraints

3.1 Probabilistic Propositional Logic Constraints

We assume that a normalized valid PC has been produced over a domain with variables $\mathbf{X}$ representing a joint distribution $p(\mathbf{X})$. We want to enforce probabilistic propositional logic (PPL) constraints upon p. We work with constraints of the form:

$$\sum_{i_c} \tau_{i_c} \cdot p(F_{i_c}) \leq \alpha_c, \tag{2}$$

where each F_{i_c} is a propositional logic formula defined over Boolean variables $\mathbf{X}_c = \{X_{j_c}\}_{\forall j_c} \subseteq \mathbf{X}$, τ_{i_c}, α_c are real numbers, j_c (and i_c) are indexes of variables (terms) of the constraint c, and $c \in C$ is an index over a set of PPL constraints.

We assume that constraints are placed in mutually exclusive buckets B. A bucket B can be simply an index set indicating the constraints it contains: we write $\mathbf{X}_B = \cup_{c \in B} \mathbf{X}_c$ the union of all variables appearing in a constraint inside bucket B. Then, two distinct buckets B, B' would be exclusive if $\mathbf{X}_B \cap \mathbf{X}_{B'} = \emptyset$. A set of buckets would be mutually exclusive if every selected pair of buckets are exclusive. During the process of creating the buckets, if any $\mathbf{X}_{c_1}$ and $\mathbf{X}_{c_2}$ of two constraints are not disjoint, then we put them together into the same bucket, so as to ensure that buckets have mutually exclusive sets of variables.

The constraints in each bucket B may obviously create dependencies among the variables $\mathbf{X}_B$. In order to avoid inconsistencies between such dependencies and those arising from the graph structure of the PC, we require that the variables in a bucket appear together in nodes of the model, that is, for any v, B, $X \in \mathbf{X}_B \cap \mathrm{sc}(v) \Rightarrow \mathbf{X}_B \subseteq \mathrm{sc}(v)$. Therefore, Eq. (1) can be recast as

$$p(\mathbf{X}) = \sum_z p(z) \prod_B p_z(\mathbf{X}_B) \prod_{X_i \in \mathbf{X} \setminus \cup \mathbf{X}_B} p_z(X_i), \tag{3}$$

where $p_z(\mathbf{X}_B)$ is a categorical distribution—note that notations $p_z(\mathbf{X}_B)$ and $p_z(X_i)$ employ a slight abuse, as the function itself is "aware" of the indexes of the variables in their arguments and may vary accordingly, for example, $p_z(X_i)$ is also a function of i and not only of X_i; the same abuse holds elsewhere, for example in Eq. (2). Equation (3) basically decomposes $p_z(\mathbf{x})$ of Eq. (1) into components that involve PPL variables (which remain together) and the other variables, which are assumed to be represented by univariate leaf-node distributions. We note that the method that we propose in this work does not need variables to be put in a bucket and kept together in the factorization. However, we stick to this scenario in order to be compare with previous work (where this is a requirement).

3.2 Enforcing Constraints

Given a PC representing $p(\mathbf{X})$ and PPL constraints, we aim to find an *efficient* approach to discover a new PC inducing a distribution $q^*(\mathbf{X})$ that is close to $p(\mathbf{X})$ while respecting the PPL constraints:

$$q^*(\mathbf{X}) = \underset{q(\mathbf{X})}{\operatorname{argmin}} \mathcal{L}(p(\mathbf{X}), q(\mathbf{X}))$$
$$\text{s.t.} \quad \forall B, \forall c \in B : \sum_{i_c} \tau_{i_c} \cdot q(F_{i_c}) \leq \alpha'_c, , \qquad q(\mathbf{X}) \in \mathcal{P}(\mathbf{X}), \tag{4}$$

where $\mathcal{L}$ measures the discrepancy between two distributions, $\mathcal{P}(\mathbf{X})$ denotes a set of probability distributions over $\mathbf{X}$ that can be represented by a PC, and B are buckets of PPL constraints. Optimization (4) is impractical, as it amounts to solving a complex optimization problem to search over $\mathcal{P}(\mathbf{X})$, even if $\mathcal{L}$ is simple enough (a nonlinear optimization with nonlinear constraints). We aim to simplify this generic optimization, so that it can be done efficiently, while still attaining good results in practice. Therefore, we exploit the PC on which $p(\mathbf{X})$ was estimated, in order to constraint the search space of $q(\mathbf{X})$: we enforce $q(\mathbf{X})$ to have a shape similar to $p(\mathbf{X})$, i.e.

$$q(\mathbf{X}) = \sum_z p(z) \prod_B q_z(\mathbf{X}_B) \prod_{X_i \in \mathbf{X} \setminus \cup \mathbf{X}_B} p_z(X_i), \tag{5}$$

that is, $q(\mathbf{X})$ will differ from the specification of $p(\mathbf{X})$ only in the parameters, but not the structure. Moreover, we use the Kullback-Leibler divergence $\mathcal{L}(p(\mathbf{X}), q(\mathbf{X})) = H(p(\mathbf{X}), q(\mathbf{X})) - H(p(\mathbf{X}))$ as discrepancy measure, where $H(\cdot)$ is the entropy and $H(\cdot, \cdot)$ the cross entropy. Clearly, we can focus on cross-entropy, as the second term does not contain $q(\mathbf{X})$.

3.3 Sample-Based Gradient Optimization

In this section, we explain how we define and implement a sample-based scheme to solve the generic optimization in (4), as a direct competitor to the previously introduced mathematical methods. Our idea is to enforce the constraints as regularization terms in the objective function, which leads to an unconstrained, regularized optimization that can be easily solved with gradient descent.

We turn a constraint of the form $\sum_{i_c} \tau_{i_c} \cdot p(F_{i_c}) \leq \alpha_c$ into a regularization term using a $\mathcal{L}_1$ or $\mathcal{L}_2$ norm, e.g. $(\sum_{i_c} \tau_{i_c} \cdot p(F_{i_c}) - s_c - \alpha_c)^2$, where s_c is a slack variable that converts the original inequality to an equality. With this change, the optimization now becomes

$$\underset{q(\mathbf{X}), s_c}{\operatorname{argmin}} \mathcal{L}(p(\mathbf{X}), q(\mathbf{X})) + \sum_B \sum_{c \in B} \gamma_c (\sum_{i_c} \tau_{i_c} \cdot p(F_{i_c}) - s_c - \alpha_c)^2 \tag{6}$$

In this optimization, γ_c indicates the regularization coefficient that corresponds to each constraint c. For each F_{i_c}, we assume a (potentially marginal) synthetic data point $\mathbf{x}_{i_c}$, which allows us to compute $p(F_{i_c}) = p(\mathbf{x}_{i_c})$ in a sample-based scheme. Assuming KL divergence as discrepancy measure (see Sect. 3.2, we can replace $\mathcal{L}(p(\mathbf{X}), q(\mathbf{X}))$ in Optimization 6 with a sample-based alternative to facilitate gradient-based learning:

$$\underset{q(\mathbf{X}), s_c}{\operatorname{argmin}} -\mathbb{E}_{\mathbf{x} \sim p(\mathbf{X})}[\log[q(x)]] + \sum_B \sum_{c \in B} \gamma_c (\sum_{i_c} \tau_{i_c} \cdot p(\mathbf{x}_{i_c}) - s_c - \alpha_c)^2. \tag{7}$$

Optimization (7) can be easily solved via gradient descent in a sample-based optimization scheme. Sample-based optimization does not limit the structure of $q(\mathbf{X})$, trains all the parameters together, directly optimizes the KL divergence (not an upper bound of KL divergence), and scales better with respect to the size of constraint buckets (as opposed to the method in [4], which scales poorly with the size of buckets because of its optimization method). On a separate note, sample-based optimization does not put any limitation on the constraints, and allows us to enforce constraints that are far more complex than the PPL constraints than previous work [4].

However, it theoretically requires an infinite number of samples drawn from $p(\mathbf{X})$ to accurately estimate and optimize the KL divergence. In practice, we rely on a static sample generated from $p(\mathbf{X})$ to solve this optimization, which makes the optimization practically feasible while slightly hurting the result. As long as the size of the constraint buckets is not too large, there is a large gap between the computation complexity of mathematical methods and the sample-based method. In addition, sample-based optimization has an extra set of hyperparameters in γ_c that need fine-tuning. Also, in sample-based optimization, we practically establish a trade-off between enforcing the constraints and optimizing the KL divergence, and as a result, there is no guarantee that the resulting model completely satisfies the constraints. Note that this is not the case for the mathematical methods, since in mathematical methods, the search space itself is limited by the constraints.

3.4 Fairness Constraints

So far, we have discussed applying linear PPL constraints to a given PC $p(\mathbf{X})$. In this section, we extend this domain of application to include fairness with a simple trick. Here, fairness refers to adjusting the PC in a way that eliminates the bias towards a sensitive variable during the prediction process of a class/target variable. For a specific problem, we assume the set of variables $\mathbf{X}$ comprises a binary class/target variable $Y \in \mathbf{X}$ and a binary protected/sensitive attribute $X' \in \mathbf{X}, X' \neq Y$. Our aim is to improve the distribution learned via a PC towards fairness for the target variable Y with respect to the protected attribute X'.

We consider statistical parity as our measure of fairness (we use this as an example; we will not debate on fairness measures, since it is not the main focus of the paper). The corresponding fairness constraint is

$$p(Y = 1|X' = 1) = p(Y = 1|X' = 0);$$

this constraint is clearly not of the form $\sum_i \tau_i \cdot p(F_i) \leq \alpha$: it actually induces a non-linear constraint which is harder to optimise (it is often useful that the constraints are linear so that a penalization can be achieved using a quadratic function—see Sect. 3.3 for more details). However, we can lift the optimization problem to a higher dimension by including a new unknown β such that

$$p(Y = 1|X' = 1) = p(Y = 1|X' = 0) = \beta,$$

giving two separate linear constraints in the desired form:

$$\begin{cases} p((y = 1) \wedge (x' = 1)) - \beta \cdot p(x' = 1) = 0, \\ p((y = 1) \wedge (x' = 0)) - \beta \cdot p(x' = 0) = 0; \end{cases} \tag{8}$$

and as long as β is fixed, the optimization can be carried out to impose the constraints in Eq. (8) to a learned PC using the proposed approach. In order to solve for β, we simply carry out an exhaustive search over candidate values between 0 and 1, retaining the best based on the performance of each resulting PC (obviously, this search procedure is reasonable for a single unknown β, or at most a few; otherwise, a smarter strategy would be required).

4 Experiments

The sample-based learning scheme we introduced makes it possible to integrate PPL constraints into the distribution represented by PCs. Theoretically, our method can be considered as an alternative to the mathematical method introduced in [4], which, to the best of our knowledge, is the only other method that integrates constraints into PCs. In this section, we conduct experiments to compare the ability of both to shape a desirable performance or behavior of a given PC. Similar to [4], we focus on two unique scenarios (each corresponding to a unique set of constraints), namely (i) constraints over marginals of the distribution $p(\mathbf{X})$ (Sect. 4.1), and (ii) constraints for enforcing fairness in this distribution (Sect. 4.2).

Throughout the experiments, we utilize RAT-SPN [14] for learning baseline PC models—one could also handcraft a PC for a purpose and use it with our approach, since this latter is not bound by the way the PC was obtained, as long as the structure and parameters are compatible with sample-based learning. We use the original implementation of RAT-SPN[1]. Hyperparameters that correspond to the region graph structure are set as follows: the number of recursive splits is 10, the depth of each recursive split is 2, the number of input distributions in each partition is 8, and the number of sum nodes per partition is 8; all the other hyperparameters are set to their defaults. For each experiment, RAT-SPN is trained via Expectation-Maximization (EM) for 20 epochs.

For each experiment, we utilize the baseline PC ($p(\mathbf{X})$) to generate 50000 samples as our offline dataset, of which 40000 samples will be used as the train data and the remaining 10000 are considered validation data. Next, we initialize $q(\mathbf{X})$ with the same structure and parameters as $q(\mathbf{X})$, and train the parameters of $q(\mathbf{X})$ to optimize (7) via gradient descent for 40 epochs. The results

[1] https://github.com/cambridge-mlg/RAT-SPN.

are averaged over 5 random initializations. The results are optimized over the regularization coefficient γ_c, which is assigned by an exhaustive search over the set $\{0.01, 0.1, 1, 10\}$. This optimization is done to find the suitable *ballpark* for the γ_c, and not to deliberately fine-tune the results in our favor.

4.1 Marginal Constraints

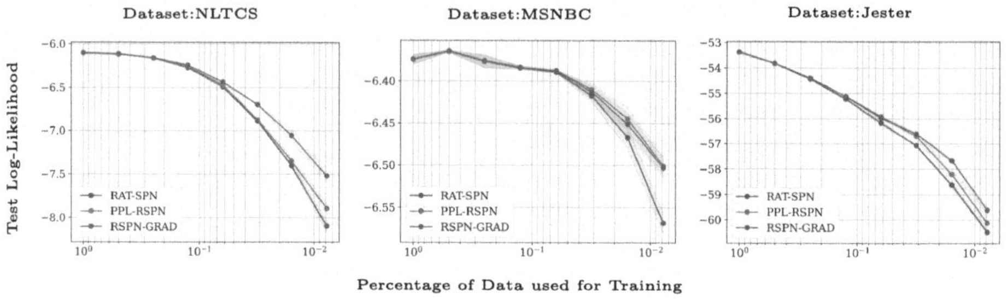

Fig. 2. Performance results on scarce datasets.

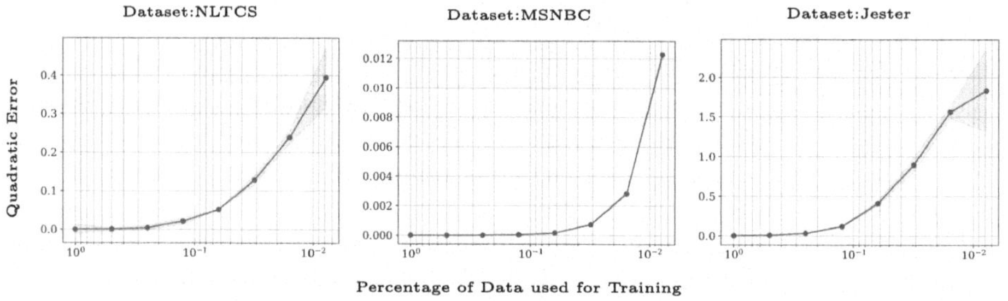

Fig. 3. Sum of quadratic differences on marginal parameters between the models with and without marginal constraints, when trained on scarce data. Constraints clearly refine the model strongly for RAT-SPNs, since standard RAT-SPN marginals are very far from matching the empirical marginal distributions when the data becomes scarce (data not shown).

As described in [4], the goal of this experiment is to illustrate how additional information pertaining to the empirical marginal distributions can be incorporated into the PC so as to compensate for data scarcity. Hence, each baseline PC in this experiment is optimized to match the empirical marginals of $\mathbf{X}$ on the training data. We carry out this experiment on three different binary datasets, namely NLTCS, MSNBC, and Jester [9].

In order to simulate scarce data, we randomly subsample each dataset $\mathcal{D}$ with a varying number of data instances. We use this subsample to train the baseline PC model using RAT-SPN. We then improve the model using the procedure described above so as to match the empirical marginal distributions: we add PPL constraints of the form $q(X_i = 1) = \alpha_i$ for every variable X_i (where α_i is the empirical marginal of X_i on the subsampled data), which we enforce globally into the model. Figure 2 gives the log-likelihood computed over the test data (this is obtained by averaging the output of the PC over the test data, and is widely used as a measure to evaluate how the model performs on the test data). The enhanced models obtained by applying the constrained optimization are called PPL-RSPN (method introduced in [4]) and RSPN-GRAD (our sample-based model using gradient optimization).

As the results in Fig. 2 suggest, both methods outperform the baseline RAT-SPN as the training data become scarcer. This performance gain (in terms of test log-likelihood) is not similar across all datasets, which we attribute to the relative amount of information captured by the marginals. Arguably, in smaller datasets (in terms of the number of samples), matching marginals should lead to larger performance gains, as marginals encode a relatively larger amount of information. More importantly, matching marginals did not harm joint accuracy. Nevertheless, RSPN-GRAD performs better than PPL-RSPN on 2 out of 3 datasets, while performing similarly on the third one, which suggests that it has an advantage over PPL-RSPN when it comes to matching marginals. Figure 3 shows the average difference between parameters of the trained model after constraints are applied. This indicates that the gradient optimization is indeed changing the model considerably.

4.2 Fairness Constraints

In this section, we investigate the impact of applying our method to a variety of fairness-specific classification tasks by adding fairness in the form of PPL constraints into PCs. As discussed in [4], most common PC learning methods are not known to be inherently compatible with fairness, and being able to apply fairness constraints to PCs opens the door to utilizing these probabilistic models in areas where fairness is a priority, thus extending their domain of applicability.

We consider six different classification datasets commonly used in fairness-aware machine learning, namely Adult, German Credit, Bank Marketing, Dutch Census, Credit Card Clients, and Law School [7]. Similar to Sect. 4.1, we refer to the variants as PPL-RSPN [4] and RSPN-GRAD (our sample-based method). The results are displayed in Table 1. Both methods are able to achieve a "fair" distribution w.r.t the protected attribute, and they also manage to do so without losing much of their representation power compared to the original RAT-SPN, that is, the test likelihood and 0-1 accuracy are barely affected while statistical parity is enforced by the use of constraints. Both methods perform comparable to one another in these fairness-related experiments. While RSPN-GRAD models always have higher accuracy across the majority of datasets (albeit by a very small margin), they tend to have worse test log-likelihood in scenarios where the statistical parity in the baseline PC is quite large (e.g. Adult, Dutch Census, and Law School datasets).

Table 1. Classification with RAT-SPN vs. PPL-RSPN vs. RSPN-GRAD enforcing statistical parity via constraints.

Dataset	Protected Attribute	Method	Test LL	Accuracy	Statistical Parity
Adult	Sex	RAT-SPN	−7.767	0.8193	0.2000
		PPL-RSPN	−7.796	0.8148	$\leq 10^{-6}$
		RSPN-GRAD	−7.801	0.8153	$\leq 10^{-4}$
German Credit	Sex	RAT-SPN	−28.752	0.745	−0.0309
		PPL-RSPN	−28.756	0.745	$\leq 10^{-6}$
		RSPN-GRAD	−28.743	0.745	$\leq 10^{-4}$
Bank Marketing	Marital Status	RAT-SPN	−13.736	0.8820	−0.0388
		PPL-RSPN	−13.739	0.8788	$\leq 10^{-6}$
		RSPN-GRAD	−13.736	0.8824	$\leq 10^{-4}$
Dutch Census	Sex	RAT-SPN	−12.880	0.7888	0.2620
		PPL-RSPN	−12.923	0.7629	$\leq 10^{-6}$
		RSPN-GRAD	−12.943	0.7728	$\leq 10^{-4}$
Cr. Card Clients	Sex	RAT-SPN	−3.998	0.7838	0.0053
		PPL-RSPN	−3.998	0.7838	$\leq 10^{-6}$
		RSPN-GRAD	−3.995	0.7838	$\leq 10^{-4}$
Law School	Race	RAT-SPN	−7.274	0.9050	−0.2054
		PPL-RSPN	−7.294	0.9034	$\leq 10^{-6}$
		RSPN-GRAD	−7.323	0.9047	$\leq 10^{-4}$

5 Conclusion

In this paper, we presented an approach for fine-tuning a PC given PPL constraints, using sample-based gradient optimization. Our strategy can be considered as a post-processing of the distribution encoded by the original PC, carried out by optimizing a penalized version of the KL divergence to the original distribution—with the penalization terms being obtained from the constraints.

We illustrate the interest of taking into account additional constraints in two scenarios: we show how requiring that the model matches empirical marginal distributions can help compensate for data scarcity or missingness, and we demonstrate that fairness requirements can be taken into account without sacrificing fitting power. Our experiments show that the sample-based approach compares favourably to a previous approach to solving the constrained optimization problem, being more versatile and efficient.

Being able to integrate a wide range of constraints opens the way to the application or PCs to many learning problems, much beyond the marginal distribution and fairness examples considered here. The investigation of such other possibilities, as well as the improvement of the strategy presented above, are left for future work.

Acknowledgments. The authors thank the Eindhoven Artificial Intelligence Systems Institute and the Department of Mathematics and Computer Science of the Eindhoven University of Technology for their support. They also thank the support of EU European Defence Fund Project KOIOS (EDF-2021-DIGIT-R-FL-KOIOS) and NWO Project NGF.1609.242.024.

Disclosure of Interests. The authors have no competing interests to declare that are relevant to the content of this article.

References

1. Van den Broeck, G., Di Mauro, N., Vergari, A.: Tractable probabilistic models: representations, algorithms, learning, and applications (2019). Tutorial at Uncertainty in Artificial Intelligence (UAI) 2019. http://web.cs.ucla.edu/~guyvdb/slides/TPMTutorialUAI19.pdf
2. Cozman, F.G., de Campos, C.P., da Rocha, J.: Probabilistic logic with independence. Int. J. Approx. Reason. **49**(1), 3–17 (2008)
3. Darwiche, A.: A differential approach to inference in Bayesian networks. J. ACM **50**(3), 280–305 (2003)
4. Ghandi, S., Quost, B., de Campos, C.: Probabilistic circuits with constraints via convex optimization. arXiv preprint arXiv:2403.13125 (2024)
5. Hansen, P., Jaumard, B.: Probabilistic satisfiability. In: Handbook of Defeasible Reasoning and Uncertainty Management Systems: Algorithms for uncertainty and defeasible reasoning, pp. 321–367. Springer (2000)
6. Karanam, A., Mathur, S., Sidheekh, S., Natarajan, S.: Bayesian learning of probabilistic circuits with domain constraints. In: The 6th Workshop on Tractable Probabilistic Modeling (2023)
7. Le Quy, T., Roy, A., Iosifidis, V., Zhang, W., Ntoutsi, E.: A survey on datasets for fairness-aware machine learning. Wiley Interdisc. Rev. Data Mining Knowl. Discov. **12**(3), e1452 (2022)
8. Liu, A., Van den Broeck, G.: Tractable regularization of probabilistic circuits. Adv. Neural. Inf. Process. Syst. **34**, 3558–3570 (2021)
9. Lowd, D., Davis, J.: Learning Markov network structure with decision trees. In: 2010 IEEE International Conference on Data Mining, pp. 334–343. IEEE (2010)
10. Peharz, R.: Foundations of sum-product networks for probabilistic modeling. Ph.D. thesis, Graz University of Technology (2015)
11. Peharz, R., Gens, R., Pernkopf, F., Domingos, P.: On the latent variable interpretation in sum-product networks. IEEE Trans. Pattern Anal. Mach. Intell. **39**(10), 2030–2044 (2016)

12. Peharz, R., et al.: Einsum networks: fast and scalable learning of tractable probabilistic circuits. In: International Conference on Machine Learning, pp. 7563–7574. PMLR (2020)
13. Peharz, R., Tschiatschek, S., Pernkopf, F., Domingos, P.: On theoretical properties of sum-product networks. In: Artificial Intelligence and Statistics, pp. 744–752 (2015)
14. Peharz, R., et al.: Random sum-product networks: a simple and effective approach to probabilistic deep learning. In: Uncertainty in Artificial Intelligence, pp. 334–344. PMLR (2020)
15. Rahman, T., Jin, S., Gogate, V.: Look ma, no latent variables: accurate cutset networks via compilation. In: International Conference on Machine Learning, pp. 5311–5320. PMLR (2019)
16. Trapp, M., Peharz, R., Ge, H., Pernkopf, F., Ghahramani, Z.: Bayesian learning of sum-product networks. In: Advances in Neural Information Processing Systems, vol. 32 (2019)
17. Vergari, A., Choi, Y., Liu, A., Teso, S., Van den Broeck, G.: A compositional atlas of tractable circuit operations for probabilistic inference. Adv. Neural. Inf. Process. Syst. **34**, 13189–13201 (2021)
18. Vergari, A., Molina, A., Peharz, R., Ghahramani, Z., Kersting, K., Valera, I.: Automatic Bayesian density analysis. In: Proceedings of the AAAI Conference on Artificial Intelligence, vol. 33, no. 1, pp. 5207–5215 (2019)
19. Zhao, H., Poupart, P., Gordon, G.J.: A unified approach for learning the parameters of sum-product networks. In: Advances in Neural Information Processing Systems, vol. 29 (2016)

Natural Language Processing

Improving Next Tokens via Second-to-Last Predictions with Generate and Refine

Johannes Schneider(✉)

Department of Computer Science and Information Systems,
University of Liechtenstein, Vaduz, Liechtenstein
johannes.schneider@uni.li

Abstract. Autoregressive language models like GPT aim to predict next tokens, while autoencoding models such as BERT are trained on tasks such as predicting masked tokens. We train a decoder-only architecture for predicting the second to last token for a sequence of tokens. Our approach yields higher computational training efficiency than BERT-style models by employing a structured deterministic approach to masking tokens. We use our model to improve the next token predictions of a standard GPT by combining both predictions in a "generate-then-refine" approach. We demonstrate on different variants of GPT-2 and different datasets that (not unexpectedly) second to last token predictions are much more accurate, i.e., more than 15% higher accuracy than standard next token predictions. The "generate-then-refine" approach also demonstrates notable improvements in next-token predictions, yielding smaller yet consistent and significant gains.

Keywords: Large Language Models · Generate-then-refine · Bidirectional decoder-only model

1 Introduction

Large language models based on transformers [23] have disrupted the field of natural language processing. In particular, autoregressive models such as the Generative Pre-trained Transformer (GPT) series, introduced in 2018 [16], marked a major step forward due to their ability to generate mostly coherent and contextually relevant text by predicting the next token solely based on preceding tokens. Conversely, autoencoding models like BERT (Bidirectional Encoder Representations from Transformers) [6] led to large improvements in understanding and representation tasks. They predict masked tokens within a sequence, thereby capturing both preceding and succeeding context. Efforts to bridge the gap between these models have led to innovations such as XLNet [27] integrating permutation-based training to capture bidirectional context while maintaining autoregressive properties. We pursue a different avenue to improve next token predictions by adopting a "generate-then-refine" approach. In our approach, a

G. Krempl et al. (Eds.): IDA 2025, LNCS 15669, pp. 223–233, 2025.
https://doi.org/10.1007/978-3-031-91398-3_17

standard autoregressive model predicts the top-k next tokens, which are provided as context, i.e., last token, to an auto-encoding model aiming to predict the second-to-last token (see Fig. 1). Thereby, we incorporate bidirectional context into a unidirectional model. Predictions of both models are then combined to improve next-token predictions. That is, we use prior token predictions as a form of feedback (or verification) to improve reliability and accuracy of next token predictions in standard autoregressive models. We believe that this approach is of great interest as the "generate-then-refine" has received considerable attention in recent years [15]. In particular, on a larger scale, our work contributes to the ongoing debate on when LLMs can self-correct during inference, where self-correction might also be triggered by external sources [9]. That is, we provide a novel approach for self-correction.

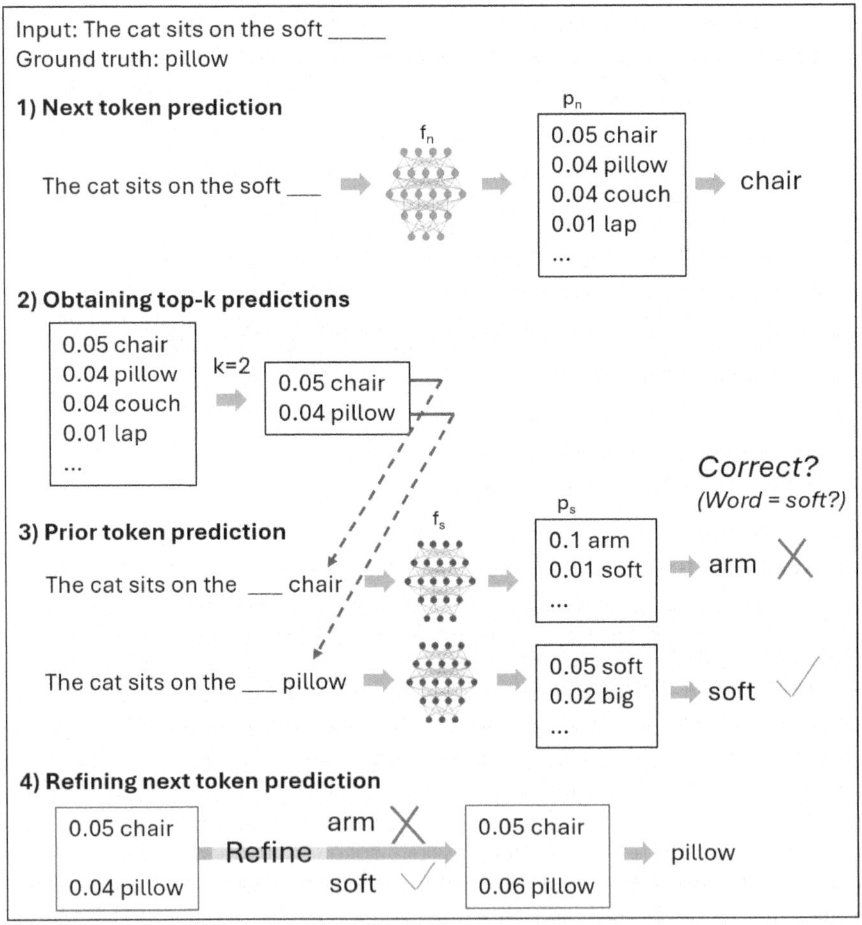

Fig. 1. Conceptual outline of our "generate-then-refine" approach using second-to-last token prediction formalized in Algorithm 1

Our implementation is also a novel way to remedy shortcomings of earlier works such as [6], e.g., eliminating the need for a dedicated mask token, and thereby increasing training efficiency due to a structured, deterministic choice of masking locations and permutation of input sequences.

We evaluate our approach on three variants of GPT-2 [17] across three datasets, showing that (as expected) leveraging bidirectional context leads to much higher prediction accuracy than next token predictions. However, translating high accuracy for second-to-last token predictions into substantial gains for next token predictions is more challenging, i.e., while we found consistent and statistically significant gains for our vanilla model using the same parameters across all datasets, the gains are small. Nevertheless, in the light of the challenges related to self-correction [9], we believe that our novel approach is of considerable interest. We also see it as a first step that can be further improved.

2 Methodology

Our approach, illustrated in Fig. 1, employs two models: A standard, unidirectional autoregressive model f_n for next token prediction, i.e., in our evaluation we use GPT-2 variants, and a second, bidirectional (autoencoding) model f_s with the same architecture trained to predict the second-to-last token. The latter model serves as an evaluator or refiner of the top-k predictions of the autoregressive model f_n by predicting the second to last token given each of the top-k predictions as last token. Our method increases the likelihood of all top-k predictions for which the second to last token was predicted correctly.

More formally, the autoregressive language model f_n computes a probability distribution p_n aiming to approximate the true conditional probability distribution p^* of a token y_t given prior tokens $y_{i<t}$:

$$f_n(y_t|y_{i<t}) = p_n(y_t|y_0, y_1, ..., y_{t-1}) \tag{1}$$

$$\approx p^*(y_t|y_0, y_1, ..., y_{t-1}) \tag{2}$$

Analogously, the autoencoding model predicts the second-to-last token y_{t-1} given tokens $y_{i<(t-1)}$ and the last token y_t:

$$f_s(y_{t-1}|y_{i<t-1}, y_t) = p_s(y_{t-1}|y_0, y_1, ..., y_{t-3}, y_{t-2}, y_t) \tag{3}$$

$$\approx p^*(y_{t-1}|y_0, y_1, ..., y_{t-3}, y_{t-2}, y_t) \tag{4}$$

To refine the next token prediction y of f_n, we compute the top-k most likely tokens Top_k (in Algorithm 1). For each token $y \in Top_k$, we compute the output of the autoencoding model f_s, i.e., $f_s(x|y_{i<t-1}, y)$. If the most likely token predicted by f_s for token y is indeed the correct second to last token y_{t-1}, i.e., $y_{t-1} = \arg\max_x f_s(x|y_{i<t-1}, y)$, we multiply the probability $f_n(y|y_{i<t})$ by a factor $(1 + w)$, where w is a parameter. We found this to work somewhat better than other approaches such as adding a fixed value. However, the essential point is that a next token prediction y of model f_n is seen as more likely correct if the

assessor model f_s correctly identifies the second to last token using the prediction y as last token. We discuss the underlying assumptions of our method in Sect. 4, underpinning them with empirical outcomes. For implementation, it is more efficient to work on logits to avoid softmax computations yielding probabilities. Code is at https://github.com/JohnTailor/AutoEncGenerateThenRefine/.

Algorithm 1. Algorithm AGR (Auto-encoding "Generate-then-Refine")

Input: Models f_n, f_s and tokens $y_{i<t}$
Output: Next token prediction y_t
$k := 15$ ▷ number of tokens to consider
$w := 0.05$ ▷ (incremental) factor if second-to-last tokens is correct
$Top_k :=$ {Tokens $y_t|y_t$ among k tokens with max probability $f_n(y|y_{i<t})$}
$y'_t := \arg\max_{y \in Top_k} p_n(y|y_{i<t}) \cdot (1 + w \cdot \mathbf{1}_{\{y_{t-1}=\arg\max_x f_s(x|y_{i<t-1},y)\}})$
Return y'_t

2.1 Implementation for Efficient Training

Previous works like BERT [6] commonly use a special token to indicate a masked token, typically masking only a small fraction of words (e.g. around 10%) for a given text, making autoencoding slow to train. We forgo the need for a mask token and increase the number of predicted tokens. We do so by eliminating randomness in choices of mask locations in the input. We decompose the input sequences into short subsequences and permute them so that they better fit existing frameworks for next token prediction. That is, the decoder of a transformer [23] employs a causal attention mask that ensures that for a given input $S = (y_0, y_1, ..., y_{t-1})$ of t tokens, to predict token y_i the decoder only sees tokens $y_{j<i}$, i.e., it cannot see the future tokens it should predict. This mechanism ensures that for training with an input sequence of length t we also obtain t predictions to compute the loss. However, our model f_s (implemented as a decoder) can effectively see the 'next' token but not the preceding one, which makes masking challenging. A simple strategy to overcome this issue would be to train without any causal attention mask, altering the input sequence a bit by removing the token to predict y_{t-1}, i.e., $S = (y_0, y_1, ..., y_{t-2}, y_t)$. The transformer predicts just a single token y_{t-1} for the entire sequence (technically, we can simply ignore all predictions except the last one for loss computation). The problem with this approach is that it slows down training by a factor of t, which is more than 100 times slower in practice, i.e., typical input lengths are significantly larger than 100 tokens. Thus, this approach is impractical. To improve training speed, we introduce a compromise. We split the entire input sequence into subsequences of length $l \ll t$, i.e., we set $l = 4$. For each of these subsequences, we obtain a prediction that is used for loss computation, effectively slowing down training by factor of $l \ll t$. More precisely, for training of f_s we replace the first subsequence of $l = 4$ tokens (y_0, y_1, y_2, y_3) by (y_0, y_1, y_2, y_4). More generally, we move tokens

at positions $j \cdot l + (l - 1)$ to position $j \cdot l + l$ leading to a permuted sequence like $S' = (y_0, y_1, y_2, y_4, y_3, y_5, y_6, y_8, y_7, y_9, y_{10}, y_{12}, ...)$. We only use the predictions for tokens $j \cdot l + l$ for loss computations. The process for altering the original input sequence for training the model for second-to-last token prediction f_s is illustrated in Fig. 2.[1]

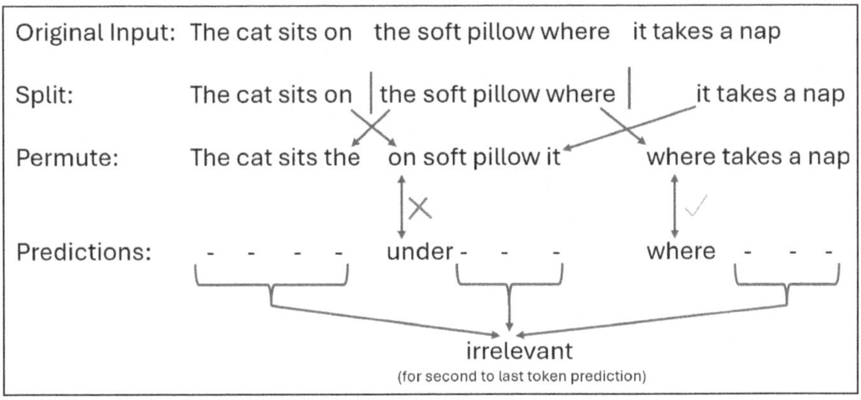

Fig. 2. Example of how input sequences are processed to obtain sequences for training the second to last token prediction model f_s

3 Evaluation

We evaluated Algorithm (1) AGR using three datasets and three variants of the GPT-2 model [17] differing mostly in terms of size, i.e., the number of parameters. Note that GPT-2 is very similar to newer models such as GPT-3 and GPT-4 (see [22] for a more detailed comparison). More specifically, we built on the NanoGPT repository by Karpathy[2]. We computed loss metrics and prediction accuracy but focused our discussion on accuracy for simplicity, as the two are highly correlated.

3.1 Datasets, Models and Training

We use three text datasets that vary in text type and size (see Table 1). Each dataset was split randomly into 90% training and 10% validation. For each dataset, we train two models, i.e., a next token prediction model f_n and prior token prediction model f_s with key parameters stated in Table 2. For data preprocessing such as tokenization model definitions and training, we leveraged the NanoGPT repository code[2]. That is, we performed warmup iterations (Column

[1] An alternative approach is to use a custom causal mask, which is more tricky to implement.

[2] https://github.com/karpathy/nanoGPT.

'WarmupIt.' in Table 2) with the minimum learning rate set to 1/10 of the starting learning rate of (see Column 'LearningRate' in Table 2) and conducted a fixed number of iterations given by column '#Iters' using a batch size of 768, with each entry having 'BlockSize' many tokens. The initial learning rate 'LearningRate' after the warmup phase is reduced until it reaches 1/10 of the initial rate at the end of training. The model architecture hyperparameters are given by the number of layers (#Layer), number of attention heads (#Head) and embedding dimension (EmbDim) yielding a total number of parameters given by '#Para.'.[3] We trained each model in Table 2 once. For evaluation of our Algorithm 1 we performed 70000 randomly chosen token predictions for both training and validation data. Each predicted token is based on an input of maximal length, i.e., of 'BlockSize' length (Table 2). We performed ten evaluation runs and reported the mean and standard deviation of accuracies of both models f_n and f_s and Algorithm AGR (1).[4]

To ensure a fair comparison in terms of the amount of training, for next token prediction we also only use tokens $j \cdot l + l$ for loss computations as for second-to-last token prediction (see Fig. 2 described in Sect. 2.1). However, we found that significantly longer training of the next token prediction model yields only limited gains without additional measures such as increasing model capacity and training data.

Computational Settings: Our experiments were executed on an Ubuntu 22.04 system equipped with Python 3.12, PyTorch 2.5, CUDA 12.5, running on a server with 512 GB of RAM, a 64-core AMD EPYC 9554 CPU, and two NVIDIA H100 NVL GPUs.

Table 1. Datasets

Dataset	#Tokens	Description / URL
simplewiki	43 Millions	Simple Wikipedia https://dumps.wikimedia.org/simplewiki/latest/ simplewiki-latest-pages-articles.xml.bz2[7]
openwebtext	1.7 Billion	Open-source replication of the WebText dataset from OpenAI http://Skylion007.github.io/OpenWebTextCorpus
Stories	364 Millions	Children stories generated by Llama3 8B https://huggingface.co/datasets/Seppel123/Stories_by_Llama3

3.2 Results

The results in Table 3 show significant gains for dataset SimpleWiki but rather small gains for other datasets. We conducted a t-test for the vanilla configuration

[3] Note that the number of parameters in our model might be smaller than those for a native GPT-2 model as we removed tokens that did not occur in the corpus and remapped the others to reduce the embedding matrix.

[4] As we chose data randomly, some data might overlap across runs, but as the validation data size of the smallest dataset simplewiki is more than a factor of 20 larger than the evaluation data, the overlap is expected to be limited.

Table 2. Model configuration for each dataset

Dataset	Model Configuration							
	#Para.	LearningRate	WarmupIt.	#Iters	BlockSize	#Layer	#Head	EmbDim
openwebtext	800 Mio	7e-4	4000	400000	208	30	16	1424
simplewiki	227 Mio	1e-3	2000	200000	112	14	4	1024
Stories	300 Mio	1e-3	2000	400000	144	16	8	1152

($k = 15$ and $w = 0.05$) and found that gains using Algorithm AGR (1), i.e., $\Delta_{AGR-f_n} := Acc^{va}_{AGR} - Acc^{va}_{s}$, are greater than zero (p-value <0.05) for each dataset. Thus, while our approach yields gains across all datasets, considering the fact that the model accuracies f_s for second-to-last token prediction are much larger than those of the next token prediction model (i.e., differing by about 20% on validation accuracy), the gains ranging from 0.03% to 0.22% for the vanilla configuration appear small. A deeper analysis identified several cases where the refiner model f_s is not of much help. First, if the second-to-last or next token is very common, e.g., a word like "a". In this case, we found, for example, that the model f_s might predict the correct token for multiple of the top k tokens of f_n, e.g., if the second to last token is 'a' and among the top k next predicted tokens are nouns like 'house' or 'car', the token 'a' might be predicted for both nouns. Second, if the token is rather uncommon and both models fail, i.e., the correct token is not among the top-k tokens. For instance, we found that the correct token is not among the top k in 50% of all predictions for $k = 2$ for the simplewiki dataset and in 75% for $k = 15$ for the openwebtext dataset. Furthermore, on this dataset, the model f_s only achieves an accuracy of about 60% meaning that its predictions are noisy. Also, we found that, if f_s is not performing well (e.g., achieving about 60%), relying too much on the prediction of f_s, i.e., choosing a large weight w ($w = 0.1$), leads to lower accuracy than not using Algorithm AGR 1, indicating that reliance on the model f_s should be limited despite its higher accuracies compared to f_n. However, if second to last token prediction f_s is fairly reliably it appears that larger weights such as $w = 0.1$ clearly outperform lower weights – A t-test shows that it performs better than any other configuration with p<0.001 for simplewiki. To provide evidence in favor of the conjecture that higher validation accuracy of the validation model leads to larger weights w being more favorable, we loaded a checkpoint of our model trained only for 20% of all iterations stated in Table 2. For the model from the checkpoint the validation accuracy was only 64% for the refiner model f_s. In turn choosing $w = 0.1$ was significantly worse (p<0.05) than the best configuration.

4 (Theoretical) Discussion of Underlying Assumptions

For our method to be useful, we need the refining model f_s to not introduce more errors than it fixes. This can be broken down into two assumptions: (i) Predicting the second-to-last token is (theoretically) easier than predicting the

Table 3. Results: Accuracies for train(tr) and validation(va) data for models f_s, f_n and Algorithm AGR varying parameters k and w; $\Delta_{AGR-f_n} > 0$ indicates gains of Algorithm AGR, which hold in all settings except for large w

Dataset	k	w	Acc_n^{tr}	Acc_s^{tr}	Acc_{AGR}^{tr}	Acc_n^{va}	Acc_s^{va}	Acc_{AGR}^{va}	Δ_{AGR-f_n}
Stories	15	0.05	77.82±0.23	93.08±0.18	78.08±0.24	65.15±0.25	83.96±0.12	65.2±0.25	0.04±0.02
Stories	5	0.05	77.84±0.39	92.97±0.14	78.1±0.4	65.09±0.15	84.01±0.15	65.11±0.16	0.03±0.03
Stories	2	0.05	77.84±0.4	93.08±0.18	78.09±0.39	65.16±0.12	83.94±0.12	65.18±0.12	0.02±0.03
Stories	5	0.01	77.86±0.32	93.15±0.12	77.93±0.32	65.11±0.15	83.94±0.12	65.13±0.15	0.02±0.01
Stories	5	0.1	77.82±0.27	93.06±0.12	78.14±0.28	65.05±0.21	83.93±0.15	64.99±0.21	-0.06±0.04
openwebtext	15	0.05	42.55±0.3	63.06±0.22	42.6±0.32	42.4±0.19	62.93±0.2	42.46±0.19	0.06±0.04
openwebtext	5	0.05	42.47±0.32	62.95±0.44	42.52±0.34	42.45±0.17	63.01±0.19	42.5±0.17	0.05±0.02
openwebtext	2	0.05	42.68±0.29	63.07±0.24	42.69±0.27	42.54±0.22	62.95±0.12	42.58±0.22	0.03±0.02
openwebtext	5	0.01	42.59±0.26	62.9±0.27	42.62±0.25	42.47±0.09	63.02±0.24	42.49±0.09	0.02±0.01
openwebtext	5	0.1	42.54±0.28	63.06±0.39	42.53±0.3	42.56±0.16	62.88±0.17	42.54±0.15	-0.02±0.04
simplewiki	5	0.1	95.9±0.14	98.7±0.08	96.57±0.14	43.44±0.18	63.8±0.16	43.78±0.19	0.35±0.03
simplewiki	5	0.05	95.92±0.1	98.66±0.05	96.39±0.1	43.46±0.14	63.77±0.15	43.68±0.14	0.22±0.02
simplewiki	15	0.05	95.88±0.11	98.7±0.08	96.33±0.11	43.46±0.17	63.77±0.18	43.68±0.17	0.21±0.02
simplewiki	2	0.05	95.91±0.1	98.72±0.05	96.34±0.09	43.46±0.2	63.85±0.19	43.67±0.21	0.21±0.01
simplewiki	5	0.01	95.81±0.11	98.66±0.08	95.93±0.12	43.44±0.16	63.84±0.13	43.5±0.16	0.06±0.02

next token. (ii) The inductive bias of (existing) models is better suited for second to last token prediction, i.e., the problem is not only theoretically easier but also that current models (e.g., transformers) actually perform better.

Let us discuss Assumption (i): While empirically we observe that the accuracy of f_s is higher than that of f_n, this is not necessarily true for all possible datasets. We formally rely on the assumption that predicting a token y_{t-1} in a bidirectional setting, i.e., "in-between", is easier than the next token y_t. We assume that the error of an optimal predictor p^* in a bidirectional setting, is less than that in a unidirectional setting, where the error is computed by comparing it to an optimal predictor p^* that has access to all the context $y_{i \neq t}$ except the token to be predicted. That is, we assume (doing a shift of indexes):

$$||p^*(y_t|y_{i\neq t}) - p^*(y_t|y_0, y_1, ..., y_{t-1})|| > \tag{5}$$

$$||p^*(y_t|y_{i\neq t}) - p(y_t|y_1, y_2, ..., y_{t-1}, y_{t+1})|| \tag{6}$$

This holds true if knowing the next token y_{t+1} is more useful than knowing a token y_0 that occurred much earlier than the token y_t to be predicted (e.g., $t-1$ tokens earlier). While this does not hold for all probability distributions (i.e., all possible datasets), it holds under the assumption of locality, i.e., nearby elements matter more than distant elements. Locality is commonly found in nature, the sciences, i.e., distributed computing [2] and specifically, also in the context of deep learning, e.g., convolutional neural networks [20].

Let us discuss Assumption (ii): Intuitively, we require that irrespective of whether a problem is theoretically easy to solve or not, a model can also approximate the problem well. We can only provide conjectures as to why next token prediction is harder than second-to-last token prediction as current explainability techniques are still emerging [21]. We observe that the next token prediction model f_n tends to overfit much more than the model f_s having the same architecture. This suggests that the standard GPT architecture generalizes better for the second to last token prediction, while showing similar training accuracies. The most illustrative case shown in Table 3 is for the simplewiki dataset, where both models achieve training accuracies of 95% to 98%, i.e. they differ only by 3% in training, while their validation accuracy differs by 20%.

5 Related Work

A large body of works [10,22] have built on the original transformer architecture [23]. The original transformer consists of an encoder and a decoder. The encoder maps input tokens to continuous representations, while the decoder uses these representations as well as inputs, along with self-attention to perform predictions such as the next token prediction in the output sequence. Encoder models like BERT [6] and its variants [11,19] focus on learning (universal) text representations and, therefore, include training tasks beyond next token prediction, i.e., identifying correct sentence order and predicting masked words, i.e., operating as auto-encoder to recover corrupted tokens. In contrast, autoregressive models for text generation, commonly employ decoder-only models as witnessed by the GPT series [3,13,14,17] and typically focus on next token prediction only – with exceptions like T5, which also employs an encoder and a decoder [18] and [1,27], which predict masked tokens. Thus, while the architecture of all of these models is similar to the original transformer [22,23], they differ significantly in their purpose and training objectives. In our work, we employ a model that predicts masked tokens like BERT using a decoder-only model like [27]. Aside from using a decoder model, we also differ from BERT in that we do not require a mask token and achieve higher training efficiency (from 15% up to 25%[8]) by using a static setup for predictions, i.e., we only predict tokens at fixed positions, such as every kth token, rather than predicting tokens at random positions. Our approach is also simpler than [27], which does not need a mask token. [27] uses attention masks to permute inputs and requires separate content and query representations, while at the same time facing challenges with conversion, effectively limiting training efficiency. That is, they only predict the last tokens in a permuted sequence. [1] (similarly to [18]) aims at a unified text language model by leveraging several training objectives, e.g., autoencoding using masking as well as autoregressive modeling. [1] follows a similar approach as BERT, only achieving a masking ratio of 15%. Furthermore, models like BERT and [1,18,27] typically aim to improve their learned representations or autoregressive modeling, while our refiner model f_s serves more to use predictions to improve outputs of a next-token model f_n, i.e., our model f_s only focuses on second-to-last token predictions, while f_n only focuses on next token predictions. Prior work has also attempted to improve outputs of LLMs, often using self-critique [15]. For example, [12] uses state-of-the-art models like GPT-4 to self-assess its outputs and improve on it iteratively. [26] trains a feedback model. Using iterative calls to the

[8] We could go as high as 50% for $l = 2$.

original LLM and the feedback model tasks such as translation can be improved. While [26] relies on language output, other methods included refinements into the architecture, e.g., [25] performs blockwise decoding to support refinements. However, claims about self-correction capabilities of LLMs have been called into question [9]. Our paper also relates to work that aims to combat hallucinations as surveyed in [8]. Many works combating hallucinations employ fundamentally different approaches; for example, [28] trained a model in a supervised manner on synthetic data to detect hallucinations at a token level. Our method falls into the generation-time category [15] or post-processing category [8,24], as we first generate a token before potentially altering it. Many prior works were specifically trained towards detecting and correcting hallucinations based on special datasets similar to [28], e.g., early works include [4,5]. One of the few generate-then-refine approaches is [7], which uses a knowledge graph to mitigate hallucinations focusing on entities like persons.

6 Conclusions

Transformer-based large language models have revolutionized natural language processing, with autoregressive models like GPT generating coherent text by predicting next tokens based on prior tokens, and autoencoding models like BERT capturing bidirectional context through masked token prediction. Our "generate-then-refine" approach constitutes a novel method for self-correction, yielding small but consistent improvements for next token predictions by combining a standard autoregressive model with an autoencoding model that predicts the second to last token using the top-k next tokens as context.

Disclosure of Interests. The authors have no competing interests to declare that are relevant to the content of this article.

References

1. Bao, H., et al.: Unilmv2: pseudo-masked language models for unified language model pre-training. In: International Conference on Machine Learning, pp. 642–652. PMLR (2020)
2. Barenboim, L., Elkin, M., Pettie, S., Schneider, J.: The locality of distributed symmetry breaking. J. ACM (JACM) **63**(3), 1–45 (2016)
3. Brown, T., et al.: Language models are few-shot learners. In: Advances in Neural Information Processing Systems (2020)
4. Cao, M., Dong, Y., Wu, J., Cheung, J.C.K.: Factual error correction for abstractive summarization models. arXiv preprint: arXiv:2010.08712 (2020)
5. Chen, S., Zhang, F., Sone, K., Roth, D.: Improving faithfulness in abstractive summarization with contrast candidate generation and selection. arXiv preprint: arXiv:2104.09061 (2021)
6. Devlin, J., Chang, M.W., Lee, K., Toutanova, K.: BERT: pre-training of deep bidirectional transformers for language understanding (2018). arXiv:1810.04805

7. Dziri, N., Madotto, A., Zaiane, O.R., Bose, A.J.: Neural path hunter: reducing hallucination in dialogue systems via path grounding. In: Proceedings of the Conference on Empirical Methods in Natural Language Processing (2021)
8. Ji, Z., et al.: Survey of hallucination in natural language generation. ACM Comput. Surv. **55**(12), 1–38 (2023)
9. Kamoi, R., Zhang, Y., Zhang, N., Han, J., Zhang, R.: When can LLMS actually correct their own mistakes? a critical survey of self-correction of LLMS. Trans. Assoc. Comput. Linguist. **12**, 1417–1440 (2024)
10. Lin, T., Wang, Y., Liu, X., Qiu, X.: A survey of transformers. Open AI **3**, 111–132 (2022)
11. Liu, Y., et al.: RoBERTa: a robustly optimized BERT pretraining approach (2019). arXiv:1907.11692
12. Madaan, A., et al.: Self-refine: iterative refinement with self-feedback. In: Advances in Neural Information Processing Systems, vol. 36 (2024)
13. OpenAI: Gpt-4 technical report (2023)
14. OpenAI: Hello gpt-4o! https://openai.com/index/hello-gpt-4o/ (2024). Accessed 19 Sep 2024
15. Pan, L., Saxon, M., Xu, W., Nathani, D., Wang, X., Wang, W.Y.: Automatically correcting large language models: surveying the landscape of diverse automated correction strategies. Trans. Assoc. Comput. Linguist. **12**, 484–506 (2024)
16. Radford, A.: Improving language understanding by generative pre-training. Open AI (2018)
17. Radford, A., et al.: Language models are unsupervised multitask learners. OpenAI blog (2019)
18. Raffel, C., et al.: Exploring the limits of transfer learning with a unified text-to-text transformer. J. Mach. Learn. Res. (2020)
19. Sanh, V., Debut, L., Chaumond, J., Wolf, T.: DistilBERT, a distilled version of BERT: smaller, faster, cheaper and lighter (2019). arXiv:1910.01108
20. Schneider, J.: Correlated initialization for correlated data. Neural Process. Lett. **54**(3), 2249–2266 (2022)
21. Schneider, J.: Explainable generative AI (GenXAI): a survey, conceptualization, and research agenda. Artif. Intell. Rev. **57**(11), 289 (2024)
22. Schneider, J.: What comes after transformers?–a selective survey connecting ideas in deep learning. arXiv preprint: arXiv:2408.00386 (2024)
23. Vaswani, A., et al.: Attention is all you need. In: Advances in Neural Information Processing Systems (2017)
24. Xiao, Z., Snoek, C.G.: Beyond model adaptation at test time: a survey. arXiv preprint: arXiv:2411.03687 (2024)
25. Xie, Y., et al.: COrAL: order-agnostic language modeling for efficient iterative refinement. arXiv preprint: arXiv:2410.09675 (2024)
26. Xu, W., et al.: LLMRefine: pinpointing and refining large language models via fine-grained actionable feedback. In: Findings of the Association for Computational Linguistics (2024)
27. Yang, Z., Dai, Z., Yang, Y., Carbonell, J., Salakhutdinov, R.R., Le, Q.V.: XLNet: generalized autoregressive pretraining for language understanding. In: Advances in Neural Information Processing Systems (2019)
28. Zhou, C., et al.: Detecting hallucinated content in conditional neural sequence generation. In: Findings of the Association for Computational Linguistics (2021)

Detection of Large Language Model Contamination with Tabular Data

Benoît Ronval[1(✉)], Pierre Dupont[1], and Siegfried Nijssen[1,2]

[1] UCLouvain, ICTEAM, Louvain-la-Neuve, Belgium
{benoit.ronval,pierre.dupont,siegfried.nijssen}@uclouvain.be
[2] KU Leuven, DTAI, Leuven, Belgium

Abstract. Large Language Models (LLMs) are increasingly used to solve machine learning tasks on tabular data, such as classification tasks. Using standard benchmarks, recent studies have shown impressive performance for such tasks. However, in this paper we study a major concern with the use of standard benchmarks: LLMs may have also been trained using these benchmarks, that is, these LLMs are contaminated. While previous work mostly focused on the GPT models, we propose a methodology to evaluate whether a large range of LLMs is contaminated. We propose new tests to detect contamination with tabular data over two aspects: knowledge and memorization. We also design an algorithm to parse the answers of the LLMs and detect hints of contamination. Our experiments conclude that the bigger and more closed-source an LLM is, the more likely it is contaminated.

Keywords: Contamination Detection · LLMs · Tabular Data

1 Introduction

Large Language Models (LLMs) have shown impressive performance on various benchmarks [26] but also on tasks involving data that is not part of their training corpus [8]. Assessment of this last point is usually done through *zero* and *few-shots* settings where the LLM sees no to few examples of the task to perform. In either case, only the examples, if any, are given to the model and no fine-tuning is done. LLMs can also be used with data other than text. In this work, we focus on tabular data due to their use in data analysis in many domains in research and industry [7]. Once converted into text with a *serialization* process, tabular data can be processed by LLMs for a given task. Thus, an LLM can be used as classifier without any further training, as done with TabLLM [15], which claimed to obtain comparable performance to classical machine learning models.

However, **how can we be certain the LLM did not see the data used for evaluation during its training?** Indeed, if LLMs are *contaminated* with tabular data, i.e. have seen test data during their training, their classification performance may be overestimated. This motivates the need for techniques to determine the level at which models are contaminated. However, this is not

G. Krempl et al. (Eds.): IDA 2025, LNCS 15669, pp. 234–245, 2025.
https://doi.org/10.1007/978-3-031-91398-3_18

straightforward as the training sets of most LLMs are not public. Models may also fail to provide hints of contamination if prompted incorrectly or even include filters limiting the output. Contamination can occur at different levels: models may have seen parts of the data, or know only the metadata of a dataset. Moreover, if prompting is used to determine contamination, an automatic analysis of the LLMs' output is needed to scale toward large number of datasets and LLMs.

Currently, companies developing LLMs have higher computing power than most universities. As part of their business plan, these companies claim to have increasingly powerful models. In this environment, we believe universities could play an objective role, study the proposed models and detect potential problems such as data contamination. We believe this area of study will grow in importance within academic research in the coming years.

In this work, we distinguish two types of contamination: *knowledge*, referring to metadata of the tabular dataset, and *memorization*, referring to the actual tabular examples [5]. We propose prompting techniques that allow to determine different forms and levels of contamination with new algorithms to analyze the LLMs' outputs. Our principal contributions are the following:

- We argue there are different forms and levels of contamination, and propose tests to detect knowledge and memorization contamination. Here we propose a test based on random perturbation of examples. A specific focus is to apply tests to a range of LLMs that is as wide as possible.
- We propose algorithms to detect hints of contamination in the output. They are designed for generalizable, automatic, fast, and reproducible testing.
- We perform a thorough study of contamination on many LLMs with many common tabular datasets. By doing so, we report which LLMs, closed and open-source, are more contaminated.
- We study the evolution of contamination between versions of LLMs and the impact of serialization.

We start this paper with the Related Work in Sect. 2 before explaining our Proposed Approach (Sect. 3) and Setup (Sect. 4) used in the work. Section 5 presents and analyzes the contamination results. We end with discussing the possible advantages of contamination in Sect. 6. The code is available on GitHub: https://github.com/bronval/LLM_tabular_contamination_detection.

2 Related Work

The field of contamination detection inside LLMs has mainly focused on text data and closed-source models, mostly GPT, with only few studies on open-source models. The most common approach is to use prompts designed to make the LLM output hints of contamination. Given the start of the sentence from a text example, the LLM is asked to complete it [13]. The output is then compared to the actual example through the BLEURT and ROUGE-L metrics [18,19] to decide if the LLM shows hints of contamination for this dataset. The authors discuss the use of another LLM which decides how similar the produced answer is to the actual example. A variant of this approach is to use a quiz format [12]. The prompt is phrased as a question with multiple choices corresponding, for example, to the remaining parts of the truncated example put in the prompt. The LLM then selects the proposition corresponding to the correct answer. This approach allows for simpler parsing of the output and may also avoid copyright filters from closed-source LLMs. Another prompt-based method is to use the probabilities of the output tokens and assume that examples seen during training will have higher probabilities [21]. However, since not all LLMs provide output probabilities, this method limits the comparisons between them.

Regarding tabular data, few studies have been done on contamination detection. Using serialization, researchers fell back on the aforementioned prompt approach [5]. Using the examples as text, they designed tests to detect knowledge or memorization contamination. For example, they ask the LLM to output the names of the features for a given dataset. Using few-shot classification with the studied LLMs, they show that there may be a link between contamination and high classification scores. They also indicated that the models generally pass the knowledge tests, i.e. contamination is detected, but have more difficulties regarding the memorization part. However, this research is limited to GPT-3.5 and GPT-4. Moreover, the parsing of the LLMs' outputs for the tests is based on the Levenstein distance. They also keep a single serialization throughout their work and do not study the impact of different templates on the results.

Previous works focus on closed-source LLMs, often overlooking the serialization impact. The contamination decision typically relies on text metrics or another model. To the best of our knowledge, no study compares closed and open-source LLMs with an algorithm to process the test outpus.

3 Proposed Approach

Our approach is prompt-based and relies on the answer produced by the LLM, aiming to apply to the widest range of models. After careful prompt design using in-context examples, context, or indicators for the beginning of the answer, we select the dataset and the LLM for which we want to test the contamination. We also select the serialization template used to transform the tabular examples. The tests are then launched independently from each other and our decision algorithms parse the output. For each output, they determine if there is a hint of contamination or not, i.e. if there is information in the output that could hardly have been produced without having seen the dataset before. Furthermore, we study the link between contamination and classification performance by using the LLMs as classifiers for the studied datasets following TabLLM's design [15].

3.1 Contamination Tests Descriptions

We detail here the different contamination tests. We distinguish the knowledge tests (K) from the memorization tests (M) and separate tests inspired by the literature [5] from the ones designed in this work.

Tests Inspired by the Literature

- **Feature List (K)**: Given the dataset name, the LLM must output its list of features. Additionally to the literature, we consider two setups: an uninformed case, where we only give the dataset name, and an informed case, where we add a subset of features to the prompt.
- **Feature Values (M)**: For the uninformed case, given the dataset name and a feature, the LLM must give the values for the feature. The informed case, new to our work, adds 4 serialized examples from the dataset. We select the examples such that there is always at least one remaining value to guess.
- **Incomplete Example Completion (M)**: Given 4 full examples and one with the last 3 features removed, the LLM must complete this last example. The test can be made more (resp. less) difficult by removing more (resp. less) features. The literature uses consecutive rows while we sample randomly all the examples making the test more difficult if the dataset is ordered.
- **Full Example Completion (M)**: Given the dataset name, its features, and the 4 first examples, the LLM must generate the next example. This assumes the dataset is ordered, as webpages may display data without shuffling them.
- **Feature Completion (M)**: Given the dataset name, features, 4 random examples, and a random example with a missing feature value, the LLM must complete it. Each example should have a different value for this feature.

New Tests. Our proposed tests have different advantages. Following the idea of the quiz format, we try to reduce the output of the LLM to a short answer that is easy to parse. We also consider the reverse case of classical tests; instead of asking to produce examples, we ask if the model recognizes them.

- **Dataset Recognition from Features (K)**: Given the list of features, the LLM must output the name of the corresponding dataset.
- **Dataset Recognition from Examples (M)**: Given 4 random serialized examples, the LLM must output the name of the corresponding dataset.
- **Example Membership (M)**: The task is to classify an example as original, i.e. taken directly from the dataset, or modified. A modified example is created by swapping the value of a feature with another value from the same feature obtained in another example, a classical perturbation seen in previous work [14]. The LLM must output "Yes" if the example to classify is original and "No" otherwise. To help the model, we put two examples in the prompt: one original, taken from the dataset as well, and its modified version, with their respective categories (original or modified). The test is done with 100 examples to classify, 50 original and 50 modified. This test does not have a binary contamination result: its score, similar to the accuracy, is used to evaluate to which extent the model can distinguish examples directly taken from the dataset from noisy examples.

3.2 Decision Algorithm for Contamination

Here, we explain the decision algorithm we propose. Given the LLMs' test outputs, the algorithm indicates if there is a hint of contamination or not. It distinguishes two scenarios: one for regular text output and one when the test output

Algorithm 1. Regular Text Decision Algorithm

Input: output (LLM's output), truth_words (the expected full answer of the test)
Output: True if contamination detected, False otherwise

```
1: unknown_words := { word ∈ truth_word | word ∉ English dictionary }
2: if unknown_words is empty then
3:     return ∀ word ∈ truth_words: word ∈ output
4: else
5:     return ∃ word ∈ unknown_words: word ∈ output
6: end if
```

involves examples. Note that in both cases, we limit the number of output tokens depending on the test to prevent false positive. A summary is provided in Table 1.

The first scenario is detailed in Algorithm 1 and is applied only to categorical features. It relies on an external English dictionary[1] to distinguish two cases: 1) there are words in the example that are specific to the dataset or 2) all the words in the example are actual English words (and thus included in the dictionary). In the first case, we consider a hint of contamination if at least one of these non-English words appears in the LLM's output as it is unlikely that a word not in the dictionary is accidentally generated by an uncontaminated LLM. For the second case, all the expected words have to appear to indicate the contamination. The test on which the algorithm is applied is passed, i.e. contamination is detected, if the algorithm returns True at least one time.

The second, simpler scenario checks that the value of every missing feature appears in the LLM's output. If one of these values does not appear with an exact match in the output, the test is failed. The test is passed if all missing values are in the output. Note that different features in the datasets we use have generally different values, reducing the risk of false positives.

For classification and membership tests, we do not use the decision algorithm as the LLMs are asked only to output "yes" or "no".

3.3 Serialization Details

To convert tabular data to text using serialization, we consider this template (whitespaces added for readability):

[specifier] [feature name] [value separator] [feature value] [feature separator]

The *feature name* and *feature value* are taken from the tabular example. The other elements have to be decided according to the sentence we aim to build. Here we consider 4 templates to assess the impact of this process: 1) "The [f. name] is [f. value]," [6], the default template for our tests, 2) "[f. name] = [f. value]," with an empty specifier, 3) "[f. name]: [f. value]", similar to a Python dictionary, and 4) "The [f. name] is equal to [f. value];", similar to a real sentence.

[1] https://pyenchant.github.io/pyenchant/.

Table 1. List of contamination tests and the algorithm used to assess their output.

Test Id	Test Name	Algorithm	Max. Output Tokens	Type
1	Feature list (inf.)	1	256	Knowledge
2	Feature list (uninf.)	1	256	Knowledge
3	Recognition (feat.)	1	16	Knowledge
4	Recognition (ex.)	1	16	Memorization
5	Feature values (inf.)	1	128	Memorization
6	Feature values (uninf.)	1	128	Memorization
7	Incomplete completion	2	256	Memorization
8	Full completion	2	512	Memorization
9	Feature completion	2	64	Memorization

4 Experiments Setup

4.1 LLMs

We separate the LLMs into 3 categories: the closed-source (no access to the model architecture and the training sets), the open-source without access to the training sets (we refer to them as semi-open-source), and the open-source where we have access to everything. For the first category, we access the models through the APIs. Otherwise, we use HuggingFace and consider the smallest version of the LLMs due to hardware limitations.

Closed-Source LLMs: We use 3 models from the GPT family: GPT-3.5 Turbo [8], GPT-4 [2], and GPT-4o[2]. More precisely, we use the checkpoints *gpt3-3.5-turbo-0125*, *gpt-4-0125-preview*, and *gpt-4o-2024-08-06*. We also use the Gemini-1.0 and Gemini-1.5 models [4].

Semi-open-source LLMs: From the Llama family, we use Llama 2 [24], Llama 3 [3], and Llama 3.1 [10], respectively with 7B, 8B, and 8B parameters. From the Gemma family, we use Gemma with 7B parameters [22] and Gemma 2 with 9B parameters [23]. We also use Phi-2 (2.7B parameters) [16] and Phi-3 (3.8B parameters) [1]. Finally, we consider Mistral (7B parameters) [17].

Open-Source LLMs: We use T0 [20] with 11B parameters. This LLM has been trained on different language modeling datasets, listed in the original paper. We also use GPT-J [25] with 6B parameters. It has been trained on The Pile [11], a dataset that combines 22 language modeling datasets.

4.2 Tabular Datasets

We consider common tabular datasets used in machine learning: Adult income, Bank marketing, California housing, German credit, Iris, Wine, and Titanic. We

[2] https://openai.com/index/hello-gpt-4o/.

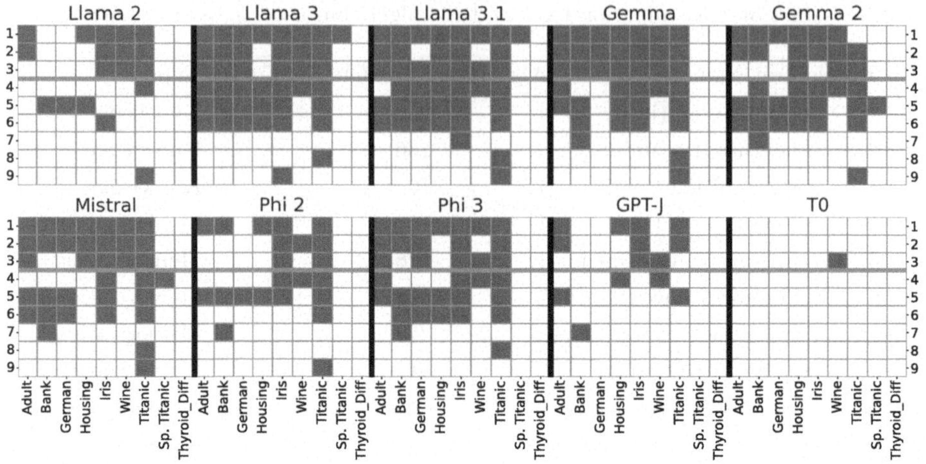

Fig. 1. Results for open-source (GPT-J, T0) and semi-open source LLMs. Green squares mark when contamination is detected. The bold, vertical black lines separate the models. The bold gray line splits the knowledge tests (above) from the memorization tests (below). (Color figure online)

also consider 2 more recent datasets: Spaceship Titanic, released just before the training cut-off of most closed-source models, and the Differentiated Thyroid Cancer Recurrence dataset, released long after the training cut-off of any LLM.

4.3 Classification Tests

We evaluate the classification accuracy of LLMs with 0 and 1-shot settings, giving one example of each class in this second case. The test is done over 100 random examples for each dataset. Following the approach of TabLLM, the prompts used are dataset specific. We ask the LLM to answer by "Yes" or "No" depending on the class of the output except in two cases: 1) for the housing dataset, we ask if the house price is higher than the median, and 2) for the iris and wine datasets, we ask the original class as they both contain 3 classes.

5 Results

In this section, we present the results of the tests described in Sect. 3. We present the contamination results and comment on the evolution between models from the same family (GPT, Llama, Gemma, Phi). Then, we discuss the impact of the serialization and display the contamination ranking based on the results.

5.1 Knowledge Tests

We observe that open-source models, GPT-J and T0 in Fig. 1, have little to no information about the datasets. On the contrary, most semi-open models (Fig. 1)

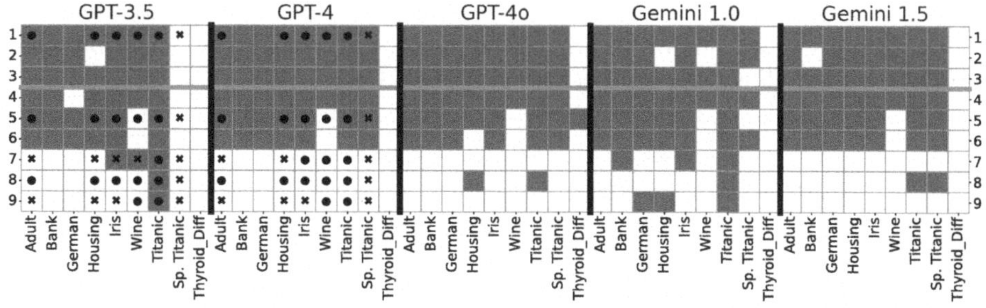

Fig. 2. Results for closed-source LLMs. Green squares mark when contamination is detected. The bold, vertical black lines separate the models. The bold gray line splits the knowledge (above) and the memorization tests (below). For GPT-3.5 and GPT-4, results from [5] are shown by a circle (contamination detected) or a cross (no detection). (Color figure online)

and all closed-source ones in Fig. 2 have extensive knowledge of the data. The informed case when the prompt contains a subset of the features tends to lead to more contamination detection. This confirms the idea that this task is not trivial and sensitive to many false negatives. Indeed, directing LLMs towards a certain dataset may result in an indication of contamination that did not appear with an uninformed prompt. The only exception is the Gemma 2 model with the Titanic dataset where we do not detect hints of contamination with the feature informed test. This shows that doing both informed and uninformed cases is useful in assessing the contamination of a model.

As expected due to their recent release dates, the Spaceship Titanic dataset is only known by the recent models (the closed-source) and the thyroid one is unknown to all models. This shows that our approach is unlikely to produce false positives and the results indicate true hints of contamination for the studied models and datasets. Compared to previous work that uses GPT models [5], we see similar results for the relevant tests and datasets. However, our experiments detect contamination in GPT-4 for the Spaceship Titanic dataset, hinting that the model contamination continues to evolve.

5.2 Memorization Tests

As it appears from the results, memorization tests are more difficult than the knowledge ones. This comes from the expected exact match to validate the completion tests as explained in Sect. 3. One exception is the Titanic dataset. We suppose these tests work well because of the *PassengerId* that has a unique value for each example and because it is a famous dataset from Kaggle[3]. Such combination of factors is difficult to have for other datasets, explaining why we detect less contamination with these completion tests. The difficulty of these tests varies with the number of features to complete and with the number of

[3] https://www.kaggle.com/competitions/titanic.

Table 2. Scores of LLMs. Mean accuracy over all datasets is reported for the membership, 0, and 1-shot tests. LLMs are ordered according to their contamination score, reported from the above figures, and scaled between 0 and 1.

	Gemini 1.0	GPT-4o	Gemini 1.5	GPT-4	GPT-3.5	Llama 3	Llama 3.1	Gemma	Mistral	Gemma 2	Phi 3	Phi 2	Llama 2	GPT-J	T0
Contamination	**0.60**	0.59	0.58	0.57	0.53	0.51	0.51	0.47	0.43	0.43	0.42	0.27	0.22	0.17	0.01
Membership	0.51	0.60	**0.70**	0.59	0.59	0.49	0.49	0.17	0.53	0.47	0.50	0.48	0.45	0.49	0.52
0-shot	0.56	0.59	0.51	**0.72**	0.64	0.47	0.49	0.44	0.34	0.15	0.41	0.02	0.23	0.03	0.58
1-shot	0.68	0.77	0.68	**0.78**	0.68	0.52	0.59	0.31	0.57	0.14	0.00	0.05	0.12	0.16	0.43

possible values for each feature. For example, the completion for the *income* feature (2 possible values) of the Adult income dataset will surely be easier than for the *education* feature (16 possible values).

The results for the membership test, shown in Table 2, are split into two parts. The open-source LLMs are close to the random baseline (0.5 accuracy) while the closed-source models perform better. This hints that these larger models can recognize easier examples that truly belong to the dataset. It also mostly matches the results of the other memorization tests where we detect more contamination for the closed-source models. We notice it is possible that LLMs sometimes output an invalid answer, i.e. other than "yes" or "no", thus obtaining a score largely below the random baseline.

As for the knowledge tests, the more an LLM is closed-source, the more it memorizes the data. Compared to previous work, the results are mostly the same [5]. However, GPT-4 fails more completion tests in our case. We recall that these tests are not performed the exact same way and that we expect an exact match while the other work is based on the Levenstein distance.

5.3 Contamination Evolution

We focus now on the differences in the contamination results between versions of models from the same family. We look at GPT-3.5, GPT-4, and GPT-4o, Gemini-1.0 and Gemini-1.5 (Fig. 2), Llama 2, Llama 3, and Llama 3.1, Gemma and Gemma 2, and Phi-2 and Phi-3 (Fig. 1). We observe that for most families, the contamination increases in the latest versions. We explain this phenomenon with three (not mutually exclusive) factors: 1) We suppose the training sets are larger for the latest versions, or at least contain new information compared to the older version; 2) Due to their increased capabilities, the newest LLMs may better model the task and thus answer more precisely; 3) The newest versions have probably been initialized with the weights of the previous model, leading to some contamination being transferred to the next version of the model.

We highlight some exceptions as the Gemini and Gemma models show less detected contamination. For the Gemini models, a look at the outputs indicates that more filters have been put in place in the model as Gemini-1.5 often claims to be unable to answer due to a lack of information while Gemini-1.0 passed the test. For Gemma models, the main difference in the results is visible in the feature recognition test. The absence of conclusive results is due to the model generating code to answer the question. We also observe that there is no significant difference between GPT-4 and GPT-4o, and Llama 3 and Llama 3.1.

5.4 Classification Scores and Ranking

Table 2 displays the ranking of the LLMs and their classification scores. We see that the most contaminated LLMs are often the largest ones. It also appears that the more a model is closed-source, the more it is contaminated. These two observations are linked but explain the result for T0, an LLM with 11B parameters with only two tests passed.

For the classification tests, the more contaminated a model is, the more likely it is to perform well. However, other aspects can impact these results. As for the membership test, LLMs may give an invalid answer and thus obtain a score below a random baseline. Since the most contaminated models are also the biggest ones, the results may be impacted by the capabilities of the model. It also depends on the inherent difficulty of the dataset for the classification task.

5.5 Influence of the Serialization

To study the impact of serialization, we use the templates described in Sect. 3 for GPT-4, Llama 3, Mistral, and Gemma 2. We only consider the memorization tests since only they require examples in the prompt. Our experiments show that there were little to no differences in the contamination results between the serialization templates. We recommend that future researchers use the template of their choice, without losing contamination results. The general template of the prompt appears to be more important than the serialization template.

6 Discussion

In this section, we comment on the results of our tests and on the contamination itself. First, a person or company with bad intentions may avoid contamination detection by tweaking the data during the training phase [9]. However, we believe this is not the case for the LLMs studied in this work given the obtained results.

Regarding the contamination results, we must be cautious about false positives and, more importantly, false negatives. False positives can be reduced by improving the decision algorithm or handling the number of maximum outputted tokens. However, our current tests may fail to detect certain forms of contamination. We would for example need to use other prompts to change the output probabilities or do the tests a sufficient number of times with high temperature values. The risk resulting from the false negative tests is to unintentionally bias the later tests (for example, classification) and report overestimated results.

We acknowledge that the best method to assess our approach would be to train LLMs in two settings, with and without contamination. However, hardware limitations prevent this method for the models used in this paper.

An interesting point is whether the contamination of LLMs is necessarily bad. As detailed before, contamination in the context of evaluating LLMs as classifier or similar tasks has to be avoided to prevent biased results. However, contamination, seen as prior knowledge of a dataset, may be useful in other cases. If we consider data generation [6], having prior knowledge about a dataset may lead to better synthetic data or reduce the time of fine-tuning.

7 Conclusion and Future Work

In this work, we study the detection of two types of tabular data contamination, knowledge, and memorization, for various LLMs with different levels of openness. To do this, we propose new tests to detect contamination and algorithms to decide if the LLM show hints of contamination in their outputs. We then apply these tests and algorithms to many different LLMs and tabular datasets.

We show that many LLMs have knowledge about the datasets we consider and some have memorized parts of the data. We highlight that the larger and the more closed-source a model is, the more likely it is to be contaminated with tabular data. This shows the need for careful use of LLMs for tasks such as classification, where the LLMs may know the benchmark under use.

As guidelines for future works, we highly recommend performing a contamination check if the task under study may be impacted by prior knowledge of the data. Any LLM could be used with the condition that our tests do not detect contamination to ensure that the results would not be biased. A possibility is also to use an LLM that is not contaminated, such as T0 used in TabLLM [15]. One important goal of future research for contamination detection is the reduction of false negative results. As mentioned in Sect. 6, this would require either new prompt designs or entirely new approaches.

Acknowledgments. Computational resources have been provided by the supercomputing facilities of the Université catholique de Louvain (CISM/UCL) and the Consortium des Équipements de Calcul Intensif en Fédération Wallonie Bruxelles (CÉCI) funded by the Fond de la Recherche Scientifique de Belgique (F.R.S.-FNRS) under convention 2.5020.11 and by the Walloon Region.

Disclosure of Interests. The authors have no competing interests to declare that are relevant to the content of this article.

References

1. Abdin, M., et al.: Phi-3 technical report: a highly capable language model locally on your phone. arXiv preprint: arXiv:2404.14219 (2024)
2. Achiam, J., et al.: GPT-4 technical report. arXiv preprint: arXiv:2303.08774 (2023)
3. AI, M.: Introducing meta Llama 3: the most capable openly available LLM to date (2024). https://ai.meta.com/blog/meta-llama-3/
4. Anil, R., et al.: Gemini: a family of highly capable multimodal models. arXiv preprint: arXiv:2312.11805 (2023)

5. Bordt, S., Nori, H., Caruana, R.: Elephants never forget: testing language models for memorization of tabular data. arXiv preprint: arXiv:2403.06644 (2024)
6. Borisov, V., Seßler, K., Leemann, T., Pawelczyk, M., Kasneci, G.: Language models are realistic tabular data generators. arXiv preprint: arXiv:2210.06280 (2022)
7. van Breugel, B., van der Schaar, M.: Why tabular foundation models should be a research priority. arXiv preprint: arXiv:2405.01147 (2024)
8. Brown, T., et al.: Language models are few-shot learners. In: Advances in Neural Information Processing Systems, vol. 33, pp. 1877–1901 (2020)
9. Dekoninck, J., Müller, M.N., Baader, M., Fischer, M., Vechev, M.: Evading data contamination detection for language models is (too) easy. arXiv preprint: arXiv:2402.02823 (2024)
10. Dubey, A., et al.: The Llama 3 herd of models. arXiv preprint: arXiv:2407.21783 (2024)
11. Gao, L., et al.: The Pile: An 800gb dataset of diverse text for language modeling. arXiv preprint: arXiv:2101.00027 (2020)
12. Golchin, S., Surdeanu, M.: Data contamination quiz: a tool to detect and estimate contamination in large language models. arXiv preprint: arXiv:2311.06233 (2023)
13. Golchin, S., Surdeanu, M.: Time travel in Llms: tracing data contamination in large language models. arXiv preprint: arXiv:2308.08493 (2023)
14. Hanhijärvi, S., Ojala, M., Vuokko, N., Puolamäki, K., Tatti, N., Mannila, H.: Tell me something i don't know: randomization strategies for iterative data mining. In: Proceedings of the 15th ACM SIGKDD International Conference on Knowledge Discovery and Data Mining, pp. 379–388 (2009)
15. Hegselmann, S., Buendia, A., Lang, H., Agrawal, M., Jiang, X., Sontag, D.: TabLLM: few-shot classification of tabular data with large language models. In: International Conference on Artificial Intelligence and Statistics, pp. 5549–5581. PMLR (2023)
16. Javaheripi, M., et al.: Phi-2: the surprising power of small language models. Microsoft Research Blog (2023)
17. Jiang, A.Q., et al.: Mistral 7B. arXiv preprint: arXiv:2310.06825 (2023)
18. Lin, C.Y.: Rouge: A package for automatic evaluation of summaries. In: Text Summarization Branches Out, pp. 74–81 (2004)
19. Papineni, K., Roukos, S., Ward, T., Zhu, W.J.: BLEU: a method for automatic evaluation of machine translation. In: Proceedings of the 40th annual meeting of the Association for Computational Linguistics, pp. 311–318 (2002)
20. Sanh, V., et al.: Multitask prompted training enables zero-shot task generalization. arXiv preprint: arXiv:2110.08207 (2021)
21. Shi, W., et al.: Detecting pretraining data from large language models. arXiv preprint: arXiv:2310.16789 (2023)
22. Team, G., et al.: Gemma: open models based on Gemini research and technology. arXiv preprint: arXiv:2403.08295 (2024)
23. Team, G., et al.: Gemma 2: improving open language models at a practical size. arXiv preprint: arXiv:2408.00118 (2024)
24. Touvron, H., et al.: Llama 2: open foundation and fine-tuned chat models. arXiv preprint: arXiv:2307.09288 (2023)
25. Wang, B., Komatsuzaki, A.: GPT-J-6B: A 6 Billion Parameter Autoregressive Language Model (2021). https://github.com/kingoflolz/mesh-transformer-jax
26. Zhao, W.X., et al.: A survey of large language models. arXiv preprint: arXiv:2303.18223 (2023)

Imbalanced Data Clustering via Targeted Data Augmentation Using GMM and LLM

Noor Khalal[1(✉)], Abdallah Alaa-Eddine Djamai[2], Imed Keraghel[1,2], and Mohamed Nadif[1]

[1] Centre Borelli UMR9010, Université Paris Cité, 75006 Paris, France
noor.khalal@etu.u-paris.fr
[2] Kernix Software, 75014 Paris, France

Abstract. In Natural Language Processing (NLP), dealing with underrepresented topics is challenging, especially in unsupervised tasks where clustering might not adequately capture minority topics. To tackle this challenge, our paper presents a novel unsupervised data augmentation method that integrates Gaussian Mixture Models (GMMs) and Large Language Models (LLMs). Due to their flexibility and robustness, GMMs can detect clusters corresponding to underrepresented areas in the data, while LLMs create synthetic documents to enrich these clusters and improve their representation. Experiments on various imbalanced text datasets demonstrate that our approach preserves clustering performance in all cases and often enhances cluster interpretability, offering a robust and scalable solution for improving data representation in unsupervised NLP tasks.

Keywords: Data Augmentation · Large Language Models · Gaussian Mixture Models · Unsupervised Learning · Clustering

1 Introduction

In unsupervised NLP tasks, the quality of data representation is crucial for effective clustering. However, real-world datasets often contain underrepresented semantic regions, leading to dramatically imbalanced clusters that are challenging for most clustering algorithms. These underrepresented clusters may contain meaningful information that remains poorly captured by traditional clustering algorithms, leading to irrelevant or difficult-to-interpret groupings.

Traditional data augmentation techniques, such as synonym replacement and random insertion [29], aim to increase data diversity by generating new samples from existing ones. Although these methods improve dataset size and variability, they typically apply augmentation uniformly across the entire dataset, without considering differences in data distribution. This uniformity does not specifically address sparsely represented regions. Moreover, many existing approaches depend on labeled data to perform class-specific augmentation or resampling [2,9,13], which makes them unsuitable for unsupervised learning scenarios where class labels are unavailable.

G. Krempl et al. (Eds.): IDA 2025, LNCS 15669, pp. 246–260, 2025.
https://doi.org/10.1007/978-3-031-91398-3_19

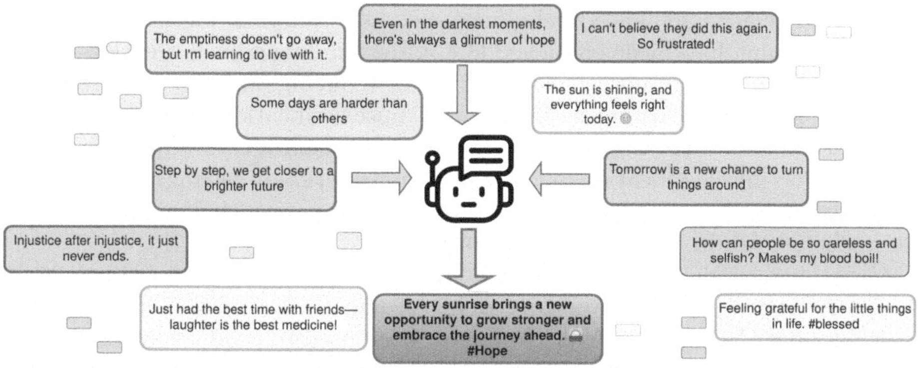

Fig. 1. Example from the Tweet Emotion dataset: four emotion clusters (joy in yellow, sadness in blue, anger in red, optimism in purple) where an LLM generates a new tweet about optimism based on three representative samples. (Color figure online)

Generative models, particularly Large Language Models (LLMs), have recently gained attention for their ability to produce high-quality synthetic text by capturing complex semantic relationships within the data. However, most of the existing work employs LLMs for uniform augmentation across datasets [31], which fails to address the imbalance problem.

To overcome these limitations, we present a new method that combines Gaussian Mixture Models (GMMs) [1] and LLMs to perform targeted data augmentation. GMMs are used to analyze the distribution of document embeddings, pinpointing sparse regions that correspond to underrepresented areas in the dataset. LLMs are then used to create synthetic documents in these sparse regions, thereby improving their representation. Figure 1 illustrates how, in the *Tweet Emotion* dataset, three tweets about optimism are given to an LLM, which generates a new tweet aligned with the same theme. This process not only adds diversity to the corpus but also enhances its representation. Through comprehensive experiments on various imbalanced text datasets, we show that this augmentation does not diminish clustering algorithm performance. In the worst-case scenario, it maintains the same performance levels, while often leading to enhancements in the interpretability of clusters.

2 Background and Related Work

Data augmentation is a widely used technique in NLP to enhance the diversity and size of training datasets by generating new data from existing samples [2,19]. In text classification tasks, traditional augmentation methods such as synonym replacement, random insertion and deletion-collectively known as Easy Data Augmentation (EDA) [29]-aim to create varied versions of existing texts to improve model robustness. While these approaches can increase data variability, they often fail to address the issue of class imbalance, as they typically apply augmentation uniformly across all classes.

To tackle class imbalance, techniques such as resampling methods and cost-sensitive learning have been employed [3,13]. Resampling adjusts the class distribution by replicating minority class samples or reducing majority class samples, while cost-sensitive learning assigns higher misclassification costs to minority classes. However, these methods rely on label information to adjust class distributions or misclassification penalties. In unsupervised settings, where class labels are unavailable, applying these techniques becomes challenging, limiting their effectiveness in balancing the dataset.

Generative models have also been explored for data augmentation in NLP. Generative Adversarial Networks (GANs) [6] and their variants, such as Deep Convolutional GANs (DCGANs) [20], have been used to produce synthetic text data by learning the data distribution and generating new samples accordingly. More recently, LLMs have been used to generate diverse textual data for augmentation purposes [15,28]. While these models can produce high-quality text, prior work often employs them to augment data uniformly across all classes without specifically targeting underrepresented ones. For instance, methods like AugGPT [4] rephrase all documents multiple times but do not address data imbalance, leaving minority classes underrepresented.

Despite these advancements, there remains a gap in the literature regarding data augmentation strategies that combine clustering methods and generative models to target underrepresented topics in unsupervised settings. Existing approaches either require labeled data or do not specifically address class imbalance. Developing effective augmentation strategies for imbalanced data in unsupervised learning remains a challenge and necessitates novel methods that can generate and utilize synthetic data without predefined labels. Our proposal addresses this gap by using GMMs to identify underrepresented clusters and employing LLMs to generate synthetic documents, providing a strategy that requires no labeled data nor prior knowledge of the class structure.

3 Contribution

In this section, we outline our approach for augmenting textual data through a combination of embedding techniques and model-based Gaussian methods [1]. The methodology involves several key steps: 1) document representation, 2) clustering with the Expectation-Maximization algorithm (EM) [5] , 3) synthetic data point generation, and 4) using an LLM for data augmentation. The approach is illustrated in Fig. 2.

3.1 Document Representation

To effectively represent documents for clustering, we use a Transformer-based embedding model to generate embeddings that capture rich semantic information. These embeddings are particularly suitable for clustering tasks because they enhance the ability to identify subtle semantic connections between unlabeled textual data [14].

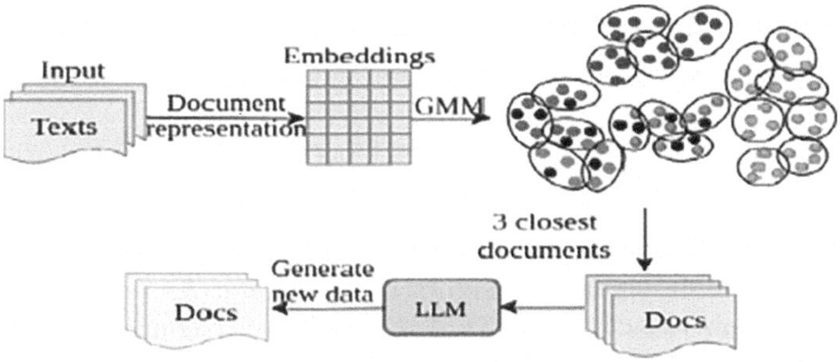

Fig. 2. Workflow for generating synthetic documents in a dataset. 1) Text inputs are processed using a Transformer-based embedding model, 2) The EM algorithm derived from GMM is used to cluster these embeddings, 3) Synthetic data points are generated in clusters with the highest volume-to-proportion ratios and their three nearest documents are identified, and 4) The three documents are given to an LLM to generate new textual instances.

We use Uniform Manifold Approximation and Projection (UMAP) [16] for dimensionality reduction, to enhance clustering efficiency. UMAP, a nonlinear technique, preserves data structures while reducing dimensions, maintaining essential relationships and reducing computational complexity. This allows clustering algorithms like GMMs to operate more efficiently by simplifying and speeding up computations.

3.2 Targeted Clustering with GMM

Clustering with Gaussian Mixture Models. In a finite GMM, the data $\mathbf{x}_1, \ldots, \mathbf{x}_n$ are assumed to be a sample of n independent instances of a random variable $\mathbf{X}$ in $\mathbb{R}^d$ where d is the dimensionality of the space. The density of the data can be expressed as:

$$f(\mathbf{x}_i; \Theta) = \sum_{k=1}^{g} \pi_k \varphi_k(\mathbf{x}_i | \mu_k, \Sigma_k), \quad \forall i \in \{1, \ldots, n\}$$

where $\Theta = (\pi_1, \ldots, \pi_g, \mu_1, \ldots, \mu_g, \Sigma_1, \ldots, \Sigma_g)$, $\varphi_k(\mathbf{x}_i|\mu_k, \Sigma_k)$ is the kth component density for observation $\mathbf{x}_i$ with parameters (μ_k, Σ_k), $(\pi_1, \ldots, \pi_g)$ are the mixing weights or probabilities (such that $\pi_k > 0, \sum_k \pi_k = 1$) and g is the number of mixture components. This means that the clusters are ellipsoidal, with the mean vector μ_k at the center, and other geometric features such as volume, shape, and orientation are determined by the covariance matrix Σ_k [1]. To estimate the parameter Θ, the log-likelihood is maximized, which is given by:

$$L(\mathbf{X}; \Theta) = \sum_{i=1}^{n} \log \left(\sum_{k=1}^{g} \pi_k \varphi_k(\mathbf{x}_i | \mu_k, \Sigma_k) \right).$$

Maximization uses the Expectation-Maximization (EM) algorithm, which iteratively maximizes the conditional expectation of the complete log-likelihood given Θ'.

$$Q(\Theta|\Theta') = \sum_i \sum_k s_{ik} \log\left(\pi_k \varphi_k(\mathbf{x}_i|\mu_k, \Sigma_k)\right).$$

where $s_{ik} = \frac{\pi_k \varphi_\ell(\mathbf{x}_i|\mu_k,\Sigma_k)}{\sum_\ell \pi_\ell \varphi_\ell(\mathbf{x}_i|\mu_\ell,\Sigma_\ell)}$ are the posterior probabilities.

The GMMs provide valuable insights into clusters. An eigen-decomposition of the covariance matrices $\Sigma_k = \lambda_k D_k A_k D_k^\top$, where λ_k controls the volume, A_k specifies the shape with $\det(A_k) = 1$, and D_k determines the orientation. This decomposition allows for the detection of clusters with different shapes, volumes, and orientations. The parameter π_k represents the proportion of each cluster.

However, GMMs can face challenges with irregular classes. By increasing the number of clusters, it is possible to capture varying shapes, orientations, sizes, and proportions, as illustrated in Fig. 3, thereby enabling more accurate data modeling. In contrast, K-means optimizes a criterion that assumes spherical classes with equal volumes and proportions. In our work, we fit a GMM to the reduced embeddings. This approach ensures that each cluster contains enough data points to accurately calculate its volume, which is essential for our subsequent analysis.

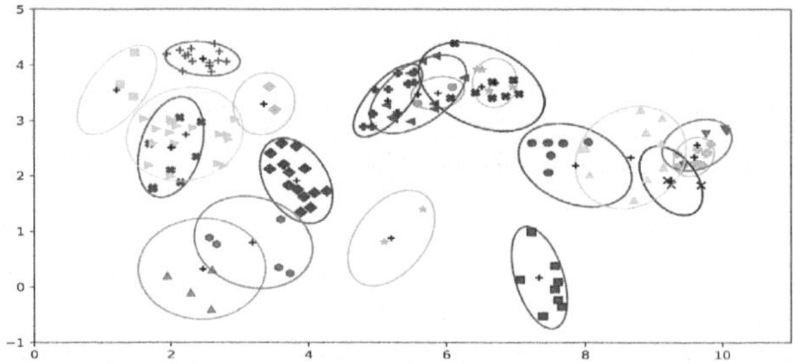

Fig. 3. Clusters with various orientations, shapes, proportions, and volumes.

Volume and Proportion Analysis. The clusters obtained by EM can be characterized by their proportion and volume. The weight of a cluster, π_k, represents the proportion of the dataset expected to belong to it, while the volume V_k is given by:

$$V_k = \frac{\pi_k^{d/2}}{d\Gamma(d/2)} \sqrt{\det(\Sigma_k)}$$

where d is the dimensionality of the data, Γ is the Gamma function and $\det(\Sigma_k)$ is the determinant of the covariance matrix.

However, when working with high-dimensional low-variance text embeddings such as in our case, the task of calculating the determinant of the covariance matrix can become ill-posed, due to the eigenvalues being close to zero, making the matrix nearly or fully singular and nullifying the determinant, and thus making the computation of cluster volumes impossible.

To stabilize the calculation and ensure a well-posed problem, Tikhonov regularization [27] is applied by adding a small positive constant to the diagonal elements of the covariance matrix [10]. Mathematically, the regularized matrix is given by $\Sigma_k^{'} = \Sigma_k + \alpha I$, where I is the identity matrix and α is the regularization parameter. This adjustment has the effect of shifting the eigenvalues away from zero, thus making the matrix invertible and the determinant (and consequently cluster volume) calculation stable.

Identifying Underrepresented Clusters. Clusters with a high volume but a low proportion highlight regions in the feature space where data points are sparse and spread out, indicating potential underrepresentation. By focusing on clusters with high volume-to-proportion ratios, we target regions where there is significant variability but relatively few data points. This is crucial for targeted data augmentation, where increasing the representation of underrepresented regions can lead to a more balanced and robust dataset.

In our work, we rank all clusters based on their volume-to-proportion ratios to identify underrepresented regions in the data. This approach highlights clusters that have a low proportion of data points (less dense) but occupy a large space in the embedding dimension (higher volumes). By focusing our augmentation on the top half of clusters with the highest volume-to-proportion ratios, we target areas where data is sparse yet potentially significant. This ensures that our data augmentation efforts are directed towards the most *critical* and *underrepresented* regions of the dataset.

3.3 Targeted Data Augmentation

Synthetic Data Point Generation. In each *underrepresented cluster* selected as described above, new data points are randomly generated using the Gaussian distribution parameters of the cluster. This approach ensures that the generated points follow the same distribution characteristics as the original data within each cluster. First, the number of samples from each Gaussian component is determined using the multinomial distribution based on the weights π_k. More specifically, the total number of samples n_{samp} is allocated among the g components using the multinomial distribution:

$$P(n_1, n_2, \ldots, n_g) = \frac{n_{\text{samp}}!}{n_1! n_2! \cdots n_g!} \pi_1^{n_1} \pi_2^{n_2} \cdots \pi_g^{n_g}.$$

This results in a set of counts $(n_1, n_2, \ldots, n_g)$ where each n_k represents the number of samples to draw from the k-th component. Then, for each component k, samples X_k are drawn from a multivariate normal distribution with mean μ_k and covariance Σ_k. This method ensures that the synthetic data points maintain the integrity and structure of each cluster.

Textual Document Generation. For each sampled point in the reduced representation space, we identify the three closest neighbors among the original documents based on Euclidean distance. These neighbors correspond to real textual documents that are most similar to the sampled point in the embedding space. Using the content of these three nearest documents, we prompt the LLM to generate a new textual document that reflects their combined themes and content. This approach ensures that the generated document is coherent and consistent with the semantic context of the original documents.

All the steps of our method are outlined in Algorithm 1. It is important to note that simply increasing the data size is not always advantageous. Instead, our approach focuses on augmenting clusters that are identified as sparse by GMMs. This ensures that the added data enhances the representation of these clusters rather than merely increasing the overall dataset size.

Algorithm 1. Document Clustering and Generation

Input: $D = \{d_1, \ldots, d_n\}$, g (number of components), k^* number of augmented clusters, $\mathcal{M}$ (an embedding model), $\mathcal{M}^*$ (an instruction-tuned model)
$X_{\text{emb}} \leftarrow \mathcal{M}(D)$
$X_R \leftarrow \text{UMAP}(X_{\text{emb}})$
$\{C_1, \ldots, C_g\} \leftarrow \text{EM}(X_R)$
for $k \leftarrow 1, g$ **do**
    $S_k \leftarrow \frac{V_k}{\pi_k}$
end for
Sort $\{C_1, \ldots, C_g\}$ according to S_k
Retain k^* clusters
for $k \leftarrow 1, k^*$ **do**
    Generate $(DA)_k$
end for
for $k = 1 \leftarrow k^*$ **do**
    $(DA)^*_k = \{\}$
    for each $i \in (DA)_k$ **do**
        Select 3 closest points of i
        $New\_Doc \leftarrow \mathcal{M}^*$(3 closest documents)
        $(DA)^*_k \leftarrow (DA)^*_k \cup \{New\_Doc\}$
    end for
end for
$(DA)^* \leftarrow \bigcup_{k=1}^{k^*} (DA)^*_k$
$X^*_{\text{emb}} \leftarrow X_{\text{emb}} + \mathcal{M}((DA)^*)$
$X^*_R \leftarrow \text{UMAP}(X^*_{\text{emb}})$
$\{C^*_1, \ldots, C^*_p\} \leftarrow \text{Clustering}(X^*_R)$ // such as Kmeans Clustering For Evaluation purposes

4 Experiments and Results

Since our approach operates in an unsupervised setting, we evaluate the effectiveness of our targeted data augmentation by applying clustering algorithms to the data. Specifically, we compare the performance of clustering on the original dataset with that on the augmented dataset using baseline methods such as KMean. In this section, we describe our experimental setup and present the results and analyses to demonstrate the effectiveness of our approach.

4.1 Datasets

We use five datasets in our experiments, four of which come from the Massive Text Embedding Benchmark (MTEB) [17]. Each dataset is represented as $D = (nClasses, nDocs, Balance, Avg)$ where $nClasses$ represents the number of groundtruth classes of the dataset, $nDocs$ is the number of total documents of the dataset, $Balance$ is the ratio of the number of documents in the minority class to that in the majority class, expressed in the power of 10, it reflects the degree of class imbalance in the dataset, and Avg is the average number of tokens per document in the datasets.

Arxiv $(12, 7000, 6.8 \times 10^{-3}, 10)$, **Biorxiv** $(26, 53787, 4.1 \times 10^{-4}, 13)$, and **Medrxiv** $(51, 17647, 4.9 \times 10^{-4}, 16)$ are sourced from MTEB [17], specifically the ArxivClusteringS2S, BiorxivClusteringS2S, and MedrxivClusteringS2S datasets[1]. Each dataset consists of article titles from their respective repositories. The cluster labels correspond to categories assigned to the papers by humans. **Reddit** $(15, 5114, 1.2 \times 10^{-3}, 11)$ comprises titles of Reddit posts, sourced from the RedditTitleBody dataset[2]. Each post is associated with a subreddit, which serves as the cluster label. The dataset reflects a variety of topics and an unbalanced distribution across subreddits. The dataset **Tweet Emotion**$(4, 3257, 2.1 \times 10^{-1}, 16)$ consists of tweets labeled with one of four emotions: anger, joy, optimism, and sadness.

4.2 Experimental Setup

Embedding Generation. We employ the *NoInstruct-Small-v0* embedding model[3], which generates embeddings of size 384. This model is selected based on its strong performance on the MTEB leaderboard[4], balancing meaningful semantic representation with computational efficiency. The 384-dimensional embeddings are sufficiently large to capture the nuances of the text while being manageable for subsequent processing.

Dimensionality Reduction. To facilitate efficient clustering, we apply UMAP to reduce the embeddings to 10 dimensions. UMAP effectively preserves the local and global structure of the data in lower dimensions. Reducing to 10 dimensions ensures that, in the GMM clustering step, the number of dimensions does not exceed the number of data points in any cluster, which is important for stable covariance matrix estimation.

[1] https://github.com/embeddings-benchmark/mteb/.
[2] https://huggingface.co/datasets/sentence-transformers/reddit-title-body.
[3] https://huggingface.co/instructor/NoInstruct-Small-v0.
[4] https://huggingface.co/spaces/mteb/leaderboard.

Clustering Parameters. We fit a Gaussian Mixture Model (GMM) to the reduced embeddings X_R. The number of components g is set to $\lceil nDocs^{1/3} \rceil$, following the recommendation in [30]. We use full covariance matrices to allow each cluster to have its own covariance, providing flexibility in modeling the data distribution. Thus, the parameters are estimated using the EM algorithm, and the clusters are determined based on the maximum a posteriori principle.

Augmentation Process. After obtaining the GMM clusters, we compute the volume-to-proportion ratio $S_k = \frac{V_k}{\pi_k}$ for each cluster, where V_k is the volume of cluster C_k and π_k is its mixture weight. Clusters are sorted in descending order based on S_k, highlighting clusters that are large in volume but contain relatively few data points, indicative of underrepresented regions in the data. We select the top k^* clusters with the highest S_k values for augmentation. For each selected cluster, we sample additional points within the cluster's distribution to generate synthetic embeddings $(DA)_k$. For each synthetic embedding, we identify the three nearest original embeddings in X_R using Euclidean distance. The corresponding documents of these neighbors are retrieved to provide context for the LLM. We use *Mistral-7B-Instruct-v0.2* [12] as our instruction-tuned LLM for text generation. This model is capable of generating coherent and contextually relevant text based on the provided prompts. By feeding the LLM with the texts of the three nearest neighbors, we generate new documents that reflect the combined content of these examples. The generated documents $(DA)^*$ are then embedded using the same embedding model $\mathcal{M}$ and added to the original embeddings X_{emb} to form the augmented dataset X^*_{emb}. UMAP is applied again to reduce the dimensionality of the augmented embeddings, resulting in X^*_R for clustering evaluation.

To assess the impact of our targeted data augmentation, we applied two baseline clustering algorithms to both the original and augmented datasets. These algorithms are chosen for their effectiveness and simplicity in clustering applications.

- **KMeans:** A widely used algorithm that partitions data into k clusters by minimizing the sum of squares within the cluster. We used the implementation from scikit-learn with 'k-means++' initialization for better convergence and clustering performance.
- **Spherical KMeans (SKmeans)** is a variant of KMeans designed for high-dimensional data, where data points are normalized to lie on the surface of a unit sphere. It uses cosine similarity instead of Euclidean distance to measure the closeness between points and cluster centroids.

4.3 Results and Analysis

To evaluate the impact of our targeted data augmentation, we examine three aspects:
(1) cluster distribution, to observe how augmentation influences the balance of

clusters; (2) keyword comparison, to assess improvements in thematic coverage and interpretability; and (3) clustering performance, evaluated using two widely used metrics: *Normalized Mutual Information* (NMI) [26], which measures the quality of the clustering against ground truth, and *Adjusted Rand Index* (ARI) [25], which evaluates the similarity between predicted and true cluster assignments while accounting for chance.

Cluster Distribution. Figure 4 illustrates the document distributions across clusters for two datasets: Reddit and Arxiv. These datasets were selected due to their manageable number of clusters (12 and 15, respectively), which allow for a clear visualization of distribution patterns. The bar graphs depict three scenarios: the original class labels (blue zigzags), the baseline KMeans clustering applied to the original non-augmented data (red bricks), and the KMeans clustering after data augmentation (pink dots).

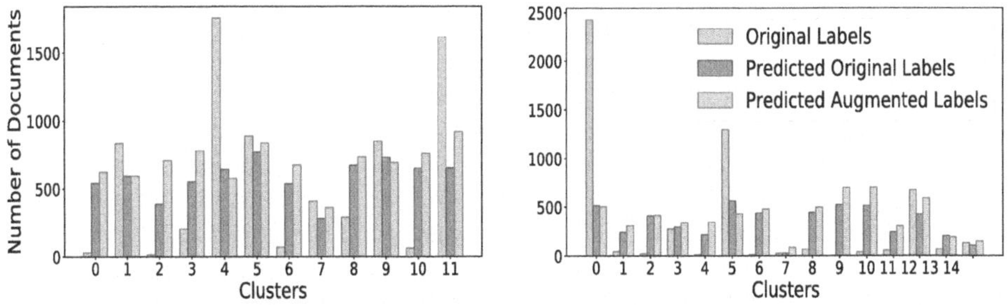

Fig. 4. Document distributions across clusters for the Arxiv (left) and Reddit (right) datasets.

In the original labels (blue zigzags), the imbalance of the datasets is evident, with some classes overrepresented and others underrepresented. The baseline KMeans (red bricks) uniformly distributes documents, reducing the dominance of overrepresented classes but not accurately reflecting the initial imbalance. After data augmentation (pink dots), KMeans continues to equalize cluster sizes. Yet, an important trend emerges: peaks in the augmented data (pink dots) often align with clusters corresponding to minority classes in the original distribution (blue zigzags).

In summary, data augmentation improves the representation of minority classes, demonstrating the effectiveness of this strategy in addressing data imbalance. However, it is now reasonable to ask two questions: a) What does our methodology contribute in terms of the interpretability of clusters? b) Does this augmentation challenge the quality of the clustering?

Keyword Comparison. To address the interpretability aspect of the classes, we first rely on the analysis of distribution of keywords in the KMeans clusters on augmented and non-augmented data. The keywords are extracted using

KeyBERT; a keyword extraction technique that uses contextual embeddings to identify the most representative terms within a cluster.

Table 1 presents the top 12 keywords (based on their frequency) for each cluster from both augmented and non-augmented data, spanning various datasets: *Tweet_Emotions*, *Reddit*, and the scientific corpora (*Arxiv*, *Biorxiv*, and *Medrxiv*). The correspondence between clusters in the augmented and non-augmented datasets is identified using the Hungarian algorithm. The analysis highlights how augmentation affects underrepresented and overrepresented clusters across different domains. In the *Tweet_Emotions* dataset, the augmentation enriched the *Optimism* cluster with new positive terms such as *happy* and *smile*, while negative words like *nervous* and *panic* were removed. This resulted in a sharper focus on optimistic themes. For the *Anger* cluster, new terms such as *game* and *racism* were introduced, reflecting broader and more intense expressions of anger. However, much of the cluster remained stable, as expected for an overrepresented category.

In the *Reddit* dataset, the underrepresented *Skincare* cluster became more specific with new terms like *cream* and *toner*, while the overrepresented *Dogecoin* cluster showed minimal changes, maintaining its focus on *doge* with slight increases in words such as *bought* and *currency*.

For the scientific corpora, the augmentation had varied impacts. In the *Arxiv* dataset, the *Economics* cluster introduced terms like *regression* and *sparsity*, broadening its methodological scope. The *Computer Science* cluster remained stable with minor additions. In the *Biorxiv* dataset, the *Animal Behavior and Cognition* (*A B&C*) cluster introduced sensory-related terms like *sleep* and *taste*, reflecting a focus on sensory mechanisms. In the *Neuroscience* cluster, new terms like *brain* emerge, while core words show a tenfold increase (e.g., *cortex*, increasing from 69 to 546). This reflects the proportional increase in generated data relative to the dataset's large size. In the *Medrxiv* dataset, the neurology cluster highlighted terms like *parkinsons* and *stroke*, emphasizing neurological conditions. The *Epidemiology* cluster introduced terms like *sarscov2* and *testing*, reinforcing diagnostics and epidemic detection.

Overall, the results demonstrate that augmentation effectively enriched underrepresented clusters by introducing domain-specific and diverse terms, improving their interpretability. Overrepresented clusters, as expected, showed minimal changes, preserving their structure while occasionally gaining some frequency shifts.

Clustering Performance. In this section, we present the clustering performance analysis summarized in Table 2. The results compare clustering metrics (NMI and ARI) across our five datasets, for two algorithms: KMeans and SKMeans, using non-augmented and augmented data. Metrics, averaged over five runs, show that augmentation often improves performance, particularly in ARI, and maintains comparable results even in worst-case scenarios, ensuring that clustering quality is not compromised.

Table 1. Comparison of top words and their frequencies in overrepresented (+) and underrepresented (-) clusters before and after data augmentation. The number at the bottom right of each word represents the frequency of the keyword in the dataset vocabulary. Words in bold are new terms introduced in the augmented data, while the rest of the words are shared between both datasets and are underlined when they have the higher frequency.

Cluster	Non-Augmented	Augmented	Changes Observed
Optimism (-)	depression$_{58}$, life$_{38}$, day$_{26}$, feel$_{24}$, nervous$_{22}$, lost$_{21}$, panic$_{16}$, optimism$_{16}$, despair$_{15}$, gloomy$_{15}$, love$_{14}$, shy$_{13}$	depression$_{58}$, **music**$_{40}$, life$_{40}$, day$_{36}$, feel$_{25}$, lost$_{22}$, **happy**$_{22}$, love$_{21}$, optimism$_{17}$, **sober**$_{17}$, **smile**$_{16}$, **birthday**$_{15}$	Augmentation expanded the focus to more optimistic and positive themes, introducing diversity while reducing emphasis on negative or neutral terms.
Anger (+)	angry$_{41}$, bully$_{35}$, outrage$_{30}$, terror$_{28}$, people$_{26}$, rage$_{25}$, dont$_{24}$, offended$_{23}$, revenge$_{17}$, irritate$_{17}$, insult$_{17}$, hate$_{13}$	angry$_{41}$, bully$_{35}$, terror$_{29}$, people$_{28}$, outrage$_{27}$, rage$_{25}$, insult$_{17}$, hate$_{13}$, **game**$_{10}$, **play**$_{10}$, revenge$_{9}$, **racism**$_{9}$	Terms didn't change significantly, but some new ones were added, expressing more intense anger (e.g., *game*, *racism*). Overall, the cluster remained largely consistent.
Skincare (-)	sellus$_{6}$, farmacy$_{2}$, moisturizer$_{1}$, brand$_{1}$, moisture$_{1}$, birthday$_{1}$, glossier$_{1}$, balm$_{1}$, rituals$_{1}$, babor$_{1}$, small$_{1}$, look$_{1}$,	sellus$_{8}$ **skincare**$_{6}$, sale$_{5}$, glossier$_{4}$, **cream**$_{4}$, **toner**$_{4}$, balm$_{3}$, **treatments**$_{3}$, **dark**$_{3}$, **spot**$_{3}$, **sun**$_{3}$, farmacy$_{2}$, moisturizer$_{2}$,	Augmentation added skincare-specific terms (e.g., *skincare, cream, toner*) and shifted focus slightly toward sales (*sale*).
DogeCoin (+)	doge$_{250}$, bought$_{17}$, currency$_{11}$, market$_{7}$, coin$_{7}$, community$_{7}$, dollar$_{6}$	doge$_{254}$, bought$_{27}$, currency$_{17}$, dollar$_{11}$, coin$_{11}$, community$_{11}$, price$_{10}$, market$_{8}$	Terms remained consistent, with minor increases in frequencies. Augmentation slightly emphasized financial aspects, but the dominant focus on doge was unchanged.
Economics (-)	sustainable$_{5}$, inequality$_{4}$, frequency$_{4}$, likelihood$_{3}$, dimensional$_{2}$,regression$_{2}$ pandemic$_{1}$,governments$_{1}$ evolutionary$_{1}$,strategy$_{1}$, approach$_{1}$,macroecon$_{1}$	**model**$_{7}$, regression$_{5}$, **series**$_{4}$, dimensional$_{4}$, **learning**$_{3}$, pandemic$_{3}$, **sparsity**$_{2}$, sustainable$_{2}$, approach$_{2}$, likelihood$_{1}$	The augmentation introduced new terms (e.g., *series*, *sparsity*) reflecting broader methodological aspects of economics. Existing terms saw increased emphasis, indicating a stronger focus on statistical applications.
CS (+)	learning$_{47}$, neural$_{21}$, networks$_{20}$, classification$_{17}$, adversarial$_{15}$, deep$_{15}$, models$_{14}$, detection$_{12}$, recognition$_{11}$	learning$_{74}$, neural$_{43}$, detection$_{35}$, networks$_{31}$, deep$_{31}$, classification$_{30}$, recognition$_{27}$, adversarial$_{22}$, models$_{21}$,	The words remain consistent, with increased emphasis on core computer science terms like *learning*, *neural*, and *detection*.
A B&C (-)	drosophila$_{14}$, olfactory$_{10}$, neurons$_{6}$, circadian$_{4}$, cortex$_{4}$, pathway$_{3}$, learning$_{3}$, dopaminergic$_{3}$, model$_{3}$, light$_{2}$, endocrine$_{2}$	drosophila$_{257}$, olfactory$_{176}$, circadian$_{166}$, **sleep**$_{66}$, **clock**$_{62}$, neurons$_{54}$, **taste**$_{47}$, light$_{30}$, **sensory**$_{26}$, **rhythms**$_{24}$, **odorant**$_{24}$	New terms like *sleep*, *clock*, and *taste* introduce broader behavioral contexts. Existing words such as *drosophila* and *olfactory* gained prominence, emphasizing sensory and neural mechanisms in animal cognition.
Neuroscience(+)	neural$_{69}$, cortex$_{58}$, learning$_{39}$, visual$_{38}$, memory$_{30}$, auditory$_{25}$, temporal$_{19}$, cortical$_{19}$, speech$_{17}$, spatial$_{16}$, dynamics$_{15}$, prefrontal$_{15}$	neural$_{546}$, cortex$_{407}$, visual$_{318}$, learning$_{280}$, **brain**$_{251}$, memory$_{244}$, cortical$_{195}$, auditory$_{168}$, **attention**$_{132}$, **perception**$_{121}$, dynamics$_{119}$, speech$_{117}$	The cluster shows a substantial increase in emphasis on core terms such as *neural* and *cortex*. New terms like *brain*, and *perception* expand the focus to include cognitive and sensory processes, complementing existing terms like *visual* and *auditory*.
Neurology (-)	alzheimers$_{35}$, cognitive$_{14}$, disease$_{12}$, dementia$_{10}$, brain$_{10}$, genetic$_{5}$, impairment$_{5}$, biomarkers$_{4}$, diagnosis$_{4}$, amyotrophic$_{4}$, trials$_{4}$, risk$_{4}$	alzheimers$_{156}$, **parkinsons**$_{115}$, cognitive$_{115}$, disease$_{107}$, brain$_{97}$, dementia$_{85}$, impairment$_{45}$, **epilepsy**$_{44}$, genetic$_{30}$, **eeg**$_{29}$, **stroke**$_{28}$, **cognition**$_{25}$	New terms like *parkinsons* and *stroke* broaden the scope to include more neurological conditions. Existing words such as *alzheimers* and *dementia* show significant increases, reinforcing the focus on degenerative diseases and brain function.
Epidemiology(+)	sarscov2$_{89}$, covid19$_{28}$, antigen$_{22}$, testing$_{19}$, diagnostic$_{16}$, rtpcr$_{15}$, saliva$_{14}$, rna$_{13}$, detection$_{10}$, test$_{9}$, nasopharyngeal$_{9}$, viral$_{8}$	sarscov2$_{329}$, antigen$_{125}$, covid19$_{104}$, detection$_{70}$, testing$_{70}$, saliva$_{67}$, diagnostic$_{54}$, rtpcr$_{44}$, test$_{39}$, rna$_{32}$, **tests**$_{31}$, **screening**$_{30}$	The focus on diagnostics and testing is reinforced with increases in terms like *sarscov2*, *antigen*, and *covid19*. New additions such as *tests* and *screening* highlight expanded approaches to epidemiological detection.

Table 2. Clustering performance results (mean ± standard deviation). Results are reported for both non-augmented (N.A) and augmented (A) datasets.

Algo.	Type	Metric	Tweet_Emo.	Reddit	Arxiv	Medrxiv	Biorxiv
KMeans	N.A	NMI	20.32 ± 2.28	54.74 ± 1.73	44.22 ± 0.91	30.07 ± 0.21	**34.03 ± 0.17**
		ARI	22.79 ± 6.23	25.69 ± 2.52	32.09 ± 2.26	6.92 ± 0.35	20.09 ± 1.79
	A	NMI	**22.10 ± 3.49**	**55.08 ± 0.58**	**44.28 ± 0.57**	**34.62 ± 0.21**	30.39 ± 0.21
		ARI	**22.81 ± 5.83**	**29.41 ± 1.78**	**34.10 ± 1.64**	**7.22 ± 0.24**	**20.59 ± 1.49**
SKmeans	N.A	NMI	20.85 ± 2.80	**55.91 ± 0.83**	43.59 ± 0.48	30.03 ± 0.30	**34.12 ± 0.16**
		ARI	19.39 ± 6.72	29.88 ± 1.83	32.75 ± 1.49	7.17 ± 0.44	22.24 ± 2.25
	A	NMI	**22.42 ± 5.04**	55.46 ± 1.21	**44.42 ± 0.72**	**30.44 ± 0.29**	33.86 ± 0.29
		ARI	**23.67 ± 9.74**	**31.53 ± 2.78**	**33.58 ± 1.60**	**7.65 ± 0.34**	**23.36 ± 3.01**

5 Conclusion

In this work, we presented a novel data augmentation framework integrating GMMs with LLMs to address the challenges of class imbalance in unsupervised natural language processing tasks. Our approach targets underrepresented regions in the dataset, employing GMMs for precise cluster identification and LLMs for generating contextually relevant synthetic documents. Through extensive experiments on multiple imbalanced text datasets, we demonstrated that our method maintains clustering performance, improves the representation of minority classes, and enriches cluster interpretability. The results show that targeted data augmentation is effective in addressing imbalance without compromising the quality of clustering algorithms.

In our contribution, we relied on GMMs; it should be interesting to test other mixtures, such as von-Mises Fisher mixture models [21–23] or latent block models [7,8,11,18]. Additionally, a potential limitation of using LLM-generated data is the risk of reinforcing biases present in the training data of the LLM itself [24]. While our approach ensures that generated text aligns with cluster themes, future work should explore ways to assess and mitigate biases in synthetic data.

Disclosure of Interests. The authors have no competing interests to declare that are relevant to the content of this article.

References

1. Banfield, J.D., Raftery, A.E.: Model-based gaussian and non-gaussian clustering. Biometrics **49**(3), 803–821 (1993)
2. Bayer, M., Kaufhold, M.A., Reuter, C.: A survey on data augmentation for text classification. ACM Comput. Surv. **55**(7), 1–39 (2022)
3. Cai, X., Xiao, M., Ning, Z., Zhou, Y.: Resolving the imbalance issue in hierarchical disciplinary topic inference via LLM-based data augmentation. In: ICDMW, pp. 1424–1429 (2023)
4. Dai, H., et al.: AugGPT: Leveraging ChatGPT for text data augmentation (2023)

5. Dempster, A.P., Laird, N.M., Rubin, D.B.: Maximum likelihood from incomplete data via the EM algorithm. J. R. Stat. Soc.: Ser. B **39**(1), 1–22
6. Goodfellow, I., et al.: Generative adversarial nets. In: Advances in Neural Information Processing Systems, vol. 27 (2014)
7. Govaert, G., Nadif, M.: Clustering with block mixture models. Pattern Recogn. **36**(2), 463–473 (2003)
8. Govaert, G., Nadif, M.: Co-clustering: Models, Algorithms and Applications. John Wiley & Sons (2013)
9. Guo, H., Mao, Y., Zhang, R.: Augmenting data with Mixup for sentence classification: an empirical study (2019)
10. Gupta, S., Kanchinadam, T., Conathan, D., Fung, G.: Task-optimized word embeddings for text classification representations. Front. Appl. Math. Stat. **5** (2020)
11. Hoseinipour, S., Aminghafari, M., Mohammadpour, A., Nadif, M.: A sparse exponential family latent block model for co-clustering. In: Advances in Data Analysis and Classification, pp. 1–37 (2024)
12. Jiang, A.Q., et al.: Mistral 7B (2023)
13. Karimi, A., Rossi, L., Prati, A.: AEDA: an easier data augmentation technique for text classification (2021)
14. Keraghel, I., Morbieu, S., Nadif, M.: Beyond words: a comparative analysis of LLM embeddings for effective clustering. In: IDA, pp. 205–216 (2024)
15. Kumar, V., Choudhary, A., Cho, E.: Data augmentation using pre-trained transformer models (2020)
16. McInnes, L., Healy, J., Melville, J.: UMAP: uniform manifold approximation and projection for dimension reduction (2018)
17. Muennighoff, N., Tazi, N., Magne, L., Reimers, N.: MTEB: massive text embedding benchmark. In: Vlachos, A., Augenstein, I. (eds.) EACL (2023)
18. Nadif, M., Govaert, G.: Block clustering of contingency table and mixture model. In: International Symposium on Intelligent Data Analysis, pp. 249–259. Springer (2005)
19. Qiu, S., et al.: EasyAug: an automatic textual data augmentation platform for classification tasks. In: Companion Proceedings of the Web Conference 2020, pp. 249–252 (2020)
20. Radford, A., Metz, L., Chintala, S.: Unsupervised representation learning with deep convolutional generative adversarial networks (2015)
21. Salah, A., Nadif, M.: Model-based von mises-fisher co-clustering with a conscience. In: Proceedings of the 2017 SIAM International Conference on Data Mining, pp. 246–254. SIAM (2017)
22. Salah, A., Nadif, M.: Social regularized von mises-fisher mixture model for item recommendation. Data Min. Knowl. Disc. **31**, 1218–1241 (2017)
23. Salah, A., Nadif, M.: Directional co-clustering. Adv. Data Anal. Classif. **13**, 591–620 (2019)
24. Shumailov, I., Shumaylov, Z., Zhao, Y., Papernot, N., Anderson, R., Gal, Y.: Ai models collapse when trained on recursively generated data. Nature **631**(8022), 755–759 (2024)
25. Steinley, D.: Properties of the Hubert-arable adjusted rand index. Psychol. Methods **9**(3), 386 (2004)
26. Strehl, A., Ghosh, J.: Cluster ensembles—a knowledge reuse framework for combining multiple partitions. J. Mach. Learn. Res. **3**(Dec), 583–617 (2002)
27. Tikhonov, A., Arsenin, V.: Solutions of Ill-posed Problems. Halsted Press book

28. Van Nooten, J., Daelemans, W.: Improving Dutch vaccine hesitancy monitoring via multi-label data augmentation with GPT-3.5. In: the 13th Workshop on Computational Approaches to Subjectivity, Sentiment, & Social Media Analysis, vol. 1, pp. 251–270 (2023)
29. Wei, J., Zou, K.: EDA: easy data augmentation techniques for boosting performance on text classification tasks (2019)
30. Wong, M.A.: A hybrid clustering method for identifying high-density clusters. J. Am. Stat. Assoc. **77**(380), 841–847 (1982)
31. Ye, J., et al.: LLM-DA: data augmentation via large language models for few-shot named entity recognition. arXiv preprint: arXiv:2402.14568 (2024)

Make Literature-Based Discovery Great Again Through Reproducible Pipelines

Bojan Cestnik[1,2], Andrej Kastrin[3], Boshko Koloski[1,4], and Nada Lavrač[1(✉)]

[1] Jožef Stefan Institute, Ljubljana, Slovenia
{boshko.koloski,nada.lavrac}@ijs.si
[2] Temida d.o.o, Ljubljana, Slovenia
bojan.cestnik@temida.si
[3] University of Ljubljana, Ljubljana, Slovenia
andrej.kastrin@mf.uni-lj.si
[4] International Postgraduate School Jožef Stefan, Ljubljana, Slovenia

Abstract. By connecting disparate sources of scientific literature, literature/based discovery (LBD) methods help to uncover new knowledge and generate new research hypotheses that cannot be found from domain-specific documents alone. Our work focuses on bisociative LBD methods that combine bisociative reasoning with LBD techniques. The paper presents LBD through the lens of reproducible science to ensure the reproducibility of LBD experiments, overcome the inconsistent use of benchmark datasets and methods, trigger collaboration, and advance the LBD field toward more robust and impactful scientific discoveries. The main novelty of this study is a collection of Jupyter Notebooks that illustrate the steps of the bisociative LBD process, including data acquisition, text preprocessing, hypothesis formulation, and evaluation. The contributed notebooks implement a selection of traditional LBD approaches, as well as our own ensemble-based, outlier-based, and link prediction-based approaches. The reader can benefit from hands-on experience with LBD through open access to benchmark datasets, code reuse, and a ready-to-run Docker recipe that ensures reproducibility of the selected LBD methods.

Keywords: literature-based discovery · bisociative knowledge discovery · open science · reusable software · Jupyter Notebooks

1 Introduction

In an era of rapidly increasing amounts of scientific publications, researchers face the challenge of keeping up with the latest findings, even within their narrow scientific fields. In a flood of information, researchers can overlook valuable segments of knowledge. Focused on their silos of information, researchers are incapable of delving into novel literature from neighboring domains or getting insights from possibly yet unconnected domains, as means for sparkling novel research insights or forming new hypotheses.

G. Krempl et al. (Eds.): IDA 2025, LNCS 15669, pp. 261–273, 2025.
https://doi.org/10.1007/978-3-031-91398-3_20

Literature-based discovery (LBD) is an area of natural language processing (NLP) research, with the first approaches reported in the mid-1980s [27]. LBD tools aim to uncover new knowledge by connecting disparate sources of literature to generate novel research hypotheses that are not evident in domain-specific documents. In his pioneering work, Swanson [27], reading articles in separate literature sources on Raynaud's disease and dietary fish oil, observed that some knowledge concepts were shared between the two separate sets of documents. This led to the discovery that fish oil could be used in the treatment of Raynaud's disease, a hypothesis that was later clinically confirmed. During the last four decades, researchers have proposed several different methods, ranging from simple latent semantic indexing [5], more complex techniques based on knowledge graphs [24], up to novel large language models [32]. A detailed overview of LBD approaches can be found in studies of several authors [9,25,30].

This paper focuses on bisociative LBD, an area of research that is the focus of our research interests [9,15]. We explore a specific aspect of LBD that is grounded in two research traditions: bisociative knowledge discovery [2] and LBD [29]. By integrating these two research traditions, bisociative reasoning can support the formation of serendipitous hypotheses by discovering the connections of seemingly unrelated concepts [12]. The main contribution of this paper is the open science approach to bisociative LBD, providing access to benchmark LBD datasets, replication of LBD discoveries, as well as a collection of Jupyter Notebooks—interactive computational narratives containing code and formatted text—that implement the main steps of the LBD process: data acquisition, text preprocessing, hypothesis discovery, and evaluation.

The paper first explains the open science approach to data and software reuse in Sect. 2. To replicate the reported discoveries, Sect. 3 provides access to several benchmark datasets used in pioneering LBD papers and more recent bisociative LBD approaches. This paper supports reproducible science through dockerized Jupyter Notebooks. Tutorials illustrating how to implement simplified LBD workflows using Jupyter Notebooks are presented in Sect. 4. The developed Python notebooks demonstrate traditional LBD approaches alongside our own bisociative LBD methods, which include (i) an ensemble-based approach for identifying cross-domain bridging terms, (ii) an outlier-based method that reduces the search space of linking terms to outlier documents, (iii) a link prediction approach for ranking associations, and (iv) a large language model (LLM)-based approach to LBD domain conceptualization. The paper concludes with a discussion in Sect. 5.

2 Approach to Reproducible Science

The development of modern programming languages, programming paradigms, and operating systems enabled the development of computer platforms for text analysis, and of modern integrated text analysis tools. Such platforms offer a high degree of abstraction and allow the user to focus on analyzing the results rather than on the methods used to obtain the results.

Originally, individual algorithms were implemented as complete solutions for specific NLP problems, while second-generation NLP systems were intended to provide integrated solutions for entire NLP pipelines including text preprocessing, transformation, and text analysis tasks, ideally also with graphical user interfaces. Many of them, such as scikit-learn [17], take advantage of operating system-independent languages, such as the Python environment, to develop complete solutions that include methods for text preprocessing and visual representation of results. These solutions also provide interfaces to call each other and thereby enable cross-solution development.

A common practice in modern scientific publishing is to make code and data publicly available, e.g., via GitHub. Today's AI software is further confronted with the challenge of how to adapt such open science solutions to big data and deep learning, which require grid computing and GPU acceleration. Hugging Face [34] exemplifies open-source collaboration within the NLP community by providing widespread access to state-of-the-art models. Similarly, ReScience [23] is a community-driven initiative that focuses on republishing and reimplementing previously published methods to promote reproducible and accessible science.

Additional means of simplifying access to NLP tools and enabling cross-solution development are modern web-based Jupyter Notebooks [19,31] that allow the creation and sharing of documents containing live code, equations, and visualizations. The open science approach to LBD implemented in this paper uses Jupyter Notebooks to practically illustrate selected concepts and LBD pipelines, allowing the reader to run the described methods on illustrative examples. The notebooks provide a set of examples that we have carefully chosen to be easy to understand but are non-trivial and can be used as recipes or starting points for developing more complex LBD scenarios. Moreover, following the recommendations and guidelines for reproducible science and FAIR benchmarking [18], we prepared dockerized versions of all methods, which provide off-the-shelf and ready-to-use environment for testing and working with the LBD methods surveyed in this paper.

3 Benchmark Datasets Enabling Experiment Replicability

A selection of datasets prepared for reuse and tool utility verification has been constructed in a way that enables the replication of the originally reported LBD experiments. To achieve this, the query enabling to download a PubMed dataset is limited by the year of the paper presenting the originally discovered bridging terms (b-terms), which is different for every dataset. The summary of statistics on the datasets made available for reuse is presented in Table 1.

We used the b-terms, which were identified in each domain pair dataset, as the gold standard to evaluate the utility of tested LBD approaches to cross-context link discovery. Table 2 presents the b-terms for Raynaud's Syndrome–Dietary Fish Oil (RS-DFO), Migraine–Magnesium (Mig-Mg), and Autism–Calcineurin (Aut-CaN) domain pair datasets used in our experiments.

Table 1. Basic properties and statistics of domain pair datasets.

Property	Dataset		
	RS-DFO	Mig-Mg	Aut-CaN
No. of docs.	1273/153	1156/7320	10,819/5320
Mean no. of words/doc.	13.1/18.3	11.3/16.1	166/259
Mean no. of terms/doc.	2.56/1.92	11.5/12.1	4.02/4.03
No. of unique terms	4554/1059	1812/6217	423,153/316,702
No. of common terms	27	228	46,473
No. of *b*-terms	3	43	13

Table 2. Bridging terms for the RS-DFO, Mig-Mg, and Aut-CaN domain pairs.

Raynaud's Syndrome–Dietary Fish Oil (RS-DFO)
Terms: blood viscosity, platelet aggregation, vascular reactivity
Migraine–Magnesium (Mig-Mg)
Terms: serotonin, spread, spread depression, seizure, calcium antagonist, vasospasm, paroxysmal, stress, prostaglandin, reactivity, spasm, inflammatory, anti inflammatory, 5 hydroxytryptamine, calcium channel, epileptic, platelet aggregation, verapamil, calcium channel blocker, nifedipine, indomethacin, prostaglandin e1, anticonvulsant, arterial spasm, coronary spasm, cerebral vasospasm, convulsion, cortical spread depression, brain serotonin, 5 hydroxytryptamine receptor, epilepsy, antimigraine, 5 ht, epileptiform, platelet function, prostacyclin, hypoxia, diltiazem, convulsive, substance p, calcium blocker, prostaglandin synthesis, anti aggregation
Autism–Calcineurin (Aut-CaN)
Terms: synaptic, synaptic plasticity, calmodulin, radiation, working memory, bcl 2, type 1 diabetes, ulcerative colitis, asbestos, deletion syndrome, 22q11 2, maternal hypothyroxinemia, bombesin

Novelty: This is the first time an interested LBD researcher can get a systematic presentation and open access to the benchmark datasets and the targeted *b*-terms. The datasets are available in the CLARIN.SI repository [8].

3.1 Raynaud's Syndrome–Dietary Fish Oil (RS-DFO) Benchmark

Swanson's [27] RS-DFO hypothesis linking RS and DFO literatures is the most commonly used baseline for evaluating novel LBD approaches. In our experiments, we used the following PubMed queries to build the RS-DFO dataset:

1. `raynaud* [TI] AND 1900/01/01:1985/11/30 [PDAT]`
2. `fish oil [TI] AND 1900/01/01:1985/11/30 [PDAT]`

Note that Swanson formulated the RS-DFO hypothesis at the end of November 1985; hence, we limited the upper bound of the time window with the PubMed "Publication Date" (PDAT) tag to reflect the body of scientific literature available at that time. The retrieved data are stored in the dataset `swanson_1986.psv.gz` and comprise 1273 records associated with Raynaud's disease (domain C) and 153 records associated with fish oil (domain A). The three most important terms on which we focus are "blood viscosity", "platelet aggregation", and "vascular reactivity".

3.2 Migraine–Magnesium (Mig-Mg) Benchmark

The second benchmark, the Migraine–Magnesium domain pair dataset, pertains to the discovery of implicit links between migraine (C) and magnesium (A). Swanson [28] described eleven neglected connections (i.e., b-terms) between sets A and C, explaining magnesium deficiency as an important cause of the onset of migraine disease in humans. In the LBD process, Swanson managed to find more than 60 pairs of articles connecting the migraine domain with magnesium deficiency through several bridging concepts. In this process, Swanson identified 43 b-terms connecting the two domains. For our research, we used two PubMed queries to construct the dataset:

1. `migraine [TI] AND 1900/01/01:1987/12/31 [PDAT]`
2. `magnesium [TI] AND 1900/01/01:1987/12/31 [PDAT]`

The gathered bibliographic data is stored in the `swanson_1988.psv.gz` dataset.

3.3 Autism–Calcineurin (Aut-CaN) Benchmark

The third dataset is the Autism–Calcineurin domain pair [22]. Autism belongs to a group of pervasive developmental disorders portrayed by an early delay and abnormal development of the cognitive, communication, and social interaction skills of a person. It is a very complex, and not yet sufficiently understood domain, where precise causes are still unknown. Like Swanson [27,28], Petrič et al. [22] also provided b-terms, 13 in total, whose importance in connecting autism to calcineurin (a protein phosphatase) was discussed and confirmed by the domain expert. The dataset was constructed using the following PubMed query:

1. `autis* [TIAB] AND 1900/01/01:2007/12/31 [PDAT]`
2. `calcineurin [TIAB] AND 1900/01/01:2007/12/31 [PDAT]`

The retrieved PubMed data are collected in the `petric_2009.psv.gz` dataset.

The programming scripts for preprocessing the described datasets are available in the GitHub repository of the project. The workflow involves several text-mining steps, including text tokenization, stopword removal, and word stemming/lemmatization, using the LemmaGen lemmatizer for English. Context-specific terms representing knowledge concepts are then determined using n-grams, which are defined as sequences of 1 to n consecutive words with a minimum support frequency. The processed text is then transformed into a bag-of-words (BoW) representation using term frequency–inverse document frequency (TF-IDF) weighting. In addition, to enhance the standard workflow, terms containing words that served as query terms during document selection were removed from the dataset to ensure unbiased preprocessing.

4 Reproduced LBD Techniques

This section starts by introducing two novel Python pipelines, illustrating the general principle of closed LBD based on Swanson's ABC discovery approach in Sect. 4.1, and open LBD based on Weeber's concept-based approach in Sect. 4.2. The section continues with brief introductions to different bisociative LBD techniques, representing prototypical approaches to closed and open LBD,[1] explaining both the intuition behind the techniques and the implementation details. The presentation of reusable LBD pipelines, aimed at enabling replication of our own LBD approaches, includes the (i) ensemble-based approach to identifying cross-domain bridging terms in Sect. 4.3, (ii) outlier-based method that reduces the search space of linking terms to outlier documents in Sect. 4.4, (iii) outlier-based open discovery method in Sect. 4.5, and (iv) network-based approach to ranking associations in Sect. 4.6. Finally, we also present an approach to LBD domain conceptualization achieved by the AHAM LLM approach to topic modeling in Sect. 4.7. Table 3 summarizes the LBD approaches implemented in Jupyter Notebooks.

Table 3. Summary of reproduced LBD techniques.

LBD technique	Section in paper	Discovery mode	Dataset	Document representation	URL
Swanson's ABC [27]	4.1	Closed	RS-DFO	TF-IDF BoW	jupyter
Concept-based [33]	4.2	Open	Mig-Mg	BoW	jupyter
CrossBee [7]	4.3	Closed	Aut-CaN	TF-IDF BoW	jupyter
OntoGen-based [4]	4.4	Closed	Aut-CaN	TF-IDF BoW	jupyter
RaJoLink [3]	4.5	Open	Aut-CaN	TF-IDF BoW	jupyter
Network-based [10]	4.6	Closed	RS-DFO	Network	jupyter
AHAM & LLM [14]	4.7	Topics	Papers	Embeddings	jupyter

Novelty: This is the first time for an interested LBD researcher to get access to prototypical LBD techniques via tutorials implemented in dockerized Jupyter Notebooks. The code is available at https://github.com/akastrin/ida2025lbd.

[1] Due to space constraints, we cannot explain the differences between the open and closed discovery settings, commonly used in LBD. For details, see Weeber et al. [33].

4.1 Swanson's ABC Closed Discovery Model

This section illustrates the general principle of LBD based on Swanson's [27] ABC closed discovery model, where A and C refer to the two domains in the A–C domain pair, and B refers to a set of potential b-terms. The goal is to replicate Swanson's link between Raynaud's syndrome and dietary fish oil (i.e., RS-DFO hypothesis) employing the closed discovery approach.

For the RS-DFO hypothesis, the term "raynaud" represents the concept c, while "fish oil" corresponds to the concept a. To simulate the closed discovery mode, we exclude any records shared between domains C and A, thereby focusing exclusively on indirect connections between the two domains. For document representation, we use article titles, while terms are represented as bigrams. Domain C comprises 4554 unique terms (uni- and bi-grams), while domain A contains 1059 distinct terms. We then identify intersecting terms shared between the two domains. This set consists of 27 common terms, including interesting b-terms such as "blood viscosity", "platelet aggregation", and "vascular reactivity", which were originally identified by Swanson [27]. Swanson's ABC closed discovery approach is illustrated in the `01_closed_discovery.ipynb` notebook.

4.2 Concept-Based Open Discovery

In this section, we extend the closed discovery approach outlined above to the open discovery mode to generate new research hypotheses. We guide the reader through reproducing Swanson's second landmark discovery, which connects migraine headaches to magnesium [28]. We aim to identify novel therapeutic candidates (A) for migraine headaches (C).

Rather than relying on phrases extracted from plain text (as in Sect. 4.1), this tutorial leverages a structured knowledge base, an approach originally introduced by Weeber et al. [33]. However, in the present implementation, we use the Medical Subject Headings (MeSH) ontology [16] to facilitate the LBD process. The use of MeSH allows for structured background knowledge to reduce textual variations and synonyms of terms to standardized knowledge concepts.

Each document in the dataset is represented as a list of MeSH headings. The LBD process begins with a dataset of 1156 PubMed articles on migraines, which map to a total of 1812 unique MeSH headings. To narrow the search space, we apply a semantic type filter, restricting MeSH headings to categories such as "Amino Acid, Peptide, or Protein", "Pathologic Function" and "Phenomenon or Process", resulting to a more manageable set of 116 MeSH headings.

Next, we rank the filtered MeSH headings using TF-IDF and select three b-concepts for further investigation: "Vasoconstriction", "Platelet Aggregation", and "Spreading Cortical Depression". For each b-concept, we perform a new PubMed query to identify associated a-concepts. This search retrieves 2898, 6273, and 180 bibliographic records for "Vasoconstriction", "Platelet Aggregation", and "Spreading Cortical Depression", respectively. Applying the semantic type filter to these records yields 11, 73, and 13 unique MeSH headings for the respective b-concepts.

The intersection of these results contains ten candidate a-concepts. After applying rank aggregation [13], two concepts, "Calcium" and "Magnesium", emerge as the top-ranked candidates. Notably, "Calcium" already has four direct connections to migraines in the literature, whereas "Magnesium" shows no prior joint reports, making it a promising novel hypothesis. The implementation of the concept-based open discovery approach can be found in the `02_open_discovery.ipynb` notebook.

4.3 Text-Mining-Based Closed Discovery

The CrossBee method [7] represents documents using a BoW document representation, with TF-IDF weights used to model the importance of words and multiword (n-gram) expressions. The method operates on a dataset spanning over two domains, Autism and Calcineurin, with 10,819 documents from domain Autism and 5320 documents from domain Calcineurin (Table 1). For each domain, TF-IDF weighted BoW representations are built. Next, each domain representation is aggregated, allowing for easier analysis of the text data categorized by domain. On top of the calculated statistics, several pre-prepared heuristics are used to estimate the importance of terms as bridging term candidates. These heuristics include elementary heuristics, such as *freqTerm*, *freqDoc*, *freqRatio*, which can be combined in an ensemble heuristic for b-term discovery. From the initial vocabulary, 46,473 terms that occur in both domains are selected as candidates for bridging terms and ranked according to a selected ensemble heuristic. The rank of the interesting b-terms from Table 2 is determined and displayed in the Receiver Operating Characteristic diagram. The Area Under the Curve is also calculated for the selected ensemble heuristic. The method is implemented in the `03_mini_crossbee.ipynb` notebook.

4.4 Outlier-Based Closed Discovery

Outlier-based LBD methods for closed discovery efficiently narrow down the search for cross-domain bridging terms to outlier documents, thus simplifying the exploration of bridging terms by domain experts. The workflow illustrates the experiments with the OntoGen tool [4], which is used for b-term discovery by detecting outlier documents in the Autism–Calcineurin domain pair [21,26].

The process starts with cleaning and structuring the text data, followed by feature extraction using a BoW model and a TF-IDF matrix, both of which are refined by filtering out less relevant terms. Dimensionality reduction with principal component analysis enables two-dimensional visualization of document clusters determined by k-means clustering. The combinations of domains and clusters are analyzed to detect overlaps and identify outlier groups (i.e., Autism [Cluster #1] and Calcineurin [Cluster #0] with 356 and 45 documents, respectively). The workflow concludes with the storage of these outliers, which are saved in a file for further processing and experimentation. The illustrated method is implemented in the `04_mini_ontogen.ipynb` notebook.

4.5 Outlier-Based Open Discovery

RaJoLink represents an outlier-based open-discovery three-step method, applied to the Autism dataset. The method requires a predefined list of terms. In the first step (Ra), we limit the terms to MeSH headings, corresponding to semantic types "Enzymes and Coenzymes" and "Amino Acids, Peptides, and Proteins" [16], resulting in 495 terms of interest. After that, TF-IDF weighting of the terms is conducted across documents, resulting in TF-IDF BoW representation over the retrieved MeSH terms. Next, the representation is aggregated per term, resulting in a single score per term. Finally, an expert examines the terms based on their positioning over the aggregated BoW representation, selecting the most interesting ones. This manual intervention identified three terms out of 495, specifically "calcium channel", "synaptophysin", and "lactoylglutathione", for further investigation of their association with autism. The positions of the three selected rare terms in the sorted list of candidate rare terms were 38, 37 and 377.

In the next step (Jo), the documents (titles and abstracts) for each rare term were retrieved from PubMed and combined into a single document list. 11,666 documents from all three domains were collected and then preprocessed similarly to step Ra. The BoW model was created and aggregated by domain. Based on the BoW model, 428 common word candidates were selected and sorted based on the TF-IDF scores. The expert selected the term "calcineurin", which was in position 18 of the list of 428 sorted joint word candidates.

The last step Link in RaJoLink implements the principle of closed discovery between two domains, in our case Autism and Calcineurin. Using Autism as the start domain of interest, the three-step RaJoLink LBD open discovery procedure [20] is illustrated in the `05_mini_rajolink.ipynb` notebook.

4.6 Network-Based Closed Discovery

The network-based approach extends the discovery of bridging terms to relation discovery, exploiting their potential to reveal bisociative relationships among concepts across different biomedical domains. The idea is to mimic the LBD discovery process as a link prediction problem [10]. In this tutorial, we exploit the co-citation network generated from the primary documents in the RS-DFO hypothesis.

The initial dataset of 1426 documents from the RS-DFO experiment is expanded into a training network, including 7128 and 2072 cited references for Raynaud's disease and fish oil, respectively, retrieved from the OpenAlex database. Additionally, the testing network consists of the first 1000 cited references published after November 30, 1985. The notebook explores unsupervised link prediction using proximity measures, including common neighbors, the Jaccard coefficient, and the Adamic/Adar index. A time-sliced evaluation demonstrates high prediction accuracy, highlighting the efficacy of this approach. The approach is illustrated in the `06_mini_linkpred.ipynb` notebook.

4.7 LLM-Based Conceptualization of the LBD Research Field

In addition to the LBD methods presented above aimed at new hypothesis generation, we recently proposed an LLM-based approach to LBD literature conceptualization [14], offering novel domain insights through topic modeling using LLMs. The method begins by using a BERTopic [6], a topic modeling procedure, on initial embeddings to identify topics covering the datasets, which are subsequently named by an LLM using an expert-designed prompt. Next, the ratio of outliers to topics is calculated and multiplied by the topic name similarity between documents, forming the AHAM score. The embeddings are then fine-tuned to the corpora, and the previous step is repeated for 50,000 steps, evaluating the AHAM scores for each 10,000 steps, selecting the embedding with the lowest AHAM score. This procedure identified 19 topics of interest that represent the field of LBD. The method is implemented in the `07_aham_lbd.ipynb` notebook.

5 Discussion

The concept of literate programming, introduced by Knuth in the mid-1980s [11], emphasizes the importance of integrating documentation with source code, enabling both human readability and computational functionality. In recent years, reproducibility has garnered significant attention from various scientific fields and communities, with numerous studies identifying the challenges and proposing solutions to address the reproducibility crisis [1]. These efforts underscore the critical need for consistent practices and tools to ensure that research outputs can be independently verified and built upon.

The reproducible programming paradigm is highly relevant to LBD as it highlights the value of creating reproducible and transparent workflows that facilitate both the theoretical understanding of implemented methods and the reuse of programming code. Key issues to address include managing dependencies effectively, resolving potential environmental conflicts, such as operating system incompatibilities, and simplifying installation processes to avoid unnecessary complexity. Failure to consider these factors can hinder the ability of others to replicate results or extend the existing work. The reproducible approach has not been explicitly addressed in the literature on LBD, although recent studies have incorporated at least some demonstrative data and code segments [25,30].

We assist LBD researchers by enabling the reproducibility of studies and by providing robust access to resources. These include benchmark datasets, computational notebooks, and a containerized development environment that can be effectively integrated at various stages of the research lifecycle. By offering these tools, we seek to mitigate barriers to reproducibility and foster a culture of scientific rigor and transparency within the LBD community.

In conclusion, making LBD "great again" involves not only advancing LBD methods but also ensuring that the underlying research practices align with the principles of reproducibility. By providing comprehensive Python notebooks wrapped in a dockerized programming environment and supported by standardized benchmark datasets, researchers can drive innovation while fostering trust and collaboration in the field.

Acknowledgments. The authors acknowledge the financial support from the Slovenian Research and Innovation Agency through the Knowledge Technologies (No. P2-0103), and Methodology for Data Analysis in Medical Sciences (No. P3-0154) core research projects, as well as Embeddings-Based Techniques for Media Monitoring Applications (No. L2-50070) research project. The young researcher grant (No. PR-12394) supported the work of Boshko Koloski.

Disclosure of Interests. The authors have no competing interests to declare that are relevant to the content of this article.

References

1. Baker, M.: 1,500 scientists lift the lid on reproducibility. Nature **533**(7604), 452–454 (2016)
2. Berthold, M.R. (ed.): Bisociative Knowledge Discovery: An Introduction To Concept, Algorithms, Tools, and Applications. Springer, Heidelberg (2012)
3. Cestnik, B., Fabbretti, E., Gubiani, D., Lavrač, N., Urbančič, T.: Exploring outlier documents to investigate potential links between gut microbiota and Alzheimer's disease. In: Ortuño, F., Rojas Ruiz, I. (eds.) Extended Abstracts of the 4th International Conference on Bioinformatics and Biomedical Engineering (IWBBIO 2016), pp. 475–486. Granada, Godel (2016)
4. Cestnik, B., Fabbretti, E., Gubiani, D., Urbančič, T., Lavrač, N.: Reducing the search space in literature-based discovery by exploring outlier documents: A case study in finding links between gut microbiome and Alzheimer's disease. Genomics Comput. Biol. **3**(3), e58 (2017)
5. Gordon, M.D., Dumais, S.: Using latent semantic indexing for literature based discovery. J. Am. Soc. Inf. Sci. **49**(8), 674–685 (1998)
6. Grootendorst, M.: BERTopic: neural topic modeling with a class-based TF-IDF procedure (2022). arXiv:2203.05794
7. Juršič, M., Cestnik, B., Urbančič, T., Lavrač, N.: Cross-domain literature mining: finding bridging concepts with CrossBee. In: Maher, M.L., Hammond, K., Pease, A., Pérez y Pérez, R., Ventura, D., Wiggins, G. (eds.) Proceedings of the Third International Conference on Computational Creativity (ICCC 2012), pp. 33–40. University College Dublin, Dublin (2012)
8. Kastrin, A., Cestnik, B., Koloski, B., Lavrač, N.: LBData: Ready-to-use datasets for evaluating literature-based discovery methods (2024). http://hdl.handle.net/11356/1999, Slovenian Language Resource Repository CLARIN.SI
9. Kastrin, A., Hristovski, D.: Scientometric analysis and knowledge mapping of literature-based discovery (1986–2020). Scientometrics **126**(2), 1415–1451 (2021). https://doi.org/10.1007/s11192-020-03811-z
10. Kastrin, A., Rindflesch, T.C., Hristovski, D.: Link prediction on a network of co-occurring MeSH terms: towards literature-based discovery. Methods Inf. Med. **55**(4), 340–346 (2016)
11. Knuth, D.E.: Literate programming. Comput. J. **27**(2), 97–111 (1984)
12. Koestler, A.: The act of creation. Hutchinson, London (1964)

13. Kolde, R., Laur, S., Adler, P., Vilo, J.: Robust rank aggregation for gene list integration and meta-analysis. Bioinformatics **28**(4), 573–580 (2012)
14. Koloski, B., Lavrač, N., Cestnik, B., Pollak, S., Škrlj, B., Kastrin, A.: AHAM: adapt, Help, Ask, Model harvesting LLMs for literature mining. In: Miliou, I., Piatkowski, N., Papapetrou, P. (eds.) Advances in Intelligent Data Analysis XXII: 22nd International Symposium on Intelligent Data Analysis (IDA 2024), pp. 254–265. Springer, Cham (2024)
15. Lavrač, N., Martinc, M., Pollak, S., Pompe Novak, M., Cestnik, B.: Bisociative literature-based discovery: Lessons learned and new word embedding approach. N. Gener. Comput. **38**(4), 773–800 (2020)
16. Lipscomb, C.E.: Medical Subject Headings (MeSH). Bull. Med. Libr. Assoc. **88**(3), 265–266 (2000)
17. Pedregosa, F., et al.: Scikit-learn: machine learning in Python. J. Mach. Learn. Res. **12**, 2825–2830 (2011)
18. Peng, R.D.: Reproducible research in computational science. Science **334**(6060), 1226–1227 (2011)
19. Perkel, J.M.: Why Jupyter is data scientists' computational notebook of choice. Nature **563**(7732), 145–147 (2018)
20. Petrič, I., Cestnik, B., Lavrač, N., Urbančič, T.: Outlier detection in cross-context link discovery for creative literature mining. Comput. J. **55**(1), 47–61 (2012)
21. Petrič, I., Cestnik, B., Lavrač, N., Urbančič, T.: Bisociative knowledge discovery by literature outlier detection. In: Berthold, M.R. (ed.) Bisociative knowledge discovery: An introduction to concept, algorithms, tools, and applications, pp. 313–324. Springer, Heidelberg (2012)
22. Petrič, I., Urbančič, T., Cestnik, B., Macedoni-Lukšič, M.: Literature mining method RaJoLink for uncovering relations between biomedical concepts. J. Biomed. Inform. **42**(2), 219–227 (2009)
23. Rougier, N.P., et al.: Sustainable computational science: the ReScience initiative. PeerJ Computer Science **3**, e142 (2017)
24. Sang, S., Yang, Z., Wang, L., Liu, X., Lin, H., Wang, J.: SemaTyP: a knowledge graph based literature mining method for drug discovery. BMC Bioinform. **19**, 193 (2018)
25. Sebastian, Y., Siew, E.G., Orimaye, S.O.: Emerging approaches in literature-based discovery: techniques and performance review. Knowl. Eng. Rev. **32**, e12 (2017)
26. Sluban, B., Juršič, M., Cestnik, B., Lavrač, N.: Exploring the power of outliers for cross-domain literature mining. In: Berthold, M.R. (ed.) Bisociative knowledge discovery: An introduction to concept, algorithms, tools, and applications, pp. 325–337. Springer, Heidelberg (2012)
27. Swanson, D.R.: Fish oil, Raynaud's syndrome, and undiscovered public knowledge. Perspect. Biol. Med. **30**(1), 7–18 (1986)
28. Swanson, D.R.: Migraine and magnesium: eleven neglected connections. Perspect. Biol. Med. **31**(4), 526–557 (1988)
29. Swanson, D.R.: Medical literature as a potential source of new knowledge. Bull. Med. Libr. Assoc. **78**(1), 29–37 (1990)
30. Thilakaratne, M., Falkner, K., Atapattu, T.: A systematic review on literature-based discovery workflow. PeerJ Computer Science **5**, e235 (2019)
31. Toomey, D.: Jupyter for Data Science: Exploratory Analysis, Statistical Modeling, Machine Learning, and Data Visualization with Jupyter. Packt Publishing (2017)
32. Wang, Q., Downey, D., Ji, H., Hope, T.: SciMON: scientific inspiration machines optimized for novelty. In: Ku, L.W., Martins, A., Srikumar, V. (eds.) Proceedings

of the 62nd Annual Meeting of the Association for Computational Linguistics (ACL 2024), pp. 279–299. Association for Computational Linguistics, Bangkok (2024)

33. Weeber, M., Klein, H., de Jong-van den Berg, L.T., Vos, R.: Using concepts in literature-based discovery: simulating Swanson's Raynaud–fish oil and migraine–magnesium discoveries. J. Am. Soc. Inf. Sci. Technol. **52**(7), 548–557 (2001)
34. Wolf, T., et al.: Transformers: State-of-the-art natural language processing. In: Liu, Q., Schlangen, D. (eds.) Proceedings of the 2020 Conference on Empirical Methods in Natural Language Processing: System Demonstrations, pp. 38–45. Association for Computational Linguistics (2020)

Extracting Information in a Low-Resource Setting: Case Study on Bioinformatics Workflows

Clémence Sebe[1](✉), Sarah Cohen-Boulakia[1], Olivier Ferret[2], and Aurélie Névéol[1]

[1] Université Paris-Saclay, LISN, CNRS, Orsay 91400, France
{clemence.sebe,sarah.cohen-boulakia,aurelie.neveol}@lisn.fr
[2] Université Paris-Saclay, CEA, List, 91120 Palaiseau, France
olivier.ferret@cea.fr

Abstract. Bioinformatics workflows are essential for complex biological data analyses and are often described in scientific articles with source code in public repositories. Extracting detailed workflow information from articles can improve accessibility and reusability but is hindered by limited annotated corpora. To address this, we framed the problem as a low-resource extraction task and tested four strategies: 1) creating a tailored annotated corpus, 2) few-shot named-entity recognition (NER) with an autoregressive language model, 3) NER using masked language models with existing and new corpora, and 4) integrating workflow knowledge into NER models. Using BioToFlow, a new corpus of 52 articles annotated with 16 entities, a SciBERT-based NER model achieved a 70.4 F-measure, comparable to inter-annotator agreement. While knowledge integration improved performance for specific entities, it was less effective across the entire information schema. Our results demonstrate that high-performance information extraction for bioinformatics workflows is achievable.

Keywords: Natural Language Processing · Information Extraction · Bioinformatics Workflows

1 Introduction

In data-intensive sciences, such as *bioinformatics*, scientific results are produced using complex data analysis pipelines taking in vast amounts of experimental data. These pipelines consist of numerous steps making calls to bioinformatics tools and require considerable computational time. They can be implemented using scripts (Bash, Python, etc.) that orchestrate tool execution and serve as the glue between data, processes, tools, and the execution environment. However, the development and use of such scripts lead to challenges in the design and execution of the processing chain, as well as maintenance and reproducibility issues. In response to these challenges, efforts have been made over the past two

G. Krempl et al. (Eds.): IDA 2025, LNCS 15669, pp. 274–287, 2025.
https://doi.org/10.1007/978-3-031-91398-3_21

decades to provide bioinformaticians with approaches that guide them toward better automation of data analysis pipelines: the *scientific workflow systems* [4]. Two systems stand out today in bioinformatics: Nextflow [5] and Snakemake [16].

Currently, we observe: (1) an increase in scientific articles describing bioinformatics workflows in a *descriptive* form (steps described without being executable) and (2) a rise in bioinformatics workflows available on repositories like GitHub that are in a *programmatic* form (implementation available but the lack of documentation makes them difficult to reuse).

In Natural Language Processing (NLP), bioinformatics workflows remain underrepresented. To our knowledge, there is no comprehensive corpus encompassing a rich and diverse set of entities describing workflows (bioinformatics tools, data, management systems, etc.). However, some corpora focus on specific types of entities, such as those centered on bioinformatics tools [6,14]. This lack of resources makes bioinformatics a low-resource context, making it difficult to structure information or analyze texts with the limited annotated data available.

To overcome the low-resource challenges in the field, we consider in this paper several strategies. First, decoder-based approaches, such as those using generative models, offer a potential solution, although they may not always outperform other methods in terms of extraction accuracy [3,18]. Next, encoder-based approaches can be explored in two ways: using larger, related corpora to enrich model training, or by developing a small, domain-specific training corpus to focus on the extraction of entities relevant to the field. Finally, another promising approach is the injection of domain-specific knowledge directly into language models.

From a methodological perspective, we address in this paper several research questions related to low-resource settings. In a specific domain, what resources should be annotated, in what quantity and how should domain experts be involved? Which existing resources can be reused or adapted for the new task? What methods are the most effective for the target task and how can they be combined? From a more application-oriented perspective, the challenge is to develop a simple and automated approach capable of extracting information about workflows described in the literature, aiming to not only systematically document workflows presented in the literature but also facilitate their registration in workflow repositories.

In preliminary work [22], we presented a corpus of scientific articles annotated with entities representing bioinformatics workflow information (BioToFlow). We also reported preliminary experiments with Named-Entity Recognition (NER) from the corpus with a specific focus on the memorization vs. generalization abilities of statistical and rule-based methods. Herein, we provide a brief description of BioToFlow and we present new contributions as follows: (a) experiments with few-shot NER using an autoregressive language model; (b) NER experiments using a pre-existing corpus and BioToFlow and (c) integration of workflow knowledge into NER models. Our code is publicly available[1].

[1] https://gitlab.liris.cnrs.fr/sharefair/ner_workflows.

2 Workflow Information Representation

2.1 Workflow Representation Schema

To systematically describe the bioinformatics workflows and their components, we relied on discussions with workflow experts (notably from the ELIXIR and BOSC International Bioinformatics communities) and the review of several articles describing workflows (e.g., [4,5,15,16,27]. Generally, bioinformatics workflows consist of data analysis steps, each containing a script that may or may not invoke a bioinformatics tool. For a workflow to be executed, it is crucial to maintain a record of its execution environment. Figure 1 presents the proposed schema, which distinguishes between three main categories of entities. The first category is *core* entities, representing major components of a workflow, such as the name of the workflow described, bioinformatics tools, and data. *Tools* can also be further classified based on the context: whether they are explicitly identified as bioinformatics tools (BioInfo), laboratory-related tools (Lab), tools mentioned but not used in the final workflow (Context), or tools referenced in a general sense without specific details (General). One entity is dedicated to methods, i.e. the descriptions of workflow steps, but without mentioning specific tool names (e.g., "mapping DNA sequences against a genome" vs. "BWA"). The data refers to the information processed by the workflow (e.g., "DNA sequence"). Articles exhibit different levels of specificity and can contain mentions of a database name or the exact filenames that were used. However, in other cases, they remain more general and only refer to broad categories of data. The second category includes *environment* entities, which capture the resources and configurations required for workflow execution, such as the execution environment (e.g., the management system used, the hardware it runs on) or programming languages used in scripts. Finally, the third category addresses more specific details, including versioning information, bibliographic references describing tools or libraries, as well as descriptions and parameters associated. This additional information can be associated with the entities present in the workflow body or be more closely related to the workflow environment, providing further context about it.

2.2 Annotation of Workflow Information in Articles: *BioToFlow*

Article Selection Criteria. Articles describing bioinformatics workflows with a link to workflow code in Github have been targeted using a query on PubMed and PubMedCentral [20][2]. During the annotation phase (February 2024), this query returned 89 and 91 articles describing workflows from Nextflow and Snakemake systems respectively, from which we randomly selected articles for annotation. As of 11/18/2024, the same query now returns 124 and 123 articles, demonstrating a continuous increase in the number of published articles on these workflows. Articles were pre-processed to select the contents of the *Material and Methods* and *Implementation* sections, which commonly contain workflow descriptions that we aim to annotate.

[2] PubMed query: (nextflow[Abstract] OR snakemake[Abstract] OR nextflow[Title] OR snakemake[Title]) AND github[All Fields].

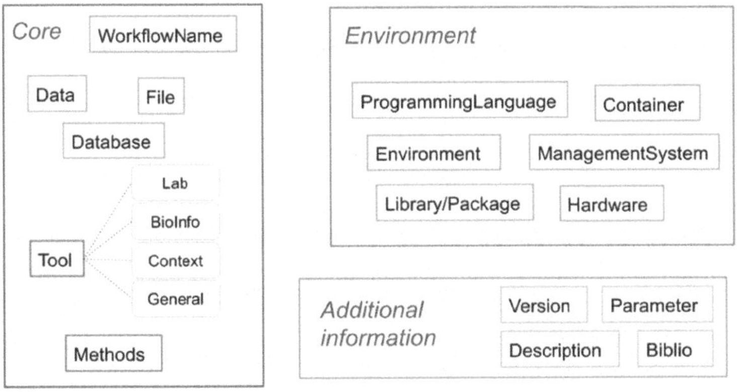

Fig. 1. Bioinformatics workflow representation schema.

Creation of an Annotated Corpus. To validate the information representation schema described in Sect. refs21 and create a resource for evaluating the extraction of such information, we developed an annotated corpus using a collaborative method [7] with the help of an annotation guide that we designed. Annotations were performed using BRAT (Brat Rapid Annotation Tool [23], v1.3 p1), recognized for its qualities including the ability to process nested entities, an existing annotation schema, and pre-annotations [19].

Seven annotators participated in the annotation campaigns, which were conducted in multiple stages: the annotators first worked on common texts, then on different texts. Inter-annotator agreement (IAA) was calculated using the F-measure[3] with BRAT-Eval [25](v0.0.2), a tool that compares two sets of annotations and analyzes divergences. The evaluation of agreement can be strict (agreement between two annotations only if identical text portions are annotated with the same label) or relaxed (agreement if two identical or overlapping text portions are annotated with the same label). The score ranges from 0 (no agreement) to 1 (perfect agreement). IAA obtained were all higher than 70% in relaxed mode. When we look in more detail at the extraction scores for each entity, we found a heterogeneity in the scores showing that some are easier to annotate than others. There are several causes for these differences, including ambiguity in the tagset, criteria for discrimination [8], and, finally, the expertise domain of the annotators. Figure 2 shows entity annotations using our schema on an excerpt from an article describing a Nextflow workflow (PMID 35171290).

Overall, our corpus, named *BioToFlow*[4], is a small corpus composed of 52 articles (26 articles related to Nextflow workflows and 26 on Snakemake workflows, randomly selected from PubMed while ensuring a balanced representation

[3] For a detailed review of inter-annotator agreement measures and their usage context, we refer the reader to [1]. In the case of named entity annotation, [10] demonstrates that the F-measure and κ can be considered equivalent.

[4] https://doi.org/10.5281/zenodo.14900544.

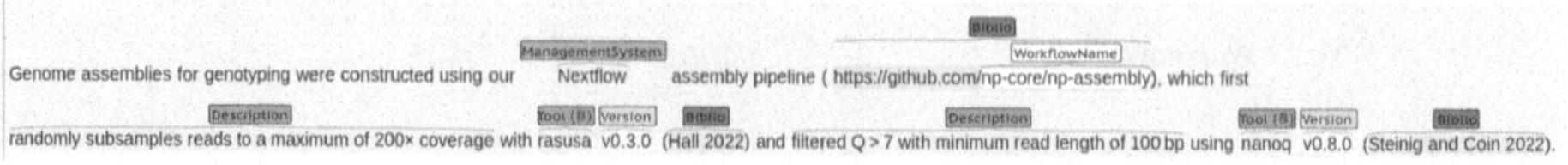

Fig. 2. Excerpt from the annotated corpus using the BRAT software.

of both workflow management systems) with a total of 78,419 tokens (27,786 annotated tokens). However, the entities are diversified and distributed according to Table 1. In our corpus, we allow the presence of nested entities (~8% of entities in BioToFlow; an example can be seen in Figure 2, where a WorkflowName entity is nested within a Biblio entity).

Table 1. BioToFlow: number of occurrences per entity.

Entities	Occurrences	Entities	Occurrences
Data	2,434	Version	454
Tool	1,482	Hardware	429
Description	1,300	Database	288
Biblio	1,251	ManagementSystem	243
Method	936	Container	108
WorkflowName	851	ProgrammingLanguage	104
File	780	LibraryPackage	101
Parameter	464	Environment	83

Using this corpus and its annotations as a starting point, we explored different extraction approaches, starting with decoder-based approaches and then moving on to encoder-based approaches.

3 Decoder-Based Approach for Named Entity Recognition

Because of the small size of the BioToFlow corpus, we sought to leverage an auto-regressive language model to perform few-shot named entity recognition. Specifically, we used Llama-3-8B-Instruct, which demonstrated the best performance among generative models in a comparative study of masked language models and generative models utilizing prompting techniques [18]. Following the same methodology as in this study[5], we conducted a series of few-shot experiments, varying the number of examples provided (either 5 or 10) and testing different prompt formulations, such as providing or omitting entity definitions and task framing.

[5] The code is available at https://github.com/marconaguib/autoregressive_ner.

We use random sampling to split the BioToFlow corpus: we use 75% of the corpus (39 articles) for training and the remaining 25% (13 articles) for testing. The results obtained on the BioToFlow test set using the Llama-3-8B-Instruct model with the best hyperparameters are presented in Table 2.

Table 2. Performance of Llama-3-8B-Instruct model with the best hyperparameters (relaxed mode).

Entities	P	R	F1	Entities	P	R	F1
Biblio	93.1	82.2	87.3	LibraryPackage	3.8	5.6	4.5
Container	29.2	50.0	36.8	ManagementSystem	41.1	38.3	39.7
Data	31.2	27.6	29.3	Method	15.4	12.1	13.6
Database	28.3	24.6	26.3	Parameter	11.4	8.8	9.9
Description	26.2	19.4	22.3	ProgrammingLanguage	23.5	14.3	17.8
Environment	60.0	13.0	21.4	Tool	54.9	59.1	56.9
File	30.7	34.8	32.6	Version	67.1	39.5	49.7
Hardware	20.5	26.3	23.1	WorkflowName	29.1	37.7	32.8
Overall	40.1	36.7	38.3				

We obtained an overall performance below 40% in relaxed mode and 31.2% in strict mode (not shown). These results are consistent with findings on other biomedical corpora [18]. Bioinformatics workflows are complex, and although they are closely related to the medical domain, the entities are often very specific. This leads us to explore alternative methods leveraging encoder-based models.

4 Encoder-Based Approach for Named Entity Recognition

Encoder-based approaches for NER typically require a substantial amount of annotated corpus. In our experiments, we leverage a large corpus with annotations covering a subset of workflow information and a small corpus with comprehensive coverage of workflow information.

4.1 Experiments on a Large Corpus: *SoftCite*

To our knowledge, no corpus fully covers workflows. However, we identified corpora with annotations partially aligned with our schema, such as those focusing on datasets [21] or bioinformatics tools. We focused on bioinformatics tools, as identifying them offers crucial insights into a workflow's functionality. We can cite the CZ Software Mentions dataset [14], which includes software mentions extracted using a trained SciBERT model from diverse sources like PubMed Central and the Chan Zuckerberg Initiative. Another relevant corpus is SoftCite

[6], a gold-standard dataset where all articles were manually annotated. SoftCite comprises 4,971 articles from the fields of biomedicine and economics. We chose SoftCite due to its manual annotation process and, to assess its relevance, we selected a subset of 1,159 articles specifically related to biology.

Since SoftCite does not contain the same entity types as BioToFlow, our first step was to identify possible overlaps between the two entity schemas and perform a mapping to convert SoftCite entity names to those used in BioToFlow. Table 3 represents the correspondence between entities in SoftCite (left column) and those in BioToFlow (right column). Some entities in the SoftCite corpus, *publisher_person*, *figure*, *table*, and *formula*, have no equivalent in BioToFlow. Similarly, certain entities in BioToFlow, including *data*, *container*, *hardware*, and *description*, are not present in SoftCite. After the conversion, the number of entities and occurrences are presented in Table 4.

Table 3. Entity correspondence between SoftCite and BioToFlow.

SoftCite		BioToFlow
software		Tool
	environment	Tool
	url	Biblio
	component	LibraryPackage
	implicit	Tool_general
publisher		Biblio
	environment	Environment
bibr		Biblio
version		Version
url		Biblio
language		ProgrammingLanguage

Table 4. SoftCite: Number of occurrences per entity.

Entities	Occurrences	Entities	Occurrences
Tool	4,142	Biblio	3,946
Version	1,349	LibraryPackage	65
ProgrammingLanguage	61		

For the extraction of our target entities, we chose a biLSTM-CRF neural model (combination of a bidirectional LSTM (for sequence processing) with CRF (for prediction)) [26], implemented by the NLStruct Python library package[6] (v0.2.0), which performs multiclass classification. All the experiments with

[6] https://github.com/percevalw/NLStruct.

NLStruct were run with the hyperparameters provided in Appendix A. This library offers the dual advantage of supporting the detection of nested entities, which is essential for certain entity types in our dataset, and accepting input files in the BRAT format. Additionally, it has demonstrated strong performance in the biomedical domain, which closely aligns with our field. More specifically, we used this model with the SciBERT language model [2], more specifically trained on scientific articles. We conventionally chose to randomly divide the articles into two main sets: the first set, 75% of the corpus (927 articles), was used to create the training set, while the remaining 25% (232 articles) constituted the test set. Within the training set, two-thirds of the articles (649 articles) were used for the actual training of the models, and one-third (278 articles) for their validation. A distinctive aspect of our approach is that we constructed five splits through random sampling to minimize dependency on a specific division. For each split, five versions of the entity recognition model were produced using different random seeds, ensuring more robust and reproducible results.

Table 5. Average performance metrics on five splits and five random seeds (with standard deviation) of percentage scores obtained using SciBERT and NLStruct in relaxed mode on SoftCite corpora.

	Softcite		
	P	R	F1
Biblio	90.6±1.4	98.0±0.4	94.6±0.9
Environment	0.0±0.0	0.0±0.0	0.0±0.0
LibraryPackage	2.5±2.4	14.3±10.5	4.1±3.9
ProgLanguage	48.9±5.3	88.8±4.8	62.8±4.8
Tool	73.4±1.9	51.0±0.6	60.2±1.0
Version	49.8±2.2	93.5±0.6	65.0±1.9
Overall-focused	73.1±1.1	70.7±0.6	71.9±0.8

We test the results on 13 test articles from BioToFlow to see if the entities of the SoftCite corpus could be used to train and extract a part of the BioToFlow entities. The performance on a subset of entities (Table 5) using this corpus is significantly better than those achieved with a decoder-based approach.

4.2 Experiments on a Small Corpus: *BioToFlow*

Similar to the SoftCite corpus experiments, we conducted experiments on BioToFlow using NLStruct and divided the data as follows: 75% of the corpus (39 articles) in the training set, with two-thirds of these (26 articles) used for model training and one-third (13 articles) for validation and 25% (13 articles) formed the test set. The results obtained are presented in Table 6.

The first results show that training a model on our little corpus obtains higher results for all the entities (see Tables 2, 5, and 6).

Table 6. Average performance on five splits and five random seeds (with standard deviation) of percentage scores obtained with SciBERT and NLStruct in relaxed mode on the BioToFlow corpus. **Overall-focused** results are directly comparable to Table 5.

Entities	P	R	F1	Entities	P	R	F1
Biblio	97.9±0.4	97.4±1.0	97.7±0.4	LibraryPackage	39.8±12.3	30.9±6.7	33.9±7.4
Container	78.3±11.2	80.2±12.5	78.9±11.8	MgmtSystem	72.9±5.6	90.9±4.0	80.6±3.4
Data	70.9±2.1	57.1±2.3	63.2±1.9	Method	38.6±2.6	68.3±3.4	49.2±2.9
Database	63.8±3.6	54.8±4.3	58.7±1.8	Parameter	53.2±3.3	53.1±4.8	52.9±2.9
Description	65.4±3.9	59.3±1.8	62.1±1.5	ProgLanguage	86.9±2.1	87.2±2.9	86.9±2.4
Environment	50.3±17.9	79.4±10.1	60.3±17.3	Tool	77.7±3.2	72.2±2.8	74.8±1.0
File	84.3±0.5	82.5±1.4	83.4±0.6	Version	86.8±2.6	90.8±2.0	88.7±0.9
Hardware	65.8±1.2	79.2±4.4	71.7±1.7	WorkflowName	67.1±2.2	77.8±4.4	72.0±2.1
Overall	71.2±1.3	69.6±0.9	70.4±0.8	**Overall-focused**	85.5±1.2	83.4±1.2	84.4±0.5

4.3 Fusion of *SoftCite* and *BioToFlow* Corpora

The previous results show that SoftCite and BioToFlow, when used independently, achieve good scores (above 70%). We also merged the two corpora and trained models on the combined dataset to see the contribution of the SoftCite corpus. Two configurations of the SoftCite corpus were considered: (a) entities were only converted to match BioToFlow entity names and (b) each SoftCite article was annotated using predictions from a model trained on BioToFlow.

The scores for specific entities are provided in Table 7. The results suggest that the SoftCite corpus, in its silver form, achieves a slightly higher F1 score compared to a model trained only on BioToFlow (Table 6). These corpus fusions resulted in some entity scores increasing while others decreased.

Table 7. Average performance metrics on five splits and five random seeds (with standard deviation) of percentage scores obtained using SciBERT and NLStruct in relaxed mode on the fusion of BioToFlow and SoftCite corpora.

	With SoftCite conversion			With SoftCite silver		
	P	R	F1	P	R	F1
Data	56.9±2.3	64.1±1.5	60.1±0.7	75.8±2.6	56.8±2.1	64.9±0.8
Description	62.9±3.64	62.5±1.3	62.6±1.2	70.0±0.8	57.4±0.9	63.5±0.7
LibraryPackage	36.2±7.9	38.1±6.3	36.1±7.0	25.3±6.4	37.2±8.9	29.0±5.7
ProgLanguage	88.6±5.4	87.6±4.0	87.9±2.4	90.3±1.3	88.3±3.0	89.2±1.4
Tool	76.9±1.3	72.2±1.2	74.5±0.9	78.8±1.9	70.0±0.9	74.1±1.1
Overall	64.4±1.0	73.6±0.9	68.7±0.3	73.0±0.5	68.5±1.4	70.7±0.6

5 Integration of Knowledge into Language Models

Since the various tested approaches did not show any significant improvement in NER scores compared to a masked language model trained on BioToFlow, we explored one final approach: the integration of knowledge into language models.

Used of a Pre-initialized Encoder. In the spirit of work exploiting pretraining for NER [17], we trained models on BioToFlow following the same methodology as in Section 4.2 but with an encoder initialized with one of the models trained during the NER experiments on the SoftCite corpus (see Section 4.1). The scores of Table 8 for a subset of entities show that the results obtained are nearly identical to those achieved with a model whose encoder was not pre-initialized.

Table 8. Average performance metrics on five splits and five random seeds (with standard deviation) of percentage scores for a subset of entities obtained using SciBERT and NLStruct in relaxed mode on BioToFlow corpus with the encoder initialization.

	P	R	F1		P	R	F1
Database	63.9±3.2	56.2±3.3	59.7±1.3	Method	38.2±1.2	68.1±3.3	48.9±1.6
Environment	47.7±18.9	80.0±14.3	58.5±20.1	ProgLanguage	87.1±1.0	85.9±5.0	86.2±2.5
LibraryPackage	40.2±11.5	32.4±5.6	35.1±5.9	Tool	77.3±3.1	72.1±2.2	74.5±0.6
MgmtSystem	72.7±5.9	94.4±5.7	82.0±4.1	**Overall**	71.3±0.8	69.6±1.3	70.4±0.8

Adding Vocabulary. *Tools* are considered a key entity in our work because each tool has a specific function in the workflow, and it is essential for the overall understanding of the workflow task that all tools are extracted properly. To improve the extraction of bioinformatics tool names, another strategy involves incorporating external knowledge into the encoder. Since the knowledge to inject in our case is made of lists of names, we chose to perform this injection at the level of its vocabulary, similarly to work such as [9, 12]. Various knowledge bases of bioinformatics tools exist, from which tool and binary names (the executable files or commands used to run the tools in a program) can be extracted. We utilized four knowledge bases: *Biotools* [13], *Bioconda* [11], *Biocontainers* [24] and *Bioweb*[7]. From these knowledge bases, we aggregated the names of tools and binaries to create an enriched vocabulary. This new vocabulary was then injected into the SciBERT language model to improve its ability to recognize bioinformatics tool names. We evaluated this strategy by conducting two experiments on BioToFlow, testing two approaches: (a) fine-tuning the SciBERT model with the new vocabulary and (b) using the enriched vocabulary without additional fine-tuning. In both cases, the embeddings of the added tokens were initialized by averaging the embeddings of their subtokens.

[7] https://bioweb.pasteur.fr/welcome.

Adding bioinformatics tool vocabulary appears to improve the extraction of tool names, particularly when fine-tuning the SciBERT model after incorporating the new vocabulary, with an F1 score increasing from 74.8±1.0% to 77.0±2.8%. As observed in previous experiments, some entities benefit from the added knowledge, while others experience slight decreases in performance (see Table 9). Specifically, the *Data* entity shows improvements with vocabulary addition in both fine-tuning and non-fine-tuning scenarios. Entities like *LibraryPackage* and *Tool* also benefit from fine-tuning with the added vocabulary. Conversely, scores for *ManagementSystem* and *ProgrammingLanguage* entities show slight declines.

Table 9. Average performance metrics on five splits and five random seeds (with standard deviation) of percentage scores obtained using SciBERT and NLStruct in relaxed mode with the addition of bioinformatics tool vocabulary, with or without fine-tuning.

	Add voc - finetune			Add voc - no finetune		
	P	R	F1	P	R	F1
Data	70.0±7.2	63.4±10.9	66.4±8.9	71.7±2.4	56.9±2.8	63.4±2.3
LibraryPackage	52.5±7.1	57.4±15.6	52.3±10.0	37.1±8.0	30.4±7.2	32.7±6.0
ManagementSystem	75.6±7.9	76.8±6.7	75.5±7.0	69.7±3.9	93.4±3.3	79.7±1.7
ProgrammingLanguage	81.9±12.6	82.1±11.5	81.9±12.0	86.0±2.1	85.8±2.3	85.8±2.1
Tool	77.2±3.7	76.9±2.7	77.0±2.8	74.5±2.5	73.9±2.6	74.12±0.3
Overall	70.4±1.7	70.2±1.5	70.3±0.4	70.6±1.2	69.5±1.2	70.1±0.9

6 Conclusion

Based on a workflow representation schema, we developed a small entity-rich corpus, BioToFlow. Using this corpus, we experimented with various entity extraction techniques, including decoder-based models, which achieved scores below 50%, and encoder-based methods. These included leveraging a large corpus with annotations covering a subset of the schema and our BioToFlow corpus. A final approach sought to incorporate additional knowledge into the models. The most effective strategy involved training a model on our small corpus while integrating a vocabulary list specific to bioinformatics tools, which improved the extraction of this entity type. All these experiments and combinations of approaches demonstrate that bioinformatics workflows represent a highly specialized branch of biology. This implies that named entity extraction requires a large amount of domain-specific annotated data, as well as the use of language models tailored to bioinformatics. To improve entity extraction, additional few-shot learning approaches should be explored, such as using NER models trained to identify any type of entity, like GLiNER [28]. To further extend the approach of incorporating domain knowledge, we will also investigate various ways to enhance the embeddings of the added vocabulary, starting with their adaptation during the training of the NER models, similarly to [12]. More globally, the entity linking between the components extracted from workflows described in scientific literature and those found in code from public repositories is an essential task that we will consider to merge information from both sources.

Acknowledgments. We sincerely thank N. Bossut, A. Gaignard, G. Marchment, M. Schmit, H. Menager, F. Lemoine for their help with the annotation process. We also thank M. Naguib for his help with the decoder-based experiments. This work has received support from the French government (Agence Nationale pour la Recherche) under the France 2030 program grant agreement ANR-22-PESN-0007 (ShareFAIR).

Disclosure of Interests. The authors have no competing interests.

A Hyperparameters Used with NLStruct

(See Table 10).

Table 10. Hyperparameters used with NLStruct.

Parameter	Value	Parameter	Value
Base encoder	BERT-base-uncased SciBERT_scivocab_uncased	Learning rate	1e-3
Sequence length	256	Dropout	0.1
Train iteration	4,000	Warmup ratio	0.1
Optimizer	AdamW	Random seed	1 - 8 - 22 - 42 -100

References

1. Artstein, R., Poesio, M.: Inter-coder agreement for computational linguistics. Comput. Linguist. **34**(4), 555–596 (2008)
2. Beltagy, I., Lo, K., Cohan, A.: SciBERT: a pretrained language model for scientific text. In: Inui, K., Jiang, J., Ng, V., Wan, X. (eds.) Proceedings of the 2019 Conference on Empirical Methods in Natural Language Processing and the 9th International Joint Conference on Natural Language Processing (EMNLP-IJCNLP), pp. 3615–3620. Association for Computational Linguistics, Hong Kong, China (November 2019). https://doi.org/10.18653/v1/D19-1371, https://aclanthology.org/D19-1371
3. Chen, Q., et al.: A systematic evaluation of large language models for biomedical natural language processing: benchmarks, baselines, and recommendations (September 2024). https://doi.org/10.48550/arXiv.2305.16326, http://arxiv.org/abs/2305.16326
4. Cohen-Boulakia, S., Belhajjame, K., Collin, O., Chopard, J., Froidevaux, C., et al.: Scientific workflows for computational reproducibility in the life sciences: status, challenges and opportunities. Futur. Gener. Comput. Syst. **75**, 284–298 (2017)

5. Di Tommaso, P., Chatzou, M., Floden, E.W., Barja, P., Palumbo, E., Notredame, C.: Nextflow enables reproducible computational workflows. Nature Biotech **35**, 316–319 (2017). https://doi.org/10.1038/nbt.3820
6. Du, C., Cohoon, J., Lopez, P., Howison, J.: Softcite dataset: a dataset of software mentions in biomedical and economic research publications. J. Assoc. Inf. Sci. Technol. **72**(7), 870–884 (2021). https://onlinelibrary.wiley.com/doi/abs/10.1002/asi.24454
7. Fort, K.: Collaborative Annotation for Reliable Natural Language Processing: Technical and Sociological Aspects. Wiley-ISTE, Hoboken, July 2016. https://hal.archives-ouvertes.fr/hal-01324322
8. Fort, K., Nazarenko, A., Rosset, S.: Modeling the complexity of manual annotation tasks: a grid of analysis. In: Kay, M., Boitet, C. (eds.) Proceedings of COLING 2012, pp. 895–910. The COLING 2012 Organizing Committee, Mumbai, India (December 2012). https://aclanthology.org/C12-1055
9. Gee, L., Zugarini, A., Rigutini, L., Torroni, P.: Fast vocabulary transfer for language model compression. In: Proceedings of the 2022 Conference on Empirical Methods in Natural Language Processing: Industry Track, pp. 409–416. Association for Computational Linguistics, Abu Dhabi, UAE (December 2022). https://aclanthology.org/2022.emnlp-industry.41
10. Grouin, C., Rosset, S., Zweigenbaum, P., Fort, K., Galibert, O., Quintard, L.: Proposal for an extension of traditional named entities: from guidelines to evaluation, an overview. In: Proceedings of the Linguistic Annotation Workshop (LAW-V), pp. 92–100. Portland, OR, 23–24 June 2011. http://www.aclweb.org/anthology/W11-0411
11. Grüning, B., et al.: Bioconda: sustainable and comprehensive software distribution for the life sciences. Nat. Methods **15**(7), 475–476 (2018)
12. Hong, J., Kim, T., Lim, H., Choo, J.: AVocaDo: strategy for adapting vocabulary to downstream domain. In: Proceedings of the 2021 Conference on Empirical Methods in Natural Language Processing, pp. 4692–4700. Association for Computational Linguistics, Online and Punta Cana, Dominican Republic (November 2021). https://doi.org/10.18653/v1/2021.emnlp-main.385
13. Ison, J., et al.: Tools and data services registry: a community effort to document bioinformatics resources. Nucl. Acids Res. **44**(D1), D38–D47 (2016)
14. Istrate, A.M., Li, D., Taraborelli, D., Torkar, M., Veytsman, B., Williams, I.: A large dataset of software mentions in the biomedical literature (September 2022). http://arxiv.org/abs/2209.00693
15. Kieser, S., Brown, J., Zdobnov, E.M., Trajkovski, M., McCue, L.A.: ATLAS: a Snakemake workflow for assembly, annotation, and genomic binning of metagenome sequence data. BMC Bioinformatics **21**(1), 257 (2020)
16. Köster, J., Rahmann, S.: Snakemake - a scalable bioinformatics workflow engine. Bioinformatics (Oxford, England) **28**, 2520–2 (2012). https://doi.org/10.1093/bioinformatics/bts480
17. Liu, Z., Jiang, F., Hu, Y., Shi, C., Fung, P.: NER-BERT: a pre-trained model for low-resource entity tagging. arXiv preprint arXiv:2112.00405 (2021)
18. Naguib, M., Tannier, X., Névéol, A.: Few-shot clinical entity recognition in English, French and Spanish: masked language models outperform generative model prompting. In: Al-Onaizan, Y., Bansal, M., Chen, Y.N. (eds.) Findings of the Association for Computational Linguistics: EMNLP 2024, pp. 6829–6852. Association for Computational Linguistics, Miami, Florida, USA (November 2024). https://aclanthology.org/2024.findings-emnlp.400

19. Neves, M., Ševa, J.: An extensive review of tools for manual annotation of documents. Brief. Bioinform. **22**, 146–163 (2019). https://doi.org/10.1093/bib/bbz130
20. NLM: MEDLINE, PubMed, and PMC (PubMed Central): How are they different? (2021). https://www.nlm.nih.gov/bsd/difference.html
21. Pan, H., Zhang, Q., Dragut, E., Caragea, C., Latecki, L.J.: DMDD: a large-scale dataset for dataset mentions detection. Trans. Assoc. Comput. Linguist. **11**, 1132–1146 (2023)
22. Sebe, C., Cohen-Boulakia, S., Ferret, O., Névéol, A.: Extraction d'entités nommées décrivant des chaînes de traitement bioinformatiques dans des articles scientifiques en anglais. In: Balaguer, M., Bendahman, N., Ho-dac, L.M., Mauclair, J., G Moreno, J., Pinquier, J. (eds.) Actes de la 31ème Conférence sur le Traitement Automatique des Langues Naturelles, volume 1 : articles longs et prises de position, pp. 422–434. ATALA and AFPC, Toulouse, France (July 2024). https://aclanthology.org/2024.jeptalnrecital-taln.30/
23. Stenetorp, P., Pyysalo, S., Topic, G., Ohta, T., Ananiadou, S., Tsujii, J.: Brat: a web-based tool for NLP-assisted text annotation. In: Proceedings of the Demonstrations at the 13th Conference of the European Chapter of the Association for Computational Linguistics, pp. 102–107 (April 2012)
24. da Veiga Leprevost, F., et al.: BioContainers: an open-source and community-driven framework for software standardization. Bioinformatics **33**(16), 2580–2582 (2017)
25. Verspoor, K., et al.: Annotating the biomedical literature for the Human Variome. Database J. Biol. Databases Curation **2013**, bat019 (2013). https://doi.org/10.1093/database/bat019
26. Wajsbürt, P.: Extraction et normalisation d'entités simples et structurées dans les documents médicaux. Thèse de doctorat, Sorbonne université (December 2021). https://www.theses.fr/2021SORUS541
27. Yates, J.A.F., et al.: Reproducible, portable, and efficient ancient genome reconstruction with nf-core/eager. PeerJ **9**, e10947 (2021). https://peerj.com/articles/10947
28. Zaratiana, U., Tomeh, N., Holat, P., Charnois, T.: GLiNER: generalist model for named entity recognition using bidirectional transformer. In: Duh, K., Gomez, H., Bethard, S. (eds.) Proceedings of the 2024 Conference of the North American Chapter of the Association for Computational Linguistics: Human Language Technologies (Volume 1: Long Papers), pp. 5364–5376. Association for Computational Linguistics, Mexico City, Mexico (June 2024). https://doi.org/10.18653/v1/2024.naacl-long.300, https://aclanthology.org/2024.naacl-long.300/

Vocabulary Quality in NLP Datasets: An Autoencoder-Based Framework Across Domains and Languages

Vu Minh Hoang Dang(✉) and Rakesh M. Verma

University of Houston, Houston, USA
vdang9@uh.edu, rmverma2@central.uh.edu

Abstract. Linguistic richness is essential for advancing natural language processing (NLP), as dataset characteristics often directly influence model performance. However, traditional metrics such as Type-Token Ratio (TTR), Vocabulary Diversity (VOCD), and Measure of Textual Lexical Diversity (MTLD) do not adequately capture contextual relationships, semantic richness, and structural complexity. In this paper, we introduce an autoencoder-based framework that uses neural network capacity as a proxy for vocabulary richness, diversity, and complexity, enabling a dynamic assessment of the interplay between vocabulary size, sentence structure, and contextual depth. We validate our approach on two distinct datasets: the DIFrauD dataset, which spans multiple domains of deceptive and fraudulent text, and the Project Gutenberg dataset, representing diverse languages, genres, and historical periods. Experimental results highlight the robustness and adaptability of our method, offering practical guidance for dataset curation and NLP model design. By enhancing traditional vocabulary evaluation, our work fosters the development of more context-aware, linguistically adaptive NLP systems.

Keywords: Lexical Diversity · Vocabulary Richness Evaluation · Vocabulary Complexity · Autoencoder Framework · Language Evolution

1 Introduction

The growing prevalence of natural language processing (NLP) applications has heightened the need for effective measures to evaluate the quality of linguistic datasets. Linguists, however, have been interested in vocabulary measurement for over 50 years, exploring ways to capture lexical richness and diversity across different contexts. Their lexical metrics such as TTR (type-token ratio) [23], Measure of Textual Lexical Diversity (MTLD) [16] and VOCD (Vocabulary Diversity) [14] have been widely adopted, but they exhibit notable limitations. These popular measures often fail to account for critical factors such as the

G. Krempl et al. (Eds.): IDA 2025, LNCS 15669, pp. 288–301, 2025.
https://doi.org/10.1007/978-3-031-91398-3_22

contextual relationship between tokens, the impact of duplication, and the variation in sentence length distributions across datasets [15,16]. Consequently, their applicability is constrained, particularly when addressing datasets with diverse linguistic and structural characteristics.

This problem is particularly challenging due to the inherent complexity of language. Datasets vary in vocabulary size, sentence structure, and contextual depth. While some approaches normalize vocabulary measures or incorporate sentence-length metrics [17], they often rely on oversimplified assumptions. These metrics fail to address key dimensions of dataset complexity, leading to potentially flawed evaluations in NLP tasks.

Our research addresses these gaps by focusing on vocabulary quality assessment through an autoencoder-based methodology. We demonstrate that reconstruction complexity can effectively measure vocabulary quality, providing insights into the underlying dataset characteristics. Through extensive validation, our framework establishes a robust approach for evaluating vocabulary quality in NLP datasets. Our contributions include:

1. A systematic critique of existing metrics, highlighting their limitations in real-world scenarios.
2. An autoencoder-based framework that leverages neural network capacity as a proxy for vocabulary diversity and complexity.
3. Evaluation of the framework using two datasets: DIFrauD [2] for deceptive text and Project Gutenberg[1] for diverse domains and languages.

2 Related Work

2.1 Lexical Metrics

Lexical metrics like TTR, VOCD, and MTLD assess lexical diversity in linguistic datasets [14,16,23]. Each has strengths and limitations. TTR, simple and efficient, is unreliable for longer texts [21]. VOCD uses a probabilistic model but is complex and hard to interpret [15]. MTLD is robust to text length but introduces inconsistencies and lacks semantic considerations [16].

Newer approaches like HD-D and MTLD+ address some shortcomings but still have limitations [8,11]. These metrics overlook sentence-level variations, contextual relationships, and duplication impact. They treat all tokens equally, disregarding semantic and syntactic roles, and are sensitive to dataset size and domain-specific characteristics [10,22]. This highlights the need for more robust, context-aware metrics in modern NLP applications.

2.2 Vocabulary Quality in NLP

Vocabulary quality is critical for dataset evaluation, directly influencing language model performance. Existing metrics like token counts or frequency distributions [20] often neglect semantic richness, syntactic variability, and cross-domain generalizability. Normalizing vocabulary size by dataset size as in Data

[1] https://huggingface.co/datasets/manu/project_gutenberg.

Quality Index 1 (DQI 1) [17] fails to account for sentence length differences, while duplication inflates scores without adding diversity. Research highlights biases in quality metrics due to oversampling or poor curation [7]. A comprehensive taxonomy of NLP metrics, including vocabulary quality, diversity, and inter-class differences, is provided in [5].

Cross-genre and multilingual datasets pose challenges due to distinct linguistic features [13] and language-specific variations [4]. Recent advancements address these issues by evaluating contextual and structural properties. Tools assess domain diversity and lexical richness [9], while token-level attention metrics measure vocabulary informativeness [24]. Metrics like overlap percentage and token recall [3] align vocabulary evaluation with downstream requirements, aiding domain-specific applications.

Despite these advancements, comprehensive vocabulary quality evaluation remains a challenge. Many metrics focus on surface-level characteristics, neglecting contextual relationships, semantic richness, and structural complexity. Our work proposes a novel autoencoder-based framework that evaluates vocabulary quality across multiple linguistic dimensions, providing insights for dataset development and analysis.

3 Critique of DQI 1 and Other Vocabulary Metrics

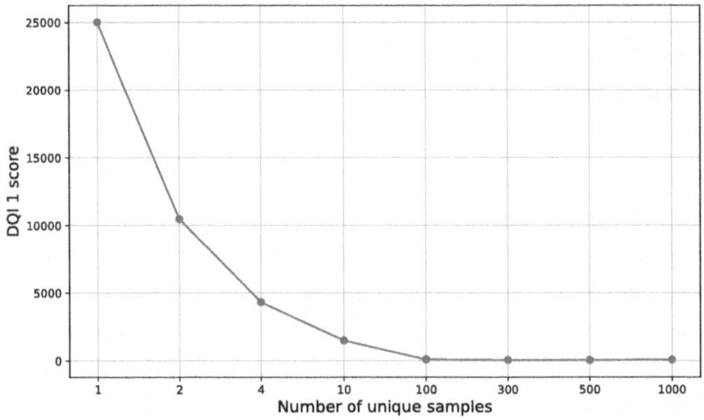

Fig. 1. This plot shows the change in the DQI 1 metric (Y-axis), which measures vocabulary quality, across different numbers of unique rows in a fixed sample size of 1000 rows (X-axis). The DQI 1 value decreases as the number of unique rows increases.

Vocabulary quality is a critical aspect of evaluating NLP datasets. Several metrics have been proposed to measure vocabulary richness and diversity, including DQI 1, VOCD, and MTLD. While these metrics offer valuable insights, they exhibit notable limitations. This section provides a detailed critique of DQI 1 and compares it with the shortcomings of other metrics, such as TTR, VOCD, and MTLD.

3.1 Critique of DQI 1

The authors of [17] propose DQI 1, which aims to measure dataset quality through a combination of two components: vocabulary and sentence length. However, the formulation of DQI 1 exhibits several critical issues, which limit its applicability and interpretability.

Let X be the dataset, v be the vocabulary, s be the sentence length, S be the set of all sentences in the dataset, a and b be the lower and upper thresholds of sentence length, and $size$ be the total number of samples, then the DQI 1 formula is as follows:

$$\text{Vocabulary} = \frac{v(X)}{size(X)} + \sigma(s(X)) \cdot \frac{\sum_S(sgn((s-a)(b-s))}{size(S)}$$

Firstly, the name "Vocabulary" in the DQI 1 metric is misleading. While the metric claims to focus solely on vocabulary, it incorporates sentence length into the formula, causing ambiguity. A more precise name, such as "Vocabulary and Sentence-Length Composite," would better reflect its scope.

The vocabulary component oversimplifies richness by ignoring the saturation of unique tokens as datasets grow. Its normalization by dataset size assumes uniformity across domains, which is invalid for domains like social media with inherently shorter texts. Similarly, the sentence length component overlooks the impact of duplication, as normalizing by unique sentences distorts diversity representation. Figure 1 demonstrates this flaw, showing that duplication paradoxically increases the DQI 1 score, contrary to expectations. While removing the sentence length component might seem like a solution, this would not address the fundamental issues with vocabulary normalization and would leave us with an incomplete measure of dataset quality.

3.2 Critique of TTR, VOCD, and MTLD

Metrics for lexical diversity, such as TTR, VOCD, and MTLD, are commonly used to evaluate dataset vocabulary quality. However, these metrics have limitations, especially for large-scale datasets.

TTR, the ratio of unique tokens to total tokens, is highly sensitive to dataset size, as shown in Fig. 2. As datasets grow, TTR values decline rapidly, making it unreliable for evaluating vocabulary diversity at scale [21]. Domain-specific biases further impact TTR: technical texts yield lower TTRs, while creative writing inflates them.

VOCD uses a probabilistic model to estimate lexical diversity, addressing some of TTR's limitations. However, it struggles with stability in short texts due to limited token counts affecting curve fitting and ignores deeper contextual relationships between words [15]. Its lack of interpretability also reduces its practical utility.

MTLD segments text, calculates TTR for each segment, and averages the results. This reduces sensitivity to dataset size, making it more reliable for cross-domain comparisons. However, MTLD is dependent on segment length choice and fails to capture deeper linguistic properties [12].

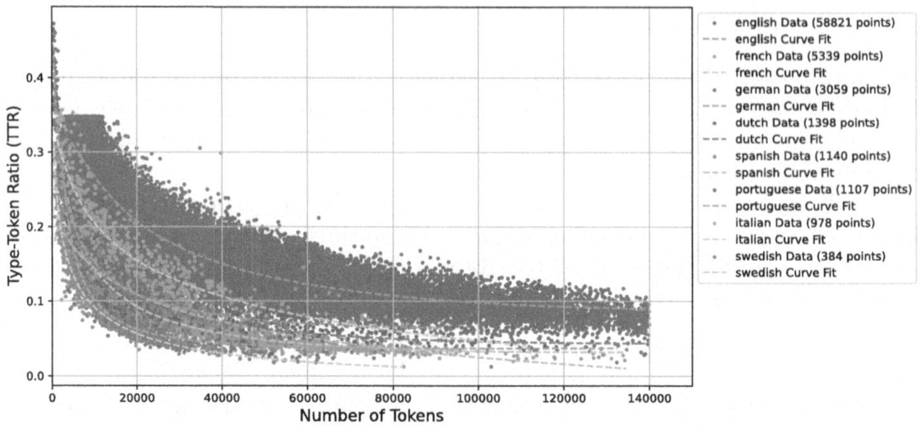

Fig. 2. Scatter plot of Type-Token Ratios (TTR) vs. token counts for multilingual Gutenberg datasets, showing TTR's sensitivity to dataset size and its limitations for comparing vocabulary diversity in large corpora. Curve fits highlight consistent trends across languages.

In summary, current vocabulary metrics like DQI 1, TTR, VOCD, and MTLD have limitations in dataset size sensitivity, domain biases, and linguistic depth. To address these issues, we propose a neural network-based autoencoder methodology that evaluates vocabulary richness through model complexity, providing a more dynamic and comprehensive assessment framework. Due to space restrictions, we cannot include all experiments. An extended version, which includes additional experiments for VOCD and MTLD, is available on arXiv [6].

4 Methodology

In this paper, we distinguish four key attributes of a dataset's vocabulary. First, vocabulary richness measures the raw count and range of unique words, indicating how extensive the vocabulary is. Second, vocabulary diversity examines how uniformly words are distributed across the text, revealing whether certain terms dominate or if usage is balanced. Third, vocabulary complexity analyzes surface-level linguistic patterns, including factors like sentence structure and word length. Finally, vocabulary quality evaluates the overall fitness and utility of the vocabulary for specific applications, considering factors like domain relevance and coverage. Our analysis focuses on intrinsic vocabulary characteristics, independent of pre-trained embeddings to avoid potential biases.

This section outlines our proposed methodology for evaluating these vocabulary attributes in NLP datasets. Addressing the limitations of metrics like DQI 1, TTR, VOCD, and MTLD, we propose an autoencoder-based approach that uses neural network complexity as a proxy for overall vocabulary quality – particularly capturing facets of richness, diversity, and complexity. By analyzing autoencoder capacity requirements, our framework captures deeper insights into lexical characteristics, dynamically evaluating the interplay of vocabulary size, token distribution, and structural patterns without relying on static assumptions.

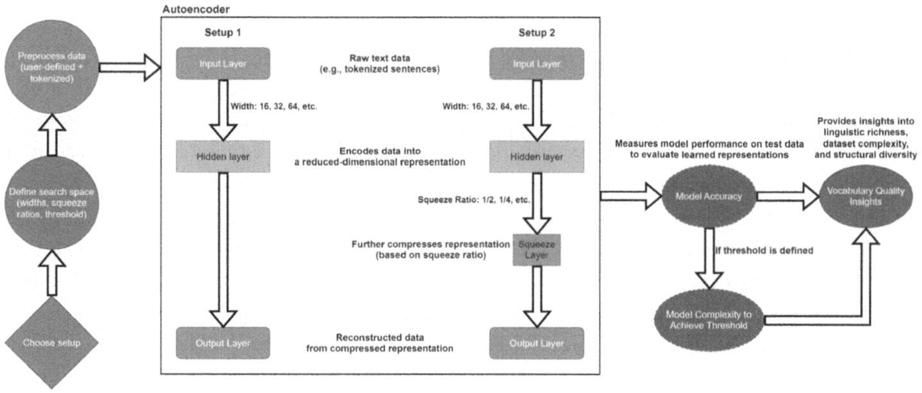

Fig. 3. Framework for Evaluating Vocabulary Quality Using an Autoencoder Model. The process begins with preprocessing and setup definition, followed by autoencoder training and evaluation, resulting in model accuracy and insights into vocabulary richness, diversity, and complexity.

4.1 Autoencoder Framework

As illustrated in Fig. 3, the proposed framework uses autoencoders to evaluate vocabulary quality. Autoencoders compress input data into lower-dimensional representations and reconstruct it, with the evaluation focusing on the model's capacity—the network size and configuration needed to achieve target accuracy. A dataset with a broad range of distinct tokens or more intricate token distributions (high richness, diversity, or complexity) will require a wider hidden layer to accurately reconstruct the text, whereas a repetitive dataset with fewer unique tokens can be reconstructed with a smaller network width, indicating lower diversity.

The input to the framework is any text, which undergoes preprocessing, including tokenization and user-defined transformations, as outlined in Fig. 3. The model aims to reconstruct the entire input text accurately. The output is the model's reconstruction accuracy given a specific configuration, including network widths and squeeze ratios. We selected a reconstruction accuracy threshold of 51% for some experiments as a baseline to surpass random performance (50%). While useful for consistent evaluation, this threshold is not always necessary, as trends in model performance and relative dataset complexity often provide sufficient insights without relying on fixed accuracy thresholds.

Setup 1: Basic Non-bottlenecked Autoencoder. In the first setup, we employ a non-bottlenecked autoencoder architecture with an input layer, a single hidden layer, and an output layer. The hidden layer width is varied across 64, 128, 256, and 512 neurons to assess its impact on reconstruction performance. ReLU activation [1] is used. While we initially experimented with masking tokens from a Named Entity Recognition (NER) model to reduce bias, we ultimately decided against using this approach as it did not significantly impact relative performance.

We use the Keras Tokenizer[2] for text preprocessing, focusing on surface-level linguistic features to evaluate vocabulary quality without biases from pretrained embeddings. The output of the tokenizer serves as the input to the autoencoder, configured with a learning rate of 0.001, trained for 100 epochs with a batch size of 32. The input dimension is set to the vocabulary size, and the output layer uses a softmax activation function. Reconstruction loss is computed using sparse categorical crossentropy, and the optimizer is Adam. Dataset complexity is inferred from the hidden layer width required to achieve a predefined reconstruction error threshold, with higher quality vocabularies necessitating wider layers for comparable performance.

Setup 2: Squeezed Autoencoder. To explore the relationship between compression and reconstruction, we add a squeeze layer between the hidden and output layers, forcing the model to condense information before reconstruction. The squeeze ratio, defined as the neurons in the squeeze layer relative to the hidden layer, is varied across 1/2, 1/4, 1/8, and 1/16 to evaluate how well the model compresses vocabulary and structural diversity.

This squeezed autoencoder provides a detailed assessment of dataset complexity, balancing compression with reconstruction fidelity. Richer vocabularies with intricate contextual relationships are expected to require larger squeeze ratios (closer to 1) to maintain accuracy, reflecting higher informational density.

4.2 Rationale for Using Multi-Layer Perceptron (MLP)

We employ Multi-Layer Perceptrons (MLPs) [18] as the foundational architecture for our autoencoders due to their simplicity, flexibility, and interpretability. MLPs are well-suited for systematically varying model capacity by adjusting layer widths and depths, making them an ideal choice for exploring the relationship between vocabulary quality and model complexity. Additionally, their straightforward architecture ensures that the results are not confounded by domain-specific nuances or task-specific optimizations inherent in more complex models such as transformers. By focusing on MLPs, we maintain a consistent baseline for evaluating datasets across different domains.

[2] https://www.tensorflow.org/api_docs/python/tf/keras/preprocessing/text.

4.3 Datasets

We evaluate our methodology using two datasets: the DIFrauD dataset and the Project Gutenberg dataset, selected for their complementary linguistic and structural diversity. DIFrauD focuses on domain-specific challenges with five subsets: *political statements*, *job scams*, *product reviews*, *phishing*, and *fake news*. These subsets represent a wide range of lexical patterns, such as rhetorical ambiguity in political statements, manipulative language in job scams and phishing, and linguistic differences in authentic versus fake reviews.

The Project Gutenberg dataset offers a multilingual benchmark, including English, Swedish, Italian, German, French, Portuguese, Spanish, and Dutch, while spanning diverse genres and historical periods. It is curated with a rigorous proofreading framework where multiple proofreaders meticulously verify each book page by page to ensure minimal OCR errors[3], making it an ideal dataset for evaluating vocabulary quality. Together, these datasets facilitate a comprehensive analysis of our framework across domain-specific, multilingual, and historically diverse corpora.

5 Results and Discussions

5.1 DIFrauD Datasets - Setup 1

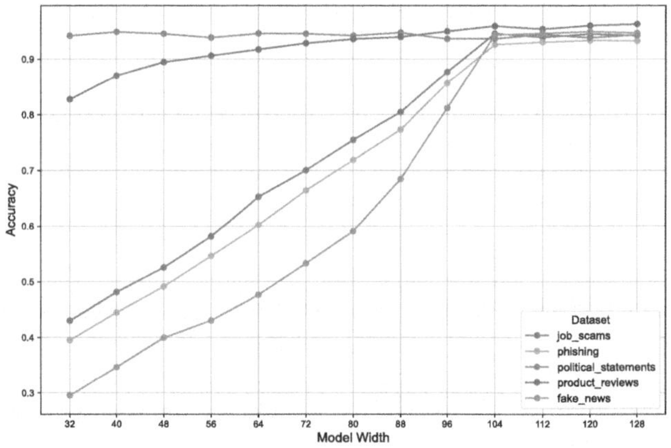

Fig. 4. Accuracy Across Different Model Widths for DIFrauD Datasets: The plot highlights accuracy variations across configurations, showing how different datasets respond to model complexity, with some requiring greater width for optimal performance.

The results depicted in Fig. 4 support our hypothesis that datasets with larger and more complex vocabularies require greater model width to achieve comparable performance to those of simpler datasets. Using a 51% accuracy threshold,

[3] https://www.pgdp.net/c/.

the *political statements* and *job scams* datasets meet this threshold at a width of 32, though *job scams* exhibit lower performance and benefit more from increased widths, indicating greater vocabulary and structural complexity. In contrast, *phishing* and *fake news* datasets require higher widths, 56 and 72 respectively, to achieve the same threshold, reflecting their richer and more complex vocabularies. These findings highlight how datasets with complex linguistic properties demand greater model capacity for effective representation.

5.2 Duplication Results

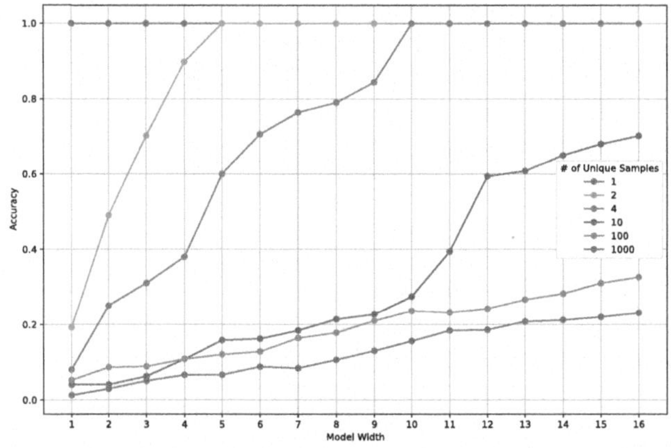

Fig. 5. Impact of Model Width on Accuracy for Datasets with Varying Unique Sample Counts. Datasets with fewer unique samples achieve high accuracy with smaller widths, while richer vocabularies require wider models to capture greater linguistic diversity.

Figure 5 illustrates that our approach maintains robust performance in assessing vocabulary characteristics, even when datasets contain many duplicate entries. For instance, datasets with only 1 or 2 unique samples reach high accuracy at minimal autoencoder widths, indicating limited linguistic diversity and lower overall vocabulary quality. This underscores the framework's capacity to recognize such datasets without demanding unnecessary model complexity.

Meanwhile, datasets featuring more unique samples (e.g., 10, 100, or 1000) require larger widths to achieve comparable accuracy, reflecting their richer and more varied vocabularies. Notably, datasets at intermediate levels of duplication (e.g., 10 or 100 unique samples) benefit from particular width configurations, highlighting the framework's flexibility in handling a range of complexities. In this way, our method accurately represents vocabulary characteristics while remaining resilient to duplication artifacts.

5.3 Robustness Against Languages and Text Length

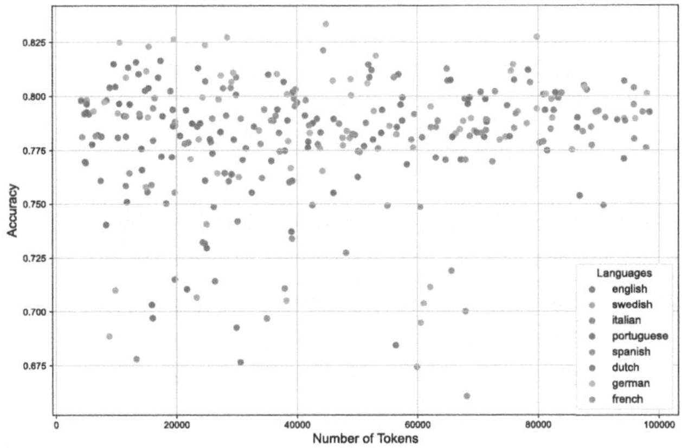

Fig. 6. Accuracy vs Text Length across Languages. The figure demonstrates that performance is primarily influenced by vocabulary quality rather than text length or the language of the dataset. This is evidenced by the consistent distribution of accuracy values across languages, regardless of the text length.

The results presented in Fig. 6 reveal that the performance of our framework is primarily determined by vocabulary quality, as opposed to text length or the language of the dataset. Due to resource constraints, 40 works were randomly sampled from each language to ensure equal representation across the datasets. All evaluations were conducted with a fixed model width of 64, demonstrating consistent accuracy across languages and indicating that performance is independent of specific languages.

Statistical analyses further support this conclusion. A Pearson correlation analysis between accuracy and text length yielded a weak, non-significant correlation ($r = 0.1219$, $p = 0.2814$), suggesting text length has minimal influence on performance. Pairwise t-tests between languages revealed no significant differences, reinforcing that the framework evaluates datasets independently of linguistic features.

These findings demonstrate the robustness of our framework in assessing vocabulary characteristics across diverse datasets. By remaining unaffected by text length or language, the method provides a fair and reliable evaluation of vocabulary richness in multilingual contexts.

Table 1. Impact of Model Width and Squeeze Ratio on Accuracy for 18th- and 20th-Century Datasets. Bold values indicate the highest accuracy for each row, reflecting optimal performance at specific width.

		Squeeze Ratio			
Century	Width	1/2	1/4	1/8	1/16
18th	64	**0.801**	0.731	0.608	0.406
18th	128	**0.814**	0.790	0.739	0.620
18th	256	0.803	**0.808**	0.790	0.745
18th	512	0.795	0.800	**0.802**	0.786
20th	64	**0.855**	0.831	0.783	0.663
20th	128	**0.845**	0.839	0.832	0.788
20th	256	0.836	**0.840**	0.836	0.831
20th	512	0.838	**0.839**	0.837	0.836

5.4 18th-Century Vs 20th-Century - Language Evolution

The results in Table 1 show that model width and squeeze ratio significantly impact accuracy for 18th-century and 20th-century datasets, each consisting of 50 works selected from Project Gutenberg and spanning multiple genres and authors. Narrower models (e.g., 64 neurons) exhibit sharp performance drops as the squeeze ratio decreases, particularly for 18th-century works, reflecting their richer vocabulary and complex linguistic structures. Wider models (256 and 512 neurons) demonstrate robustness to smaller squeeze ratios, effectively adapting to the increased complexity.

In contrast, the 20th-century dataset maintains consistently higher accuracy for all widths and squeeze ratios, suggesting simpler vocabularies and less intricate structures. This stability is supported by a pooled t-test comparing the 18th and 20th centuries across the configurations, which yields a t-statistic of -2.6609 and a p-value of 0.0081, confirming that the higher accuracy of the 20th-century dataset is statistically significant. This highlights their reduced modeling demands compared to 18th-century datasets.

Results Summary. Overall, these results demonstrate the framework's versatility in assessing vocabulary attributes across diverse datasets. In the DIFrauD experiments, datasets with more complex and diverse vocabularies demanded wider autoencoder configurations, while duplication tests underscored the model's sensitivity to linguistic diversity. Multilingual evaluations further confirmed its resilience across varying languages and text lengths. Lastly, the historical experiment showcased the method's adaptability: 18th-century texts required greater capacity and exhibited higher sensitivity to reduced squeeze ratios compared to 20th-century texts.

6 Limitations

While our methodology demonstrates the effectiveness of autoencoder-based approaches for assessing vocabulary richness, diversity, and complexity, several limitations merit attention. First, the computational overhead of training multiple autoencoder configurations can challenge real-time applications or resource-constrained settings. Additionally, although our approach successfully identifies structural and lexical intricacies, its performance in noisy or low-resource datasets remains underexplored, highlighting the need for further research to enhance robustness and generalizability across diverse real-world scenarios.

Our use of learnability—measured by how well an autoencoder reconstructs text—as a proxy for these vocabulary attributes assumes that data exhibiting richer, more diverse, and structurally coherent token distributions are inherently more predictable and thus more accurately reconstructed. Empirical evidence supports this premise, indicating that autoencoders yield higher reconstruction errors when faced with noisy or low-quality inputs [19]. However, relying solely on reconstruction quality has clear drawbacks. It does not directly capture semantic context, label correctness, or broader communicative intent. Consequently, while unsupervised learnability offers valuable insights into lexical richness, distributional diversity, and structural complexity, it should be used alongside other metrics for a well-rounded evaluation of dataset quality.

7 Conclusion and Future Works

This paper presents a new autoencoder-based methodology for evaluating vocabulary quality in NLP datasets. Our framework treats model capacity as a stand-in for linguistic complexity, providing a flexible assessment that exceeds the capabilities of traditional metrics. Through extensive experiments, we show that richer vocabularies demand wider models, linguistic diversity has a pronounced effect on model behavior, and our method remains robust across multiple languages and text lengths. Notably, the historical analysis highlights that 18th-century texts require larger model capacities and exhibit greater sensitivity to reduced squeeze ratios than those from the 20th century, demonstrating the framework's adaptability to varying linguistic complexities. Future work includes extending the method to noisy and low-resource datasets and incorporating contextual embeddings to enrich the evaluation.

Acknowledgments. Research partly supported by NSF grants 2210198 and 2244279, and ARO grants W911NF-20-1-0254 and W911NF-23-1-0191. Verma is the founder of Everest Cyber Security and Analytics, Inc.

References

1. Agarap, A.F.: Deep learning using rectified linear units (ReLU). arXiv preprint arXiv:1803.08375 (2018)
2. Boumber, D.A., Qachfar, F.Z., Verma, R.: Domain-agnostic adapter architecture for deception detection: extensive evaluations with the DIFrauD benchmark. In: Calzolari, N., Kan, M.Y., Hoste, V., Lenci, A., Sakti, S., Xue, N. (eds.) Proceedings of the 2024 Joint International Conference on Computational Linguistics, Language Resources and Evaluation (LREC-COLING 2024), pp. 5260–5274. ELRA and ICCL, Torino, Italia (May 2024). https://aclanthology.org/2024.lrec-main.468
3. Caswell, I., Chelba, C., Clark, J.: Quality at a glance: an audit of web-crawled multilingual datasets. In: Proceedings of the 2021 Conference of the North American Chapter of the Association for Computational Linguistics (NAACL), pp. 1336–1350 (2021)
4. Conneau, A., Khandelwal, K., Goyal, N., Chaudhary, V., Wenzek, G., Guzmán, F.: Unsupervised cross-lingual representation learning at scale. In: Proceedings of the 58th Annual Meeting of the Association for Computational Linguistics (ACL), pp. 8440–8451 (2020). https://doi.org/10.18653/v1/2020.acl-main.747
5. Dang, V.M.H., Verma, R.M.: Data Quality in NLP: metrics and a comprehensive taxonomy. In: Miliou, I., Piatkowski, N., Papapetrou, P. (eds.) Advances in Intelligent Data Analysis XXII. IDA 2024. Lecture Notes in Computer Science, vol. 14641, pp. 217–229. Springer, Cham (2024). https://doi.org/10.1007/978-3-031-58547-0_18
6. Dang, V.M.H., Verma, R.M.: Autoencoder-based framework to capture vocabulary quality in NLP. arXiv preprint (2025)
7. Dodge, J., Ilharco, G., Schwartz, R., Farhadi, A., Hajishirzi, H., Smith, N.A.: Documenting the English colossal clean crawled corpus. In: Proceedings of the 2021 Conference on Empirical Methods in Natural Language Processing (EMNLP), pp. 8780–8794 (2021). https://doi.org/10.18653/v1/2021.emnlp-main.694
8. Egbert, J., Biber, D.: Developing a more effective measure of lexical diversity for automated text analysis. Text Talk **35**(6), 667–699 (2015). https://doi.org/10.1515/text-2015-0012
9. Gururangan, S., et al.: Don't stop pretraining: adapt language models to domains and tasks. In: Proceedings of the 58th Annual Meeting of the Association for Computational Linguistics (ACL), pp. 8342–8360 (2020). https://doi.org/10.18653/v1/2020.acl-main.740
10. Jarvis, S.: Capturing the diversity in lexical diversity. Lang. Learn. **63**(S1), 87–106 (2013). https://doi.org/10.1111/j.1467-9922.2012.00739.x
11. Jarvis, S., McCarthy, P.M.: A comparison of sophisticated approaches for measuring lexical diversity in texts. J. Second Lang. Res. **26**(3), 235–265 (2010). https://doi.org/10.1177/0267658310377374
12. Koizumi, R., In'nami, Y.: Effects of text length on lexical diversity measures: using short texts with less than 200 tokens. TESOL Q. **46**(4), 765–774 (2012). https://doi.org/10.1002/tesq.46
13. Lin, B.Y., Tan, C., Yang, L., Ren, X.: DIFFPREP: a unified approach to pretrained language model tuning for data with high domain diversity. In: Proceedings of the 2020 Conference on Empirical Methods in Natural Language Processing (EMNLP), pp. 5580–5593 (2020)
14. Malvern, D., Richards, B., Chipere, N., Durán, P.: Lexical Diversity and Language Development: Quantification and Assessment. Palgrave Macmillan, Basingstoke, UK (2004)

15. McCarthy, P.M., Jarvis, S.: VOCD: a theoretical and empirical evaluation. Lang. Test. **24**(4), 459–488 (2007). https://doi.org/10.1177/0265532207080767
16. McCarthy, P.M., Jarvis, S.: MTLD, vocd-D, and HD-D: a validation study of sophisticated approaches to lexical diversity assessment. Behav. Res. Methods **42**(2), 381–392 (2010). https://doi.org/10.3758/BRM.42.2.381
17. Mishra, S., Arunkumar, A., Sachdeva, B.S., Bryan, C., Baral, C.: DQI: measuring data quality in NLP. CoRR **abs/2005.00816** (2020)
18. Rumelhart, D.E., Hinton, G.E., Williams, R.J.: Learning representations by back-propagating errors. Nature **323**, 533–536 (1986). https://doi.org/10.1038/323533a0
19. Salekshahrezaee, Z., Leevy, J.L., Khoshgoftaar, T.M.: A reconstruction error-based framework for label noise detection. J. Big Data **8** (2021). https://doi.org/10.1186/s40537-021-00447-5
20. Sharoff, S.: Know thy corpus! Robust methods for digital curation of web corpora. In: Proceedings of the Twelfth Language Resources and Evaluation Conference. pp. 2453–2460. European Language Resources Association, Marseille, France (May 2020). https://aclanthology.org/2020.lrec-1.298
21. Tweedie, F.J., Baayen, R.H.: How variable may a constant be? Measures of lexical richness in perspective. Comput. Humanit. **32**(5), 323–352 (1998). https://doi.org/10.1023/A:1001749303137
22. Wang, X., Zhang, Y.: Rethinking lexical diversity metrics: implications of duplication in linguistic datasets. Trans. Assoc. Comput. Linguist. **9**, 355–372 (2021). https://doi.org/10.1162/tacl_a_00369
23. Yule, G.U.: The Statistical Study of Literary Vocabulary. Cambridge University Press, Cambridge, UK (1944). originally introduced the Type-Token Ratio concept as a measure of lexical richness
24. Zhang, X., Hashimoto, T.B.: On the inductive bias of pre-trained representations for text classification. In: Proceedings of the 2021 Conference on Empirical Methods in Natural Language Processing (EMNLP), pp. 103–120 (2021)

Temporal and Streaming Data

Expertise Prediction of Tetris Players Using Eye Tracking Information

Stijn J. Rotman(✉), Gianluca Guglielmo, Boris Cule, and Michal Klincewicz

Tilburg University, Warandelaan 2, 5037 AB Tilburg, The Netherlands
{s.j.rotman,g.guglielmo,b.cule,m.w.klincewicz}@tilburguniversity.edu

Abstract. We report on an application of three multivariate time series classification methods, HIVE-COTE 2.0, MINIROCKET, and MR-PETSC, to gaze and eyelid movement data to classify expertise. Our methods can be used to noninvasively monitor performance and identify experts using low-grade equipment. The test case was Tetris, which is a video game in which players arrange falling blocks to clear horizontal lines with increasing points and difficulty as the game advances. In addition to being able to classify the expert players, we can attribute the patterns within a time series that led to a prediction with MR-PETSC. This allows us to describe eye behavior that is associated with expertise. This method can be used in any performance classification that involves screen-based activity that is accompanied by eye movement recordings, for example with a webcam.

Keywords: Time series classification · Expertise · Eye tracking

1 Introduction

Experts can use their attentional resources more efficiently than non-experts [7, 20]. This can be captured by a variety of methods, including eye tracking. Expert chess players and radiologists, for example, direct their gaze and visual attention differently compared to novices [5,8]. Unfortunately, identifying attentional features of expertise with eye behavior data typically relies on invasive or costly laboratory equipment, unsuitable for many real-world applications.

This study looks specifically at Tetris expertise as a real-world test case. We recorded players' eye and eyelid movements with a webcam, combining horizontal and vertical gaze data with eye-aspect-ratio (eyelid movement) signals. These simple, accessible signals were sufficient to identify the experts. Because we do not focus on specific game mechanics, our methods can be easily applied to other domains of expertise, whenever they involve monitors or computer use.

At the core of our approach are time series classification (TSC) methods: HIVE-COTE 2.0 [29], MINIROCKET [10], and MR-PETSC [13]. While HIVE-COTE 2.0 and MINIROCKET represent state-of-the-art TSC methods [30], MR-PETSC provides competitive performance and offers greater interpretability.

G. Krempl et al. (Eds.): IDA 2025, LNCS 15669, pp. 305–317, 2025.
https://doi.org/10.1007/978-3-031-91398-3_23

The rest of this paper is outlined as follows: Sect. 2 situates our work in the existing literature by addressing previous work. Then, Sect. 3 describes the data collected for this study. The TSC methods used in this study, as well as the classification baselines, are defined in Sect. 4. Subsequently, Sect. 5 presents and discusses our results. The results of our analyses are further interpreted in Sect. 6, providing the context of earlier findings. Finally, we conclude the paper in Sect. 7.

2 Theoretical Background

Similar to most real-life expertise, expertise in video games is gained through practice over many hours. In video games, it is manifested through the adaptive use of game mechanics and optimizing strategies. Expert video game play has been extensively studied within the context of online competitive games, such as DOTA II [21], Hearthstone [18], but also single-player games, such as Tetris [15, 25,32]. Tetris has drawn researchers' attention since the early 90s [20], just a few years after it was developed by Alexey Pajitnov. It is ideal for research questions that aim to take advantage of a video game environment to study basic cognitive and perceptual mechanisms. However, up to date, no studies with Tetris focused on time series classification of Tetris expertise with physiological markers. Our study aims to fill this gap and thereby also provide a general method for expertise classification by addressing the following research questions (RQs):

RQ1 To what extent can gaze- and eyelid-related time series be used to discriminate Tetris players with different levels of proficiency in Tetris?
RQ2 What are the key differences in performance across interpretable and opaque classifiers?
RQ3 Which factors contribute to the classification of experts?

We choose to focus on eye movements because previous studies demonstrated significant differences in eye movements between experts and novices in action video games. For instance, experts tend to fixate on targets in these games less often than novices, but for a longer time [34]. Expert players have higher blinks per minute (blinks/m) when playing competitive Hearthstone [18], while in Tetris novices have a higher decrease in baseline-corrected blinks/m (blinks at rest minus blinks during the task) when compared to experts [16]. Significant differences between eye behavior of experts and novices in those studies motivated our thinking that the video recording of the eyes would be sufficient to *identify* expert players. We may not need any other data about their performance, including their in-game decisions.

There are also independent reasons to think that eye behavior is a good candidate to identify expertise. First, features of gaze direction and trajectory of eye movement have been demonstrated to be a proxy for working memory involvement [9,14]. Secondly, we know that experts are more effective in retrieving encoded task-specific information [12] and effectively allocating their attentive resources [14]. In other words, their working memory functions better in tasks they are experts in. Given this evidence, we expect that information contained in the players' eye aspect ratio signal and their eye movements—as a proxy for working memory performance—can be used as an important feature that identifies expert players.

3 Data

3.1 Collection Procedure

A total of 80 participants (39 females, 40 males, 1 undisclosed; mean age = 22.27, SD = 5.92) were recruited from Tilburg University using, SONA, Tilburg University's online subject recruitment system. The study was approved under REDC2021.35 by the Research Ethics and Data Management Committee of Tilburg University School of Humanities and Digital Sciences. Data collection involved 40-min sessions that started with the collection of informed consent, demographic information, and self-assessed Tetris experience (1–5 Likert scale). Following a brief introduction to NES Tetris and a 2-min practice session with the game controls, participants then played the game for 13-min, without knowledge of the exact duration to prevent distraction. Their faces and game play were simultaneously recorded using Open Broadcaster Software (OBS) at 30 frames per second (fps).

3.2 Expertise Definition

On average, participants completed 2.76 matches (SD = 1.88) with an average score of 4 291.93 (min = 84.75, max = 32 141, SD = 6 771). As in prior studies [15–17,25,32], we define expertise by players' score and match count. Proficient players are expected to score higher and complete fewer, longer matches [15]. To that end we used a k-means algorithm for clustering. In our sample, two players played without losing, so their total scores for a single match were used.

We assessed skewness, finding the number of matches relatively normal (1.06) but the score highly skewed (2.09) [35]. This is typical for Tetris [15,32]. Taking into consideration that k-means is sensitive to highly skewed variables [22], a log transformation was applied, as suggested by previous work [15]. Finally, we used the elbow method [15,36] to identify three clusters, which we then used as ground truths: Novices, Intermediates, and Experts. Experts (20 players) averaged 1.1 matches, Intermediates (38 players) 2.32 matches, and Novices (22 players) 5.22 matches.

An ANOVA with Holm correction ($F(2,77) = 9.32$, $p < 0.001$) confirmed significant differences in self-assessed Tetris experience across groups: Novices (M=1.73, SD=0.69) had a lower self-assessed experience than Intermediates (M = 2.21, SD = 0.87; $p < 0.05$) and Experts (M. = 2.80, SD = 0.75; $p < 0.001$), while Experts had a higher self-assessed experience than Intermediates ($p < 0.05$) and Novices ($p < 0.001$).

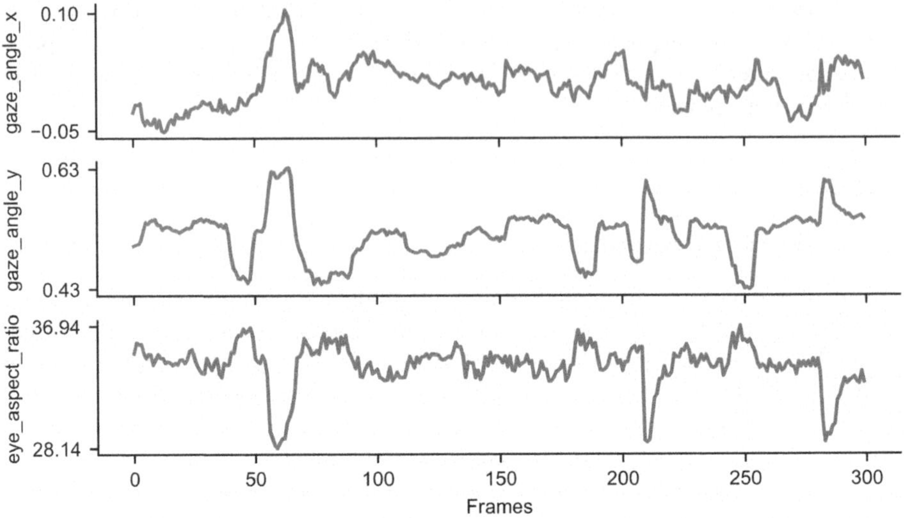

Fig. 1. 300 frames (10 s) of signal for all three channels of one participant.

3.3 Eye Movement Extraction and Preprocessing

After data collection, videos were cropped to contain only faces. From these, two time series were created: the eye-aspect ratio and eye movement. Eye aspect ratio was extracted using the *FaceMeshDetector* available in the *cvzone* Python library. This measure is derived from the distance between upper and lower eyelids and captures blinks as troughs. The same method can be used to identify eyelid movements caused by eye movements [19]. Eye movements were extracted using OpenFace [2] software as the *gaze_ angle_ x* (negative values = left, positive = right) variable and the *gaze_ angle_ y* (lower values = up, higher = down) variable in a recording. Each measure, combined as input channels, produced 13 minutes of data per participant at 30 fps, totaling 23,400 observations per participant. Excluding one corrupted recording of an intermediate player, the full data set contained 5,545,800 data points for 79 participants. Each individual signal is scaled to a mean of 0 and a standard deviation of 1. Figure 1 shows recorded signals of one participant before normalization over 10 seconds of play as an example.

As becomes apparent in Sect. 4, most time series classifiers work by extracting domain-agnostic features. We extract no other features from the time series than those extracted by the time series classification models we employ. Our goal is to propose a framework for expertise classification that is applicable to a wide variety of screen-based tasks. Domain-specific features would reduce the adaptability of this framework to different tasks, thus would be inappropriate.

4 Methodology

4.1 Baseline Algorithms

This study uses two baseline algorithms: naive majority class classification and one-nearest-neighbor dynamic time warping (1-NN DTW). Majority class classification assigns the most common class label to all test instances, ignoring feature patterns. The 1-NN DTW algorithm assigns the class of the closest time series neighbor based on dynamic time warping (DTW) distance and is regarded as a robust baseline for time series classification tasks [1].

4.2 HIVE-COTE 2.0

The central idea behind HIVE-COTE 2.0 is a hierarchy of classifiers [29]. A hierarchical ensemble of base classifiers is created by organizing multiple classifiers into a structured hierarchy, where each level specializes in recognizing distinct patterns or characteristics within time series data.

The most recent implementation of HIVE-COTE 2.0 uses four classifiers, one dictionary based (Temporal Dictionary Ensemble [28]), one interval based (Diverse Representation Canonical Interval Forest [27]), one convolution based (ARSENAL, which is based on the ROCKET classification model), and one shapelet based (Shapelet Transform Classifier [6]). All models in the ensemble are ensemble methods themselves [29]. Cross-validation accuracy weighted probabilistic ensemble (CAWPE) is used to combine predictions [23]. CAWPE weighs the probability estimates for each class by each classifier's training performance, giving more attention to the estimates of models that performed better on the training set.

4.3 MINIROCKET

MINIROCKET is another state-of-the-art algorithm for TSC. The main steps are the generation of random convolutional kernels, extracting features by transforming the time series with the kernels, and training a linear model [11].

MINIROCKET starts by selecting random subsequences from the time series, creating fixed-length subseries from random starting points. These subsequences serve as the basis for feature extraction. Unlike deep neural networks, which learn convolutional kernels through iterative tuning, MINIROCKET uses a large set of randomly generated kernels. These are then applied to the time series to generate transformed features for classification, typically using a linear model, which has shown the best performance. While more interpretable than HIVE-COTE 2.0 due to its linear model, MINIROCKET's interpretability remains limited to finding which convolutions are most discriminative. Interpreting the meaning of these kernels remains a challenge.

4.4 Mr-Petsc

Mr-Petsc extends the PETSC (Pattern-Based Embedding for Time Series Classification) method [13], which consists of segmentation, pattern mining, embedding, and linear classification steps. Initially, time series are segmented and transformed via Sequential Aggregate approXimation (SAX) [24], which compresses data by reducing series length and discretizing sequences. Mr-Petsc extends PETSC by recursively segmenting the time series from coarser to finer representations in the SAX representation.

Next, cohesive sequential patterns are mined. Cohesive patterns are characterized by minimal gaps to improve their interpretability. The sequential pattern mining approach directly mines top-k frequent patterns, eliminating the need for a minimum support threshold. These patterns are then used to create an embedding, counting occurrences for each time series. For multivariate data, patterns are mined separately in each dimension and merged into a single embedding. As with MiniRocket, a tabular classifier is then applied to the embedding. We follow the implementation of Feremans et al. [13], fitting a logistic regression with L1 and L2 regularization with stochastic gradient descent.

Though like MiniRocket, Mr-Petsc extracts features from the time series, which it feeds into a linear classifier, the classification attributions are much easier to interpret for Mr-Petsc than for MiniRocket. For both algorithms, the weights of the linear model rank the extracted features in their importance. However, the patterns extracted by Mr-Petsc are much more interpretable than the convolutional kernels generated by MiniRocket. The specific extracted patterns are represented as abstract shapes with discrete values. Moreover, the occurrences of specific patterns can be used to attribute what aspects of a time series led to a certain prediction.

4.5 Experimental Setup

Instead of performing a single train-test split, we split the data in five stratified folds for cross-validation. This prevents data leakage just as a single train-test split would. Additionally, cross-validation makes our evaluations more robust, as it increases the test set size. To evaluate classifier performance, we use weighted and unweighted accuracy, weighted F1 score, and area under the receiver operating characteristic curve (AUROC). Given the unbalanced multi-class target, we compute the one-vs-one AUROC score weighted by class support.

In the time series classification community, much effort has gone into reducing the need for data-dependent hyperparameters. Thus, we train the models using default parameters to minimize overfitting to our particular training data. Both baseline models are parameter-free. Though kNN can have larger values for k, 1-NN DTW is standard for baseline comparisons [1]. Hive-Cote 2.0 uses the CAWPE algorithm to construct ensemble weights, using the default $\alpha = 4$ suggested by the authors [23]. For MiniRocket, we follow the authors' advice to use a ridge regression classifier instead of unregularized logistic regression,

Table 1. Classification results on 13 minutes of game-play

	Accuracy	Weighted accuracy	Weighted F1	AUROC
Majority class	0.468	0.333	0.299	0.487
1-NN DTW	0.443	0.449	0.445	0.584
HIVE-COTE 2.0	0.519	0.478	0.508	0.692
MINIROCKET	0.532	0.501	0.526	0.621
MR-PETSC	**0.633**	**0.662**	**0.630**	**0.775**

because we have fewer training examples than features [10]. Finally, as MR-PETSC combines the discretization results of SAX using different settings, there is no need for learning the optimal discretization parameters.

Majority classification, 1-NN DTW, HIVE-COTE 2.0, and MINIROCKET use *aeon v0.11.1* [26]. For MR-PETSC, we use an adaptation of the Python and Java code provided by the authors [13]. Evaluation metrics use the implementation from *scikit-learn v1.2.2* [31]. We use Python v3.11, Java v11.0.20.1, and Apache Maven v3.6.3. All experiments are run on a Linux Mint 21 machine with an AMD Ryzen 7 16-thread processor and 32GB of RAM. Our data and code are available at `github.com/steenrotsman/tetris-expertise-classification`.

5 Results

5.1 Comparison of Model Performance

Table 1 shows the classification results. All three classifiers outperform both baseline models on all metrics. In terms of AUROC, MR-PETSC outperforms all other methods, followed by HIVE-COTE 2.0 and MINIROCKET. For the other metrics, applying MR-PETSC to the problem results in the best performance as well. However, the relative ranking of HIVE-COTE 2.0 and MINIROCKET differs. For the weighted F1, weighted accuracy, and unweighted accuracy, HIVE-COTE 2.0 is outperformed by MINIROCKET.

The confusion matrices displayed in Fig. 2 give more context to the results presented in Table 1. Each confusion matrix shows the predicted labels against the true labels for each model trained on the full 13 minutes of recording. Most TSC models struggle with the Intermediate class. 1-NN DTW and MR-PETSC both have misclassified a relatively high number of Intermediate players as Expert players, whereas the HIVE-COTE 2.0 and MINIROCKET show a tendency toward mostly predicting players as being Intermediate. Both these models have misclassified at least as many Novices and Experts as Intermediates as they have correctly classified players to the other classes.

5.2 Interpreting Performance of MR-PETSC

As described in Sect. 4, MR-PETSC is the most interpretable method out of the TSC algorithms. To better understand its superior performance we looked at the

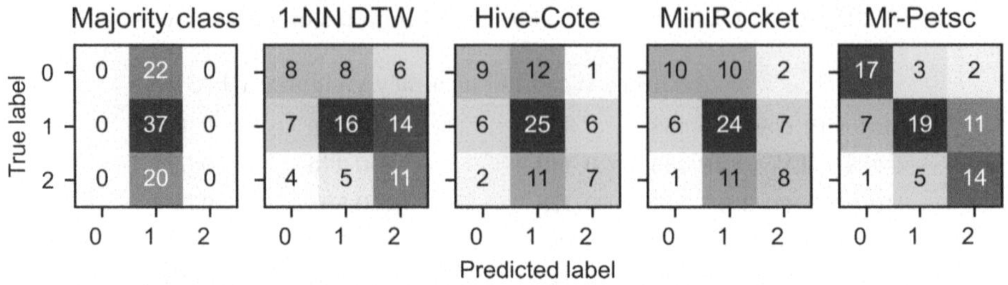

Fig. 2. Confusion matrices of results on 13 minutes of game-play. Labels correspond to 0: 'Novice', 1: 'Intermediate', 2: 'Expert'.

top 10 most important patterns based on their logistic regression weights. These patterns appear as abstract shapes, but nevertheless their interpretability is highly context-dependent. Here we look at the relationship between the patterns and ocular behavior and speculate about how they led to better predictions. Our context is domain knowledge about eye behavior and the nature of our dataset.

The three signals of the first minute of play of a participant are shown in Fig. 3. Notably, only one of the top patterns occurs in this minute of the *gaze_angle_x*. In other words, the patterns found in this signal tend to be less important for classification than ones in the other signals. The *gaze_angle_y* signal contains multiple rapid up-and-down movement of the eyes that are strongly associated with experts. We speculate that vertical movement can either correspond to tracking the falling zoid or looking at the next one, indicating strategic planning. What supports our interpretation is that Tetris involves vertical movements on screen, so less horizontal gaze change is needed to play. This means that there is less variation in horizontal eye movements between categories of players than in the other signals. Moving on to the *eye_aspect_ratio*, we find troughs highlighted in red. These patterns correspond to blinks. From the attribution, we can conclude that the model interprets patterns that correspond to blinks as reducing the probability of being an expert. In plain terms, the more someone blinks, the less likely are they to be an expert. Though the *gaze_angle_y* and *eye_aspect_ratio* seem conflicting, we stress that Mr-Petsc bases its classification on the relative frequency of pattern occurrences.

6 Discussion

Our first research question was *To what extent can gaze- and eyelid-related time series be used to discriminate Tetris players with different levels of proficiency in Tetris?* To study this question, we compared the Hive-Cote 2.0, MiniRocket, and Mr-Petsc time series classifiers alongside two baselines: majority classification and 1-NN DTW. Among these, Hive-Cote 2.0 and MiniRocket can be considered state-of-the-art, as they typically outperform other methods on most data sets [30]. For multivariate data sets, Mr-Petsc performs competitively,

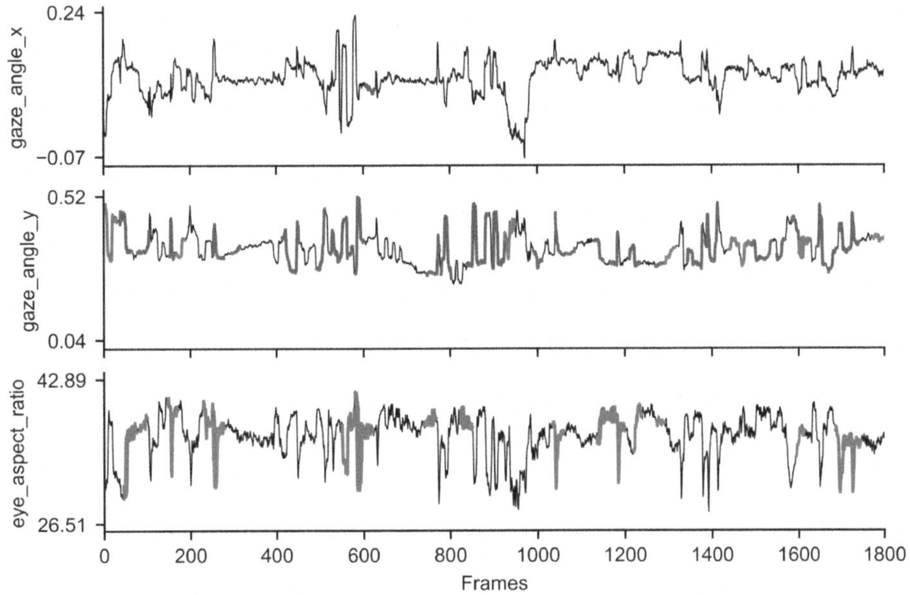

Fig. 3. 1800 frames (first minute) of signal for all three channels for one participant. Top 10 most important patterns are highlighted. Patterns that contributed towards and against classification as an expert are marked blue and red, respectively. Thicker lines correspond to higher weights. (Color figure online)

but on average is outperformed by the state-of-the-art [13]. Interestingly, MR-PETSC outperformed the state-of-the-art classifiers on this task. Unexpectedly, HIVE-COTE 2.0 showed poorer performance.

The strong performance of MR-PETSC can be attributed to the algorithm's ability to identify where expertise manifests as recurring patterns, e.g., tracking falling zoids or less frequent blinking. While this result indicates a potential for pattern-based TSC on other expertise classification problems using eye-tracking data, we should be wary not to overgeneralize our findings to avoid the Texas sharpshooter fallacy [3]. The Texas sharpshooter fallacy occurs when conclusions are drawn by focusing only on data sets where a method outperforms another, suggesting that the classifier is especially appropriate for aspects that these data sets have in common. In this context, while MR-PETSC did well on this eye-tracking dataset, claiming it is universally superior for eye-tracking data or expertise classification would be premature unless we can predict when and why it will outperform other models. In further research, we plan to further study the appropriateness of pattern-based classification for this problem setting.

Regarding the risk of overgeneralization, it should be taken into account that the sample contained 79 participants. While we have a lot of observations per participant, we need a larger sample over a variety of screen-based tasks before we can make any generalizations about the potential applicability of our conclusions outside of Tetris. This is what we aim to focus on going forward.

Regarding our second research question, *What are the key differences in performance across interpretable and opaque classifiers?* we expected a trade-off between accuracy and interpretability. However, we found the reverse, with more interpretable classifiers also performing better. In this case, the interpretability of MR-PETSC strengthens our confidence of the classifier learning how to generally classify expertise, as part of the attributions consist of patterns that we expected to find.

This relates to the last research question *Which factors contribute to the classification of experts?* One of the justifications for using the eye signals of participants to classify expertise is the relation between blinks and expertise found in prior work [18]. Our findings show that MR-PETSC was indeed able to identify a 'blink' pattern in the *eye_aspect_ratio* signal. MR-PETSC found domain-specific patterns as well, for example the rapid and strong vertical eye-movements contributing to expertise.

Beyond Tetris, our methodology's potential extends to other fields. In esports as well as research fields beyond the gaming world, our methodology for expertise detection can inform training and performance studies. For example, prior research has shown that experts and novices exhibit different blink patterns in microsurgery [4] and different eye movement patterns in radiology [5]. By adapting our approach to screen-based simulations of these tasks, expertise could be detected with the methods used in this study, using data collected noninvasively and with widely available technology. The adaptation of our approach to other screen-based tasks does not require knowledge of advanced tools or domain-specific features. Rather, a laptop with a webcam and access to open-source software is sufficient. For future work, we plan to study the applicability of our approach in other domains. Additionally, we will compare the use of frequent motifs instead of sequential patterns in time series classification problems [33].

7 Conclusion

We are able to discriminate different levels of proficiency in Tetris using time series with a duration of 13 minutes, extracted from gaze and eyelid movements. The three models we apply to the problem, HIVE-COTE 2.0, MINIROCKET, and MR-PETSC, are all able to beat majority class and 1-NN DTW baseline models. The classification performance of the most interpretable classifier, MR-PETSC, was higher on this classification problem than the more opaque classifiers HIVE-COTE 2.0 and MINIROCKET. We reported the important features that MR-PETSC identified and provided general speculation about their relationship to eye behavior of expert Tetris players.

References

1. Bagnall, A., Lines, J., Bostrom, A., Large, J., Keogh, E.: The great time series classification bake off: a review and experimental evaluation of recent algorithmic advances. Data Min. Knowl. Disc. **31**, 606–660 (2017)
2. Baltrušaitis, T., Robinson, P., Morency, L.P.: Openface: an open source facial behavior analysis toolkit. In: 2016 IEEE Winter Conference on Applications of Computer Vision (WACV), pp. 1–10. IEEE (2016)
3. Batista, G.E., Keogh, E.J., Tataw, O.M., De Souza, V.M.: Cid: an efficient complexity-invariant distance for time series. Data Min. Knowl. Disc. **28**, 634–669 (2014)
4. Bednarik, R., Koskinen, J., Vrzakova, H., Bartczak, P., Elomaa, A.P.: Blink-based estimation of suturing task workload and expertise in microsurgery. In: Proceedings: 31st IEEE International Symposium on Computer-Based Medical Systems-CBMS 2018 (2018)
5. Bertram, R., Helle, L., Kaakinen, J.K., Svedström, E.: The effect of expertise on eye movement behaviour in medical image perception. PLoS ONE **8**(6), e66169 (2013)
6. Bostrom, A., Bagnall, A.: Binary shapelet transform for multiclass time series classification. Transactions on Large-Scale Data-and Knowledge-Centered Systems XXXII: Special Issue on Big Data Analytics and Knowledge Discovery, pp. 24–46 (2017)
7. Breivik, G., Roth, W.T., Jørgensen, P.E.: Personality, psychological states and heart rate in novice and expert parachutists. Persona. Individ. Differ. **25**(2), 365–380 (1998)
8. Bruder, C., Hasse, C.: Differences between experts and novices in the monitoring of automated systems. Int. J. Ind. Ergon. **72**, 1–11 (2019)
9. Colzato, L.S., Spapé, M.M., Pannebakker, M.M., Hommel, B.: Working memory and the attentional blink: blink size is predicted by individual differences in operation span. Psychon. Bull. Rev. **14**, 1051–1057 (2007)
10. Dempster, A., Petitjean, F., Webb, G.I.: Rocket: exceptionally fast and accurate time series classification using random convolutional kernels. Data Min. Knowl. Disc. **34**(5), 1454–1495 (2020)
11. Dempster, A., Schmidt, D.F., Webb, G.I.: Minirocket: a very fast (almost) deterministic transform for time series classification. In: Proceedings of the 27th ACM SIGKDD Conference on Knowledge Discovery & Data Mining, pp. 248–257 (2021)
12. Ericsson, K.A., Kintsch, W.: Long-term working memory. Psychol. Rev. **102**(2), 211 (1995)
13. Feremans, L., Cule, B., Goethals, B.: Petsc: pattern-based embedding for time series classification. Data Min. Knowl. Disc. **36**(3), 1015–1061 (2022)
14. Gegenfurtner, A., Lehtinen, E., Säljö, R.: Expertise differences in the comprehension of visualizations: a meta-analysis of eye-tracking research in professional domains. Educ. Psychol. Rev. **23**, 523–552 (2011)
15. Gray, W.D., Banerjee, S.: Constructing expertise: surmounting performance plateaus by tasks, by tools, and by techniques. Top. Cogn. Sci. **13**(4), 610–665 (2021)
16. Guglielmo, G., Klincewicz, M., in'Veld, E.H., Spronck, P.: Tracking early differences in tetris performance using eye aspect ratio extracted blinks. IEEE Trans. Games (2023)

17. Guglielmo, G., Klincewicz, M., Huis in't Veld, E., Spronck, P.: Predicting tetris performance using early keystrokes. In: Proceedings of the 18th International Conference on the Foundations of Digital Games, pp. 1–4 (2023)
18. Guglielmo, G., Mavromoustakos Blom, P., Klincewicz, M., Huis in't Veld, E., Spronck, P.: Blink to win: blink patterns of video game players are connected to expertise. In: Proceedings of the 17th International Conference on the Foundations of Digital Games, pp. 1–7 (2022)
19. Helmchen, C., Rambold, H.: The eyelid and its contribution to eye movements. Neuro-Ophthalmology **40**, 110–131 (2007)
20. Kirsh, D., Maglio, P.: Reaction and reflection in tetris. In: Artificial Intelligence Planning Systems, pp. 283–284. Elsevier (1992)
21. Kokkinakis, A.V., Cowling, P.I., Drachen, A., Wade, A.R.: Exploring the relationship between video game expertise and fluid intelligence. PLoS ONE **12**(11), e0186621 (2017)
22. Kumar, C.N.S., Rao, K.N., Govardhan, A., Sandhya, N.: Subset k-means approach for handling imbalanced-distributed data. In: Emerging ICT for Bridging the Future-Proceedings of the 49th Annual Convention of the Computer Society of India CSI, vol. 2, pp. 497–508. Springer (2015)
23. Large, J., Lines, J., Bagnall, A.: A probabilistic classifier ensemble weighting scheme based on cross-validated accuracy estimates. Data Min. Knowl. Disc. **33**(6), 1674–1709 (2019). https://doi.org/10.1007/s10618-019-00638-y
24. Lin, J., Keogh, E., Wei, L., Lonardi, S.: Experiencing sax: a novel symbolic representation of time series. Data Min. Knowl. Disc. **15**, 107–144 (2007)
25. Lindstedt, J.K., Gray, W.D.: Distinguishing experts from novices by the mind's hand and mind's eye. Cogn. Psychol. **109**, 1–25 (2019)
26. Middlehurst, M., et al.: aeon: a python toolkit for learning from time series. J. Mach. Learn. Res. **25**(289), 1–10 (2024). http://jmlr.org/papers/v25/23-1444.html
27. Middlehurst, M., Large, J., Bagnall, A.: The canonical interval forest (CIF) classifier for time series classification. In: 2020 IEEE International Conference on Big Data (Big Data), pp. 188–195. IEEE (2020)
28. Middlehurst, M., Large, J., Cawley, G., Bagnall, A.: The temporal dictionary ensemble (tde) classifier for time series classification. In: Machine Learning and Knowledge Discovery in Databases: European Conference, ECML PKDD 2020, Ghent, Belgium, 14–18 September 2020, Proceedings, Part I, pp. 660–676. Springer (2021)
29. Middlehurst, M., Large, J., Flynn, M., Lines, J., Bostrom, A., Bagnall, A.: Hivecote 2.0: a new meta ensemble for time series classification. Mach. Learn. **110**(11-12), 3211–3243 (2021)
30. Middlehurst, M., Schäfer, P., Bagnall, A.: Bake off redux: a review and experimental evaluation of recent time series classification algorithms. Data Min. Knowl. Discov. 1–74 (2024)
31. Pedregosa, F., et al.: Scikit-learn: machine learning in Python. J. Mach. Learn. Res. **12**, 2825–2830 (2011)
32. Qi, Y.: Random forest for bioinformatics. Ensemble Mach. Learn. Methods Appl. 307–323 (2012)
33. Rotman, S.J., Cule, B., Feremans, L.: Efficiently mining frequent representative motifs in large collections of time series. In: 2023 IEEE International Conference on Big Data (BigData), pp. 66–75 (2023)

34. Smith, C.F., Tsai, Y., Wong, J.H., Brooks, D.T., Peterson, M.S.: More than meets the eye: investigating expert and novice differences in action video games. J. Vis. **8**(6), 105–105 (2008)
35. Tabachnick, B.G., Fidell, L.S.: Using Multivariate Statistics (6. baskı). Pearson, MA (2013)
36. Thorndike, R.L.: Who belongs in the family? Psychometrika **18**(4), 267–276 (1953)

Integrating Inverse and Forward Modeling for Sparse Temporal Data from Sensor Networks

Julian Vexler[1](✉), Björn Vieten[2], Martin Nelke[3], and Stefan Kramer[1]

[1] Johannes Gutenberg University Mainz, Staudingerweg 9, 55218 Mainz, Germany
j.vexler@gmx.net, kramer@informatik.uni-mainz.de
[2] Fraport AG, 60547 Frankfurt am Main, Germany
B.Vieten@Fraport.de
[3] MIT GmbH, Pascalstraße 69, 52076 Aachen, Germany
Martin.Nelke@mitgmbh.de

Abstract. We present CavePerception, a framework for the analysis of sparse data from sensor networks that incorporates elements of inverse modeling and forward modeling. By integrating machine learning with physical modeling in a hypotheses space, we aim to improve the interpretability of sparse, noisy, and potentially incomplete sensor data. The framework assumes data from a two-dimensional sensor network laid out in a graph structure that detects certain objects, with certain motion patterns. Examples of such sensors are magnetometers. Given knowledge about the objects and the way they act on the sensors, one can develop a data generator that produces data from simulated motions of the objects across the sensor field. The framework uses the simulated data to infer object behaviors across the sensor network. The approach is experimentally tested on real-world data, where magnetometers are used on an airport to detect and identify aircraft motions. Experiments demonstrate the value of integrating inverse and forward modeling, enabling intelligent systems to better understand and predict complex, sensor-driven events.

Keywords: Inverse modeling · Forward modeling · Sensor networks · Magnetometers · Aviation

1 Introduction

Sensor networks can collect an almost arbitrary amount of multivariate data in a fraction of a second. However, sensors only measure the effect or outcome of physical phenomena. The true parameters that cause these outcomes are usually unknown. *Inverse modeling* is the approach to estimate the unknown causal parameters from the available observations. In other words, models are trained with the obtained data, in order to estimate the unknown parameters. Machine Learning (ML) approaches are the state of the art for making inferences from data. However, they frequently achieve their peak performance when the true labels are known and they can be trained in a supervised fashion. In contrast to

G. Krempl et al. (Eds.): IDA 2025, LNCS 15669, pp. 318–329, 2025.
https://doi.org/10.1007/978-3-031-91398-3_24

Fig. 1. Left: An aircraft leaving a parking area, while trespassing the perpendicular bus line. Right: Sensor field set-up with one bus line along the yellow center guiding line (68 sensors) and one bus line perpendicular to it (53 sensors).

inverse modeling, *forward modeling* uses well-defined parameters to generate synthetic data from simulations in, usually, controlled environments. In this work, we present CavePerception, a framework for the analysis of sparse data from sensor networks that incorporates both inverse modeling and forward modeling. We assume a two-dimensional sensor network, laid out in a graph structure, where the sensors deliver multi-variate data about the surrounding area. The data is influenced by objects of certain categories, which are able to move in different directions at certain speeds and angles. The framework is able to handle any stationary-installed sensors, as long as they deliver time-series data. In this work, we consider *magnetometers*, which can be used to measure the distortion of the Earth's magnetic field by metallic objects, components, and external magnetic fields. While the measurements of the sensors themselves are not giving information about the category of the object or its motion vector, knowledge about the overall setting can be used to derive valid options for object categories and motion vector assignments. This knowledge may be used to develop a synthetic data generator, which mimics the environment and sensor set-up, to simulate a motion for each of the valid options. In CavePerception, we integrate ML-based inverse modeling and forward modeling to develop multiple hypotheses about the events detected by the sensors. The inverse modeling part predicts a category and a motion vector for the observation, while the forward modeling part aims at improving the predictions and filling in missing forecasts, in case the inverse modeling was not able to predict a parameter. Our framework is tested on real-world data, where 121 magnetometers are used at Frankfurt airport in Germany, to detect and identify aircraft motions. The sensor field is shown in Fig. 1. The experiments demonstrate the value of integrating inverse and forward modeling, in comparison to conventional ML approaches. In future work, CavePerception shall be integrated into novel magnetometer-based traffic surveillance systems on airport aprons, which aim to replace outdated systems, as these often fail to prevent hazardous situations[1].

[1] https://www.aerotime.aero/articles/air-europa-b737-collides-with-condor-b757-at-palma-de-mallorca-airport.

In summary, the main contributions of this paper are: (i) CavePerception, a framework for object classification and motion estimation from sensory networks, (ii) a combination of inverse and forward modeling, for an improved recognition of events, (iii) a solution to handle sparse temporal sensor data, and (iv) the validation of the framework in a real-world scenario.

2 Related Work

Inverse modeling is currently the predominant paradigm in ML and Data Mining (DM). Work on the integration of forward and inverse modeling, however, is currently found mainly outside ML and DM. A paper from applied physics [8] presents a deep learning approach to the forward and inverse designs of plasmonic metasurface structural color. Another applied physics paper describes a deep learning framework for forward and inverse problems in time-domain electromagnetics [4]. Closer in spirit are robotics approaches for tracking and prediction based on deep forward and inverse perceptual models [7]. Deriving data from simulations of theoretical models and then training classification, regression, and ranking models is quite common in ML applied to particle physics [6]. One has to note that the integration of forward and inverse modeling is similar to abduction [5], a currently underrepresented topic in DM [13].

In case of traffic surveillance on airport grounds, there is no actual standard for sensor systems for the tasks of object detection, speed estimation, tracking, and classification. Consequently, different technological systems are explored in the literature, where most of them focus on trajectory optimization during flight [9], mid-air classification of aircraft using radar [14] or aircraft classification based on top-view images [1,3]. However, in case of surface traffic surveillance on airports, aircraft need to be classified on the ground. Hence, visual-based technologies may struggle with obstacles and visually bad weather conditions like heavy rain, snow or fog [15]. On the other hand, magnetometers are not affected as they measure the Earth's magnetic field. They may also be installed directly below the surface of the holding point positions that need to be monitored, functioning as on-site sensors. Consequently, magnetometers have a high potential to be a part of traffic surveillance systems on airport grounds.

3 Simulation-Assisted Forward Modeling for Sparse Temporal Sensor Data

Our solution to the challenge of object detection, motion estimation, and classification of a moving object based on multivariate time-series data, is mirrored in our framework, named CavePerception. Inspired by Plato's allegory of the cave, where humans only see shadows of objects, which are inaccurate representations of ideal objects, we interpret our problem as follows: The sensors measurements are reflections of object motions, yet, they are not able to accurately represent the objects themselves. Consequently, the goal of CavePerception is to perceive

Algorithm 1: CavePerception

Input: Initial hypotheses set $\mathcal{H}_0$
Output: Predicted hypotheses $\tilde{\mathcal{H}}$
Data: Detected real-world event E
// Inverse Modeling
$\overrightarrow{v} \longleftarrow$ estimate motion vector
$\phi(E) \longleftarrow$ classify object category
$\mathcal{H}_{real} \subseteq \mathcal{H}_0$, inversely modeled hypotheses
// Forward Modeling
$\mathcal{H}_{sim} \longleftarrow$ simulation hypotheses generated by a synthetic data generator
// Inverse and Forward Modeling Matching
$\tilde{\mathcal{H}} \in \mathcal{H}_{real} \cap \mathcal{H}_{sim}$, retrieve the most likely hypotheses (see Algorithm 2)

real-world observations as imperfect representations of forwardly modeled synthetic patterns that are generated from known objects. Despite the imprecision in the data, the matching between observed and well-defined patterns shall reveal the true object.

A top-level view of the processing pipeline of CavePerception is shown in Algorithm 1 and explained in detail below. The inverse modeling aims to classify an object and estimate its motion vector based on the information retrieved from the sensory data and geometric domain knowledge. This part of the framework is based on previous work [11,12], briefly summarized below. The forward modeling and hypotheses matching parts are the novel aspects of CavePerception. Overall, the aim is to improve the results from inverse modeling, by assisting the data analysis with well-defined, knowledge-infused simulations.

Inverse Modeling. In the context of signal processing, it is often required to first employ a signal detection algorithm, before applying subsequent algorithms. Let $F = \{s_1, ..., s_k\}$ denote a sensor field, where each sensor $s_i \in F$ delivers a multivariate vector reading. In case of 3-axis magnetometers, it is a three-dimensional vector of the Earth's magnetic field at each time point t: $s_i^t = \{X_i^t, Y_i^t, Z_i^t\}$. We are only interested in sensor signals, when an object ρ is present. Hence, an *event* detection algorithm based on a Z-score approach[2] is applied, to calculate a moving threshold $\tau(s_i^t)$ for each sensor $s_i \in F$ at time t. A sensor s is *activated* if its measurements exceed the threshold: $|s^t| \geq \tau(s^t)$. Furthermore, we define an *event* $E(t_0, t_k)$ as a sequence of activated sensors over time, from t_0 to t_k, as:

$$E(t_0, t_k) := \{s^{t_m} | \forall s \in F, m \in [0, k] : |s^{t_m}| \geq \tau(s^{t_m})\}, \tag{1}$$

where at least one sensor is activated at each time point t_m. For simplification, event $E(t_0, t_k)$ is abbreviated as E. In real-world scenarios, an event E captures a temporal frame of a moving object ρ, where the object activates one or multiple sensors at each time point in E. Ideally, the complete object motion is observed in

[2] https://stackoverflow.com/questions/22583391/peak-signal-detection-in-realtime-timeseries-data/22640362.

one event E. However, depending on the sensor field set-up, the sensor type, and the object itself, a crossing may be not continuously observable by the sensors, leading to a separation into multiple coherent events, e.g., E_1, E_2, and E_3.

The inverse modeling approach of CavePerception processes the sensors measurements by applying an object detection algorithm. As soon as an object is detected within an event E, CavePerception aims to reduce an initially user-defined set of hypotheses $\mathcal{H}_0 = \{h_1, h_2, ..., h_n\}$. $\mathcal{H}_0$ contains all hypotheses $h_i = (\rho_i, \overrightarrow{v}_i)$ with object ρ_i and motion vector assignments $\overrightarrow{v}_i$ that may be observed in the set up environment, i.e. $\mathcal{H}_0$ is defined based on background knowledge. The activated sensors within E are clustered using ST-DBScan [2] to retrieve a first geometrical pattern of the object. A cluster fusion approach [12] combines clusters to obtain more accurate object representations. Furthermore, a data-driven event fusion algorithm [12] is incorporated, to fuse consecutive events E_i if they were triggered by the same object. Based on the clustering results, the motion vector $\overrightarrow{v}$ of the object ρ is estimated. Finally, a Transformer-based [10] classification of the object is performed to categorize the object [11]. At the end of the inverse modeling procedure, CavePerception returns a set of hypotheses $\mathcal{H}_{real} \subseteq \mathcal{H}_0$, where $h_i = (\phi(E), \overrightarrow{v}_i) \in \mathcal{H}_{real}$ contains the Transformer's classification $\phi(E)$ and the estimated motion vectors $\overrightarrow{v}_i$.

Forward Modeling. The idea behind forward modeling is to have a synthetic data generator, which generates data similar to the real-world data. Therefore, we make use of geometrical knowledge about the sensor field installation and the geometric pattern of polygonal-shaped objects. The user can then define which objects shall be simulated, how the sensors are positioned in a two-dimensional grid, and specify the motion behaviour of the objects to be simulated. Additional parameters like the starting position, the sensor detection range, and the data noise level may be defined. Based on the application field, background knowledge is used to perform arbitrarily many simulations of varying complexities to retrieve all simulation hypotheses $\mathcal{H}_{sim}$ that would be expected as valid observations for the considered sensor field. At the same time, a synthetic data set $\mathcal{D}_{S_i}$ is obtained for each simulation S_i, or equivalently, each hypothesis $h_i \in \mathcal{H}_{sim}$.

Inverse and Forward Modeling Matching. In real-world domains, a data-driven inverse modeling approach may not always deliver the full set of required or accurate results. The idea of the forward modeling approach is to fill the gap of missing information and to improve the derived results by matching realistic simulations with real-world observations.

After CavePerception's inverse modeling pass, each hypothesis $h \in \mathcal{H}_{real}$ contains two values: $h = (\phi(E), \overrightarrow{v})$, where $\overrightarrow{v} = \{d, \nu, \alpha\}$ is a motion vector with motion direction d, motion velocity ν, and motion angle α. If the inverse pass is able to predict any of these parameters, then all hypotheses $h \in \mathcal{H}_{real}$ have the same value for the respective parameter. For example, if $d = right$, then $\forall h = (\phi(E), \overrightarrow{v}) \in \mathcal{H}_{real} : \overrightarrow{v} = \{right, \nu, \alpha\}$. After the inverse pass, there are four different scenarios: (i) the object was classified and a motion was detected (fully or partially), (ii) the object was classified, but no motion was detected,

Algorithm 2: Inverse and Forward Model Matching

Output: Final hypotheses $\tilde{\mathcal{H}}$
Data: Hypotheses derived from real-world measurements $\mathcal{H}_{real}$, Synthetic hypotheses $\mathcal{H}_{sim}$, Real-world observed event E

// Hypotheses matching
$\mathcal{H}_{red} \longleftarrow \mathcal{H}_{sim} \cap \mathcal{H}_{real}$
// Time-invariant hypotheses reduction
$B(\mathcal{D}_{red}) \longleftarrow \{B(\mathcal{D}_{S_i}) | \forall \mathcal{D}_{S_i} \in \mathcal{D}_{red}\}$, binary activation matrices of matched synthetic datasets
$B(E) \longleftarrow$ binary activation matrix of event E
$Y_{red} \longleftarrow$ initialize empty set for scalar distances
for *each* $S_i \in \mathcal{H}_{red}$ **do**
  $M_{E,S_i} \longleftarrow pdist(B(E), B(\mathcal{D}_{S_i}))$ // see Eq. 5
  $Y_{red} \longleftarrow Y_{red} \cup \{sdist(M_{E,S_i})\}$ // see Eq. 6
end
$\tilde{\mathcal{H}} \longleftarrow \{S_i \in \mathcal{H}_{red} | sdist(M_{E,S_i}) = \min(Y_{red})\}$ // see Eq. 7

(iii) a motion was detected, but the object could not be classified, and (iv) there is not enough information to detect a motion or to perform a classification.

Our forward modeling approach aims to fill the gaps of missing information in $\mathcal{H}_{real}$, including the identification of the specific object type ρ, by employing a time-invariant hypotheses matching algorithm within CavePerception. The pseudocode is shown in Algorithm 2 and explained below.

Hypotheses Matching. The first step of the algorithm is the matching of the inversely derived hypotheses $\mathcal{H}_{real}$ of event E with the simulated hypotheses $\mathcal{H}_{sim}$. In contrast to $\mathcal{H}_{real}$, the parameters of all hypotheses $h = (\hat{\rho}, \overrightarrow{v}) \in \mathcal{H}_{sim}$ are known. Therefore, only those simulations are considered where the simulation parameters correspond to the estimated parameters $\phi(E)$ and $\overrightarrow{v}$:

$$\mathcal{H}_{red} \longleftarrow \mathcal{H}_{sim} \cap \mathcal{H}_{real}. \tag{2}$$

The hypotheses reduction to $\mathcal{H}_{red}$ results in hypotheses $h_i \in \mathcal{H}_{red}$, where each parameter is known, as the missing information from the inverse modeling in $\mathcal{H}_{real}$ is filled with the available information in $\mathcal{H}_{sim}$. For example, if the category $\phi(E)$ of object ρ is successfully predicted during inverse modeling, then all simulations of object types $\{\rho_1, ..., \rho_n\}$ of category $\phi(E)$ are extracted from $\mathcal{H}_{sim}$. This way, not only the categories of the objects are known, but also the specific object types $\hat{\rho}_i$, which were unknown in $\mathcal{H}_{real}$. Here, we use the notation $\hat{\rho}_i$ for the predictions of the true type ρ based on the predicted category $\phi(E)$. With the hypotheses matching, the set of simulation datasets $\mathcal{D}_S$ is reduced to:

$$\mathcal{D}_{red} = \{\mathcal{D}_{S_i} \in \mathcal{D}_S | h_{S_i} = (\hat{\rho}, \overrightarrow{v}) \in \mathcal{H}_{red}\}, \tag{3}$$

where $\mathcal{D}_{red}$ only contains those simulated datasets where the simulations were performed according to the estimated parameter combinations in $\mathcal{H}_{real}$.

	0	0.13	0.26	0.39	0.52	0.65	0.78
0	0	1	0	1	2	0	1
0.13	1	2	3	0	1	2	3
0.26	0	3	4	0	1	2	3
0.39	1	0	0	4	5	0	1
0.52	2	1	1	5	2	3	4

$M =$ (matrix above)

Fig. 2. An example of a distance matrix M. The rows and columns specify temporal distances to the event start, whereas the entries represent the pairwise distance values.

Time-Invariant Hypotheses Reduction. After the hypotheses matching, it remains to retrieve the most likely hypotheses from $\mathcal{H}_{red}$ to describe the observed event E. Therefore, binary activation matrices of the simulated datasets $\mathcal{D}_{red}$ are compared with the binary activation matrix of event E. This way, we aim to find an optimal geometry-based match. Hence, each synthetic dataset $\mathcal{D}_{S_i} \in \mathcal{D}_{red}$ is converted from a continuous data representation to a binary activation matrix $B(\mathcal{D}_{S_i})$, where each column represents a sensor and the rows represent the time:

$$B_{i,j} = \begin{cases} 1 & \text{, if } s_j^{t_i} \geq \tau(s_j^{t_i}), \\ 0 & \text{, otherwise.} \end{cases} \tag{4}$$

Subsequently, each binary activation matrix $B(\mathcal{D}_{S_i})$ needs to be matched with $B(E)$. The matches are then used to derive dissimilarity scores that describe to which extent a simulation coincides with the observation. The most similar simulation hypotheses are then returned as the most likely explanations for the observed event. The matching of the binary activation matrices can be performed with any pair-wise distance measure *pdist*. The matrices themselves, $B(E)$ and $B(\mathcal{D}_{S_i})$, need to have the same number of columns n_c (the number of sensors), but can vary in the number of rows n_k and n_l, i.e. $|B(E)| = (n_k, n_c)$ and $|B(\mathcal{D}_{S_i})| = (n_l, n_c)$, as the lengths of different events may vary. The pair-wise distance between the two matrices results in a distance matrix:

$$\begin{aligned} M_{E,S_i} &\longleftarrow pdist(B(E), B(\mathcal{D}_{S_i})) \\ &= (dist(v_i, v_j))_{i,j}, \forall i \in \{1, ..., n_k\}, j \in \{1, ..., n_l\}, \end{aligned} \tag{5}$$

where v_i and v_j are the i-th and j-th row vectors of matrices $B(E)$ and $B(\mathcal{D}_{S_i})$, respectively. In this work, we apply the Hamming distance, which counts at how many positions the entries in two vectors differ. Applied to the sensor activation matrices, it counts how many sensors differ in their activation in event E and simulation S_i. Figure 2 shows an example of such a distance matrix M, where the rows and columns represent the temporal distances of both matrices $B(E)$ and $B(\mathcal{D}_{S_i})$. However, we are interested in a scalar score to describe the dissimilarity between both matrices. Therefore, the diagonals within matrix M are considered (framed in red). Each diagonal is a full-match of observation E within the simulation S_i. We restrict the diagonals to mirror the full path of the smaller matrix. Hence, each diagonal needs to have $min(n_k, n_l)$ values. If a full match is not required, i.e. both crossings can also match partially, then the length of the diagonals could be set to be less restrictive. We can assure n_k to be smaller than n_l by transposing the resulting matrix, if $n_k > n_l$. To obtain a scalar distance $sdist(M_{E,S_i})$, we search for the minimal dissimilarity in the diagonal sums of all diagonals $diag(M) = \{diag_1(M), ..., diag_k(M)\}$:

$$sdist(M_{E,S_i}) = min_{i \in \{1,...,|diag(M)|\}} \sum_{v \in diag_i(M)} v. \tag{6}$$

Equation 6 returns the optimal time-invariant match of event E with simulation S_i. Applying Eq. 6 to every simulation $S_i \in \mathcal{H}_{red}$, a set of optimal time-invariant matches is obtained and collected in Y_{red}. Finally, Algorithm 2 concludes the search for the optimal match in the simulations by choosing the simulations with the lowest scalar dissimilarity:

$$\tilde{\mathcal{H}} \longleftarrow \{S_i \in \mathcal{H}_{red} | sdist(M_{E,S_i}) = \min(Y_{red})\}. \tag{7}$$

Notice that synthetic data generation for sensors may be a challenging task. However, data normalization techniques may simplify the task, as the data generator then only has to mimic the normalized data. To further reduce the dependence on an exact data representation, our solution matches the hypotheses parameters h during the hypothesis matching, and afterwards, only the binary activation matrices B are compared to each other. Hence, it does not harm the solution, if the synthetic data does not perfectly match the real-world sensor data.

4 Experimental Results

The real-world data are multivariate time-series data covering a period of two years. The objects to be identified are aircraft, of which only the gears and engines are detectable by magnetometers. The real-world data has a high level of noise and latent factors that may reduce the data quality. Further, the data is sparse as many aircraft types occur rarely compared to a few types, which are often observed. On the other hand, the synthetic data (generated by forward modeling) has a larger aircraft type variety due to simulations of more aircraft types. The simulations were performed according to the background knowledge available. That is, for each aircraft type that passed the considered sensor field during the two-year observation period and whose aircraft dimension is known, motions were simulated with different motion vector assignments: The sensor field was passed in two directions at varying speeds and with small variations in the motion angles, such that aircraft do not always exactly follow the guiding line. All experiments were performed on a computer with OS Ubuntu 22 with 32 GB memory, an Intel Core i7-4770K CPU and a GeForce GTX 1080 TI GPU. The GPU was used for the ML algorithms in the model comparison section, which are also a part of the inverse modeling of CavePerception.

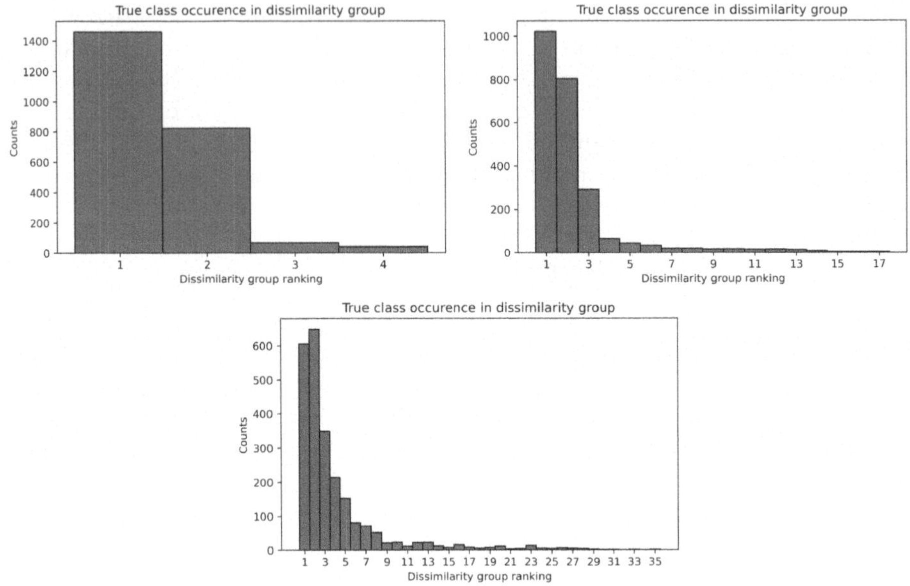

Fig. 3. Group rank assignment of the true aircraft class. Group rank 1 corresponds to the predicted group of aircraft types or categories with the smallest dissimilarity. Top-Left: ARC. Top-Right: Cat. 17. Bottom: Aircraft type.

For the performance evaluation of CavePerception we consider three types of classification, i.e. the aircraft type, the *Aerodrome Reference Code*[3] (ARC), and cat. 17. The ARC is a universal categorization for aircraft types, introduced by the *International Civil Aviation Organization* (ICAO)[4]. The ARC contains six categories, from A (smallest wing span) to F (largest wing span). The aircraft type classification considers each model of aircraft separately. For example, an Airbus A320 is considered as a different type as the Airbus A320 neo, although the geometric differences are minor. Consequently, the type classification poses the most challenging classification task. Therefore, we additionally cluster the aircraft types into 17 categories (cat. 17), where aircraft types with minor geometrical differences are grouped into one class.

Group Rank Evaluation. The predicted classes of CavePerception are contained in $\tilde{\mathcal{H}}$. Based on the dissimilarities, one can obtain a ranking of dissimilarity groups $G_i = \{S_j | sdist(M_{E,S_j}) = i\}$, where each group G_i contains those simulations S_j where the dissimilarities are equal to i. This approach allows us to measure the degree of deviation from the true result based on the dissimilarity values. Figure 3 shows the histograms of the dissimilarity group ranks that contain the true class. That means, group one is the group with the highest match

[3] www.skybrary.aero/articles/icao-aerodrome-reference-code.
[4] www.icao.int/Pages/default.aspx.

Table 1. Real-world data results: Weighted macro averaged precision, recall, and F1-score for aircraft type, ARC, and cat. 17 classifications. RF and XGBoost are trained with 75% of training data.

		Prec.	Rec.	F1
ARC	Random	0.49	0.70	0.58
	RF	0.52	0.66	0.59
	XGBoost	0.80	0.67	0.63
	Transformer	0.84	0.66	0.71
	CavePerception	**0.89**	**0.73**	**0.79**
Cat. 17	Random	0.35	**0.59**	0.44
	RF	0.45	0.54	0.46
	XGBoost	0.73	0.55	0.52
	Transformer	0.61	0.56	0.55
	CavePerception	**0.86**	0.54	**0.64**
Type	Random	0.05	0.21	0.08
	RF	0.15	0.00	0.00
	XGBoost	0.70	0.08	0.12
	Transformer	0.40	0.32	0.33
	CavePerception	**0.98**	**0.43**	**0.59**

with event E, while the largest group rank contains the unlikeliest matches. In these figures, we count how often the true class has been assigned to the respective group ranks. Optimally, it should always be the first group. As can be seen, predicting the ARC is the easiest task, as there are only four possible classes: C, D, E, and $Unknown$. The ARC values A, B, and F have not been observed at this position during the time of observation, while $Unknown$ includes all those events, where the true labels of an event are unknown, i.e. vehicles, false detections, and others. Furthermore, in all three classification approaches, the true class remains in the group ranks with the smallest dissimilarities. More than 65% of the true classes remain in the first three group ranks. Consequently, even if less than 50% of the correct targets were assigned to group 1 for the type and cat. 17 classifications, considering the difficulty of the classification task, the majority of the true classes are close to the highest matches with the observations.

Model Comparison. We also compare CavePerception to commonly used ML algorithms to validate the benefit of combining forward with inverse modeling procedures. Therefore, we consider Transformers, which are frequently considered one of the best performing learning algorithms, XGBoost, random forest, and a random guessing baseline, which predicts the majority class. The disadvantage of many standard implementations of tree-based learners is the requirement of having all data preloaded, such that the models can be trained on the whole training data at once. Due to the size of our training data and the memory size required to train these models, random forests and XGBoost were trained on 75% of the training data. Incremental learning or similar techniques may be used to train the models on the whole dataset by iteratively updating from subsamples, however, this may introduce other challenges like representation bias from successive subsamples or the loss of correlation information. Hyperparameter optimization has been performed for all models to ensure a fair comparison. Table 1 shows that CavePerception is able to outperform all other models for each classification task, except of the recall for cat. 17, where random guessing is leading due to overpredicting the majority class. Notice that despite the difficulty of the aircraft type classification, CavePerception manages to achieve an impressive F1-score compared to the inverse models.

5 Conclusion

In real-world applications, temporal data from sensor networks are frequently sparse and noisy. In this paper, we have shown that the combination of forward and inverse modeling can improve the overall results in such settings. Experiments have shown that our framework, CavePerception, does not only outperform other methods on complex data, but also that its group ranking approach based on hypothesis matching is able to capture most of the correct classes within the first two to three ranked groups. The model comparison also shows that classification results may be improved by including forwardly modeled hypotheses, compared to models that are solely inversely trained. Another advantage of our solution is that missing information from inverse modeling is filled with assignments from synthetic simulations based on the event matching with the real-world observation.

CavePerception should work, as far as possible, without homogeneity assumptions in real-world sensor network deployments. It is able to handle any two-dimensional sensor layout structure and also works independent of the sensor sampling frequency, as the temporal differences between consecutive points are considered instead of fixed frequency-based time points. However, it requires at least two sensor lines perpendicular to each other, of which one is necessary to estimate the motion vector, while the other one is important for classification. Another limitation of CavePerception is the necessity of object type geometries. If the geometry of an object type is unknown, a simulation for this type cannot be performed, resulting in a non-detection of the correct type, in case it occurs in an observation. Further, the performance of CavePerception depends on the inverse modeling results. If the inverse process yields a false classification, then it will affect the final outcome of the hypotheses matching. Therefore, it is crucial to optimize the inverse modeling procedure. In future work, this dependence on inverse modeling may be loosened by providing probabilistic results, such that the matching between inverse and forward modeling may take the probabilities into account. On our roadmap, CavePerception shall be further developed and integrated into a prototype for real-time traffic surveillance on airport aprons.

Acknowledgments. The authors thank the Federal Ministry for Economic Affairs and Climate Action and the German Aerospace Center for supporting the project.

Disclosure of Interests. The authors have no competing interests to declare that are relevant to the content of this article.

References

1. Azam, F., Rizvi, A., Khan, W.Z., Aalsalem, M.Y., Yu, H., Zikria, Y.B.: Aircraft classification based on PCA and feature fusion techniques in convolutional neural network. IEEE Access **9**, 161683–161694 (2021)
2. Birant, D., Kut, A.: St-dbscan: an algorithm for clustering spatial-temporal data. Data Knowl. Eng. **60**(1), 208–221 (2007)
3. Gao, K., He, H., Lu, D., Xu, L., Ma, L., Li, J.: Optimizing and evaluating swin transformer for aircraft classification: Analysis and generalizability of the mtarsi dataset. IEEE Access **10**, 134427–134439 (2022)
4. Hu, Y., Jin, Y., Wu, X., Chen, J.: Solving time domain electromagnetic forward and inverse problems using a differentiable programming platform. In: 2021 International Conference on Electromagnetics in Advanced Applications (ICEAA), pp. 216–221. IEEE (2021)
5. Kakas, A.C.: Abduction, pp. 1–8. Springer US, Boston, MA (2017)
6. Köppel, M., et al.: Learning to rank higgs boson candidates. Sci. Rep. **12**(1), 13094 (2022)
7. Lambert, A., Shaban, A., Raj, A., Liu, Z., Boots, B.: Deep forward and inverse perceptual models for tracking and prediction. In: 2018 IEEE International Conference on Robotics and Automation (ICRA), pp. 675–682. IEEE (2018)
8. Roberts, N.B., Keshavarz Hedayati, M.: A deep learning approach to the forward prediction and inverse design of plasmonic metasurface structural color. Appl. Phys. Lett. **119**(6) (2021)
9. Tian, Y., He, X., Xu, Y., Wan, L., Ye, B.: 4D trajectory optimization of commercial flight for green civil aviation. IEEE Access **8**, 62815–62829 (2020)
10. Vaswani, A., et al.: Attention is all you need. Adv. Neural Inf. Process. Syst. **30** (2017)
11. Vexler, J., Kramer, S.: Classifying aircraft categories from magnetometry data using a hypotheses-based multi-task framework. In: European Conference on Artificial Intelligence (2023)
12. Vexler, J., Kramer, S.: Identifying aircraft motions and patterns from magnetometry data using a knowledge-based multi-fusion approach. In: 2023 26th International Conference on Information Fusion (FUSION), pp. 1–8. IEEE (2023)
13. Wicker, J., et al.: Cinema data mining: the smell of fear. In: Proceedings of the 21th ACM SIGKDD International Conference on Knowledge Discovery and Data Mining, pp. 1295–1304 (2015)
14. Xia, S., et al.: Aircraft target classification method for conventional narrowband radar based on micro-doppler effect. Math. Probl. Eng. **2022**, 1–11 (2022)
15. Zhang, Y., Carballo, A., Yang, H., Takeda, K.: Perception and sensing for autonomous vehicles under adverse weather conditions: a survey. ISPRS J. Photogramm. Remote Sens. **196**, 146–177 (2023)

Bridging Spatial and Temporal Contexts: Sparse Transfer Learning

Daniel Persson(✉), William Wahlberg, Anna Vettoruzzo, and Sławomir Nowaczyk

Center for Applied Intelligent Systems Research, Halmstad University, Halmstad, Sweden
Daniel.p-00@hotmail.com

Abstract. This paper introduces a novel transfer learning adapter, the Bridged Attention Module (BAM), designed to enhance the performance of Spatial-Temporal Graph Convolutional Networks (ST-GCN) in data-limited forecasting scenarios. BAM improves fine-tuning efficiency by jointly capturing spatial and temporal dependencies, optimizing information flow, and significantly reducing the number of trainable parameters while preserving model accuracy. Experimental evaluations demonstrate that the BAM-enhanced ST-GCN consistently achieves competitive accuracy and, in some cases, surpasses traditional fine-tuning methods, even with limited data. The effectiveness of this approach is validated using electric vehicle (EV) charging station occupancy forecasting, highlighting the practical utility of BAM.

Keywords: Transfer learning · Time series · Electric vehicles · Deep learning · Parameter-efficient learning

1 Introduction

Accurate spatial-temporal forecasting is essential across various fields. However, achieving high accuracy in these models can be challenging, particularly in scenarios with limited access to comprehensive datasets. This paper addresses the need for parameter-efficient forecasting methods that can deliver accurate predictions while relying on only a small amount of data. To demonstrate the proposed approach, electric vehicle (EV) charging station occupancy forecasting is used as a test case, reflecting the practical importance of scalable solutions in data-sparse environments. In the EV industry, where infrastructure is still developing and historical data may be scarce, reliable occupancy forecasting is critical for optimizing charging station networks. In 2022, global EV sales surpassed 10 million, increasing their share of total vehicle sales from 4% in 2020 to 12% [18]. This shift, driven by policy initiatives like the European Green Deal [9], underscores the need for efficient, data-sparse forecasting methods that can adapt to growing demands.

Accurate forecasting in data-limited scenarios has broad relevance as it enables strategic decision-making and efficient resource management across various fields. In cases where collecting extensive data is impractical or costly,

D. Persson and W. Wahlberg—Equal contribution.

G. Krempl et al. (Eds.): IDA 2025, LNCS 15669, pp. 330–342, 2025.
https://doi.org/10.1007/978-3-031-91398-3_25

parameter-efficient models can provide reliable predictions while minimizing data requirements [11,15,45]. This capability supports more effective planning in areas such as transportation, energy consumption, urban development, and healthcare, making forecasting solutions adaptable and scalable in rapidly evolving domains.

Accurately forecasting in spatial-temporal contexts presents several unique challenges, particularly when aiming for parameter efficiency with limited data. Traditional models often struggle to maintain accuracy when data is sparse or non-continual, as they are not designed to handle incomplete datasets effectively [8]. Furthermore, spatial dependencies–where conditions or usage patterns at one location can impact nearby locations–add another layer of complexity, requiring models that can capture these interdependencies dynamically [40]. Temporal variations, influenced by patterns or unexpected external factors, further complicate reliable forecasting [36]. Additionally, scalability is crucial: in applications with numerous spatial points, retraining a separate model for each location is computationally prohibitive. Addressing these spatial and temporal complexities while maintaining parameter efficiency across large networks necessitates advanced solutions capable of achieving high accuracy without extensive retraining.

This paper presents a novel approach by introducing the Bridged Attention Module (BAM) to build upon a transformer-based Spatio-Temporal Graph Convolutional Network (ST-GCN) for efficient transfer learning. The ST-GCN's architecture addresses both spatial and temporal aspects of charging station occupancy, with a GCN modeling spatial relationships between stations [22] and a transformer network capturing temporal information [39]. The proposed BAM adapter improves the adaptability and accuracy of this architecture in data-sparse scenarios, making it particularly effective for time series forecasting of EV charging station occupancy. A key innovation of this approach is the integration of the BAM adapter, which facilitates efficient transfer learning, allowing the model to be fine-tuned for different charging networks without requiring complete retraining. By significantly reducing the number of trainable parameters, BAM improves the model's ability to generalize across stations with limited data, enhancing scalability and computational efficiency. The combination of the ST-GCN and BAM modules provides a highly effective solution for forecasting charging station occupancy across charging station networks.

In summary, the main contributions of this paper are as follows:

- We introduce the BAM adapter, which enhances the performance of ST-GCN for transfer learning in time series forecasting under data-scarce conditions, while also significantly reducing the number of trainable parameters.
- We demonstrate the effectiveness of BAM in transfer learning scenarios by enabling the model to transfer knowledge from one EV charging station network to another. This is particularly beneficial for newly established charging station networks with limited historical data, making BAM an ideal solution for large-scale applications characterized by data-limited conditions.

2 Related Work

The rapid growth in EV adoption and demand for efficient charging infrastructure have brought significant attention to forecasting EV charging station occupancy. Various approaches address challenges in predicting usage patterns and optimizing management.

Handling missing data is essential for time series forecasting with intermittent datasets. Simple univariate methods like Last Observation Carried Forward (LOCF) and Next Observation Carried Backwards (NOCB) are effective for sparse series [13], while multivariate approaches, such as k-Nearest Neighbors (k-NN), improve imputation accuracy by leveraging relationships between variables, resulting in smoother imputed data [25,41].

Previous literature often used foundational ML models for time series forecasting, such as Moving Average (MA), Autoregressive (AR), and Autoregressive Integrated Moving Average (ARIMA) [19,33,35]. While extensively explored, these models primarily handle univariate data and linear relationships, limiting their effectiveness with complex patterns [6,7,27,37,43].

Unlike traditional ML methods, deep learning approaches such as Long Short-Term Memory (LSTM) networks and transformers excel with complex, high-dimensional, nonlinear data [14,23,39]. LSTM networks effectively capture long-term dependencies in time series data, while transformer models leverage attention mechanisms to process sequential data in parallel [28,31,34]. Techniques like fractional positional encoding further enhance transformers' ability to handle intermittent series with irregular intervals, making them popular for time series forecasting [1,2,12]. While LSTMs and transformers capture temporal patterns, they struggle with spatial dependencies; GCNs address this by modeling spatial relationships in graph-structured data [22,26]. ST-GCNs extend GCNs by incorporating temporal dynamics, enabling simultaneous modeling of spatial and temporal dependencies for tasks like traffic prediction [20,21,29]. These architectures have also been employed to predict EV charging station occupancy, utilizing the spatial distribution of stations alongside their temporal usage patterns [30,38].

Another used technique in time series forecasting is transfer learning. Transfer learning enables models to generalize effectively across diverse scenarios (e.g., urban networks) even when data is limited [3,42]. By enhancing both scalability and adaptability, transfer learning makes it feasible to handle large-scale scenarios. Some widely used transfer learning techniques include fine-tuning the last layer [44] or fine-tuning all layers [16]. Recently, the focus of transfer learning was more directed towards the integrations of small adapters modules into pre-trained models for task-specific fine-tuning while keeping the original weights frozen. This approach reduces the number of trainable parameters and improves the efficiency of the model [11,15,45]. Adapter-based techniques have been successfully applied to architectures like CNNs and RNNs [5,10,24] and to transformers in large language models (LLMs) [17]. These adapters, such as the MLP-based modules introduced by [15], improve computational efficiency by reducing the number of trainable parameters. A further improvement is obtained with the Tiny-Attention Adapter, which enhances model performance by utilizing a compact multi-head attention mechanism to focus on key features [45]. However, it lacks the ability to capture spatial dependencies, being originally designed for large language models rather than spatio-temporal tasks. Although the adapter itself uses only 0.05% of the model's parameters, the decoder is fine-tuned alongside the adapter, severely increasing the total number of retrained parameters. Another famous parameter-efficient technique includes Low-Rank Adaptation (LoRA) which achieves parameter efficiency by representing weight updates as low-rank matrices [17]. While typically used in large language models, LoRA can also be applied for time-series forecasting [11] allowing efficient adaptation of pre-trained models by adding a minimal number of task-specific parameters, maintaining model accuracy while significantly reducing computational and memory overhead. This paper introduces a new parameter-efficient method named BAM, that is integrated between the GCN and the transformer components within an ST-GCN model, effectively capturing both spatial and temporal dependencies while minimizing the number of trainable parameters.

3 Method

3.1 Background

To address the spatial and temporal dependencies in EV charging station occupancy data, this paper employs an ST-GCN model [20,21,29,30,38]. The ST-GCN combines a GCN to capture spatial relationships among charging stations with a flexible temporal modeling component. The GCN output is represented in our paper as

$$H_S = \mathrm{GCN}(X, G)$$

where $G = (V, E)$ represents the graph structure of the stations, and $X = \{x_t\}_{t=1}^{T}$ denotes the sequence of occupancy data over time, with each $x_t \in \mathbb{R}^{|V'| \times d}$ capturing the occupancy state of all stations at time t. In particular, the nodes V and edges E in G represent the set of charging stations and the distance between these stations, respectively. Each edge $e_{ij} \in E$ between nodes v_i and v_j is weighted according to the distance between the stations by road, enabling the ST-GCN to incorporate spatial relationships. For the temporal component, instead, various models can be integrated into the ST-GCN architecture. This study specifically employs a transformer model which applies self-attention to the spatial encoding H_S [39], resulting in a spatio-temporal representation H_{ST} that allows the ST-GCN to account for both inter-station dependencies and temporal occupancy patterns.

Furthermore, this paper introduces the BAM adapter, a novel enhancement that improves the model's adaptability and performance of transfer learning with datasets containing limited labeled samples. To evaluate its performance, several transfer learning strategies and data percentages are considered during the evaluation. Each combination is represented as

$$\mathcal{T}(M, D_{\text{percentage}})$$

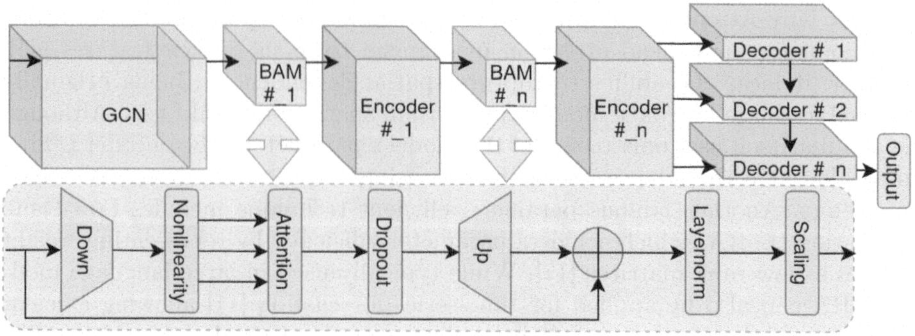

Fig. 1. Architecture of the proposed BAM adapter integrated into the ST-GCN network. Green and red represent trainable weights, while purple indicates frozen weights. (Color figure online)

where M denotes the learning strategy, and $D_{\text{percentage}}$ specifies the fraction of target data used for fine-tuning.

3.2 Bridged Attention Module (BAM)

Figure 1 illustrates the architecture and placement of the proposed BAM adapter within the ST-GCN. The BAM adapter is designed to maximize adaptability in data-sparse scenarios while minimizing trainable parameters. It integrates both spatial and temporal information to enhance the flow of data from the GCN to the final predictions. The BAM architecture begins with a downscaling bottleneck layer to reduce the input feature dimensionality, helping the model to focus on essential information, reducing parameter count and limiting overfitting risk. This is followed by a single-head attention mechanism that captures dependencies within the input sequence, offering expressiveness without the computational cost of multi-head attention [45]. The attention layer is defined as $z_t = \text{Attention}(W_q x_t, W_k x_t, W_v x_t)$, where W_q, W_k, and W_v are learnable weight matrices for queries, keys, and values, respectively [39]. The attended features are then regularized with a dropout layer to reduce overfitting, resulting in $\hat{z}_t = z_t \cdot \text{Bernoulli}(1 - p)$. The features are then upscaled and combined with a skip connection from the initial input, to maintain both the original context and the additional insights learned during adaptation. This procedure is expressed as $y_t = \text{Concat}(\hat{z}_t, x_t)$. After that, a Layer Normalization defined as $\text{LayerNorm}(y_t) = \frac{y_t - \mu}{\sigma + \epsilon} \cdot \gamma + \beta$, where μ and σ are the mean and standard deviation of y_t, and γ and β are learnable parameters, stabilizes the training. Finally, the output is scaled by a learnable parameter and is expressed as $\text{Output} = \text{Scale}(\text{LayerNorm}(y_t))$.

The BAM adapter is placed between the GCN and the transformer architecture and within the transformer's encoder-decoder layers. This positioning acts as a "bridge", harmonizing representations and ensuring smooth information flow. The BAM adapter is the sole component that is fine-tuned to the target scenario, during transfer learning, highlighting its adaptation capabilities and ensuring a low number of trainable parameters.

4 Result

4.1 Experiments

The experiments were conducted across two main scenarios, each designed to evaluate the model's ability to generalize across different networks. Three distinct EV charging-station networks, each representing small charging station areas of three different cities in Sweden—Värnamo, Varberg, and Malmö-were utilized to create the two transfer learning scenarios: Värnamo to Varberg and Varberg to Malmö. In each scenario, the model was trained on data from the source network using three distinct random seeds to ensure variability in initialization. Following this, each of these pre-trained models was fine-tuned on the target network using three new seeds, creating a total of 9 unique seed combinations per scenario, allowing a comprehensive assessment of the model's robustness and performance consistency.

The effectiveness of the BAM adapter was evaluated by comparing it against several common transfer learning strategies: No Fine-Tuning, Trained from Scratch, Fine-Tune All, Fine-Tune Last GCN Layer, and Fine-Tune Last Transformer Layer. Additionally, two adapter-based approaches were included in the comparison: the MLP Adapter, a widely used method in the time-series literature, and the Tiny-Attention Adapter, which architecture closely resembles this paper's proposed BAM adapter. Incorporating these baselines helped demonstrate that the proposed modifications in the BAM adapter provide meaningful improvements over existing approaches.

The evaluation metric used for all experiments was Mean Squared Error (MSE), and in all tables, the displayed values represent the MSE along with the standard deviation. The same experimental setup was repeated for the comparative study against the baseline models (Sect. 4.2) and for the ablation study (Sect. 4.3) to analyze the contributions of individual components in the BAM adapter.

For each city, a network of charging stations was constructed. While cities typically contain more than five stations, the selection was limited to five stations per city due to several practical considerations. First, the dataset was limited, and the available data required preprocessing to filter out stations with noisy or incomplete data. Additionally, computational constraints played a role in defining the network size, ensuring feasibility within the available resources. Most importantly, to enable a fair transfer learning setup across all investigated cities, the selected number of stations had to be consistent and present in each case. This ensured that the model could be applied uniformly across different scenarios, facilitating meaningful comparisons and reliable generalization insights.

Experiments were conducted for both scenarios (Värnamo to Varberg and Varberg to Malmö); however, only the results of the Värnamo to Varberg scenario are presented in this paper due to space constraints, as the outcomes were consistent across both cases.

Table 1. Performance of baseline models at different data usage percentages (1%, 5%, 20%, and 50%) for the Värnamo to Varberg scenario. The "Parameters" column indicates the percentage of trainable parameters in each model. All values are scaled by 10^4. Bold numbers highlight the best-performing model at each data usage percentage, while in the "Parameters" column, they indicate a low percentage (below 10%) of parameters used for fine-tuning.

Värnamo to Varberg					
Model	1%	5%	20%	50%	Parameters
No fine-tuning	126.4 ± 99.08	126.4 ± 99.08	126.4 ± 99.08	126.4 ± 99.08	**0.00%**
Trained from scratch	122.49 ± 28.16	82.59 ± 14.1	55.75 ± 1.73	52.55 ± 2.68	100.00%
Fine-tune all	**58.37 ± 5.27**	54.45 ± 2.21	**50.65 ± 1.26**	50.2 ± 1.21	100.00%
Last GCN Layer	95.89 ± 49.74	68.25 ± 16.17	57.56 ± 5.50	56.94 ± 6.16	**0.22%**
Last transformer Layer	62.26 ± 1.05	57.26 ± 4.53	53.17 ± 2.80	53.49 ± 2.74	20.99%
MLP Adapter	71.53 ± 10.70	58.03 ± 3.88	52.44 ± 2.01	51.65 ± 1.53	**2.69%**
Tiny-Attention Adapter	63.07 ± 5.39	58.24 ± 3.22	52.49 ± 1.59	51.43 ± 2.78	57.01%
BAM	63.14 ± 7.63	**53.7 ± 1.81**	51.22 ± 1.39	**50.09 ± 1.31**	**3.02%**

Data The data for this project was provided by ChargeFinder [4], and it consists of occupancy information from 29 EV charging stations across three Swedish cities: Värnamo, Varberg, and Malmö. Each dataset includes essential features: station ID, outlet count, occupied count, available count, offline count, and timestamps for each data point. As the data was logged intermittently based on user queries, an imputation method was applied to resample it into a continuous time series, making it suitable for time series forecasting.

To address this, the k-NN imputation method [32] was implemented for its capability to capture complex relationships among multiple features. By identifying patterns based on the similarity of data points in the temporal space, k-NN effectively fills in missing values.

Following the transformation from an intermittent to a continuous time series, the data was structured as a graph, denoted by $G = (V, E)$, as detailed in Section 3.1. For this study, a subset of five stations was chosen from each city, resulting in a total of 15 stations. These stations were selected based on the quality and quantity of data, which was necessary for training our ST-GCN model. Indeed, several stations had extremely few entries, or presented noisy patterns in the data, making it necessary to discard them for this paper's experiments. For each city, a network of charging stations is constructed, denoted by $N = (V', E')$, with $V' \subset V$ nodes and $E' \subset E$ edges. This subset enables an efficient demonstration of the model's ability to leverage spatial relationships while managing computational complexity.

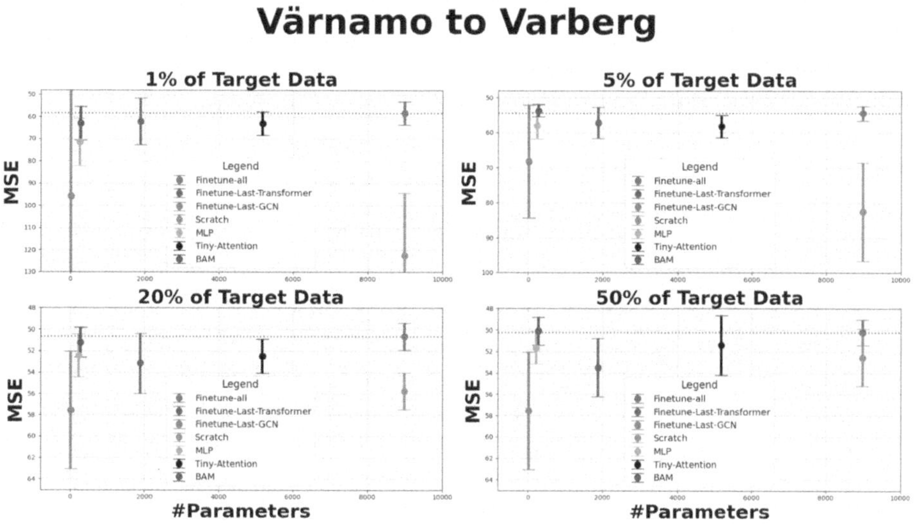

Fig. 2. MSE scores for different models at 1%, 5%, 20%, and 50% data usage, displayed across four Pareto plots. Each plot reports the number of trainable parameters on the X-axis and the MSE values on the Y-axis, with error bars representing the standard deviations. Results are shown for the Värnamo to Varberg scenario. All MSE values are scaled by 10^4.

4.2 Comparative Study

The results of the comparative study, presented in Table 1, highlight the strong performance of the BAM adapter. BAM's performance, with different percentage of data used, is comparable, or even outperform (in $\mathcal{T}(\text{BAM}, D_{5\%})$ and $\mathcal{T}(\text{BAM}, D_{50\%})$) the Fine-tune all baseline, which requires fine-tuning all model parameters. BAM, on the other hand, is efficient in terms of the number of fine-tuned parameters, as reported in the last column of the table.

It is also worth noting the high standard deviation observed in several models, particularly at 1% data usage in Table 1. Models such as $\mathcal{T}(\text{No fine-tuning}, D_{1-50\%})$, $\mathcal{T}(\text{Trained from scratch}, D_{1\%})$, and $\mathcal{T}(\text{Last GCN layer}, D_{1\%})$ exhibit significant variability, also due to the scaling factor introduced for results presentation (i.e., all results are scaled by 10^4), which further amplify relative fluctuations. Furthermore, the low data regime causes overfitting and poor generalization, as demonstrated by the performance of $\mathcal{T}(\text{Trained from scratch}, D_{1-5\%})$. Also, fine-tuning only the last GCN layer is insufficient for stable performance, especially with limited data (see $\mathcal{T}(\text{Last GCN layer}, D_{1-50\%})$). In contrast, $\mathcal{T}(\text{BAM}, D_{1-50\%})$ maintains consistent performance with low variability, showcasing its adaptability across different data percentages.

To qualitatively assess the efficiency of the BAM adapter compared to previous baselines in different data regimes, Figure 2 shows a Pareto plot with the MSE and the number of parameters trained (or fine-tuned) when transfer-

Table 2. Ablation study on the proposed BAM adapter in the Värnamo-Varberg scenario. Each row of the table reports the results when one component is selectively removed from the original BAM architecture. All values are scaled by 10^4. Bold numbers highlight the best-performing model at each data usage percentage.

Ablation study (Värnamo to Varberg)				
Model	1%	5%	20%	50%
BAM	63.14 ± 7.63	$\mathbf{53.7 \pm 1.81}$	$\mathbf{51.22 \pm 1.39}$	$\mathbf{50.09 \pm 1.31}$
- *GCN Fine-tuning*	65.28 ± 9.98	55.46 ± 1.90	52.04 ± 1.43	50.51 ± 1.25
- *ST-GCN Bridge*	$\mathbf{55.39 \pm 3.86}$	54.53 ± 1.91	53.23 ± 1.10	51.77 ± 1.63
- *Transformer Bridge*	67.03 ± 7.12	57.67 ± 2.84	53.83 ± 2.97	52.41 ± 2.16
- *Attention*	61.55 ± 7.33	54.42 ± 2.01	52.54 ± 1.33	51.18 ± 1.44
- *Dropout*	64.77 ± 7.54	54.88 ± 2.05	51.30 ± 1.44	50.53 ± 1.39
- *Layer normalization*	61.47 ± 9.36	54.80 ± 2.17	52.79 ± 1.36	51.46 ± 1.62
- *Scalar*	65.82 ± 9.37	55.45 ± 1.93	52.35 ± 1.22	50.21 ± 2.34

ring knowledge among different stations. The performance of the BAM adapter (blue color) outstands when considering its efficiency in utilizing fewer trainable parameters compared to other models. The only comparable baselines in terms of parameter count is the MLP adapter, which operates in a similar parameter range, and the fine-tuning of the last GCN layer, which uses even fewer parameters. However, when considering together both the MSE and the number of parameters, the BAM adapter outperforms all other methods when using 5% and 50% of the data. Notably, BAM achieves these results with only 3.02% of trainable parameters, compared to 100%, 100%, 20.99%, and 57.01% used by Trained from scratch, Fine-tune all, Fine-tune last transformer, and Tiny-Attention, respectively. This analysis emphasizes the ability of the BAM adapter to maintain strong performance while using a minimal number of trainable parameters, demonstrating the model's efficiency and effectiveness in leveraging limited data and hardware resources to obtain competitive results. Similar conclusions were also drawn from experiments conducted in the Varberg to Malmö transfer learning scenario, further validating the generalizability of the BAM adapter across different charging station networks.

4.3 Ablation Study

The ablation study was conducted to assess the impact of each component of the BAM adapter. By systematically removing or altering individual elements, the study identifies which components are essential for optimal performance. The results, presented in Table 2, show that each component is a necessary contribution to the BAM architecture, and its removal would lead to an increase in MSE. The only exception is with $\mathcal{T}(\text{BAM}, D_{1\%})$. This might be attributed to the data-hungry nature of attention mechanisms, which may overfit and amplify irrelevant features at lower data percentages. Additionally, the improved performance when removing the layer normalization could be due to the fact that, with limited data, the model may benefit from the instability of training, which allows for occasional "lucky jumps" in optimization that lead to better performance. Similarly, removing the ST-GCN bridge reduces the model's complexity, which can be advantageous when working with very limited data, as it simplifies the feature transformations and reduces the risk of overfitting, thus improving performance at 1%.

$\mathcal{T}(\text{-Transformer Bridge}, D_{1-50\%})$, which removes the bridges inside the transformer, exhibits the highest MSE, overall. Additionally, $\mathcal{T}(\text{-Scalar}, D_{1-50\%})$ demonstrates that the scalar component offers notable performance improvements with minimal additional parameters. These findings highlight how different components contribute to the overall effectiveness of the BAM adapter and assure high performance in a low data regime with a low number of trainable parameters, which is crucial in many industrial applications.

5 Conclusion

This paper introduces the BAM adapter, a novel and highly effective solution for overcoming the challenges associated with data-sparse time-series prediction tasks. By enhancing the performance of ST-GCN through efficient transfer learning, BAM demonstrates a unique ability to capture both spatial and temporal dependencies while maintaining exceptional parameter efficiency. The results highlight the scalability and efficiency of the BAM adapter, which, together with an ST-GCN model, outperforms or matches the performance of previous baselines, while significantly reducing the number of trainable parameters. Extensive experiments were conducted on various EV charging station networks in Sweden, demonstrating that BAM offers an efficient transfer learning solution when new charging stations are introduced with only a few days of historical data available. This adaptability underscores BAM's potential to address critical forecasting needs across a wide range of domains.

Looking ahead, future work is expected to focus on extending the BAM adapter's application to larger and more diverse datasets, providing a deeper evaluation of its scalability and robustness across varied contexts. Furthermore, investigating BAM's potential in a wider array of domains, such as healthcare, energy management, and urban planning, could uncover valuable opportunities to apply its parameter-efficient design to domain-specific forecasting challenges. Additionally, a comparative analysis with other parameter-efficient transfer learning techniques, such as LoRA, would be valuable to further assess BAM's effectiveness. Finally, integrating BAM with other advanced spatio-temporal architectures holds promise for unlocking additional performance improvements and expanding its applicability to more complex and demanding scenarios.

Acknowledgment. The work was carried out with support from The Knowledge Foundation and from Vinnova (Sweden's innovation agency) through the Vehicle Strategic Research and Innovation Programme, FFI.

Disclosure of Interests. The authors have no competing interests to declare that are relevant to the content of this article.

References

1. Akita, R., Yoshihara, A., Matsubara, T., Uehara, K.: Deep learning for stock prediction using numerical and textual information. In: 2016 IEEE/ACIS 15th International Conference on Computer and Information Science (ICIS), pp. 1–6. IEEE (2016)
2. Boulanger-Lewandowski, N., Bengio, Y., Vincent, P.: Modeling temporal dependencies in high-dimensional sequences: Application to polyphonic music generation and transcription. In: Proceedings of the 29th International Conference on Machine Learning (ICML), pp. 1159–1166 (2012)
3. Caruana, R., Silver, D.L., Baxter, J., Mitchell, T.M., Pratt, L.Y., Thrun, S.: Learning to learn: knowledge consolidation and transfer in inductive systems (1995)
4. ChargeFinder: Data Provided for this project. https://chargefinder.com/se
5. Chen, H., et al.: Conv-adapter: exploring parameter efficient transfer learning for convnets. In: Proceedings of the IEEE/CVF Conference on Computer Vision and Pattern Recognition (CVPR) Workshops, pp. 1551–1561 (June 2024)
6. Clark, S., Rodriguez, P.: Limitations of autoregressive models for nonlinear time series data. J. Appl. Stat. **42**(3), 341–355 (2019)
7. Clark, S., Rodriguez, P.: Challenges in modeling nonlinear patterns in time series data. J. Appl. Stat. **41**(9), 1230–1245 (2021)
8. Dou, B., et al.: Machine learning methods for small data challenges in molecular science. Chem. Rev. **123**(13), 8736–8780 (2023)
9. European Union: The European Green Deal (11 December 2019). https://commission.europa.eu/strategy-and-policy/priorities-2019-2024/european-green-deal_en. Accessed 24 Nov 2023
10. Guo, D., Rush, A., Kim, Y.: Parameter-efficient transfer learning with diff pruning. In: Zong, C., Xia, F., Li, W., Navigli, R. (eds.) Proceedings of the 59th Annual Meeting of the Association for Computational Linguistics and the 11th International Joint Conference on Natural Language Processing (Volume 1: Long Papers). pp. 4884–4896. Association for Computational Linguistics, Online (August 2021). https://doi.org/10.18653/v1/2021.acl-long.378, https://aclanthology.org/2021.acl-long.378
11. Gupta, D., Bhatti, A., Parmar, S., Dan, C., Liu, Y., Shen, B., Lee, S.: Low-rank adaptation of time series foundational models for out-of-domain modality forecasting. In: Proceedings of the 26th International Conference on Multimodal Interaction, pp. 382–386 (2024)
12. Harzig, P., Einfalt, M., Lienhart, R.: Synchronized audio-visual frames with fractional positional encoding for transformers in video-to-text translation. In: 2022 IEEE International Conference on Image Processing (ICIP), pp. 2041–2045. IEEE (2022)
13. Hassani, H., Kalantari, M., Ghodsi, Z.: Evaluating the performance of multiple imputation methods for handling missing values in time series data: a study focused on East Africa, soil-carbonate-stable isotope data. Stats **2**(4), 457–467 (2019)
14. Hochreiter, S., Schmidhuber, J.: Long short-term memory. Neural Comput. **9**(8), 1735–1780 (1997)

15. Houlsby, N., et al.: Parameter-efficient transfer learning for NLP (2019). https://arxiv.org/abs/1902.00751
16. Howard, J., Ruder, S.: Universal language model fine-tuning for text classification. arXiv preprint arXiv:1801.06146 (2018)
17. Hu, E.J., et al.: Lora: low-rank adaptation of large language models. arXiv preprint arXiv:2106.09685 (2021)
18. IEA: Electric Vehicles (2023). https://www.iea.org/energy-system/transport/electric-vehicles. Accessed 24 Nov 2023
19. Johnson, K., Lee, M.: The use of arima models in time series forecasting. J. Forecast. **40**(4), 401–420 (2021)
20. Johnson, M., Wang, L.: Applications of st-gcns in urban planning and transportation. J. Urban Mobil. **25**(3), 101–115 (2022)
21. Kim, A., Lee, S.: Advantages of spatio-temporal graph convolutional networks in forecasting. J. Mach. Learn. Res. **34**(4), 345–359 (2021)
22. Kipf, T.N., Welling, M.: Semi-supervised classification with graph convolutional networks. arXiv preprint arXiv:1609.02907 (2016)
23. Lara-Benítez, P., Gallego-Ledesma, L., Carranza-García, M., Luna-Romera, J.M.: Evaluation of the transformer architecture for univariate time series forecasting. In: Alba, E., et al. (eds.) Advances in Artificial Intelligence, pp. 106–115. Springer International Publishing, Cham (2021)
24. Le, H., Pino, J., Wang, C., Gu, J., Schwab, D., Besacier, L.: Lightweight adapter tuning for multilingual speech translation. arXiv preprint arXiv:2106.01463 (2021)
25. Lee, B., Lee, H., Ahn, H.: Improving load forecasting of electric vehicle charging stations through missing data imputation. Energies **13**(18), 4893 (2020)
26. Lee, J., Zhang, M.: Advantages of graph convolutional networks in capturing local neighborhood information. J. Graph Data Anal. **11**(3), 123–136 (2018)
27. Lee, M., Johnson, K.: Limitations of moving average models in time series prediction. J. Mach. Learn. **50**(4), 567–580 (2020)
28. Leeraksakiat, P., Pora, W.: Occupancy forecasting using LSTM neural networkand transfer learning. In: 2020 17th International Conference on Electrical Engineering/Electronics, Computer, Telecommunications and Information Technology (ECTI-CON). IEEE (June 2020)
29. Lovanshi, M., Tiwari, V.: Human skeleton pose and spatio-temporal feature-based activity recognition using st-gcn. Multimed. Tools Appl. **83**(5), 12705–12730 (2024)
30. Luo, R., Song, Y., Huang, L., Zhang, Y., Su, R.: AST-GIN: attribute-augmented spatiotemporal graph informer network for electric vehicle charging station availability forecasting. Sensors (Basel) **23**(4) (2023)
31. Ma, T.Y., Faye, S.: Multistep electric vehicle charging station occupancy prediction using hybrid lstm neural networks. Energy **244**, 123217 (2022)
32. Murti, D.M.P., Pujianto, U., Wibawa, A.P., Akbar, M.I.: K-nearest neighbor (k-nn) based missing data imputation. In: 2019 5th International Conference on Science in Information Technology (ICSITech), pp. 83–88 (2019). https://doi.org/10.1109/ICSITech46713.2019.8987530
33. Nie, S.Y., Wu, X.Q.: A historical study about the developing process of the classical linear time series models. In: Yin, Z., Pan, L., Fang, X. (eds.) Proceedings of The Eighth International Conference on Bio-Inspired Computing: Theories and Applications (BIC-TA), 2013, pp. 425–433. Springer Berlin Heidelberg, Berlin, Heidelberg (2013)
34. Parikh, A.P., Täckström, O., Das, D., Uszkoreit, J.: A decomposable attention model for natural language inference. arXiv preprint arXiv:1606.01933 (2016)

35. S., H.L.: A study in the analysis of stationary time series. by herman wold. [pp. 214 + viii. almqvist and wiksells boktryckeri-a.-b., uppsala. 1938. price kr. 6.]. J. Inst. Actuaries **70**(1), 113–115 (1939). https://doi.org/10.1017/S0020268100011574
36. Saxena, D., Cao, J.: D-gan: deep generative adversarial nets for spatio-temporal prediction (2021). https://arxiv.org/abs/1907.08556
37. Smith, R., Green, A.: Strengths of autoregressive models in time series analysis. J. Stat. Model. **30**(1), 45–57 (2018)
38. Su, S., Li, Y., Chen, Q., Xia, M., Yamashita, K., Jurasz, J.: Operating status prediction model at EV charging stations with fusing spatiotemporal graph convolutional network. IEEE Trans. Transp. Electrif. 1 (2022)
39. Vaswani, A., et al.: Attention is all you need. Adv. Neural Inf. Process. Syst. (2017)
40. Vyas, A., Bandyopadhyay, S.: Dynamic structure learning through graph neural network for forecasting soil moisture in precision agriculture (2022). https://arxiv.org/abs/2012.03506
41. Wafaa Mustafa Hameed, N.A.A.: Comparison of seventeen missing value imputation techniques. J. Hun. Univ. Nat. Sci. **49**(7) (2022)
42. Weiss, K., Khoshgoftaar, T.M., Wang, D.D.: A survey of transfer learning. J. Big Data **3**(1), 1–40 (2016). https://doi.org/10.1186/s40537-016-0043-6
43. White, L., Black, G.: Challenges and limitations of arima models in time series prediction. J. Mach. Learn. **51**(5), 601–615 (2022)
44. Yosinski, J., Clune, J., Bengio, Y., Lipson, H.: How transferable are features in deep neural networks? (2014). https://arxiv.org/abs/1411.1792
45. Zhao, H., Tan, H., Mei, H.: Tiny-attention adapter: contexts are more important than the number of parameters. arXiv preprint arXiv:2211.01979 (2022)

Meta-learning and Data Augmentation for Stress Testing Forecasting Models

Ricardo Inácio[1,2(✉)], Vitor Cerqueira[1,2], Marília Barandas[3], and Carlos Soares[1,2,3]

[1] Faculdade de Engenharia da Universidade do Porto, Porto, Portugal
{rcinacio,vcerqueira,csoares}@fe.up.pt
[2] Laboratory for Artificial Intelligence and Computer Science (LIACC), Porto, Portugal
[3] Fraunhofer Portugal AICOS, Porto, Portugal
marilia.barandas@aicos.fraunhofer.pt

Abstract. The effectiveness of time series forecasting models can be hampered by conditions in the input space that lead them to underperform. When those are met, negative behaviours, such as higher-than-usual errors or increased uncertainty are shown. Traditionally, stress testing is applied to assess how models respond to adverse, but plausible scenarios, providing insights on how to improve their robustness and reliability. This paper builds upon this technique by contributing with a novel framework called MAST (Meta-learning and data Augmentation for Stress Testing). In particular, MAST is a meta-learning approach that predicts the probability that a given model will perform poorly on a given time series based on a set of statistical features. This way, instead of designing new stress scenarios, this method uses the information provided by instances that led to decreases in forecasting performance. An additional contribution is made, a novel time series data augmentation technique based on oversampling, that improves the information about stress factors in the input space, which elevates the classification capabilities of the method. We conducted experiments using 6 benchmark datasets containing a total of 97.829 time series. The results suggest that MAST is able to identify conditions that lead to large errors effectively.

This work was partially funded by projects AISym4Med (101095387) supported by Horizon Europe Cluster 1: Health, ConnectedHealth (n. 46858), supported by Competitiveness and Internationalisation Operational Programme (POCI) and Lisbon Regional Operational Programme (LISBOA 2020), under the PORTUGAL 2020 Partnership Agreement, through the European Regional Development Fund (ERDF) and Agenda "Center for Responsible AI", nr. C645008882-00000055, investment project nr. 62, financed by the Recovery and Resilience Plan (PRR) and by European Union - NextGeneration EU, and also by FCT plurianual funding for 2020–2023 of LIACC (UIDB/00027/2020 UIDP/00027/2020). Funded by the European Union - NextGenerationEU. Views and opinions expressed are however those of the author(s) only and do not necessarily reflect those of the European Union or the European Commission. Neither the European Union nor the European Commission can be held responsible for them.

G. Krempl et al. (Eds.): IDA 2025, LNCS 15669, pp. 343–357, 2025.
https://doi.org/10.1007/978-3-031-91398-3_26

Keywords: Time series · Forecasting · Stress testing · Meta-learning · Data augmentation

1 Introduction

Time series forecasting is a relevant problem in many application domains, such as industry, healthcare, and finance. Accurate forecasts help reduce uncertainty about the future and foster data-driven decision-making within organizations [19]. However, the effective application of forecasting models is sometimes hampered by extreme conditions that reduce performance and reliability [12].

In machine learning, such conditions can refer to aspects of the input space that lead a model to exhibit an abnormal behaviour, such as higher-than-usual errors or increased uncertainty [11,12], resulting in rare but remarkably unreliable outputs, causing end-users to distrust models. In the context of forecasting, these can originate from factors such as i) data-related challenges, such as outliers or missing data [20], or ii) an inadequate model that fails to capture relevant patterns in specific, uncommon time series [22]. While traditional stress testing focuses on building scenarios with adverse, but plausible input conditions to strain systems beyond their normal operating parameters, to assess the integrity of the development process, our approach aims to first identify, simulate, and model challenging instances, based on output manifestations (e.g. performance degradation), to then build a metamodel that reliably recognises such conditions on novel instances. This process compliments the traditional strategy by providing data-driven insights about potentially unknown stress factors, more efficiently assessing what characteristics undermine robustness. This process fosters a responsible deployment of machine learning algorithms, as it enables end users to defer forecasts performed on data that meets stress-inducing criteria.

To give an example of how stress-inducing data manifests in practice, Fig. 1 illustrates the forecasting performance of a model across several univariate time series. It shows the distribution of SMAPE (symmetric mean absolute percentage error) incurred by a forecasting model across 1428 time series. The distribution is right-skewed, where the long-tail denotes time series where the model is under stress and exhibits larger errors.

We propose a novel method to characterise and predict factors of stress in a given forecasting model, thus revealing the conditions that lead to a large error in a given time series. Understanding these factors is essential to uncover the inner workings of forecasting models, increasing transparency in the use of such systems [18]. We accomplish this goal by proposing a new approach for modelling stress in univariate time series forecasting models called `MAST` (short for Meta-learning and data Augmentation for Stress Testing).

The method involves two vital concepts: using meta-learning to model the structural properties of time series; and adopting resampling methods to generate synthetic time series data that lead to a degradation of performance. Meta-learning applies machine learning to improve the efficiency of models, with tasks such as algorithm selection [6] or dynamic model combination [8]. `MAST` uses

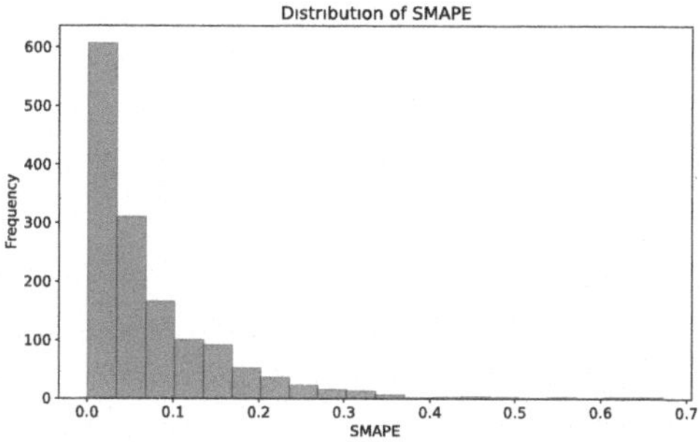

Fig. 1. SMAPE performance of a forecasting model across several time series.

meta-learning to model the occurrence of large errors based on structural properties of time series (e.g., seasonal strength). These properties are computed using time series feature extraction methods [4,29].

Large errors represent a small fraction of the dataset (c.f. Fig. 1). A limited number of samples complicates the process of modelling such instances. In this context, we frame the above meta-learning problem as an imbalanced learning task [28]. Imbalanced domain learning is the field of machine learning where the target distribution is skewed, which means most observations belong to a single class. However, the minority class is often the most relevant, which, in our case, represents instances where a model incurs a large error. Resampling techniques are typically used to tackle class imbalance in binary classification problems. We explore resampling methods (e.g., `ADASYN` [16]) to generate synthetic samples in the feature space concerning large errors. We hypothesize that the augmented meta-dataset improves the performance of the meta-learning model.

We validate the proposed approach in six benchmark datasets that comprise a total of 97.829 time series. The experiments suggest that our method is able to identify the conditions that lead to large errors by a forecasting model. Using Shapley values [24], we provide a concrete example about the conditions where a given model is expected to underperform. The proposed method and experiments are available online.[1] In summary, the contributions of this work are:

- A novel methodology to identify stress conditions in the inputs of forecasting models based on meta-learning and data augmentation called `MAST`;
- A data augmentation approach based on oversampling strategies that creates synthetic observations at the feature level based on large forecasting errors;
- A set of extensive experiments that validate the proposed method in benchmark datasets, including a usage example based on Shapley values.

[1] https://github.com/ricardoinaciopt/mast.

2 Background

2.1 Problem Definition

A univariate time series is a time-ordered sequence of observations defined as $Y = \{y_1, y_2, \ldots, y_t\}$, where $y_i \in \mathbb{R}$ is the numeric value of Y collected at time i and t is the length of Y. The goal is to predict the value of the next h observations of a given time series, $y_{t+1}, \ldots, y_{t+h}$, based on historical data.

Machine learning approaches based on auto-regression are increasingly adopted to tackle forecasting problems [8]. We use sliding windows to reconstruct a time series Y for supervised learning. Each observation of a time series is modelled as a function of its past p lags [5]. This results in a dataset $D = \{< X_i, y_i >\}_{i=p+1}^{t}$ where y_i represents the i-th observation from Y, and $X_i \in \mathbb{R}^p$ is the i-th corresponding set of p lags: $X_i = \{y_{i-1}, y_{i-2}, \ldots, y_{i-p}\}$. The goal is to build a regression model f with the form $y_i = f(X_i)$.

We focus on forecasting problems that involve a collection of multiple univariate time series: $\mathcal{Y} = \{Y_1, Y_2, \ldots, Y_n\}$, where Y_j is the j-th time series and n is the number of time series in the collection. Recent machine learning methods applied to forecasting tend to follow a global approach and build a model for a given collection using the historical data of all available time series. Global forecasting models tend to outperform local approaches [14], which build an independent model for each time series in a collection.

2.2 Data Augmentation

Data augmentation techniques are useful when a large or diverse enough dataset is not available for training a model [3]. In most cases, the goal of existing time series data augmentation techniques is to obtain an enriched dataset that leads to better forecasting performance [7]. We focus on well-established methods in the literature of imbalanced domain learning, such as `SMOTE`, which interpolates between minority class samples and their nearest neighbours to create new samples, or `ADASYN`, which uses the density distribution of minority class instances, focusing on complex regions, to dynamically generate synthetic samples [28].

2.3 Time Series Feature Extraction

Time series features, such as mean and variance, are statistical summaries extracted from time series data, that capture important characteristics and patterns, allowing for subsequent analysis [17]. Several frameworks have been proposed for this task. Examples include the works by Montero-Manso et al. [29], Barandas et al. [4], or Christ et al. [10]. This technique has been shown to improve performance in various tasks, such as forecasting [21] or classification [4].

There are several approaches in the literature for feature extraction, including TSFEL [4], tsfeatures [29], or catch22 [23]. These involve several statistical operations, such as the mean or standard deviation of the time series.

2.4 Stress Testing

Building reliable predictive models is a key step towards the responsible use of machine learning [13]. This task is challenging because real-world data is susceptible to conditions that undermine the performance of models [12]. Such conditions are traditionally identified using stress testing [11], where realistic, but challenging scenarios are designed to push models farther from normal responses, as a way to gauge reliability and robustness in the development process.

Establishing in which circumstances a model is prone to under-perform is a crucial aspect for supporting the adoption of machine learning systems, especially in sensitive domains such as healthcare [18]. Stress testing is an emerging topic that addresses this problem. The goal is to evaluate the consistency of machine learning approaches by applying challenging but plausible scenarios. Hence, we adopt a meta-learning approach to both identify and simulate the weaknesses of a model. To our knowledge, this is the first work that explores meta-learning for stress testing, in the sense that, instead of adjusting the input as to reflect such factors, we leverage output manifestations related to large errors, as potential indicators of stress properties, in the corresponding input data. This allows us to not only discern it, but also to characterise stress in terms of features.

3 Stress Testing Forecasting Models Using Meta-learning

The workflow of `MAST`, illustrated in Fig. 2, is split into: i) a development stage and; ii) an inference stage. In the development stage, we conduct performance estimation to identify instances of large errors. Then, we extract statistical features from all time series to build the meta-dataset, in conjunction with the results from the performance estimation step. As it is unbalanced by nature, data augmentation is applied to increase the representativity of stress factors. We use this set to develop the metamodel. The inference stage simply involves applying the metamodel on previously unseen data.

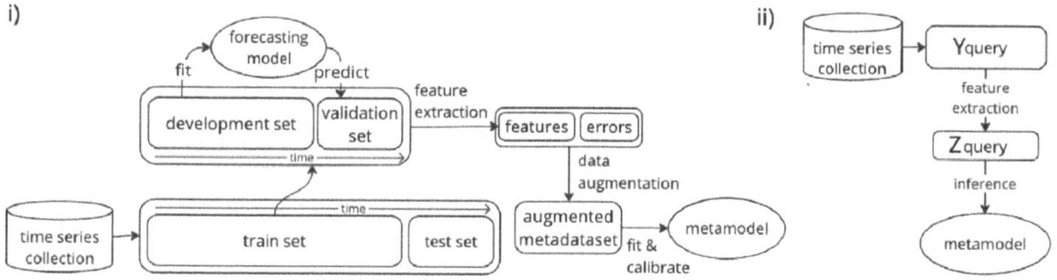

Fig. 2. Workflow behind `MAST`, split into a i) development and ii) inference stage.

3.1 Development Stage

In the development stage, we aim at building a metamodel that predicts the probability that a time series will induce a large error on a forecasting model. This goal is split into four tasks:

- Performance estimation: Conducting a base-level learning cycle to estimate the performance of a given forecasting model in input data;
- Feature extraction: Distil each time series in the train set into a feature set;
- Data augmentation: Create semisynthetic feature data with large errors;
- Meta-learning: Build a metamodel with calibrated probabilities, that predicts the performance of a base-level model based on time series features.

Performance Estimation. We conduct a procedure to estimate the performance $\hat{P}_{\mathcal{Y}}$ of a forecasting model f on each time series in collection $\mathcal{Y}$: $\hat{P}_{\mathcal{Y}} = \{\hat{P}(Y) : Y \in \mathcal{Y}\}$. Let $\mathcal{D}$ denote the dataset built from $\mathcal{Y}$, which is prepared for auto-regression. We split $\mathcal{D}$ into training and testing sets, $\mathcal{D}_{\text{train}}$ and $\mathcal{D}_{\text{test}}$, and fit a model f using $\mathcal{D}_{\text{train}}$. The performance of f is evaluated on $\mathcal{D}_{\text{test}}$, resulting in error scores according to SMAPE metric. This leads to a set of performance estimates $E = \{e_1, e_2, \dots, e_n\}$, where e_i is the estimate of f for time series Y_i.

Feature Extraction and Meta-dataset. We break down the training set of each time series $Y \in \mathcal{Y}$ using feature extraction. Let g denote a feature extraction function that transforms a time series Y_i into a set of features $Z_i = \{z_{i,1}, z_{i,2}, \dots, z_{i,m}\}$, where m is the number of features and $z_{i,k}$ is the k-th feature of Y_i. This transformation is written as $Z_i = g(Y_i)$. While this methodology is method-agnostic, we focus on *tsfeatures* [29], as mentioned in Sect. 2.3.

The results from the performance estimation, E, and feature extraction Z are concatenated into a meta-dataset $\mathcal{D}' = \{< Z_i, e_i >\}_{i=1}^{n}$. Each element in $\mathcal{D}'$ includes the features of a given time series, along with the error score of a given forecasting model f on that data. Note that the features are extracted using the training data, while the error is computed on the test data. This process preserves the temporal order of data and avoids leakage.

Our goal is to use the meta-dataset to develop a metamodel f' that classifies where the forecasting model f will incur a large error in a given time series. We define a binary variable for large errors based on the error scores E, which is determined as follows:

$$b_i = \begin{cases} 1 & \text{if } e_i > \tau, \\ 0 & \text{otherwise.} \end{cases} \tag{1}$$

For a given time series, b_i takes the value of 1 if e_i is above τ, which represents the threshold above which an error is considered large. The threshold τ can be defined by domain expertise or in a data-driven way, e.g. using percentiles. In effect, the meta-dataset $\mathcal{D}' = \{< Z_i, b_i >\}_{i=1}^{n}$ defines a binary classification task. The goal is to model upcoming large errors (b) based on time series features (Z) extracted from historical data from new samples.

Data Augmentation. We work under the assumption that large errors are rare. Thus, the distribution of b is imbalanced, with $b = 1$ being the minority class. The meta-class imbalance poses an additional challenge for learning a model for large errors. We tackle this issue by framing the task as an imbalanced domain learning problem [28], and use resampling methods for data augmentation.

We employ an oversampling algorithm (e.g. `SMOTE` [9]) where the imbalanced meta-dataset is used to create new $(Z_{i,synthetic}, b_{i,synthetic})$ samples concerning time series that lead to large errors ($b = 1$). The idea is to use existing data to simulate conditions that lead to large errors, representing factors of stress.

Meta-learning. The created synthetic samples are represented in a dataset $\mathcal{D}'_{synthetic}$, which is concatenated with the original dataset $\mathcal{D}'$ to augment it, leading to the combined meta-dataset $\mathcal{D}' \cup \mathcal{D}'_{synthetic}$. Then, a metamodel f' is trained on the meta-dataset, as described by the following equation: $\hat{b}_i = f'(Z_i)$. The training process includes a calibration step, using 30% of the meta-data. The output of the metamodel f' provides information about the probability that the forecasting model f will incur a large error on Y_i.

3.2 Inference Stage

At this stage, the metamodel f' tries to predict the probability that f will incur a large error for a query time series Y_{query}. We conduct this process as follows:

1. Feature extraction: Transform Y_{query} into a complete set of features Z_{query}.
2. Compute the probability of a large error using the metamodel: $\hat{b} = f'(Z_{query})$.

The probabilities related to large errors can be used in various ways, for example deferring the forecasts by not accepting the predictions made by f. Another possibility is to use the metamodel output for a descriptive analysis by characterizing the feature space in conditions with high probability of a large error.

4 Experiments

We carried out extensive experiments to evaluate the proposed method. The experiments focus on the following research questions:

- **RQ1:** How does the base forecasting model compare with benchmark forecasting approaches?
- **RQ2:** Can the metamodel predict large errors by a given forecasting model?
- **RQ3:** What is the impact of data augmentation via meta-features resampling in the performance of the metamodel?
- **RQ4:** Which argumentation techniques lead to better scores across datasets?
- **RQ5:** How can the results of the metamodel be used to describe the conditions where a forecasting model is expected to underperform?

4.1 Data

We use the following 6 datasets in the experiments: M3 Monthly, M3 Quarterly, M3 Yearly [25], and M4 Monthly, M4 Quarterly, M4 Yearly [26]. These datasets were part of different competitions and represent long-standing univariate forecasting benchmarks. Table 1 provides a brief summary of the data. In datasets with a monthly frequency, the forecasting horizon (h), and the number of lags (p), are set to 12. For quarterly or yearly frequencies, we set h and p to 4.

Table 1. Brief overview of the datasets, included in the experiments.

	# Time series	# Observations	Avg. length
M3 Monthly	1428	167562	117
M3 Quarterly	756	37004	48
M3 Yearly	645	18319	28
M4 Monthly	48000	11246411	234
M4 Quarterly	24000	2406108	100
M4 Yearly	23000	858458	37

4.2 Experimental Design

Performance Estimation. The data partitioning process is illustrated in Fig. 3. We split the available data into training and testing sets for performance estimation. The test set is composed of the last h observations of each time series Y in $\mathcal{Y}$. The training set is further split into a development set and a validation set, where the latter is composed of the last h observations of each time series before the test set.

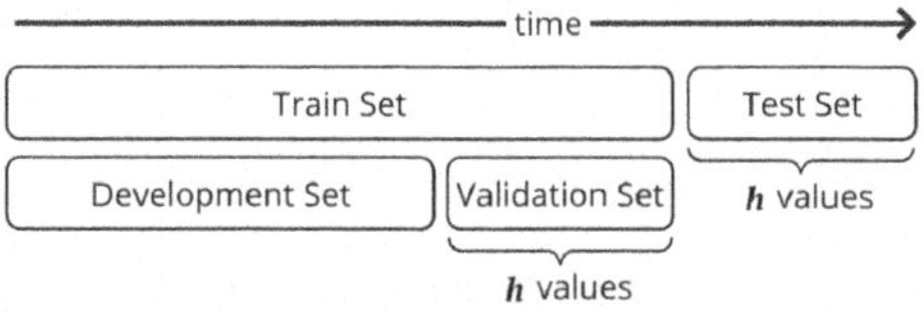

Fig. 3. Data preparation, where for each time series, the segment at the end of the train set, of length (h), becomes the test set. The preceding segment of the same length forms the validation set, and the remaining data defines the development set.

We carry out the methodology for building the metamodel using the development set and the validation set (c.f. Fig. 2). Before metamodel training, we leave 30% of the meta-dataset entries out, to then calibrate the probabilities of the predictions, thus reflecting the true class likelihood. We employ *Isotonic Regression* for calibration [34]. Using the training set, we re-train the forecasting model and carry out the inference stage described in Sect. 3.2. Finally, the performance of the forecasting model and the metamodel are evaluated using the test set. For each time series, we evaluate the forecasting performance of f. Using those results, we evaluate the ability of the metamodel f' to predict large errors. We evaluate forecasting performance using SMAPE, a popular benchmarking metric [26]. While we aim at identifying the weaknesses of a forecasting model, ensuring that it outperforms a baseline is important for its applicability. Concerning the metamodel, we resort to Log Loss for evaluation.

Model Training. The forecasting model is built using a `lightgbm` (`LGBM`) regression algorithm. This approach has been shown to provide state-of-the-art forecasting performance. For instance, it is the algorithm used in the winning solution of the M5 forecasting competition [27]. Following Bandara et al. [3], we train a global forecasting model for each dataset listed in Table 1 with the approach described in Sect. 2.1. Hyperparameter tuning was conducted with Bayesian optimization using `optuna` [1]. For each dataset, seasonal differencing was carried out to handle seasonal effects and stabilize the level of time series. We used 12, 4, and 1 periods for monthly, quarterly and yearly datasets, respectively. We include the seasonal naive method in the experiments as a baseline to validate the forecasting accuracy of the employed model. Seasonal naive forecasts the upcoming values of a time series using the last observation of the same period, providing a simple, but strong standard.

The metamodel is also trained using a `lightgbm` algorithm, but with an objective function tailored for binary classification. The metamodel is optimized using random search.

Meta-target Definition and Parameter Setting. We define large errors in a data-driven way using percentiles. The threshold τ is set to the percentile 80 of the SMAPE performance estimates from the validation set: $\tau = q_{0.80}(E)$.

As mentioned beforehand, we leverage the *tsfeatures* framework for time series feature extraction. The resulting set contains a total of 42 features, including `entropy` and `linearity`. We refer the reader to the work by Montero-Manso et al. [29] for a comprehensive description of the complete feature set.

Concerning data augmentation, we test several methods, namely: `ADASYN` [16], `SMOTE` [9], and variants: `BorderlineSMOTE` [15] and `SVMSMOTE` [31].

4.3 Results

The selected forecasting model must attain reasonable predictive performance, by capturing relevant time series patterns, thus making it interesting to investigate when it underperforms. To assess this, we compare it with a simple baseline model. The results are reported in Table 2, which shows the SMAPE scores of each approach on every dataset. The developed model outperforms seasonal naive in all cases, which validates its forecasting performance (**RQ1**). The statistical significance of the differences in output across models is verified via the Wilcoxon signed-rank test [33]. The result ($W = 0.0, p = 0.03125$) suggests a statistically significant difference in forecasting performance, given $p < 0.05$, indicating that it should not be attributed to random variation. Additionally, as $W = 0.0$, it implies that all differences ($SMAPE_{LGBM} - SMAPE_{SeasonalNaive}$) pointed in the same (negative) direction, corroborating the superiority of LGBM.

The results concerning the metamodel are reported in Table 3. The metamodel is able to predict large errors with a reasonable performance (**RQ2**), corroborating the proposed hypothesis. When no data augmentation is applied, the average Log Loss is the highest across all datasets, which emphasizes the importance of the synthetic samples (**RQ3**). Overall, it performs better when coupled with data augmentation. On average, `SMOTE` attains the lowest Log Loss across all datasets, while `ADASYN` achieves the lowest average rank (**RQ4**). Conversely, `BorderlineSMOTE` seems to be the worst performing augmentation method, as it never attains best or second-best rankings. This may stem from the fact that this technique changes the density distribution of the minority favouring samples in

Table 2. Forecasting performance of each method using SMAPE. Bold font denotes the best approach for each dataset.

	M3 M	M3 Q	M3 Y	M4 M	M4 Q	M4 Y
`LGBM`	**7.41%**	**3.70%**	**7.54%**	**6.23%**	**4.47%**	**5.42%**
`Seasonal Naive`	7.98%	4.50%	7.85%	7.17%	5.19%	6.49%

Table 3. Metamodel performance in each dataset according to Log Loss. Bold font denotes the best method in each dataset, while underlined font denotes second best. An average Log Loss and Rank analysis are included.

	`ADASYN`	`SMOTE`	`SVMSMOTE`	`BorderlineSMOTE`	`RandomUnderSampler`	`No Resampling`
M3 M	**0.396**	<u>0.404</u>	0.779	0.474	0.421	0.606
M3 Q	0.537	<u>0.460</u>	**0.459**	0.817	0.519	0.689
M3 Y	<u>0.462</u>	**0.454**	0.495	0.466	0.502	0.511
M4 M	<u>0.342</u>	0.347	**0.337**	0.354	0.354	0.359
M4 Q	0.375	0.379	**0.348**	0.364	0.359	<u>0.349</u>
M4 Y	**0.445**	0.471	0.476	0.463	<u>0.455</u>	0.475
Avg. Log Loss	<u>0.426</u>	**0.419**	0.482	0.490	0.435	0.498
Avg. Rank	**2.50**	<u>3.00</u>	3.17	4.00	3.33	4.83

the outermost regions, which could difficult the process of identifying instances in the innermost regions, while inadvertently leading to a stronger class mixture [30]. It is worth noting that the advantages of oversampling are independent of the sample size of the dataset. The three variants of the M4 dataset (M, Q, Y) contain thousands of time series, but creating additional samples leads to improvements, reflected by the lower Log Loss values.

4.4 Metamodel Explanations

The results presented in the previous section provide evidence that `MAST` is able to detect large errors, which supports and validates the postulated approach. We can leverage Shapley values [24] to explore how time series features affect the output of the metamodel. For illustrative purposes, we showcase results regarding the M3 Monthly dataset only, on the five most important features on this task, according to the metamodel. These are the following: `linearity`, `lumpiness`, `hw_beta`, `spike`, and `hw_gamma`. Figure 4 shows a summary plot of the Shapley values from the above features. This plot shows interesting insights about the performance of the forecasting model based on the predictions of the metamodel (**RQ5**). For example, low `linearity` values increase the probability of classifying a time series as large error inducing. This suggests that the forecasting model may struggle in time series with a low `linearity` value. The opposite is demonstrated by `lumpiness`, as higher values suggest an increased likelihood of inducing large errors. Moreover, to formally assess for collinearity or multicollinearity across the features of the metadataset, we compute the correlation matrix and the VIF (Variance Inflation Factor) [32] to check if any absolute value is significant, given the following thresholds: ($|r| \geq 0.7$ or $|VIF| \geq 2$).

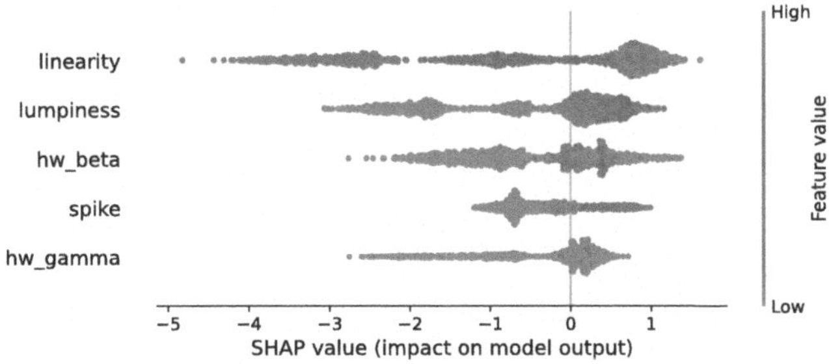

Fig. 4. Shapley values of the 5 most relevant features for the M3 monthly metamodel. Values to the right of the vertical line indicate a positive influence, while the ones at the left indicate a negative one. The colour indicates the magnitude of each feature.

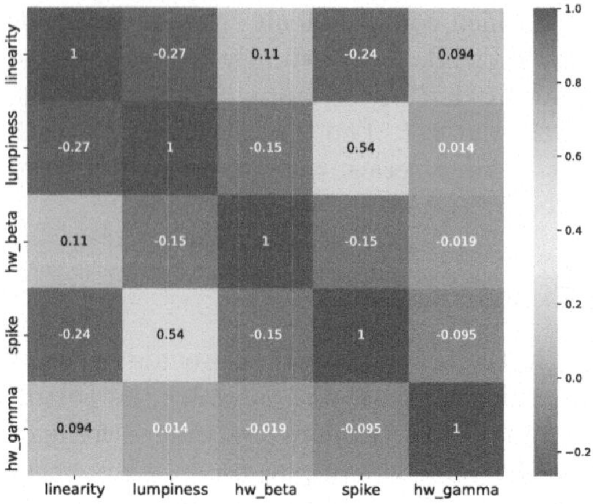

Fig. 5. Correlation matrix of the 5 most relevant features for the M3 monthly metamodel. Since no absolute value is above 0.7, no significant correlations are noted.

Since the results outlined in Fig. 5 show that no absolute value is significant, as the largest is observed between `lumpiness` and `spike` (0.54), and since all VIF values in Table 4 are below 2, we can establish that no substantial linear dependencies are present across the most important features in the metadataset.

Table 4. Variance Inflation Factor for the 5 most relevant features for the M3 monthly metamodel. As it is always below 2, significant linear dependencies are not established.

Feature	`linearity`	`lumpiness`	`hw_beta`	`spike`	`hw_gamma`
VIF	1.099	1.687	1.075	1.650	1.057

5 Discussion and Future Work

This work addresses the problem of evaluating the robustness of univariate time series forecasting models to time series with varying characteristics, specifically focusing on adverse conditions. We use meta-learning and data augmentation to model the relationship between large errors and the structural properties of time series, such as `linearity` or `spike`. The results indicate that our approach is able to identify input conditions that lead to large errors. Those can then be explored using techniques such as Shapley values, indirectly revealing hints about the shortcomings of forecasting models, and how to possibly improve them. Overall, the experiments suggest that using meta-learning to model large errors is promising for improving the transparency and reliability of forecasting models. Nonetheless, the scope is still limited, as only one definition for stress-induction is used: $\tau = q_{0.80}(E)$, which could benefit from a sensitivity analysis. Also, other manifestations of stress could be explored, such as uncertainty. We plan to extend the proposed method in future work, considering:

- Modelling stress in the form of high uncertainty, as opposed (or a complement) to stress in terms of large errors;
- Examine what is the sensitivity of the approach to distinct problems, such as calibration, stress threshold definition or synthetic data sample size.
- Test different data augmentation approaches, based on synthetic time series generation, such as TSMixup [2].
- Apply the developed approach in a wider range of time series datasets.

6 Conclusion

Understanding what aspects in the inputs of forecasting models lead them to undesirable behaviours, such as large errors, is important to improve their transparency and reliability. This paper introduces a novel framework for modelling and characterising such aspects, as to anticipate abnormal responses, based on meta-learning and data augmentation. We demonstrated how features extracted from time series, and manifestations of decreases in forecasting performance, can be used to identify which patterns in the input space contribute to large errors. This compliments traditional stress testing settings, by enabling uncovering potential stress conditions in a data-driven way. Furthermore, by generating time series meta-features based on the most challenging instances, we enhance the effectiveness of the method in predicting the probability of an observation leading to performance degradations. We validated our approach through benchmark datasets containing several univariate time series.

Disclosure of Interests. The authors have no competing interests to declare that are relevant to the content of this article.

References

1. Akiba, T., Sano, S., Yanase, T., Ohta, T., Koyama, M.: Optuna: a next-generation hyperparameter optimization framework. In: Proceedings of the 25th ACM SIGKDD International Conference on Knowledge Discovery and Data Mining (2019)
2. Ansari, A.F., et al.: Chronos: learning the language of time series. arXiv preprint arXiv:2403.07815 (2024)
3. Bandara, K., Hewamalage, H., Liu, Y.H., Kang, Y., Bergmeir, C.: Improving the accuracy of global forecasting models using time series data augmentation. Pattern Recogn. **120**, 108148 (2021)
4. Barandas, M., et al.: TSFEL: time series feature extraction library. SoftwareX **11**, 100456 (2020)
5. Bontempi, G., Ben Taieb, S., Le Borgne, Y.A.: Machine learning strategies for time series forecasting. In: Business Intelligence: Second European Summer School, eBISS 2012, Brussels, Belgium, 15–21 July 2012, Tutorial Lectures 2, pp. 62–77 (2013)

6. Brazdil, P., van Rijn, J.N., Soares, C., Vanschoren, J.: Metalearning approaches for algorithm selection II. In: Metalearning: Applications to Automated Machine Learning and Data Mining, pp. 77–102. Springer (2022)
7. Cerqueira, V., Santos, M., Baghoussi, Y., Soares, C.: On-the-fly data augmentation for forecasting with deep learning. arXiv preprint arXiv:2404.16918 (2024)
8. Cerqueira, V., Torgo, L., Pinto, F., Soares, C.: Arbitrage of forecasting experts. Mach. Learn. **108**, 913–944 (2019)
9. Chawla, N.V., Bowyer, K.W., Hall, L.O., Kegelmeyer, W.P.: SMOTE: synthetic minority over-sampling technique. J. Artif. Intell. Res. **16**, 321–357 (2002)
10. Christ, M., Kempa-Liehr, A.W., Feindt, M.: Distributed and parallel time series feature extraction for industrial big data applications. arXiv preprint arXiv:1610.07717 (2016)
11. Cunha, L., Soares, C., Restivo, A., Teixeira, L.F.: GASTeN: Generative adversarial stress test networks. In: Crémilleux, B., Hess, S., Nijssen, S. (eds.) Advances in Intelligent Data Analysis XXI, pp. 91–102. Springer, Cham (2023)
12. D'Amour, A., et al.: Underspecification presents challenges for credibility in modern machine learning. J. Mach. Learn. Res. **23**(226), 1–61 (2022)
13. Freiesleben, T., Grote, T.: Beyond generalization: a theory of robustness in machine learning. Synthese **202**(4), 109 (2023)
14. Godahewa, R., Bandara, K., Webb, G.I., Smyl, S., Bergmeir, C.: Ensembles of localised models for time series forecasting. Knowl.-Based Syst. **233** (2021)
15. Han, H., Wang, W.Y., Mao, B.H.: Borderline-SMOTE: a new over-sampling method in imbalanced data sets learning. In: International Conference on Intelligent Computing, pp. 878–887. Springer (2005)
16. He, H., Bai, Y., Garcia, E.A., Li, S.: ADASYN: adaptive synthetic sampling approach for imbalanced learning. In: 2008 IEEE International Joint Conference on Neural Networks (IEEE World Congress on Computational Intelligence), pp. 1322–1328. IEEE (2008)
17. Henderson, T., Fulcher, B.D.: Feature-based time-series analysis in r using the theft package. arXiv preprint arXiv:2208.06146 (2022)
18. Jacovi, A., Marasović, A., Miller, T., Goldberg, Y.: Formalizing trust in artificial intelligence: prerequisites, causes and goals of human trust in AI. In: Proceedings of the 2021 ACM Conference on Fairness, Accountability, and Transparency, pp. 624–635 (2021)
19. Jain, C.L.: Answers to your forecasting questions. J. Bus. Forecast. **36**(1), 3 (2017)
20. Kolambe, M., Arora, S.: Forecasting the future: a comprehensive review of time series prediction techniques. J. Electr. Syst. **20**(2s), 575–586 (2024)
21. Lemke, C., Gabrys, B.: Meta-learning for time series forecasting and forecast combination. Neurocomputing **73**(10–12), 2006–2016 (2010)
22. Li, Y., Xu, J., Anastasiu, D.C.: SEED: an effective model for highly-skewed streamflow time series data forecasting. In: 2023 IEEE International Conference on Big Data (BigData), pp. 728–737. IEEE (2023)
23. Lubba, C.H., Sethi, S.S., Knaute, P., Schultz, S.R., Fulcher, B.D., Jones, N.S.: Catch22: canonical time-series characteristics. Data Min. Knowl. Disc. **33**(6), 1821–1852 (2019)
24. Lundberg, S.M., Lee, S.I.: A unified approach to interpreting model predictions. Adv. Neural Inf. Process. Syst. **30** (2017)
25. Makridakis, S., Hibon, M.: The M3-competition: results, conclusions and implications. Int. J. Forecast. **16**(4), 451–476 (2000)
26. Makridakis, S., Spiliotis, E., Assimakopoulos, V.: The M4 competition: 100,000 time series and 61 forecasting methods. Int. J. Forecast. **36**(1), 54–74 (2020)

27. Makridakis, S., Spiliotis, E., Assimakopoulos, V.: M5 accuracy competition: results, findings, and conclusions. Int. J. Forecast. **38**(4), 1346–1364 (2022)
28. Moniz, N., Cerqueira, V.: Automated imbalanced classification via meta-learning. Expert Syst. Appl. **178**, 115011 (2021)
29. Montero-Manso, P., Athanasopoulos, G., Hyndman, R.J., Talagala, T.S.: FFORMA: feature-based forecast model averaging. Int. J. Forecast. **36**(1), 86–92 (2020)
30. Naseriparsa, M., Al-Shammari, A., Sheng, M., Zhang, Y., Zhou, R.: RSMOTE: improving classification performance over imbalanced medical datasets. Health inf. sci. syst. **8**, 1–13 (2020)
31. Nguyen, H.M., Cooper, E.W., Kamei, K.: Borderline over-sampling for imbalanced data classification. Int. J. Knowl. Eng. Soft Data Paradig. **3**(1), 4–21 (2011)
32. O'brien, R.M.: A caution regarding rules of thumb for variance inflation factors. Qual. Quant. **41**, 673–690 (2007)
33. Woolson, R.F.: Wilcoxon signed-rank test. Encycl. Biostatistics **8** (2005)
34. Zadrozny, B., Elkan, C.: Transforming classifier scores into accurate multiclass probability estimates. In: Proceedings of the Eighth ACM SIGKDD International Conference on Knowledge Discovery and Data Mining, pp. 694–699 (2002)

Pragmatic Paradigm for Multi-stream Regression

Nuwan Gunasekara(✉), Sławomir Nowaczyk, and Sepideh Pashami

Center for Applied Intelligent Systems Research, Halmstad University, Halmstad, Sweden
{nuwan.gunasekara,slawomir.nowaczyk,sepideh.pashami}@hh.se

Abstract. Deep learning methods have become increasingly popular for time series nowcasting; however, their effectiveness is limited when making predictions on out-of-distribution data in dynamic streaming environments. This limitation is mainly caused by concept drifts, which occur when the underlying data distribution changes over time – and proves particularly challenging in multi-stream settings, where numerous streams of data exist without necessarily exhibiting the same dynamic behavior. When confronted with concept drift, since many parts of the neural network may still contain useful knowledge about the relevant domain, re-training only on new data poses the risk of catastrophic forgetting. On the other hand, re-training on old and new data can incur significant computational overhead and latency, which is not desirable in a streaming context. Conversely, stream learning allows models to adapt quickly to the new distribution; however, it often fails to achieve efficiency comparable to batch learning. This work aims to combine the strengths of neural networks and streaming regressors for data stream nowcasting in two key ways. First, a streaming regressor utilizes the higher-order representations learned by the neural network, which is set up as a feature extractor and a regressor. Second, the streaming regressor predicts the residual of the neural network. Experiments conducted on challenging real-world dynamic multi-streaming data indicate that this hybrid approach provides better results than the traditional static neural network.

Keywords: Data stream regression · Online nowcasting

1 Introduction

Various approaches, including deep learning techniques, have been proposed for time series prediction. Literature categorizes those approaches into discriminative, generative, and hybrid methods [14]. Discriminative Models focus on learning the conditional probability of target outputs given inputs. On the other hand, Generative Models capture the joint probability of inputs and outputs. Hybrid Models combine features of both discriminative and generative models to leverage their strengths [14]. In practical settings, these methods face challenges such

G. Krempl et al. (Eds.): IDA 2025, LNCS 15669, pp. 358–372, 2025.
https://doi.org/10.1007/978-3-031-91398-3_27

as high-dimensional data, interpretability, slow training processes [14], and susceptibility to concept drifts (evolving data distributions) [27]. Neural Network based methods suffer from catastrophic forgetting (forgets past knowledge) when confronted with concept drifts [12,22].

Unlike traditional batch learning, Stream Learning (SL) must handle concept drifts using limited memory and processing power [3,6,9]. This makes it an interesting alternative to traditional batch-learned models for time series prediction. SL can be used for time series nowcasting (predicting the current time step given input data for current time step and past time steps) [23] and forecasting (predicting the future time steps given input and label data for current and past time steps) [23]. Data stream regression is crucial for applications where data continuously flows and predictions must be made in real-time, such as financial markets [17], and online energy demand prediction [20,21,29]. Data stream regression has been used for nowcasting in energy demand prediction by integrating previous time steps' ground truth into the next step as an input feature [25].

Streaming data is often captured in a multi-stream fashion in real-world settings: traffic information captured from multiple geographic locations [31], weather data captured from multiple stations [13], and energy demand information captured from multiple connection points [25]. Often, there are strong relationships among streams: traffic data captured at a certain crossing tends to be similar to nearby streets, and electricity demand for specific types of connection points tends to share key patterns. Multi-stream learning attempts to capture these relationships [18,31].

The data stream community often overlooks the availability of historical streaming data in practical settings. Hence, they strictly adhere to a fully online training approach. However, extensive historical data can be utilized in various ways before model deployment: to initialize an incremental learner, train a base model, or train a feature extractor. The key contributions of this work are:

- An incremental regressor is trained using the embeddings of a pre-trained network along with other stream features. The network embeddings were learned from the categorical features of historical data that assist in identifying a stream. The network structure additionally includes a prediction head.
- The incremental regressor can function as either a standalone predictor or a residual predictor for the pre-trained network.
- Empirical results on two challenging demand prediction real-world data streams further verify the validity of this approach.

The remainder of this paper is organized as follows. The next section reviews the current state-of-the-art techniques in online time series forecasting, multi-stream learning, and streaming regression. The following section introduces the proposed method, which utilizes batch-learned embeddings to train streaming regressors for multi-stream regression. The experiments section outlines the experimental setup, where the proposed method is compared to a static neural network using two real-world demand prediction data streams. The final section concludes with a discussion of the findings and suggestions for future work.

2 Related Work

Online time series forecasting [1,19,22,27] is challenging due to the data's streaming nature and the evolution of concepts. Training deep neural forecasters in real-time is difficult due to their limited capacity to adapt to changing environments while retaining previous knowledge [22,27]. Fast and Slow Learning Network (FSNet) [22] enhances Neural Networks by using per-layer adapters to facilitate rapid adaptation to recent changes while accessing past knowledge through associative memory. Each adapter maps contexts to transformation coefficients, which are stored as memories and retrieved as needed, with updates written back later. The latest literature also highlights the need for exploiting both the temporal dependency and the cross-variable dependency under a unified framework for online multivariate time series forecasting [27]. OneNet [27] builds independent models to leverage time dependency and variable dependency in a multivariate time series. The models are combined using weights from Online Convex Programming (OCP). OCP produces ensembling weights by utilizing both exponential gradient descent to preserve long-term history and offline reinforcement learning to adjust to short-term history. In the empirical evaluation, OneNet outperformed FSNet and Informer [30] in challenging online multivariate time series forecasting problems [27].

Many real-world scenarios involve multiple streams captured from similar sources, resulting in multi-stream data. A multi-stream feature fusion approach was proposed for traffic flow prediction [18]. Here, a Graph Convolutional Network (GCN) was utilized for capturing spatial features, Gated Recurrent Units (GRUs) were employed for temporal features, and a Fully Connected Network (FNN) was used for additional features. Finally, a soft-attention mechanism was applied to integrate these features. Notably, the GCN utilized a data-driven matrix instead of a distance-based matrix to construct the graphs. Another study employs GNNs to address concept drift in multi-stream environments [31]. It treats prediction tasks as online data streaming tasks within sub-graphs to forecast traffic speed. Each sub-graph corresponds to a specific adaptation target and is updated over time. Local adaptation techniques are used to manage drift within each sub-graph. The goal is to effectively address drift in each sub-graph through a designed adaptation strategy. The global correlations between the streams are maintained as a graph structure.

Many streaming regression methods are explained in the literature [5]. Hoeffding Tree Regressor (HTR) is an adaptation of the incremental tree algorithm Hoeffding Tree (HT) for regression. Like HT, HTR uses the Hoeffding bound [15] to control its split decisions. HTR relies on calculating the reduction of variance in the target space to find a split candidate and ADaptive sliding-WINdow (ADWIN) [2] for drift detection. Fast Incremental Model Trees with Drift Detection (FIMT-DD) learn model trees from an evolving data stream with drift detection [16]. It uses the variance reduction split criterion for splitting and the Page-Hinckley test for drift detection. Its regression counterpart is Fast Incremental Regression Trees with Drift Detection (FIRT-DD). When a drift warning is detected, the above methods create a background tree and replace the foreground tree if the warning surpasses an out-of-control limit. Recently, ensemble methods have outperformed single tree regressors for streaming regression [7,26]. Different ensemble generation, aggregation, and update methods for datastream regression are discussed in literature [9]. The empirical evaluation suggests that using the mean of predictions instead of the median and resetting the base models positively affect predictive performance [9]. Adaptive Random Forest Regressor (ArfReg) [26] combines k Nearest Neighbours and the Adaptive Random Forest Regressor (ArfReg) [7]. Here, a representative data point (called a centroid) is generated at each leaf by compressing the information from all instances in that leaf. At prediction, instead of letting all trees in the forest participate, only k trees with the smallest distances between the input instance and the relevant leaf's centroid are used for the prediction. Both ArfReg and SOKNL use variants of FIRT-DD as the base learner. ArfReg uses ADWIN to monitor drifts on each base learner, builds a background learner on warning, and replaces the foreground learner with the background base learner when the drift signal reaches an out-of-control limit. SOKNL has built-in drift adaption methodology of ArfReg as well as implicit adaptation of the centroids. It assumes that the centroids at the leaves adjust to recent trends. SOKNL have superior accuracy compared to ArfReg [26].

Incremental learning has been applied alongside batch models in some scenarios [10,11]. In this context, incremental models were used to predict task IDs in an online continual learning environment, allowing the appropriate frozen network to be retrieved for predictions on a given task. However, in those instances, incremental models were not utilized to predict the residuals of the batch models as they are in our case.

3 Method

This work proposes a method to address the challenges of deploying machine learning models in dynamic streaming environments where multiple concept drifts are expected. The setting assumes the availability of a considerable amount of historical streaming data, which is used to train an initial (base) machine learning model. The data comprises multiple multivariate streams, each identified by a unique combination of categorical features. For example, the store-id and product-id combination represents the demand for a specific product at a particular store [24], while the Point of Connection (PoC) and trading hour combination represents the energy demand during a specific hour at a given PoC [25].

Once the base model is trained on historical data, it is deployed to operate on future streaming data. The deployment environment is characterized by concept drifts (i.e., distributional changes in the data over time). Since the data contains numerous streams, frequent retraining of the base model to adapt to these drifts is computationally infeasible. Instead, retraining is expected to occur rarely and at intervals much longer than the rate at which the data changes. Our proposed method aims to enable the model to operate effectively in a dynamic environment by utilizing mechanisms that allow it to adapt to changing trends in the data without the need for frequent retraining. This ensures that the model remains effective despite the presence of concept drifts, supporting its application in large-scale, multi-stream data scenarios.

3.1 Batch Learned Model

For a data stream instance i with features $X^i = \{x_1, x_2, \ldots, x_n\}$ and label y^i, we denote the identifier of a particular stream from which i comes as S. Such an identifier typically consists of one or more categorical features: $S = \{s \in X_i \mid s \text{is a stream identifiable categorical feature}\}$, making it important to learn an encoding for s, denoted E, such that $\mathrm{E}(s) \in \mathbb{R}^d$. The encoded identifiers can be concatenated with remaining features: $X^i \setminus S = \{x \in X^i \mid x \notin S\}$ to yield:

$$q^i = [E(s_1), E(s_2), \ldots, E(s_{|S|}),\ X^i \setminus S] \ where\ q^i \in \mathbb{R}^{|S| \times d + |X^i \setminus S|} \quad (1)$$

and passed into an MLP to produce z^i by its last layer:

$$z^i = MLP(q^i) \quad (2)$$

The z^i is ultimately passed into a perceptron R to yield the final regression value, namely $\hat{y^i} = R(z^i)$. The whole setup can be summarized into: $F_{batch}(X^i) = R(MLP(q^i))$, and is visualized in 1.

Root mean squared loss of $\hat{y^i}$ is back-propagated to train the F_{batch} which includes R, MLP and the embeddings. Using the network structure described above enables us to learn embeddings that group similar streams closer together while pushing dissimilar ones further apart. Additionally, it allows us to extract features from various network layers. This structure also provides us with a prediction head that has been trained on a substantial amount of historical data.

In a streaming scenario where models need to quickly adapt to concept drifts in underlying data, the above network performs sub-optimally since it is not able to adapt to new trends once deployed. Re-training the entire network whenever a change is detected is not feasible, considering the streaming nature of the incoming data and the number of such streams.

3.2 Incremental Model

SL offers an attractive alternative where models are learned incrementally from a stream of data while adjusting to concept drifts online [12]. Also, their test-then-train paradigm where testing happens prior to training on a given instance [12] offers incremental models with efficient use of available data. This allows one

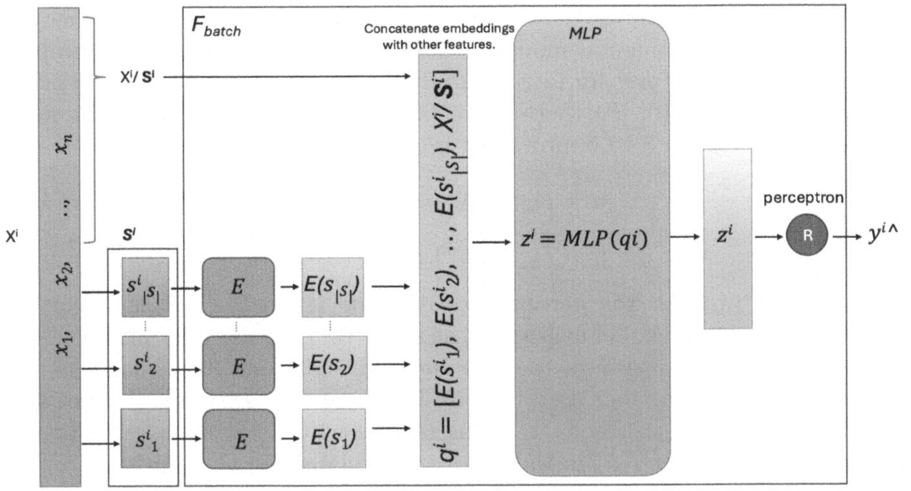

Fig. 1. Architecture of F_{batch}

to use the training and validation data utilized to train F_{batch} for initializing an incremental model prior to deployment. The above advantages allow the incremental model to constantly adjust to changes in underlying data even once they are deployed.

Incremental model F_{inc} can be leveraged in many different ways on our setting:

a) train F_{inc} from scratch using *training data* and continuously update the model once deployed:

$$\hat{y^i} = F_{inc}(X^i) \tag{3}$$

b) train F_{inc} from scratch using *extracted features* of *training data* via a *pre-trained network* F and continuously update the model once deployed:

$$\hat{y^i} = F_{inc}(F(X^i)) \tag{4}$$

c) train F_{inc} to predict the *residuals* of a batch learned model $F_{batch}(X^i)$. Here F_{inc} can be trained using the *initial features* X^i as in Eq. 3:

$$\hat{y^i} = F_{batch}(X^i) + F_{inc}(X^i) \tag{5}$$

Incremental learner F_{inc} in 4 can be trained on the q^i (the embeddings concatenated with the other features) and z^i (the final layer output of the MLP) to yield:

$$\hat{y^i} = F_{inc}(q^i) \tag{6}$$

and

$$\hat{y^i} = F_{inc}(z^i) \tag{7}$$

Although using embeddings that relate to a stream, combined with other features to train an incremental model, is novel to our knowledge, the concept of utilizing last-layer features for incrementally training a model has been explored in previous research [10]. Furthermore, q^i and z^i can be used to train the incremental learner F_{inc} in 5 to improve the predictions of $F_{batch}(X)$:

$$\hat{y} = F_{batch}(X^i) + F_{inc}(q^i) \tag{8}$$

$$\hat{y} = F_{batch}(X^i) + F_{inc}(z^i) \tag{9}$$

For simplicity, we omit the per-instance notation (i) in the rest of the paper. Since the incremental model utilizes a test-then-train sequence for model testing and training on each instance, the *training* and *validation* data that was used to train $F_{batch}(X)$ is further used to train the incremental model. Current stat-of-the-art random forest-based streaming regressor SOKNL [26] is considered as the incremental learner F_{inc} in this work. As explained in the previous work, it contains the inbuilt drift adaptability of ArfReg and the implicit adaptability of the centroids at the leaves. This allows SOKNL to produce superior results compared to ArfReg on evolving data streams [26]. In the experiments, we validate the effectiveness of approaches: 3, 5, 6, 7, 8 and 9.

Table 1. Train, Validation, and Test splits on each dataset.

	Train	Validation	Test
Demand Forecasting	57,750	60,060	32,340
NZ Energy Pricing	1,582,799	262,800	1,255,067

4 Experiments

The experiments attempt to validate the effectiveness of training the incremental learner F_{inc} using the *initial features* X (Eq.3) the *embeddings concatenated with the other features*: q (Eq.6), the *final layer output of the MLP*: z (Eq.7). They also attempt to validate the effectiveness of incremental learner F_{inc} predicting the residuals of the batch learner F_{batch}. Here F_{inc} can also be trained using X, q and z as per Eqs. 5, 8 and 9. The batch learner F_{batch} is considered as the baseline in the experiments. The above approaches were compared on two demand prediction datasets using a current state-of-the-art incremental regressor.

NZ Energy Pricing. dataset represents the real-time, time-series and streaming data of energy prices in New Zealand [25]. The data is sourced from the Electricity Market Information (EMI) website maintained by the New Zealand government. In this dataset, each instance contains energy price information for half an hour:

- *Trading Date*: yyyy-mm-dd formatted trading date.
- *Trading Period*: half an hour trading period indices, usually 1 to 48
- *Point of Connection (PoC)*: the point of connection on the grid.
- *Island*: a binary value indicating the North Island or South Island of NZ.
- *IsProxyPriceFlag*: boolean indicating whether the price is a proxy price. This can happen when the PoC is disconnected from the grid.
- *DollarsPerMWh*: the published dispatch energy price in dollars per MWh.

for roughly two years (1/11/2022 - 15/10/2024) across 91 PoCs. In the original dataset *Trading Period* and *Trading Date* were normalized to $[0, 2\pi]$ via $2\pi\frac{X_i - X_{min}}{X_{max} - X_{min}}$ and Sine and Cosine functions were applied to the normalized values yielding two more columns for each *Trading Period* and *Trading Date*. The original data were kept unchanged to preserve the seasonal information. Since the trading system allows multiple price announcements per given trading period, the previous trading period's mean and median were also included in the current period. We used 2022 data for training, 2023 data for validation, and 2024 data for testing. Here *Trading Period* and *PoC* were considered as categorical features to learn embeddings.

Demand Forecasting Kaggle Competition Dataset. is a Kaggle competition dataset published by one of the largest retail chains in the world for building a forecasting model to predict the sales for 28 Stock Keeping Units (SKU)s in its portfolio at 76 different stores using historical sales data such as: *week*, *store_id*, *sku_id*, *total price*, *base price*, *is featured sku*, *is displayed sku* and *units sold* (target variable) for the past 3 years on a week-on-week basis. Weekly sales, promotional information, and past weeks' sales are also available for each SKU-store combination. Here, the first year's data was used for training, the second year's data for validation, and the last year's data for testing. F_{batch} used categorical features *sku_id* and *store_id* to learn embeddings.

Table 1 highlights the number of instances in training, validation, and testing splits in each of the datasets. Training and validation data were used at the training step of the batch model F_{batch} while it was tested on test data. The test-then-train paradigm allows incremental models to leverage even the test data for model updates since testing the model happens prior to the model training using a given instance.

The output dimensionality of each of the embedders was set to 5 ($k = 5$) in F_{batch}. C was set to 2, since F_{batch} learned embeddings for two categorical features on each dataset. The input and output dimensions of each of the MLP layers are $(2 \times 5 + o, 512)$, $(512, 512)$, $(512, 256)$, $(256, 64)$. Here o is the size of all the other input features except the interested categorical features on each dataset. Each layer of the MLP had a dropout of 0.5 and a ReLU activation. The final output of the MLP was fed into a perception which predicted the final value.

In the previous work, SOKNL performed best with the Fast Incremental Regression Trees with Drift Detection (FIRT-DD) [16] as the base learner with *VarianceReductionSplitCriterion*, a grace period of 50 instances, split confidence of 0.01, and 100 trees [26]. The CapyMOA [8] implementation of SOKNL uses the above best hyperparameter configurations and other hyperparameter settings as its default parameters. We used the same default hyperparameters as in CapyMOA. Each incremental approach was run five times with different random seeds. All the experiments used the same F_{batch} for each dataset.

The Evaluation Metric. Coefficient of Determination, which is also known as R-squared (R^2) [28], is favored in literature against Mean Squared Error (MSE) or Mean Absolute Error (MAE) when comparing different algorithms over multiple datasets for regression [4]. R^2 is formulated as: $R^2 = 1 - \frac{\sum_{i=1}^{n}(m(x_i)-y_i)^2}{\sum_{i=1}^{n}(y_i-\bar{y})^2}$ where $m(x_i)$ is the prediction given by model m for the instance x_i, y_i and $\bar{y}$ represents mean of all true values y_i. R^2 ranges from $-\infty$ to 1.0. We also show Root Mean Squared Error (RMSE) which ranges from 0 to ∞ on periodic plots to get an understanding of the loss of the learning algorithm.

4.1 Results

Table 2 shows average R^2 and RMSE for all the incremental approaches against the batch learned Neural network F_{batch} on both Demand Forecasting and NZ Energy Pricing datasets. The results are rounded to two decimal places. We can see that $F_{batch}(X) + F_{inc}(X)$ has the lowest RMSE with the highest R^2 on the Demand Forecasting Kaggle Competition dataset. On the same dataset residual predicting approaches: $F_{batch}(X) + F_{inc}(X)$, $F_{batch}(X) + F_{inc}(q)$, and $F_{batch}(X) + F_{inc}(z)$ always seem to produce better results than $F_{batch}(X)$ with very low variance both in RMSE and in R^2. Using the initial features X seems to yield better results when predicting residuals on the Demand Forecasting Kaggle Competition dataset. On this dataset, just using the incremental learner F_{inc}

Table 2. R^2 and RMSE for each incremental approach on each dataset against baseline $F_{batch}(X)$. X: Initial input features, q: Categorical embeddings concatenated with rest of the features. z: Final layer logits of the MLP. The results are rounded to two decimal places. **Higher** R^2 is better. **Lower** RMSE is better.

Learner	Demand Forecasting		NZ Energy Pricing	
	RMSE	R^2	RMSE	R^2
$F_{batch}(X)$	31.86 ± 0.00	0.64 ± 0.00	121.11 ± 0.00	0.59 ± 0.00
$F_{inc}(X)$	39.64 ± 9.09	0.42 ± 0.30	43.28 ± 1.43	0.95 ± 0.00
$F_{inc}(q)$	38.23 ± 9.70	0.45 ± 0.31	$\mathbf{34.16 \pm 0.18}$	$\mathbf{0.97 \pm 0.00}$
$F_{inc}(z)$	49.85 ± 1.67	0.13 ± 0.06	63.47 ± 0.44	0.89 ± 0.00
$F_{batch}(X) + F_{inc}(X)$	$\mathbf{28.79 \pm 0.07}$	$\mathbf{0.71 \pm 0.00}$	42.33 ± 1.29	0.95 ± 0.00
$F_{batch}(X) + F_{inc}(q)$	29.05 ± 0.15	0.70 ± 0.00	35.21 ± 0.61	$\mathbf{0.97 \pm 0.00}$
$F_{batch}(X) + F_{inc}(z)$	30.82 ± 0.02	0.67 ± 0.00	61.09 ± 0.77	0.90 ± 0.00

(irrespective of the input features) seems to produce suboptimal results than the baseline F_{batch}.

On the other hand, baseline F_{batch} produced the worst results than all the other approaches on the NZ Energy Pricing dataset. Since F_{batch}'s performance is poor in this dataset, most of the predictive performance seems to come from the incremental models for the residual prediction approaches. The size of the dataset (over 1M instances) could also contribute to the better performance of the incremental learners since they have more instances to train. Here, the approaches that use categorical embeddings concatenated with the rest of the features: $F_{inc}(q)$, $F_{batch}(X) + F_{inc}(q)$ produced better results than the other approaches.

Considering the results on both datasets, we can see that predicting the residuals by the incremental models improves the accuracy when the base batch learner is better than the incremental models. In other situations, where batch learner is worse than the incremental model, most of the performance comes from the incremental model. In that case, the batch learner does not seem to make a meaningful contribution towards the final prediction. Though categorical embeddings concatenated with the rest of the features (q) seem to improve or not hinder the performance of the incremental models compared to all the features (X), using final logits (z) of the MLP seems to hinder the performance of the incremental model compared to using X or q.

Performance Evaluation Over Time. was conducted to further understand the behavior of different approaches against the batch learned Neural network F_{batch} on the different periods of the stream. The periodic statistics were collected on each dataset for each approach. On the Demand Forecasting dataset, these statistics were collected every 500 instances, while on the NZ Energy Pricing dataset, the statistics were collected for every 25000 instances. The results are averaged over five runs of each approach.

Figures 2 and 3 further explain these results showing the periodic evolution of each approach on Demand Forecasting and NZ Energy Pricing datasets. In a given figure, the top plot shows the periodic R^2 scores for the selected best approaches, and the middle plot shows the periodic RMSE for the same. The bottom plot shows the average prediction of multiple runs by the best incremental approach, prediction by the $F_{batch}(X)$, and the ground truth for each instance on a given dataset. Each method's legend shows its cumulative R^2 or RMSE and standard deviation.

When one considers the periodic statistics on the NZ Energy Pricing dataset in Fig. 2, it is clear that the batch learner performs badly on periods where there are drifts in the underlying distribution. This is evident by $F_{batch}(X)$ having high RMSE (middle plot) and low predictive performance (top and bottom plots) in periods where we see a concept drift in energy pricing. In those periods, incremental models seem to be more responsive to the latest trends in the data since they can adjust to the latest trends. On the other hand, when one looks at ground truth values (bottom plot) of the Demand Forecasting dataset in Fig. 3,

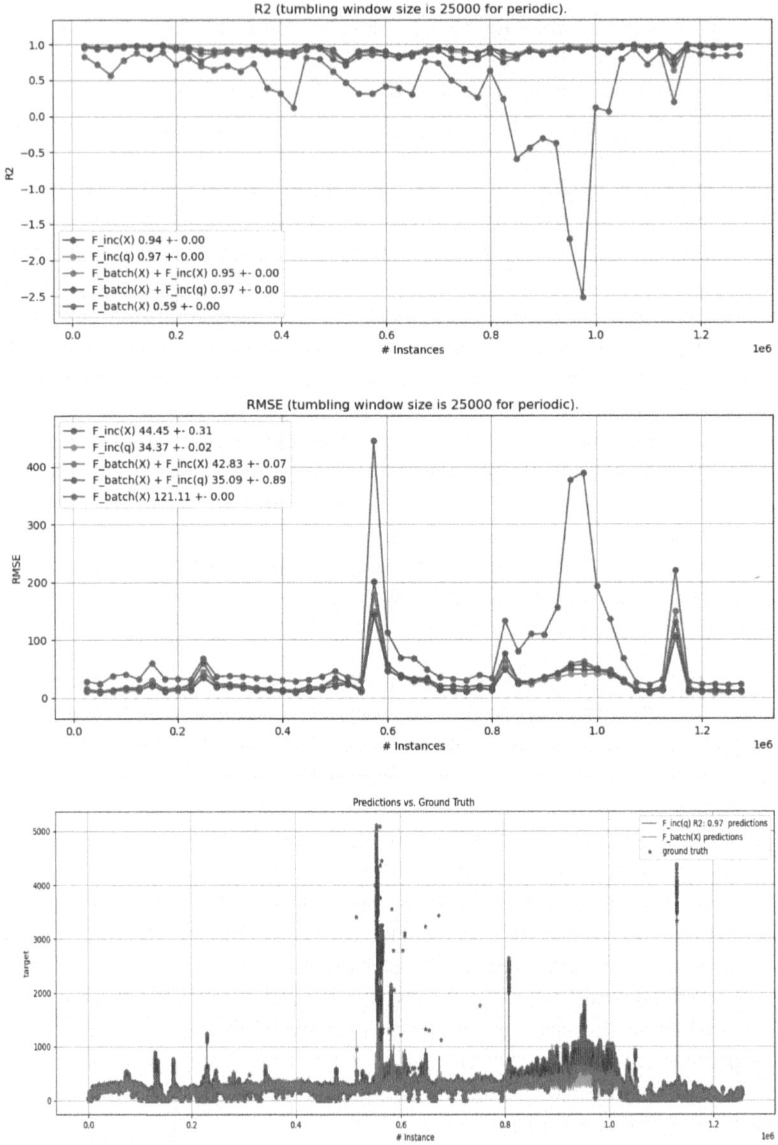

Fig. 2. NZ Energy Pricing Dataset. The **top** plot shows the periodic R^2 $(-\infty, 1.0]$ scores for the selected best approaches, and the **middle** plot shows the periodic RMSE $[0, \infty)$ for the same. The **bottom** plot shows the average prediction of multiple runs by the best incremental approach, prediction by the $F_{batch}(X)$, and the ground truth for each instance. Each method's legend shows its cumulative R^2 or RMSE and standard deviation.

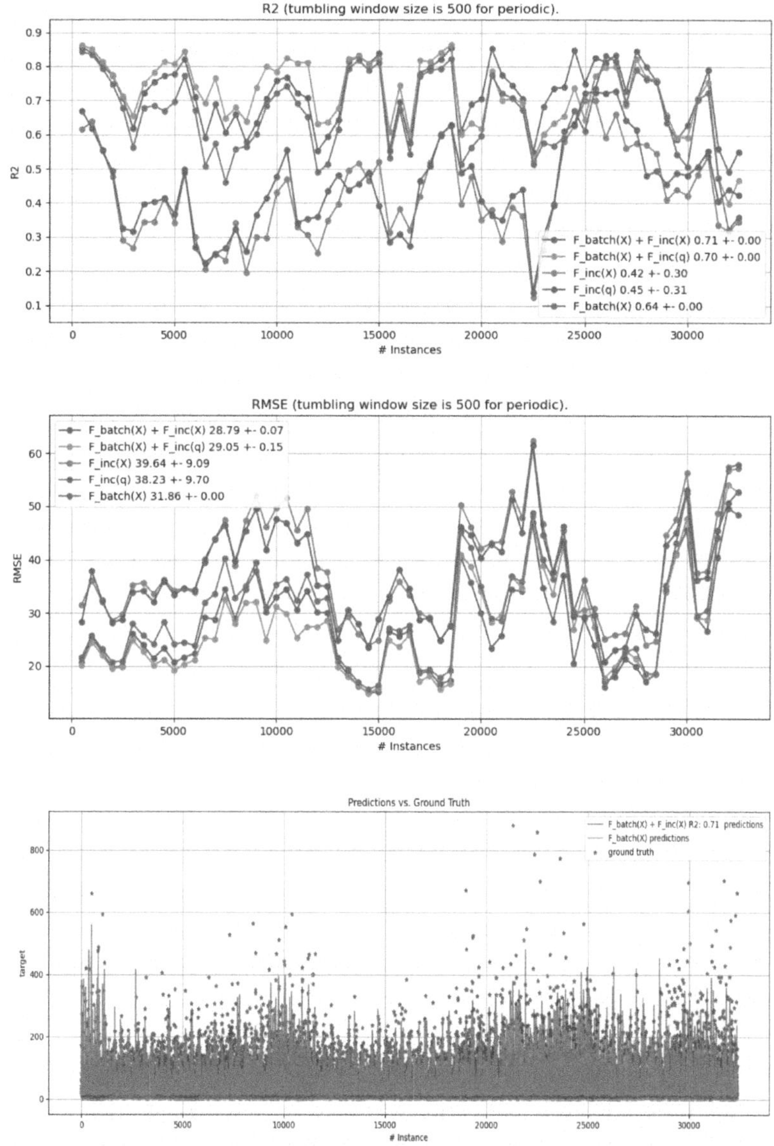

Fig. 3. Demand Forecasting Dataset. The **top** plot shows the periodic R^2 $(-\infty, 1.0]$ scores for the selected best approaches, and the **middle** plot shows the periodic RMSE $[0, \infty)$ for the same. The **bottom** plot shows the average prediction of multiple runs by the best incremental approach, prediction by the $F_{batch}(X)$, and the ground truth for each instance. Each method's legend shows its cumulative R^2 or RMSE and standard deviation.

two points can be observed compared to NZ Energy Pricing data. First, concept drifts in the data are less obvious (possible concepts around 10000, 15000, 25000, and 30000) compared to the NZ Energy Pricing dataset. Second, there are fewer instances available for a given possible concept. In such situations, even though the incremental model by itself failed to surpass F_{batch}, possibly due to scarcity of labeled data, they can improve on the predictions of F_{batch} when predicting residuals.

5 Conclusions

The paper proposes a novel learning paradigm that uses historical streaming data to learn embeddings for multiple streams. These learned embeddings are subsequently used to train an incremental regressor, which can function either as a standalone predictor or as a residual predictor of a pre-trained network. Experimental results indicate that using these embeddings enhances predictive power on datasets containing many streams without negatively impacting performance in other scenarios. The experiments also confirm that the residual prediction by the incremental learner further improves the final prediction when the base batch learner is performing well. On the other hand, in residual prediction, most of the predictive power comes from the incremental learner when the batch learner is performing poorly. In future work, we aim to compare our approach with the current state-of-the-art online time series forecasting method, OneNet, and other online multi-streaming techniques [18,31]. Additionally, we plan to explore various aspects of the proposed method using the latest synthetic data generators [22,26]. Furthermore, we would like to conduct a deeper investigation of the performance of each individual stream.

Acknowledgments. This study was funded by the KEEPER project.

Disclosure of Interests. The authors have no competing interests to declare that are relevant to the content of this article.

References

1. Anava, O., Hazan, E., Mannor, S., Shamir, O.: Online learning for time series prediction. In: Conference on Learning Theory, pp. 172–184. PMLR (2013)
2. Bifet, A., Gavalda, R.: Learning from time-changing data with adaptive windowing. In: SIAM (SDM), pp. 443–448. SIAM (2007)
3. Bifet, A., Gavaldà, R., Holmes, G., Pfahringer, B.: Machine Learning for Data Streams: With Practical Examples in MOA. MIT Press (2018)
4. Chicco, D., Warrens, M.J., Jurman, G.: The coefficient of determination R-squared is more informative than SMAPE, MAE, MAPE, MSE AND RMSE in regression analysis evaluation. Peerj Comput. Sci. **7**, e623 (2021)
5. Choudhary, A., Jha, P., Tiwari, A., Bharill, N.: A brief survey on concept drifted data stream regression. In: SocProS, pp. 733–744 (2021)

6. Gomes, H.M., Barddal, J.P., Enembreck, F., Bifet, A.: A survey on ensemble learning for data stream classification. ACM (CSUR) **50**(2), 1–36 (2017)
7. Gomes, H.M., Barddal, J.P., Ferreira, L.E.B., Bifet, A.: Adaptive random forests for data stream regression. In: ESANN (2018)
8. Gomes, H.M., et al.: CapyMOA: efficient machine learning for data streams in python. arXiv preprint arXiv:2502.07432 (2025)
9. Gomes, H.M., Montiel, J., Mastelini, S.M., Pfahringer, B., Bifet, A.: On ensemble techniques for data stream regression. In: 2020 International Joint Conference on Neural Networks (IJCNN), pp. 1–8. IEEE (2020)
10. Gunasekara, N., Gomes, H., Bifet, A., Pfahringer, B.: Adaptive neural networks for online domain incremental continual learning. In: DS, pp. 89–103. Springer (2022)
11. Gunasekara, N., Gomes, H., Bifet, A., Pfahringer, B.: Adaptive online domain incremental continual learning. In: ICANN 2022, pp. 491–502. Springer (2022)
12. Gunasekara, N., Pfahringer, B., Gomes, H.M., Bifet, A.: Survey on online streaming continual learning. In: Elkind, E. (ed.) IJCAI, pp. 6628–6637. IJCAI (2023)
13. Halstead, B., et al.: Analyzing and repairing concept drift adaptation in data stream classification. Mach. Learn. 3489–3523 (2022)
14. Han, Z., Zhao, J., Leung, H., Ma, K.F., Wang, W.: A review of deep learning models for time series prediction. IEEE Sens. J. **21**(6), 7833–7848 (2019)
15. Hoeffding, W.: Probability inequalities for sums of bounded random variables. The Collected Works of Wassily Hoeffding, pp. 409–426 (1994)
16. Ikonomovska, E., Gama, J., Džeroski, S.: Learning model trees from evolving data streams. Data Min. Knowl. Discov. **23**(1), 128–168 (2011)
17. Kovacs, A., Bogdandy, B., Toth, Z.: Predict stock market prices with recurrent neural networks using NASDAQ data stream. In: SACI, pp. 000449–000454. IEEE (2021)
18. Li, Z., et al.: A multi-stream feature fusion approach for traffic prediction. IEEE Trans. Intell. Transp. Syst. **23**(2), 1456–1466 (2020)
19. Liu, C., Hoi, S.C., Zhao, P., Sun, J.: Online ARIMA algorithms for time series prediction. In: Proceedings of the AAAI Conference on Artificial Intelligence, vol. 30 (2016)
20. Lobo, J.L., Ballesteros, I., Oregi, I., Del Ser, J., Salcedo-Sanz, S.: Stream learning in energy IoT systems: a case study in combined cycle power plants. Energies **13**(3), 740 (2020)
21. Oprea, S.V., Bâra, A., Diaconita, V.: Sliding time window electricity consumption optimization algorithm for communities in the context of big data processing. IEEE Access **7**, 13050–13067 (2019)
22. Pham, Q., Liu, C., Sahoo, D., Hoi, S.: Learning fast and slow for online time series forecasting. In: The Eleventh International Conference on Learning Representations (2023)
23. Read, J., Žliobaitė, I.: Learning from data streams: an overview and update (2023)
24. S, A.R.V.: Demand Forecasting—kaggle.com. https://www.kaggle.com/datasets/aswathrao/demand-forecasting. Accessed 31 Oct 2024
25. Sun, Y., Gomes, H.M., Pfahringer, B., Bifet, A.: Real-time energy pricing in New Zealand: an evolving stream analysis. arXiv preprint arXiv:2408.16187 (2024)
26. Sun, Y., Pfahringer, B., Gomes, H.M., Bifet, A.: SOKNL: a novel way of integrating k-nearest neighbours with adaptive random forest regression for data streams. Data Min. Knowl. Discov. **36**(5), 2006–2032 (2022)
27. Wen, Q., et al.: OneNet: enhancing time series forecasting models under concept drift by online ensembling. Adv. Neural Inf. Process. Syst. **36** (2024)

28. Wright, S.: Correlation and causation. J. Agric. Res. **20**(7), 557 (1921)
29. Wu, H., Elizaveta, D., Zhadan, A., Petrosian, O.: Forecasting online adaptation methods for energy domain. Eng. Appl. Artif. Intell. 106499 (2023)
30. Zhou, et al.: Informer: beyond efficient transformer for long sequence time-series forecasting. In: AAAI, vol. 35, pp. 11106–11115 (2021)
31. Zhou, M., Lu, J., Song, Y., Zhang, G.: Multi-stream concept drift self-adaptation using graph neural network. IEEE Trans. Knowl. Data Eng. **35**(12), 12828–12841 (2023)

Two-in-One Models for Event Prediction and Time Series Forecasting. Comparison of Four Deep Learning Approaches to Simulate a Digital Patient Under Anesthesia

Quentin Victor[1], Ianis Clavier[1], Hugo Boisaubert[1], Fabien Picarougne[1], Corinne Lejus-Bourdeau[2,3], and Christine Sinoquet[1](✉)

[1] LS2N/UMR CNRS 6004, Nantes University, Nantes, France
christine.sinoquet@univ-nantes.fr
[2] LE SiMU, Nantes University, Nantes, France
[3] Nantes University Hospital, Nantes, France

Abstract. The digitization of data in hospitals has enabled the collection of mixed temporal data for surgical procedures. These datasets describe patients through an event trace and a multivariate time series. Leveraging this digitized data presents a unique opportunity to develop innovative anesthesia training methods, enabling real-time scenario generation driven by machine learning models.

This paper aims to develop a model that (i) simulates the evolution of the physiological variables of a digital patient under anesthesia, in response to medical actions, and (ii) predicts the actions performed by the virtual medical team, with the simulator user being the sole human agent executing actions through the simulator interface. The challenge lies in achieving two goals within a single model: modeling asynchronous data streams for time series forecasting and event prediction. A key difficulty in our medical case is the sparsity of the event sequences.

We found no models addressing the combined prediction of event-contextualized time series and time-series-contextualized events. To bridge this gap, we explore four deep learning architectures that we have tailored for unsupervised representation learning and joint prediction tasks: (i) a stacked LSTM (Long Short-Term Memory) network, (ii) a Graph Neural Network (GNN), (iii) the Vanilla Transformer, and (iv) its variant, the Informer.

We evaluate these models on a real-world anesthesia dataset, focusing on overall predictive performance and computational efficiency. Our proof-of-concept study demonstrates that the Transformer outperforms the three other models which show significant shortcomings in predicting medical events.

Keywords: Deep joint modeling · Mixed asynchronous data · Time series prediction · Event prediction · Embedding · Simulation · Digital patient

G. Krempl et al. (Eds.): IDA 2025, LNCS 15669, pp. 373–388, 2025.
https://doi.org/10.1007/978-3-031-91398-3_28

1 Introduction

The present work evaluates and compares four deep learning models customized to model jointly time series and event temporal data for the double purpose of event-based contextualized time series forecasting and time series-based contextualized event prediction. Our framework is the context of a digital patient undergoing anesthesia.

Digital simulation-based training has revolutionized medicine over the past decade. However, anesthesia training has lagged behind this trend, with high-fidelity mannequin-based simulation remaining the predominant method used in hospitals. Meanwhile, the digitization of data in hospitals has enabled the collection of mixed temporal data for surgical procedures. These datasets typically comprise an event trace and a multivariate time series (MTS). Harnessing these digitized data offers an unprecedented opportunity to develop innovative anesthesia training methods, allowing for real-time scenario generation powered by machine learning models. Such models promise greater realism and variety in training scenarios.

The short-term objective of this paper is two-fold. We aim to design a model to simulate the evolution of a digital patient's physiological parameters in response to medical actions during surgery. During a training session, as the sole human member of the medical team, the user, an anesthesia resident, will perform anesthesia procedures (*e.g.*, intubation, administration of an analgesic) using the digital simulator interface. In contrast, we need to forecast all other actions throughout the surgery (*e.g.* pneumoperitoneum inflation, prosthesis placement), as they are performed by the rest of the medical team, which is virtual. The model to be built must therefore predict the next medical action to occur that is not performed by the trainee, within the context of the ongoing surgery. This challenge requires joint modeling of mixed asynchronous data streams to achieve these two goals: time series forecasting and event prediction. A critical obstacle in our medical framework lies in the sparsity of the event sequences.

In the last decades, the analysis of events [1] and time series [2] has attracted substantial attention from distinct communities, particularly for modeling with a forecasting purpose. Surpringly, despite the potential unlocked by the ubiquity of sensors in modern life, we have found no models that address the joint prediction of event-contextualized time series and time-series-contextualized events.

To tackle this gap, we have specifically tailored four deep learning architectures for the purposes of unsupervised representation learning and joint prediction tasks. These customized models include a stacked LSTM (Long Short-Term Memory), a Graph Neural Network (GNN), the Vanilla Transformer, and one of its variants, the Informer. In a proof of concept focused on the medical simulation problem that motivated this work, we explore the applicability of these models.

2 Related Work

The past two decades have seen a surge in research on the joint modeling of time series and *time-to-event* data in the very specific area of survival analysis. Other works are divided into those utilizing events to provide context for time series forecasting, and those using contextual time series data for event prediction.

Event-Contextualized Time Series Forecasting. Various approaches have explored improving time series forecasting by integrating continuous exogenous variables as additional time series. These approaches span statistical techniques (*e.g.*, [3,4]) and neural network-based models [5]. However, integrating discrete event-related variables into time series forecasting frameworks remains an under-explored area. The approaches outlined in [6] and [7] are hindered by an overly specialized focus, which hampers their reuse in other applicative frameworks. Methods that rely on extreme, highly specific events to boost time series forecasting face similar limitations (*e.g.*, [8]). Recently, the model described in [9] uses contextual variables to synchronize a Hawkes temporal point process with an autoregressive first-order hidden Markov model; another recent work developed an encoder-decoder built on stacked Recurrent Neural Networks [10].

Time Series-Contextualized Event Prediction. Multivariate temporal point processes (TPPs) are widely used for event prediction tasks [11]. The fundamental concept behind all TPPs is the conditional intensity function, which quantifies event occurrence rates based on past events. Extensions of standard TPP models have been developed in survival data analysis to integrate time-varying covariates (*e.g.*, [12]). The work reported in [13] incorporates context into the intensity function of a TPP through a linear combination of past time-series values. The method in [14] implements the intensity function of a TPP by combining two Recurrent Neural Networks: one aligns with an event history, and the other aligns with the data points of a time series.

3 Methodology

We have specifically tailored four deep learning architectures to explore their potential as unified frameworks for unsupervised representation learning of hybrid asynchronous data, and joint contextualized prediction tasks.

3.1 Handling Hybrid Asynchronous Data

Input Data. The data for our modeling problem consist of "anesthetic profiles" $\{S, E\}$. The timeframe for the MTS $S = [s_1, s_2, \ldots, \ldots s_{\ell_S}]$ of length ℓ_S and dimension d_v, the number of variables, is discrete: the observations in S are regularly sampled, with $\tau_t \in \mathbb{R}^+$ denoting the continuous timestamp of observation s_t at discrete timestamp t. In the continuous timeframe, the trace of timestamped events $E = [(e_1, t_1), \cdots (e_{\ell_E}, t_{\ell_E})]$ is defined over a finite set of n_e event categories, $\mathscr{C}$. The same event may appear multiple times in E. The events

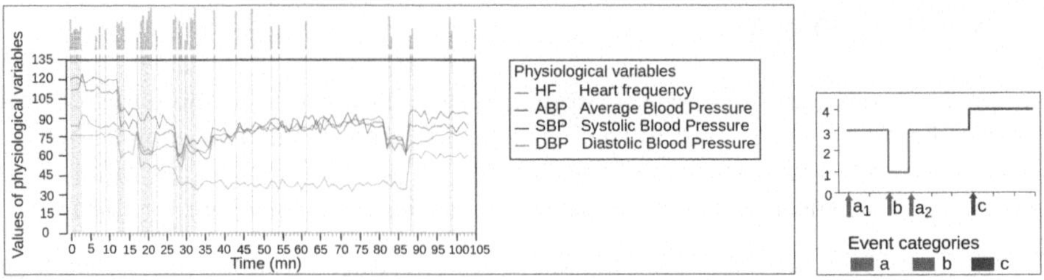

(a) (b)

Fig. 1. (a) Anesthetic profile. The events are marked in red in the upper part of the subfigure. (b) Sparse event trace converted into a piecewise plateau time series. (Color figure online)

in E may impact the dynamics of S. Figure 1(a) presents an illustration of an anesthetic profile.

We first synchronize the trace of events E on its associated MTS S. We consider ϵ, the "No Event" category. We assume that at most one event can occur in time interval $[\tau_t, \tau_{t+1}[$. If it exists, this single event is aligned with t; otherwise, ϵ is aligned with t. We thus obtain a univariate categorical time series Z defined on $\mathscr{C} \cup \{\epsilon\}$, of same length as S. We then randomly assign an integer value from $[1, n_e]$ to each event category in $\mathscr{C}$, we assign 0 to the ϵ category, and we map Z to $[0, n_e]$, to obtain the novel univariate time series Z^{num}.

Piecewise Plateau Time Series. In the **PWP** modality, we address the sparsity issue by converting Z^{num} into a piecewise plateau time series, Z^{pwp} (see Fig. 1(b)). Normalization to zero mean and unit standard deviation is applied on the $d_v + 1$ variables in S and Z^{pwp}, prior to any of the four model inputs.

Event Embedding. In the **EMB** modality, the d_v variables of S are first normalized upstream model dataflow. Neural networks offer a framework to learn embeddings of categorical data. Given a specified embedding dimension, d_e, the data in Z^{pwp} are fed into an embedding layer specifically integrated into each of the four models. The embedding is learned concurrently with the rest of the model trained for the joint prediction tasks. The resulting embedded data are then concatenated to the normalized data from S. The MTS of dimension d_v+d_e thus obtained is fed into each of the four models considered.

3.2 Models

ST-LSTM. In the class of Recurrent Neural Networks, the ST-LSTM model here refers to a stack of LSTM layers, where Long Short-Term Memory units [15] recurrently update two hidden states, c_t and h_t, designed to capture dependencies within the data. Figure 2 illustrates the operation of the model. We have chosen this model for its simplicity.

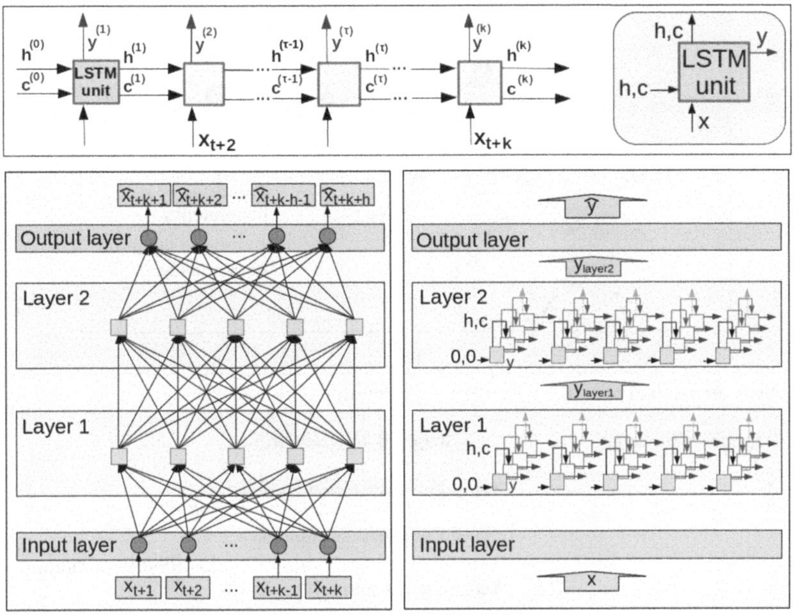

Fig. 2. Model with two stacked recurrent layers. Top section: LSTM unit (bright yellow) unfolded in time (light yellow). Bottom section: on the left, highlight on model architecture where each recurrent layer is composed of 5 LSTM units; dimension of hidden vectors h_t and c_t: 5; k: past size; h: prediction horizon. On the right, model workflow with temporal dimension. The y outputs of layer 1 serve as inputs for layer 2, in the fully connected scheme. $\hat{\mathbf{y}} = (\hat{x}_{t+k+1}, \hat{x}_{t+k+2}, \cdots, \hat{x}_{t+k+h})$. (Color figure online)

MTGNN. Deep learning models have achieved significant success in time series forecasting by more effectively capturing nonlinear temporal and spatial patterns compared to linear models [16]. However, recent advances in Graph Neural Networks (GNNs) [17] have played a pivotal role in addressing spatio-temporal graph modeling challenges (see Fig. 3(a)), particularly for time series tasks such as forecasting [18]. In the GNN class, Graph Convolutional Networks generalize the concept of convolution from regular grid data, such as images, to graph data [19]. The network progressively increases the range of nodes that influence a node representation as more graph convolutional layers are stacked. We selected the Multivariate Time-series Graph Neural Network (MTGNN) framework, specifically designed for MTS, for its ability to learn the graph structure directly from the data [20]. Figure 3(b) shows how MTGNN interweaves temporal and spatial convolutional modules, making it well-suited for modeling spatio-temporal relationships.

Transformer

Transformer. Unlike recurrent models, Transformers process entire sequences simultaneously. In a Transformer, an encoder processes the input sequence using

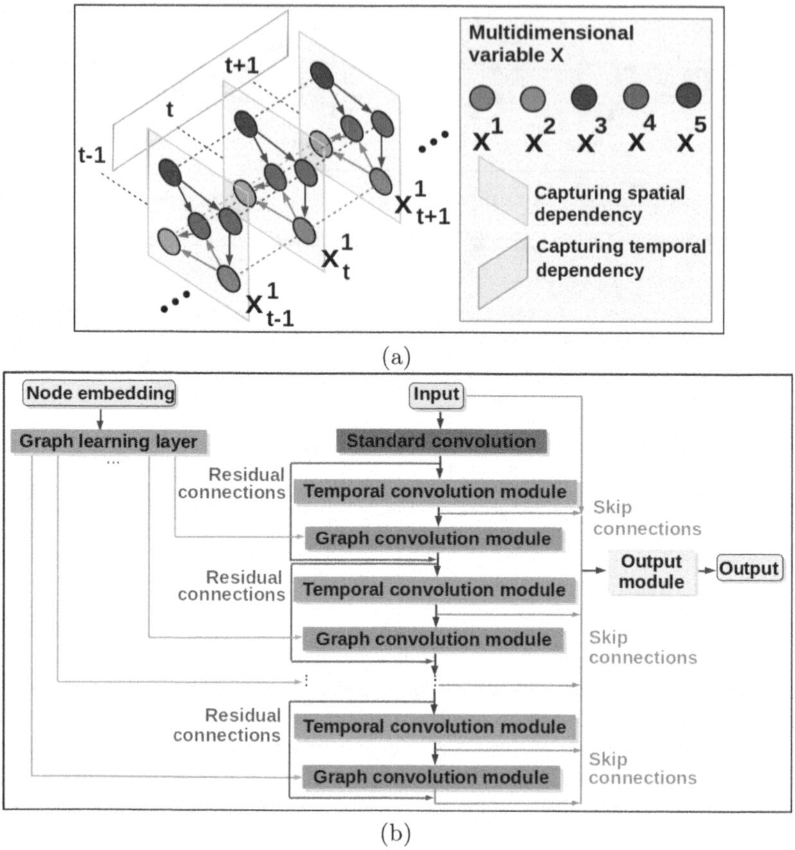

Fig. 3. (a) Graph Neural Network. Spatial dependency is captured at each timestamp t; temporal dependency is captured across the different timestamps, for each dimension of the variable. (b) MTGNN model. The graph learning layer derives a latent graph adjacency matrix, which is then provided to the graph convolution modules. Temporal and spatial convolutional modules are interwoven to capture both spatial and temporal dependencies, while residual and skip connections prevent gradient vanishing.

self-attention to generate context-rich representations, while a decoder generates the output sequence by using cross-attention to attend to the encoder's output, and self-attention to focus on its own previously generated tokens. Since Transformers lack a natural sequential structure, positional encodings are added to input embeddings to maintain the order of the sequence elements. We therefore selected the Transformer architecture, where the self-attention mechanism can alleviate performance decline caused by long-range dependencies. Among the different variants, we chose the simplest one: the seminal Vanilla Transformer [21] (see Fig. 4).

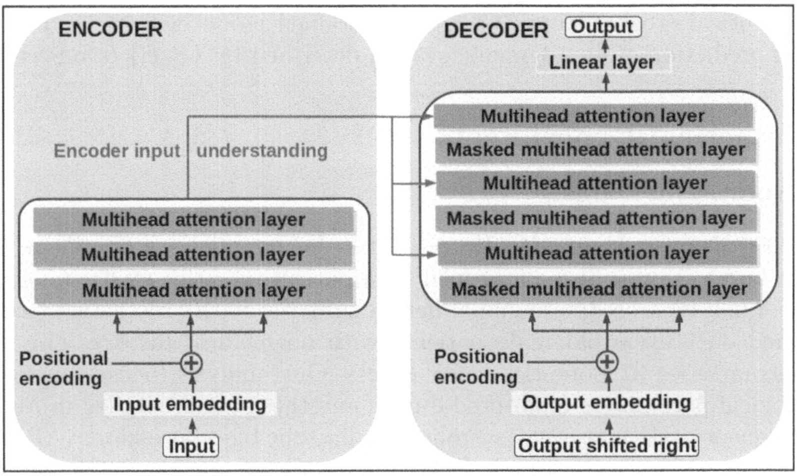

Fig. 4. Transformer. This architecture combines an encoder stack and a decoder stack, whose structures are very similar. As the data progress from layer to layer in both encoder and decoder, each multihead attention (MHA) layer pays attention to more general features in the encoder/decoder input data. The decoder predicts the next token from past tokens, based on masked-attention (all the future tokens are masked and not accessible). After each masked MHA layer of the decoder stack, an MHA layer is added. The output of the encoder stack (the encoder input understanding, or embedding) is fed into each MHA layer of the decoder stack.

Informer. The Informer was specially developed to address the scalability issue in Transformers, for the MTS forecasting task [22]. This model introduces a probabilistic self-attention mechanism that dynamically selects the most critical elements in the sequence, based on their relevance at each specific time step. The Informer model employs distillation techniques to run faster and be less memory-consuming than the seminal Transformer. Distillation in this context refers to a down-sampling technique that progressively compresses the time-series data, keeping only the most relevant information. Moreover, the generative-style decoder predicts outputs in a single forward operation, unlike the Transformer, which uses time-consuming step-by-step "dynamic decoding" prone to cumulative error propagation.

3.3 Model Learning

Each architecture will be trained on pairs $\{x, y\}$, where x and y are sequences of lengths k and h, respectively, and both have dimension $d_v + 1$ for the PWP modality, and $d_v + d_e$ for the EMB modality. Parameter k defines the length of the past context used to make an h-step ahead prediction during training, by minimizing a loss function. The loss is defined as $\mathscr{L}oss = \text{MSE} + \lambda\ \text{CCE}$, where λ compensates for the disparity in magnitude between mean squared error (MSE) and categorical cross-entropy (CCE). The trained model will be used for joint h-ahead prediction of any dynamic system described by $\{S, E\}$ (see Sect. 3.1).

4 Numerical Experiments

4.1 Anesthesia Dataset

Since 2000, Nantes University Hospital has documented an *anesthetic profile* for each surgical procedure performed on a patient. For this study, we accessed data on 1,000 laparoscopic inguinal hernia surgeries with prosthesis placement, performed on 30-year-old male patients with no medical history. This procedure encompasses 37 potential medical acts. Our analysis focuses on four key physiological parameters monitored during anesthesia, and detailed in Fig. 1(a): heart frequency (HF); average, systolic and diastolic blood pressures (ABP, SBP, DBP). In this dataset, the length of the time series ranges from 290 to 370 timestamps. We split the dataset into 3 subsets dedicated to training (60%), validation (20%) and test (20%). With hyperparameters k (past size) and h (prediction horizon) fixed to 30 and 10 respectively, we obtained 29,680 training pairs.

4.2 Implementation Details

Intensive computing resources were provided by a Bi-CPU setup (Bi-Intel Xeon Silver 4216: 2 × 16 cores (2 × 32 threads); 2.10 GHz; 32 × 64 Gb). The code for the ST-LSTM and Transformer models was adapted from the PyTorch 2.5.0 library. The MTGNN and Informer model codes were adapted from the MvTS library [23]. In addition, we used the PyTorch library to implement the EMB modality in all four models. Finally, we used the MvTS framework to load data and conduct training, validation, and testing for these four models.

4.3 Metrics for Prediction Evaluation

We selected four standard metrics for continuous variables (MSE, MAE, MAPE, SMAPE). For categorical variables, we chose the global accuracy (GA), and the arithmetic and geometric means of the F-score, denoted as AFS and GFS. Each metric is evaluated as the average performance obtained from horizons 1 to the targeted horizon. Metrics have strengths and limitations. Thus, for each of the n_s hyperparameter settings of a model, we ranked each metric from 1 to n_s (with n_s the worst performance). The model's overall score, rk, was computed as the average of rk_v and rk_e, the average ranks for continuous and categorical metrics.

4.4 Experimental Settings

Description of the Hyperparameters. Table 1 outlines the architecture-specific hyperparameters adjusted for model selection in our experiments.

Table 1. Model-specific hyperparameters adjusted for model selection.

ST-LSTM	n_ℓ: number of stacked hidden layers, n_n: number of nodes (*i.e.*, LSTM units) common to the hidden layers; defining the dimension of both hidden state vectors h_t and c_t
MTGNN	n_ℓ: total number of hidden layers, $n_{c\ell}$: number of stacked convolutional layers, n_{cc}: number of channels in the model's convolutional layers, n_{rc}: number of channels used in the residual connections of the model
Transformer and Informer	n_ℓ: total number of hidden layers in either the encoder or the decoder, d_t: embedding size used to represent the input data in a latent space

Preliminary Analysis. To assess the impact of the PWP and EMB modalities on predictive performance, we included the NOEV (No Event) modality in our experimental protocol, where only time series are input to the models. We conducted a frugal analysis prior to a more intensive grid search focused on the best event-encoding modality. In this frugal analysis, we modified one hyperparameter at a time from a reference setting, while holding all others constant for each model and modality (see Table 2). For instance, for ST-LSTM with NOEV modality, we started from $(n_\ell, n_n) = (\mathbf{2}, \mathbf{64})$ and derived the configurations $(\mathbf{2}, 128)$, $(\mathbf{2}, 256)$, $(1, \mathbf{64})$, and $(3, \mathbf{64})$. Overall, we trained 80 models.

Table 2. Hyperparameter settings for the preliminary analysis. For each architecture, the hyperparameters of the reference setting are highlighted in bold. All other settings are derived from this reference by modifying a single hyperparameter at a time.

	Modalities			
Architectures	NOE, PWP and EMB		EMB	
ST-LSTM	$n_\ell \in \{1, \mathbf{2}, 3\}$	$n_n \in \{\mathbf{64}, 128, 256\}$	$d_e \in \{\mathbf{5}, 10, 15\}$	$\lambda \in \{0.25, \mathbf{0.50}, 0.75\}$
MTGNN	$n_\ell \in \{2, \mathbf{3}, 4\}$ $n_{cc} \in \{16, \mathbf{32}, 64\}$	$n_{c\ell} \in \{1, \mathbf{2}, 3\}$ $n_{rc} \in \{\mathbf{8}, 16, 32\}$	$d_e \in \{\mathbf{5}, 10, 15\}$	$\lambda \in \{0.25, \mathbf{0{\cdot}50}, 0.75\}$
Transformer Informer	$n_\ell \in \{2, 4, \mathbf{6}, 8\}$ $d_t \in \{64, 128, 256, \mathbf{512}, 1024\}$		$d_e \in \{\mathbf{5}, 10, 15, 20, 25\}$	$\lambda \in \{0.25, \mathbf{0.50}, 0.75\}$
# models trained	ST-LSTM: 5 Transformer: 8	MTGNN: 9 Informer: 8	ST-LSTM: 9 Transformer: 14	MTGNN: 13 Informer: 14
Hyperparameters shared by the four models considered				
$k = 30$; $h = 10$; Batch size $= 32$; $n_{epochs} = 200$ (controled by early stopping); Dropout: enabled; Gradient descent optimization algorithm: Adam; Network weight initialization: Glorot Uniform; Learning rate $= 10^{-4}$				

Table 3. Hyperparameter variation range in grid search. #mt: number of models trained.

Architectures	#mt	Hyperparameter values			
ST-LSTM	81	$n_\ell \in \{2,3,4\}$	$n_n \in \{128,256,512\}$	$d_e \in \{10,15,20\}$	$\lambda \in \{0.2,0.3,0.4\}$
MTGNN	729	$n_\ell \in \{2,3,4\}$	$n_{c\ell} \in \{2,3,4\}$	$n_{cc} \in \{32,64,128\}$	$n_{rc} \in \{8,16,32\}$
		$d_e \in \{8,10,12\}$	$\lambda \in \{0.25,0.35,0.45\}$		
Transformer	81	$n_\ell \in \{3,4,5\}$	$d_t \in \{128,256,512\}$	$d_e \in \{18,20,22\}$	$\lambda \in \{0.4,0.5,0.6\}$
Informer	81	$n_\ell \in \{5,6,7\}$	$d_t \in \{256,512,1024\}$	$d_e \in \{3,5,7\}$	$\lambda \in \{0.6,0.7,0.8\}$

Refined Analysis. The preliminary analysis showed that the models generally perform better in event prediction in the EMB modality. Based on the preliminary results, we varied the hyperparameters as shown in Table 3, to run a grid search under this modality. We used the same values as in Table 2 for the hyperparameters shared across the 4 architectures. Overall, we trained 972 models.

4.5 Results and Discussion

Preliminary Analysis. Table 4 summarizes the performance evaluation, focusing on the MAPE and AFS metrics. Four key conclusions can be drawn: (i) for all models, the PWP and EMB modalities respectively worsen and improve MTS forecasting performance compared to the NOEV modality; (ii) all models, except the Transformer, show poor or failed event prediction, even under the EMB modality; (iii) under this modality, MTGNN ranks second after the Transformer; (iv) in this experimental setup, no clear trend is observed, except for ST-LSTM, in the impact of hyperparameters on MTS or event prediction performance.

Table 4. Performance ranges obtained at horizon 10 during preliminary analysis.

Architectures	Modalities				
	NOEV	PWP	EMB	PWP	EMB
	MAPE			AFS	
ST-LSTM	1.6–1.8	2.5–3.4	1.4–1.6	0.11–0.18	0.03–0.25
MTGNN	1.3–1.4	1.5–1.7	1.2–1.3	0.13–0.15	0.35–0.55
Transformer	≤0.2	0.4–2	0.01–0.1	0.78–0.99	0.90–0.96
Informer	1.5–1.6	1.8–3.7	1.9–2.0	0.08–0.16	0.31–0.56

Refined Analysis. Figure 5, which focuses on the MAPE and AFS metrics, illustrates without loss of generality the trends observed for the two prediction tasks. We can draw additional insights from the performance landscape explored by the grid search.

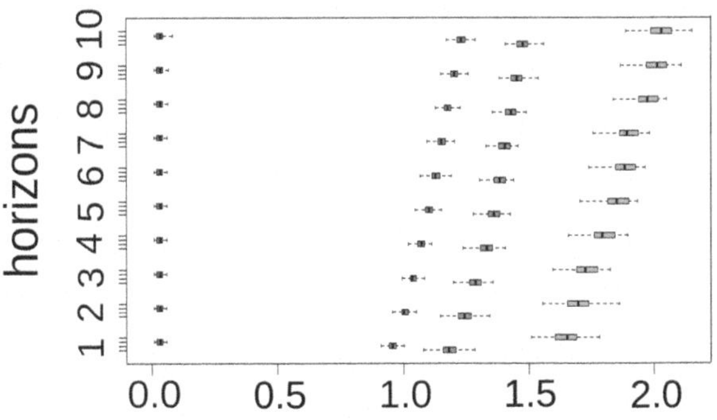

(a) Time series prediction (MAPE)

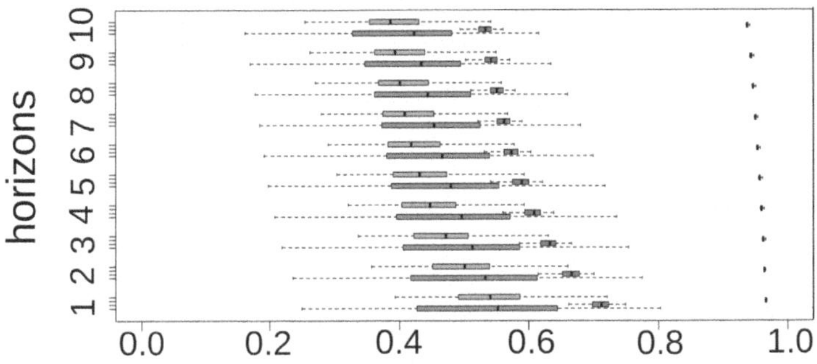

(b) Event prediction (AFS)

Fig. 5. Performance distributions across the 4 architectures and horizons, for all models trained during grid search. Performances are measured on the testing set. Boxplot display order: ST-LSTM (brown), MTGNN (pink), Transformer (blue), Informer (green). (Color figure online)

1. We observe dense distributions of very good MTS prediction performances across the four architectures. Regarding event forecasting, we observe wider distributions for ST-LSTM and the Informer, but a dense distribution for MTGNN, which are the least performing architectures. In contrast, the Transformer shows a highly concentrated distribution of excellent performances. Thus, we learn that the hyperparameter adjustment of MTGNN across 729 settings results in no diversification, let alone towards an optimal event forecasting performance, for the considered dataset. On a smaller scale of 81 settings, the diversification in event forecasting performance observed for ST-LSTM and the Informer demonstrates an inability of these models to improve for the dataset considered.
2. We identify as many distinct performance levels as there are architectures in the MTS forecasting task: all architectures show dense distributions.
3. We confirm the ranking of the architectures by decreasing MTS forecasting performance (Transformer, MTGNN, ST-LSTM, Informer), as noted in the preliminary analysis. Regarding event forecasting, ST-LSTM and the

Informer noticeably lag behind MTGNN, in turn significantly trailing the Transformer. ST-LSTM and the Informer perform similarly, with a slight edge for ST-LSTM.
4. As regards the Transformer, any hyperparameter setting tested yields high event forecasting performance, suggesting that the performance landscape is simple enough for any such setting to suffice for this model.
5. We note the outstanding stability of the Transformer for MTS and event performance through the ten prediction horizons. The relative performance decrease between horizons 1 and 10 is reduced to 4%, while it reaches 23%, 24%, and 27% for ST-LSTM, MTGNN, and the Informer, respectively.

Training Times. Table 5 outlines the ranking of ST-LSTM, MTGNN, the Transformer and the Informer by increasing training time.

Table 5. Training time distributions across the 4 architectures, for all models trained during grid search. The training set comprises data from 600 patients.

Architectures	Training time in seconds			
	Average	Median	Minimum	Maximum
ST-LSTM	1,458 ± 275 (24 h)	1,444	967	2,518
MTGNN	10,331 ± 2,363 (2.9 days)	10.159	5,905	17,827
Transformer	16,692 ± 4,347 (4.6 days)	16,868	9,184	23,738
Informer	22,092 ± 4,680 (6.1 days)	21,572	12,715	32,325

Best Models. Table 6 provides the performances of the top-10 models obtained for each of the four architectures. In line with our previous remarks, we cannot identify any trends regarding the bias of these models towards specific hyperparameter values. Figure 6 summarizes the performances obtained by the top model for each architecture. It is worth noting that the four neural models tested in this work improve contextualized MTS forecasting over the models presented in [9,10], where MAPE values are at least 3 and 4, at horizon 10, with (k,h) equal to (30, 10) on the anesthesia dataset. All MAPE values obtained from the four models we investigated are below 2.

Key Takeaways. In summary, MTS prediction is relatively straightforward for all four architectures, on the anesthesia dataset. However, the Transformer is the only architecture among the four that achieves joint performance outcomes aligned with our application objective. The spatial and temporal convolution modules of MTGNN fail to provide satisfactory performance for our future simulator. Despite being designed for fast training, the Informer is the slowest to train among the four architectures. Notably, its reported ability for long-term time series prediction does not necessarily correlate with strong short-term results.

Table 6. Ten best hyperparameter settings identified for each of the four architectures. For each architecture, underlined numbers denote the ranks of the metrics across the set of settings (the higher the worse). rk_v and rk_e: average ranks for continuous and categorical variables, respectively. rk: overall score, average of rk_v and rk_e (see Sect. 4.3)

LSTM										
n_ℓ	3	4	3	4	2	2	4	4	3	4
n_n	512	512	256	128	128	512	128	256	256	512
d_e	15	10	15	20	20	20	20	15	20	15
λ	0.2	0.2	0.2	0.2	0.3	0.4	0.3	0.2	0.2	0.3
rk	15.0	17.3	18.0	18.4	19.0	19.5	20.1	20.3	21.5	22.5
rk_v	11.0±8.7	14.3±14.7	21.8±17.6	2.7±1.7	21.3±19.6	29.3±6.8	27.5±10.7	25.5±4.4	32.3±13.5	41.8±16.6
rk_e	19.0±1.7	20.3±3.2	14.3±1.0	34.3±1.0	16.7±1.0	9.7±1.5	12.7±1.0	15.0±1.0	10.7±1.0	3.3±1.0
MSE	0.070 24	0.068 8	0.071 47	0.068 5	0.067 2	0.071 36	0.070 32	0.070 29	0.071 48	0.071 38
MAE	0.169 6	0.168 4	0.170 17	0.168 2	0.170 13	0.171 29	0.171 27	0.170 19	0.171 32	0.171 34
MAPE	1.426 6	1.472 36	1.449 17	1.407 2	1.483 48	1.452 20	1.438 13	1.460 27	1.443 15	1.501 66
SMAPE	0.397 8	0.398 9	0.397 6	0.394 1	0.400 22	0.402 32	0.402 38	0.400 27	0.402 34	0.401 29
GA	0.613 18	0.605 19	0.624 14	0.565 34	0.615 16	0.656 10	0.640 13	0.619 15	0.651 11	0.705 3
AFS	0.481 21	0.489 18	0.501 14	0.434 35	0.493 17	0.515 11	0.507 12	0.498 16	0.521 10	0.554 4
GFS	0.611 18	0.598 24	0.616 15	0.566 34	0.612 17	0.660 8	0.633 13	0.617 14	0.648 11	0.703 3
MTGNN										
n_ℓ	4	3	4	2	3	3	4	3	3	2
$n_{c\ell}$	4	3	2	2	2	2	3	4	4	3
n_{cc}	128	128	128	128	128	128	128	128	128	128
n_{rc}	32	16	32	32	32	16	16	32	8	32
d_e	10	12	12	10	12	12	12	12	10	10
λ	0.35	0.35	0.45	0.35	0.35	0.35	0.45	0.45	0.25	0.45
rk	22.1	34.8	35.5	35.9	37.6	48.6	53.5	55.4	59.5	59.6
rk_v	13.3±15.4	39.3±17.0	43.0±22.9	32.5±53.7	57.5±45.3	89.3±62.5	99.3±78.6	74.8±46.2	38.3±17.3	72.5±68.6
rk_e	31.0±14.7	30.3±8.1	28.0±2.7	39.3±9.3	17.7±5.7	8.0±4.4	7.7±6.4	36.0±26.0	80.7±46.6	46.7±34.1
MSE	0.050 1	0.052 45	0.052 38	0.051 5	0.052 41	0.053 174	0.052 94	0.052 39	0.052 50	0.050 2
MAE	0.147 4	0.148 34	0.148 49	0.147 9	0.148 56	0.148 58	0.148 44	0.148 71	0.148 25	0.148 32
MAPE	1.170 13	1.173 19	1.172 15	1.196 113	1.197 120	1.180 30	1.208 212	1.185 48	1.174 22	1.194 102
SMAPE	0.360 35	0.360 59	0.360 70	0.358 3	0.359 13	0.361 95	0.360 47	0.362 141	0.360 56	0.362 154
GA	0.825 22	0.823 35	0.824 27	0.822 47	0.826 13	0.826 10	0.827 4	0.825 21	0.820 111	0.824 25
AFS	0.563 48	0.567 21	0.565 31	0.566 29	0.567 24	0.576 3	0.568 15	0.560 66	0.566 27	0.558 86
GFS	0.797 23	0.796 35	0.796 26	0.795 42	0.798 16	0.798 11	0.800 4	0.797 21	0.792 104	0.796 29
Transformer										
n_ℓ	5	4	4	5	5	4	5	5	4	5
d_t	128	256	256	128	256	128	256	256	256	256
d_e	20	18	20	20	20	18	18	20	18	18
λ	0.6	0.6	0.4	0.5	0.4	0.5	0.6	0.5	0.5	0.4
rk	10.8	14.7	19.9	20.2	20.4	20.6	21.4	22.0	22.2	24.1
rk_v	16.3±5.7	1.8±1.0	12.3±10.3	21±3.4	10.3±6.5	27±18.8	13.5±1.7	15.8±5.1	9.3±3.3	7.8±1.0
rk_e	6.3±5.0	23.3±6.8	27.0±5.0	20.7±10.2	27.0±2.7	14.0±16.7	28.3±4.9	26.3±7.8	33.3± 6.7	38.0±4.6
MSE	10^{-5} 9	10^{-5} 3	7.10^{-6} 2	10^{-5} 16	10^{-5} 18	10^{-4} 54	10^{-5} 15	10^{-5} 14	10^{-5} 6	10^{-5} 7
MAE	0.002 15	0.002 2	0.002 6	0.003 23	0.002 13	0.003 25	0.002 11	0.003 22	0.002 7	0.002 9
MAPE	0.019 19	0.011 1	0.022 25	0.020 22	0.014 6	0.017 12	0.018 14	0.019 17	0.017 13	0.016 7
SMAPE	0.012 22	0.007 1	0.011 16	0.013 23	0.009 4	0.011 17	0.011 14	0.010 10	0.010 11	0.010 8
GA	0.979 1	0.979 31	0.979 32	0.979 9	0.979 30	0.979 33	0.979 34	0.979 35	0.979 41	0.979 37
AFS	0.937 7	0.937 18	0.937 27	0.937 28	0.937 25	0.937 2	0.937 26	0.937 24	0.937 30	0.938 43
GFS	0.978 11	0.978 21	0.978 22	0.978 25	0.978 26	0.978 7	0.978 25	0.978 20	0.978 29	0.978 34

(continued)

Table 6. (*continued*)

Informer										
n_ℓ	6	7	7	5	6	5	6	7	5	6
d_t	256	256	512	256	512	1024	256	512	512	256
d_e	3	3	5	3	3	7	3	3	3	3
λ	0.7	0.7	0.7	0.6	0.6	0.7	0.6	0.7	0.6	0.8
rk	8.3	15.2	15.3	18.3	18.8	23.2	23.5	23.9	24.5	25.4
rk_v	10.3±5.9	13.0±7.5	17±7.2	16.5±21.2	12.5±5.3	37.8±5.9	24.3±6.1	20.8±11.6	35.3±12.7	35.9±6.7
rk_e	6.3±1.0	17.3±11.0	13.7±1.0	20.0±4.4	25.0±7.0	8.7±1.0	22.7±5.7	27.0±6.0	13.7±2.5	15.0±2.0
MSE	0.118 8	0.117 4	0.118 22	0.118 10	0.118 17	0.121 45	0.118 18	0.118 23	0.119 32	0.120 41
MAE	0.248 18	0.247 11	0.248 24	0.247 5	0.247 7	0.252 40	0.248 21	0.248 22	0.249 28	0.251 39
MAPE	1.954 4	1.974 15	1.965 9	2.046 48	1.984 17	2.005 34	2.002 32	1.955 5	2.058 54	1.995 26
SMAPE	0.633 11	0.634 22	0.633 13	0.629 3	0.633 9	0.640 32	0.635 26	0.640 33	0.636 27	0.642 37
GA	0.612 6	0.655 10	0.658 13	0.683 18	0.689 22	0.644 8	0.686 21	0.694 27	0.663 14	0.672 15
AFS	0.278 7	0.369 30	0.327 14	0.358 25	0.376 33	0.289 9	0.341 18	0.354 21	0.329 16	0.341 17
GFS	0.525 6	0.582 12	0.590 14	0.606 17	0.610 20	0.565 9	0.635 29	0.641 33	0.581 11	0.587 13

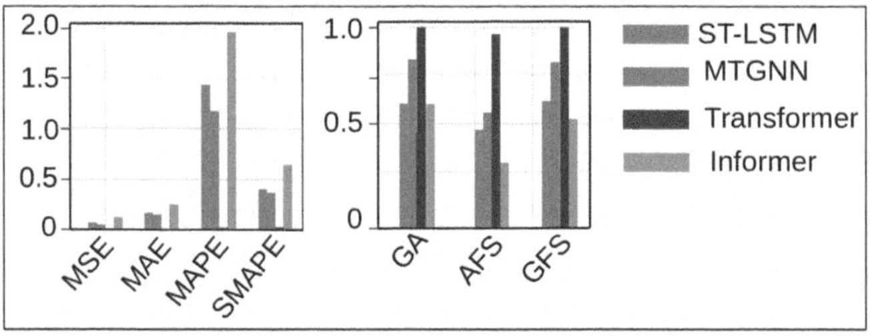

Fig. 6. Performances of the four best models at horizon 10.

5 Conclusion and Future Work

In this work, our aim was to evaluate four deep learning architectures that we have tailored for unsupervised learning of hybrid asynchronous data representation and joint forecasting tasks. We showcase the effectiveness of the Transformer model for joint prediction, with a specific emphasis on data-driven simulation of patients under anesthesia. Future work will investigate surgeries involving more variables and events. In personalized medicine, a Transformer-based solution will predict the risk associated with a medical action performed on the digital twin of a real patient, before the actual surgery is performed. In addition to simulating the next medical action not performed by the user, event forecasting will also help identify when the user strays from the expected procedure.

Disclosure of Interests. The authors have no competing interests to declare.

References

1. Zhao, L.: Event prediction in the big data era: a systematic survey. ACM Comput. Surv. **54**(5), 1–37 (2021)
2. Li, W., Eddie Law, K.L.: Deep learning models for time series forecasting: a review. IEEE Access **12**, 92306–92327 (2024)
3. Manigandan, P., et al.: Forecasting natural gas production and consumption in United States - evidence from SARIMA and SARIMAX models. Energies **14**(19), 6021 (2021)
4. Grossi, L., Nan, F.: Robust forecasting of electricity prices: simulations, models and the impact of renewable sources. Technol. Forecast. Soc. Chang. **141**, 305–318 (2019)
5. Massaoudi, M., et al.: An effective hybrid NARX-LSTM model for point and interval PV power forecasting. IEEE Access **9**, 36571–36588 (2021)
6. Bunescu, R., Struble, N., Marling, C., Shubrook, J., Schwartz, F.: Blood glucose level prediction using physiological models and support vector regression. In: 12th International Conference on Machine Learning and Applications (ICMLA), pp. 135–140 (2013)
7. Liu, Y., Ghasemkhani, A., Yang, L., Zhao, J., Zhang, J., Vittal, V.: Seasonal self-evolving neural networks based short-term wind farm generation forecast. In: IEEE International Conference on Communications, Control, and Computing Technologies for Smart Grids (SmartGridComm), pp. 1–6 (2020)
8. Ding, D., Zhang, M., Pan, X., Yang, M., He, X.: Modeling extreme events in time series prediction. In: 25th ACM SIGKDD International Conference on Knowledge Discovery and Data Mining (KDD), pp. 1114–1122 (2019)
9. Dama, F., Sinoquet, C., Lejus-Bourdeau, C.: A framework for context-sensitive prediction in time series - feasibility study for data-driven simulation in medicine. In: 10th IEEE International Conference on Data Science and Advanced Analytics (DSAA), pp. 1–12 (2023)
10. Paris, J., Sinoquet, C., Taia-Alaoui, F., Lejus-Bourdeau, C.: Deep joint modelling of mixed asynchronous streams - proof of concept for data-driven simulation of a digital patient under anaesthesia. In: 28th International Conference on Knowledge-Based and Intelligent Information and Engineering Systems (KES), pp. 2597–2606 (2024)
11. Jang, J., Oh, R.: A review on Poisson, Cox, Hawkes, shot-noise Poisson and dynamic contagion process and their compound processes. Ann. Actuarial Sci. **15**, 623–644 (2021)
12. Papageorgiou, G., Mauff, K., Tomer, A., Rizopoulos, D.: An overview of joint modeling of time-to-event and longitudinal outcomes. Annu. Rev. Stat. Appl. **6**, 223–240 (2019)
13. Choi, E., Du, N., Chen, R., Song, L., Sun, J.: Constructing disease network and temporal progression model via context-sensitive Hawkes process. In: 15th IEEE International Conference on Data Mining (ICDM), pp. 721–726 (2015)
14. Xiao, S., Yan, J., Farajtabar, M., Song, L., Yang, X., Zha, H.: Learning time series associated event sequences with recurrent point process networks. IEEE Trans. Neural Netw. Learn. Syst. **30**, 3124–3136 (2019)
15. Hochreiter, S., Schmidhuber, J.: Long short-term memory. Neural Comput. **9**(8), 1735–1780 (1997)
16. Benidis, K., Rangapuram, S.S., Flunkert, V., Wang, Y., Maddix D.C. et al.: Deep learning for time series forecasting: tutorial and literature survey. ACM Comput. Surv. **55**(6), 121:1–121:36 (2023)

17. Wu, Z., Pan, S., Chen, F., Long, G., Zhang, C., Yu, P.S.: A comprehensive survey on graph neural networks. IEEE Trans. Neural Netw Learn. Syst. **32**(1), 4–24 (2021)
18. Jin, M., et al.: A survey on Graph Neural Networks for time series: forecasting, classification, imputation, and anomaly detection. IEEE Trans. Pattern Anal. Mach. Intell. **46**(12), 10466–10485 (2024)
19. Kipf, T. N., Welling, M. : Semi-supervised classification with graph convolutional networks. In: 5th International Conference on Learning Representations (ICLR) (2017)
20. Wu, Z., Pan, S., Long, G., Jiang, J., Chang, X., Zhang, C.: Connecting the dots: multivariate time series forecasting with graph neural networks. In: 26th ACM SIGKDD International Conference on Knowledge Discovery and Data Mining (KDD), pp. 753–763 (2020)
21. Vaswani, A., et al.: Attention is all you need. In: Annual Conference on Neural Information Processing Systems (NeurIPS), pp. 5998–6008 (2017)
22. Zhou, H., et al.: Informer: beyond efficient transformer for long sequence time-series forecasting. In: 35th AAAI Conference on Artificial Intelligence (AAAI), pp. 11106–11115 (2021)
23. Ye, J., Li, W., Zhang, Z., Zhu, T., et al.: MvTS-library: an open library for deep multivariate time series forecasting. Knowl.-Based Syst. **283**, 111170 (2024)

An Analysis of Temporal Dropout in Earth Observation Time Series for Regression Tasks

Miro Miranda[1,2(✉)], Francisco Mena[1,2], and Andreas Dengel[1,2]

[1] University of Kaiserslautern-Landau, Kaiserslautern, Germany
{miro.miranda_lorenz,francisco.mena,andreas.dengel}@dfki.de
[2] German Research Center for Artificial Intelligence, Kaiserslautern, Germany

Abstract. Missing instances in time series data impose a significant challenge to deep learning models, particularly in regression tasks. In the Earth Observation field, satellite failure or cloud occlusion frequently results in missing time-steps, introducing uncertainties in the predicted output and causing a decline in predictive performance. While many studies address missing time-steps through data augmentation to improve model robustness, the uncertainty arising at the input level is commonly overlooked. To address this gap, we introduce Monte Carlo Temporal Dropout (MC-TD), a method that explicitly accounts for input-level uncertainty by randomly dropping time-steps during inference using a predefined dropout ratio, thereby simulating the effect of missing data. To bypass the need for costly searches for the optimal dropout ratio, we extend this approach with Monte Carlo Concrete Temporal Dropout (MC-ConcTD), a method that learns the optimal dropout distribution directly. Both MC-TD and MC-ConcTD are applied during inference, leveraging Monte Carlo sampling for uncertainty quantification. Experiments on three EO time-series datasets demonstrate that MC-ConcTD improves predictive performance and uncertainty calibration compared to existing approaches. Additionally, we highlight the advantages of adaptive dropout tuning over manual selection, making uncertainty quantification more robust and accessible for EO applications.

Keywords: Dropout · Earth Observation · Time Series · Regression · Uncertainty Quantification

1 Introduction

Recently, Deep Learning (DL) models have been widely used in the Earth Observation (EO) field to find optimal data-driven solutions in different applications [3]. DL effectively processes complex and heterogeneous sensor data, allowing for accurate analysis of environmental patterns [27]. For instance, predicting continuous values such as crop yield [38], water content in plants [39], and surface forecast [40] involves processing complex data. In the EO field, processing

G. Krempl et al. (Eds.): IDA 2025, LNCS 15669, pp. 389–402, 2025.
https://doi.org/10.1007/978-3-031-91398-3_29

time series sensor data is essential for understanding the changes and dynamics of our planet [29]. However, sensors may experience anomalies and occlusions, leading to missing data over certain time-steps [11,15,43]. For instance, clouds obstruct the sunlight in optical images, as on average, 67% of Earth's surface is covered by clouds [22], often resulting in uncertain predicted outputs. Addressing Missing Data in Time Series is a prevalent challenge in the DL field, with various modeling solutions and pre-processing techniques developed to mitigate its impact and ensure accurate predictions. While various techniques exist to mitigate missing data [15,26], few explicitly quantify the uncertainty it introduces, particularly the impact of input-level uncertainty. Addressing this gap is crucial for improving model reliability in EO regression applications.

In this work, we analyze two models that simulate missing data across time series during training and inference for a dual purpose. During training, it acts as an augmentation technique to increase model generalization to missing data in temporal EO data. During inference, by generating multiple Monte Carlo (MC) samples with different missing patterns, it acts as an Uncertainty Quantification (UQ) mechanism, thereby increasing the trustworthiness of a prediction. The first model is built on the dropout variational distribution applied during inference [13] and over time, referred to as Monte Carlo Temporal Dropout (MC-TD). However, the optimal dropout value is a difficult hyperparameter to tune, requiring a resource-intensive process that depends on each dataset and missing data type [26]. Instead of manually searching, we propose using Concrete Dropout (ConcD) [14] to learn the optimal dropout ratio using standard gradient descent. We refer to this method as Monte Carlo Concrete Temporal Dropout (MC-ConcTD), where ConcD is applied across time and during inference. The ConcD [14] approximates a Bernoulli distribution using a continuous relaxation, enabling the dropout probability to be optimized through gradient-based learning [24]. Unlike traditional dropout-based approaches, this model automatically learns an optimal dropout distribution, eliminating costly hyperparameter tuning.

We validate MC-TD and MC-ConcTD on regression tasks spanning three EO datasets with temporal sensor data. Experimental results demonstrate that MC-ConcTD improves both predictive accuracy and uncertainty calibration, while also eliminating the need for manual dropout tuning. The Concrete Temporal Dropout (ConcTD) not only enhances predictive performance but also improves the accessibility and practicality of UQ in EO regression applications. The code is publicly available at https://github.com/mmiranda-l/Temporal-Dropout/.

2 Related Work

Missing Data in time Series. Irregular sampled time series are common in signal processing [36]. Temporal observations can easily suffer from problems in their collection, causing irregular observations with missing data. Numerous research in DL has focused on learning to impute time series, such as BRITS [4], mTAN [44], and SAITS [10], while others focused on adapting models to ignore the missing data, such as D-GRU [5] and MissFormer [1]. In the EO field, missing spatial and temporal data are a common phenomenon due to real-world data collection constraints [43]. This includes, sensor noise, sensor failure, and cloud occlusion, affecting observations and introducing uncertainty. This problem negatively affects the performance of predictive models, where more missing data translates to worse predictions [12,20]. However, some strategies in the EO field mitigate the negative effect of missing data, such as including features from different sensors or using dropout techniques [15,35]. Recently, the Temporal Dropout (TD) technique, which involves randomly dropping time steps, has been used to enhance the model performance [15,18,47]. Furthermore, some studies leverage missing data as an augmentation technique to enhance generalization. For instance, Weitland et al. [50] randomly mask the end of a time-series for generalization to an early crop classification. On the other hand, some studies have focused on reconstruction tasks that recover the missing time-steps. For instance, Chen et al. [6] use a polynomial fit based on the Savitzky-Golay filter, while others use DL models based on Multi-layer Perceptron (MLP) [8], or convolutions [42].

Regression with Time Series Data. Regression tasks have been lesser explored than classification tasks with time-series data [32]. One reason might be that DL models are not as effective for regression as for classification tasks, as shown by Tan et al. [46]. The intrinsic challenge of predicting a continuous value instead of a categorical value is often overlooked. Nevertheless, in the EO field, there are numerous applications involving pixel-level regression with time series data. For instance, Nguyen et al. [33] use multi-spectral data and weather time series for pixel-wise crop yield prediction. They use a MLP model for processing the time series data as individual features. However, it has been shown recently that Recurrent Neural Network (RNN) models obtain better results. This is underlined by [30,37]. Furthermore, Maimaitijiang et al. [25] consider the use of drone-based optical time series data for the same task, obtaining images unaffected by cloud occlusion. Another pixel-wise task studied in the literature is cloud removal. This task has been explored with multi-spectral optical time series data [41], as well as including radar time series [11]. Similarly, in the Earth surface forecast, multi-spectral optical and weather time series have been used with spatio-temporal models to obtain accurate predictions [40], such as using ConvLSTM models [9].

Uncertainty Estimation. Uncertainty Estimation holds particular promise in safety-critical applications, including the EO field [16]. Reliable uncertainty estimates improve decision-making by identifying predictions with high confidence and those that should be treated with caution [19,21]. Uncertainty in predictive models is typically classified into two categories: epistemic and aleatoric [19,21]. While the former captures the uncertainty coming from the model itself, the latter represents the uncertainty coming from the data exclusively, such as sensor noise. In regression, the aleatoric uncertainty is often directly learned from the data using proper scoring rules such as the Gaussian Negative Log-Likelihood

(NLL) loss [34]. In contrast, epistemic uncertainty (EU) is typically estimated by modeling a distribution over the weight space, often based on Bayes' Theorem. This includes approximate Bayesian inference [2], or dropout techniques [13,14,31]. In the EO field, numerous studies have assessed the effectiveness of UQ methods, particularly in applications such as crop yield prediction with time series data, as evidenced in [23]. However, most of these studies overlook the uncertainty arising at the input level, which can be crucial in improving model reliability and robustness. In our work, we explore the simulation of missing data in EO time series for regression tasks. To the best of our knowledge, there is no method in the EO literature that accounts for the uncertainty arising from the input time series using missing time steps.

3 Temporal Dropout for Uncertainty Estimation

3.1 Notation and Preliminaries

Let $\mathbf{x} \in \mathcal{X}$ and $y \in \mathcal{Y}$ denote samples from the input and target spaces, respectively. In the case of multivariate time series data with T time-steps, $\mathbf{x}$ is given as $\mathbf{x} = (x_0, ..., x_T)$. We further denote data pairs as $(\mathbf{x}, y)$, and define the goal to predict the target y from $\mathbf{x}$. Thus, we aim to find a model $f^\theta(\cdot) : \mathcal{X} \longmapsto \mathcal{Y}$ such that $f^\theta(\mathbf{x}) = \hat{y}$, by optimizing over the model's parameters θ.

Uncertainty Estimation: We define the problem of learning the uncertainty in the predicted output ($\hat{y} = \{\mu, \sigma^2\}$) propagated through the uncertainty in the input and the model itself. To estimate the epistemic uncertainty (EU), we model the expected output and variance over multiple stochastic forward passes, by placing a distribution over the network. Given independent input samples $x^{(l)}$ drawn from the input distribution, EU is approximated as follows:

$$\mathbb{E}[f^\theta(x)] = \mu_{EU} \approx L^{-1} \sum_l f^\theta(x^{(l)})$$

$$Var[f^\theta(x)] = \sigma^2_{EU} \approx (L-1)^{-1} \sum_l (f^\theta(x^{(l)}) - \mathbb{E}[f^\theta(x)])^2.$$

To estimate the aleatoric uncertainty (AU), which arises from the data, we assume a Normal distribution parameterized by two output heads: $\mu(x), \sigma^2(x) = f^\theta(x)$. Both terms can be optimized by minimizing the NLL [34]:

$$\mathcal{L}_{NLL}(y, x) = \log \sigma^2(x) + \frac{(\mu(x) - y)^2}{\sigma^2(x)}. \tag{1}$$

This approach captures AU in $\sigma^2(x)$, allowing the model to capture data related uncertainty. In Valdenegro-Toro et al. [49] a relationship between input and output uncertainty is given, showing that input uncertainty is propagated through the model, finally resulting in model, or epistemic uncertainty (EU) by leveraging stochastic MC sampling. We follow this argumentation. Finally, the predictive uncertainty (PU) is the combined effect of all uncertainties.

3.2 Temporal Dropout

We leverage TD to estimate uncertainty. The dropout technique [45] is commonly used in hidden layers of DL models during training. Nevertheless, applying this technique at the input-level across time has shown good regularization performance in the EO field [15,18,26,28,47]. This prevents the model from focusing on individual time-steps and simultaneously handling missing time-steps, which randomly occur in EO data with sensor failure and cloud coverage. Thus, the multivariate time series data $\mathbf{x}$ is masked out as

$$\hat{\mathbf{x}} = \mathbf{x} \odot (1 - \mathbf{p}), \tag{2}$$

where $\mathbf{p} \in [0,1]^T$ is the binary dropout mask, which is drawn from a Bernoulli distribution, i.e. $p_i \sim \text{Bern}(\alpha), \forall i = 1, ..., T$, with α as the dropout ratio. Thus, the input $\mathbf{x}$ becomes a random variable $\hat{\mathbf{x}}$, providing additional network regularization. In contrast, we estimate the uncertainty arising from missing time steps, by enabling TD at the inference level. Gal et al. [13] demonstrated that casting dropout approximates Bayesian inference in any neural network architecture. Similarly, we extend this formulation along the temporal dimension, named **Monte Carlo Temporal Dropout (MC-TD)**. This corresponds to sampling L different dropout masks for prediction, $p_i^{(l)} \sim \text{Bern}(\alpha)$, $\forall l = 1, \ldots, L$, and applying them to the input data as $\hat{\mathbf{x}}^{(l)} = \mathbf{x} \odot (1 - \mathbf{p}^{(l)})$.

3.3 Concrete Temporal Dropout

When using the TD technique, an optimal dropout ratio value (α) has to be identified to get a well-calibrated model. This is a computationally expensive and time-consuming process, especially when working with over-parametrized models, commonly used in the EO field and time series modeling. In Gal et al. [14] a continuous relaxation of the dropout technique is proposed, allowing the dropout ratio to be a learnable parameter in combination with standard gradient descent. The soft dropout mask $\tilde{\mathbf{p}}$ is defined as:

$$\tilde{\mathbf{p}} = CS(\alpha, \mathbf{u}, \tau) = \sigma\left(\left((\log \frac{\alpha}{(1-\alpha)} + \log \frac{\mathbf{u}}{(1-\mathbf{u})}\right)\frac{1}{\tau}\right), \tag{3}$$

with σ the sigmoid function, $\tau \in \mathbb{R}^1$ a temperature value controlling the smoothness of the approximation, $\alpha \in \mathbb{R}^1$ a learnable parameter in the model, and $\mathbf{u}$ a uniform random variable, $u_i \sim \text{Uni}(0,1), \forall i = 1, ..., T$. The latter acts as the auxiliary variable in the sampling reparametrisation [24]. We use this sampling strategy in the dropout across time, named Concrete Temporal Dropout (ConcTD). Furthermore, to get the uncertainty, we use this technique at inference time, referred to as **Monte Carlo Concrete Temporal Dropout (MC-ConcTD)**. This corresponds to obtaining L samples of the Uniform distribution, $u_i^{(l)} \sim \text{Uni}(0,1), \forall l = 1, \ldots, L$, and forward over the model with data as $\hat{\mathbf{x}}^{(l)} = \mathbf{x} \odot (1 - CS(\alpha, \mathbf{u}^{(l)}, \tau))$. This technique is illustrated in Fig. 1. Additionally, in Fig. 2 we schematically illustrate TD and ConcTD.

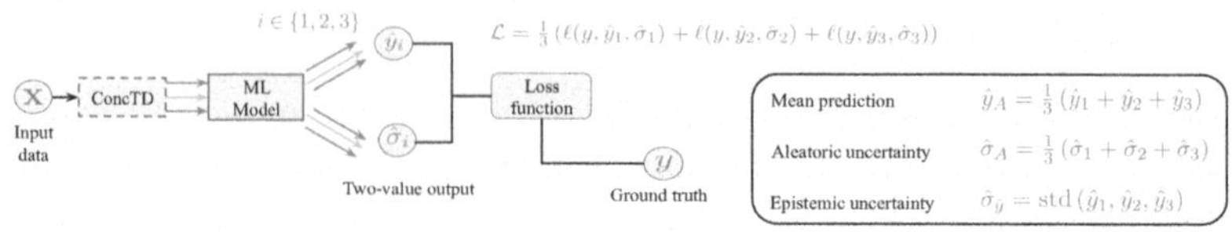

Fig. 1. Illustration of a DL model prediction with the ConcTD technique and three MC samples indexed by $i \in \{1, 2, 3\}$.

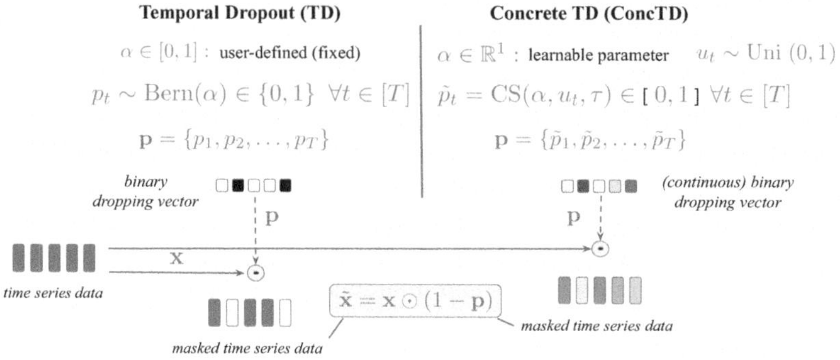

Fig. 2. Illustration of TD and ConcTD techniques over time series data.

4 Experiments

4.1 Datasets

We use three publicly available EO datasets with temporal sensor data. The characteristics of these datasets are presented in Table 1.

Swiss Crop Yield (SwissYield). We use a dataset for crop yield estimation [38], focused on cereals. This is a regression task in which the crop yield is estimated in tons per hectare (t/ha) from sensor data. The data was collected in Switzerland between 2017 and 2021. The temporal features are multi-spectral optical sensor (from Sentinel-2, with 10 bands) and weather data (with 5 bands). The time series are from seeding to harvesting, with an approximate 5 days of sampling rate. All features were spatially interpolated to a pixel resolution of 10 m.

Live Fuel Moisture Content (LFMC). We use a dataset for moisture content estimation [39]. This involves a regression task in which the vegetation water (moisture) per dry biomass (in percentage) in a given location is predicted. The data was collected between 2015 and 2019 in the western US. The temporal features are multi-spectral optical sensor (from Landsat-8, with 8 bands) and radar sensor (from Sentinel-1, with 3 bands). These features were re-sampled monthly during a four-month window before the moisture measurement, i.e. a signal of 4 time-steps is used. Additional static sensors are the topographic information, soil properties, canopy height, and land-cover class. All features were interpolated to a pixel resolution of 250 m.

Table 1. Datasets description and statistics.

Dataset	Samples	Series length	Features	Avg. target	Std. target
SwissYield	54,098	16-55	15	7.356	2.001
LFMC	2,578	4 (fixed)	61	103.987	39.562
PM2.5	167,309	120 (fixed)	9	73.673	68.546

Particles Matter 2.5 (PM2.5). We use a dataset for PM2.5 estimation [7]. This involves a regression task in which the concentration of PM2.5 (particles measure 2.5 microns or fewer in diameter) in the air (in $\mu g/m^3$) in a particular city is predicted. The data was collected between 2010 and 2015 in five Chinese cities. The temporal features are atmospheric conditions (with 3 bands), atmospheric dynamics (with 4 bands), and precipitation (with 2 bands). These are captured at hourly resolution, where we consider a three-day window for prediction.

4.2 Setup

We use an early fusion strategy that concatenates all features along the time steps. For the architecture, we adopt a RNN model with 2 layers based on long-short term memory units to extract the temporal information. Then, one additional layer is used to map the last hidden state of the RNN into a hidden dimension. Finally, two prediction heads are employed, reflecting the mean and variance, respectively. All layers consist of 128 units, including 20% of dropout. In addition, we use batch normalization layers after the RNN model. We train the models for 100 epochs with an early stopping criteria based on a patience of five. The optimization is carried out with the ADAM optimizer and a batch size of 128 over the Mean Squared Error (MSE) function. For MC sampling, 20 samples are used as in [13]. We first evaluate MC-TD and MC-ConcTD against each other and then against standard UQ methods for time series regression tasks, including MC-Dropout [13] and a Variatioinal Inference (VI) [2]. For comparison, all models employ the same architecture. If not differently specified, a TD ratio of 0.3 is used for the MC-TD model, thereby aiming for a balance between regularization and model capacity.

4.3 Results

We compare the performance of MC-ConcTD and MC-TD models across all datasets in Table 2. Both approaches demonstrate strong performance across all three datasets. However, MC-ConcTD consistently outperforms MC-TD on all regression metrics for every dataset. For instance, a maximum improvement of 6% points (p.p) is shown in R^2 on the PM2.5 dataset.

In Fig. 3(a), we evaluate the performance of various dropout ratios in the MC-TD model. The results indicate that a lower dropout ratio consistently improves performance across all datasets, with a clear decline in model performance as the ratio increases. Additionally, we observe an increase in PU with increasing TD ratio and decreasing performance. In Fig. 3(b) the calibration error (ECE) for every TD ratio is illustrated, together with the constant ECE for the MC-ConcTD model. Marker sizes indicate normalized PU. Similarly, we observe an increase in calibration error with increased uncertainty, except for the SwissYield dataset, and decreased performance. Ultimately, in Fig. 3(c), the learned dropout ratios of the MC-ConcTD model across all folds are illustrated. Notably, the dropout ratios remain below 0.25 with significant instability across folds and datasets. For instance, approaching zero for the PM2.5 dataset. The results underline the difficulty of manually tuning the dropout ratio.

In Fig. 4, we illustrate the model calibration. We define calibration as the models' capabilities to accurately reflect the true probabilities of observed outcomes. For instance, for an 80% confidence level, an event should occur approximately 80% of the time. Ideally, a perfectly calibrated model aligns with the diagonal line. In the SwissYield and LFMC datasets, the models are overconfident, though the MC-ConcTD model demonstrates relatively better calibration overall. Conversely, for the PM2.5 dataset, the models exhibit better calibration, especially for the MC-ConcTD model. In Fig. 5, prediction-versus-target plots are shown for MC-Dropout, the MC-TD, and MC-ConcTD models on the three datasets. To analyze uncertainty estimates, the predictions are scaled based on the PU. Overall, the plots show a good alignment between predicted and target values. However, the distinct characteristics of the datasets become evident,

Table 2. Results of the two variants of dropout across time. The best results are highlighted in bold. The MC-TD model uses a dropout ratio of 0.3.

Dataset	Model	R2(↑)	RMSE(↓)	MAE(↓)	ECE(↓)	PU (↓)
SwissYield	MC-TD	0.76	0.99	0.72	**2.14**	**1.00**
	MC-ConcTD	**0.79**	**0.93**	**0.66**	2.23	1.23
LFMC	MC-TD	0.69	21.92	15.67	3.77	**1.01**
	MC-ConcTD	**0.72**	**20.80**	**14.76**	**3.41**	8.01
PM2.5	MC-TD	0.71	36.36	24.56	2.92	16.00
	MC-ConcTD	**0.77**	**31.77**	**21.61**	**2.15**	**5.80**

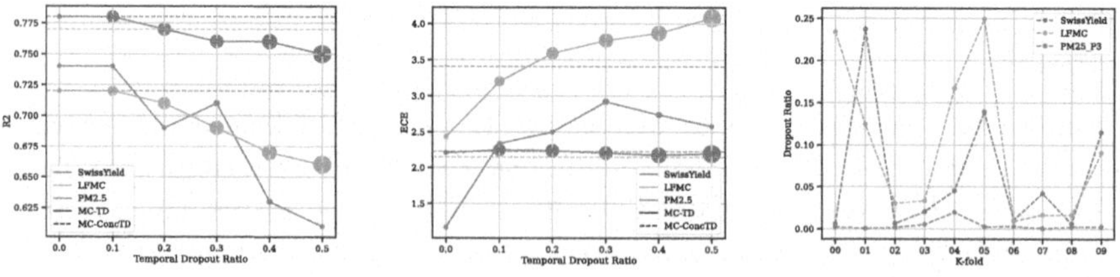

(a) R2 values (b) ECE (c) TD ratio with ConcTD

Fig. 3. Model performance in R^2 and ECE under different TD settings, where marker size indicates the normalized PU.

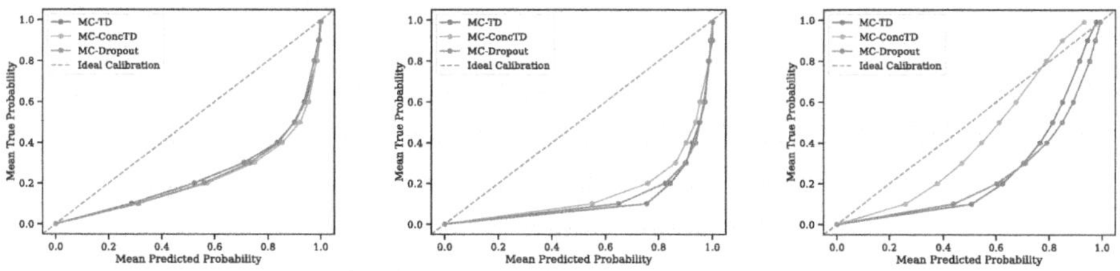

(a) SwissYield (b) LFMC (c) PM2.5

Fig. 4. Model calibration across confidence levels for all datasets.

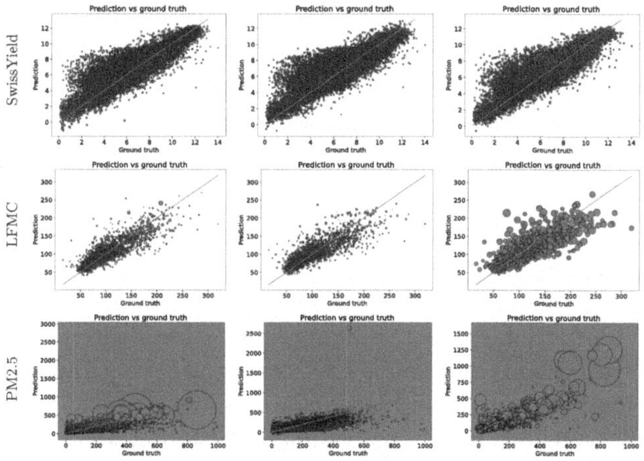

(a) MC-Dropout (b) MC-TD (c) MC-ConcTD

Fig. 5. Prediction over target plots for three regression datasets. The marker size indicates the uncertainty in each prediction.

with the PM2.5 dataset exhibiting very high uncertainties and the SwissYield dataset showing consistently low uncertainties. Furthermore, we observe that higher uncertainties are associated with lower alignment between predicted and target values, particularly for the MC-ConcTD model.

Finally, we compare the introduced models against two established UQ methods, namely MC-Dropout and VI. The results are summarized in Table 3. Notably, the VI performs poorly on regression metrics. The MC-ConcTD model achieves the best R2 scores on the SwissYield and PM2.5 datasets, while being equal to the MC-Dropout on the LFMC dataset. Considering all evaluation metrics, we find that the MC-ConcTD has the best overall results.

Table 3. Performance comparison for various UQ models.

Dataset	Model	R2 (↑)	RMSE (↓)	MAE (↓)	ECE (↓)	PU (↓)
SwissYield	MC-Dropout	0.78	0.94	0.67	2.21	1.02
	MC-TD	0.76	0.99	0.72	2.14	**1.00**
	MC-ConcTD	**0.79**	**0.93**	**0.66**	2.23	1.23
	VI	0.51	2.02	1.57	**0.73**	23.57
LFMC	MC-Dropout	**0.72**	**20.69**	**14.64**	**2.43**	1.11
	MC-TD	0.69	21.92	15.67	3.77	**1.01**
	MC-ConcTD	**0.72**	20.80	14.76	3.41	8.01
	VI	0.53	27.00	20.00	4.48	4.01
PM2.5	MC-Dropout	0.74	33.89	22.45	**1.17**	3.99
	MC-TD	0.71	36.36	24.56	2.92	16.00
	MC-ConcTD	**0.77**	**31.77**	**21.61**	2.15	5.80
	VI	0.30	57.00	39.00	3.57	**2.49**

5 Discussion

We empirically prove that TD at the inference level improves model performance over common UQ methods on various EO regression tasks (Table 2 and Table 3), thereby confirming related studies [18]. However, calibrating the dropout ratio in any DL model is challenging as it requires exploring a continuous hyperparameter space. The same applies to the MC-TD method. Consequently, MC-TD can only find the optimal dropout ratio at high computational search costs. Interestingly, higher dropout ratios or missing time-steps consistently result in reduced performance with increased uncertainty and calibration error, with an optimum value of around 0.1. The optimal value must balance predictive performance, uncertainty, and calibration error. This underlines the need for effective management of missing instances. Particularly, when time windows are small, determining the optimal ratio remains challenging. Nevertheless, MC-ConcTD demonstrates adaptability by learning different values based on the validation scenario (fold). This adaptability arises from the model's ability to dynamically adjust the dropout value using the ConcD [14], leading to improved performance and calibration. Nevertheless, we observe high variability across datasets. Moreover, we notice that there may be multiple optimal dropout ratios, as illustrated in Fig. 3(c). For instance, for the SwissYield dataset, the learned ratio of the second fold is approximately 0.25, whereas in the sixth fold, it is approximately 0.01. We attribute this variability to the distinct characteristics of each dataset, including varying time series length, number of features, and dataset size (Table 1). For instance, the LFMC dataset is characterized by only four time steps, which could potentially explain the poor performance of TD-based models on this dataset. In contrast, the PM2.5 dataset has 120 time steps, but has only short temporal dependencies compared to the SwissYield dataset. Therefore, dropping the previous time-steps can significantly reduce performance. As a result, TD is less effective in this context, potentially explaining the low TD ratios in Fig. 3(c). Overall, ConcTD enhances the robustness of UQ across different EO datasets, by continuously adapting to unique data characteristics while reducing the computational burden of optimal ratio search, making UQ more accessible for EO applications with time series data.

Limitations. The proposed methods and experimental setup have limitations that must be considered. We validate the models using EO data without real missing values in the time series, which could potentially differ from scenarios involving actual missing data. This limitation may affect how the models perform in real-world situations where missing values are present. Additionally, we use only an RNN encoder to learn the temporal patterns and only few UQ methods for comparison. However, the primary goal of this study is to demonstrate the effectiveness of TD and ConcTD in enhancing predictive performance and UQ, rather than identifying the optimal architecture for each dataset and use case. Nevertheless, additional models and architectures must be evaluated in the future, including Transformers, and other UQ methods. Moreover, while TD has shown significant improvements in related literature [18], we observe only small improvements in this study on individual datasets. This may be attributed to the task and the unique data characteristics, including short time series length (LFMC), short temporal dependencies (PM2.5), or noise in remote sensing data. More datasets must be considered to better understand the robustness of the proposed methods. Finally, while we demonstrate improvements in our models regarding the literature, we emphasize that poor calibration remains an ongoing challenge that requires careful evaluation, an issue commonly encountered in DL [17,48]. Specifically, EO data is often impacted by noisy und uncertain measurements and spatio-temporal distribution shifts, which can lead to poor calibration when the model is applied to unknown environments. Further exploring model calibration, including post-hoc calibration, will be required before deploying models into practice.

6 Conclusion

This work underscores the importance of input uncertainty in time-series data for EO regression applications. To address this, we introduced two novel uncertainty estimation methods, namely MC-TD and MC-ConcTD. Both methods account for input uncertainty in time series data by applying Temporal Dropout (TD) at the inference level using Monte Carlo sampling. While MC-TD requires manual and expensive tuning of the dropout ratio, MC-ConcTD learns this automatically as a free parameter. We empirically demonstrate their effectiveness in enhancing predictive performance in various EO datasets. While these models improve the accessibility of uncertainty estimation in EO applications, they also present challenges that require further investigation. Future research will focus on validating these approaches across diverse EO applications and by considering model calibration.

Acknowledgments. We acknowledge the support of the University of Kaiserslautern-Landau (RPTU) and the German Research Center for Artificial Intelligence (DFKI).

Disclosure of Interests. The authors have no competing interests to declare that are relevant to the content of this article.

References

1. Becker, S., Hug, R., Huebner, W., Arens, M., Morris, B.T.: Missformer: (in-) attention-based handling of missing observations for trajectory filtering and prediction. In: International Symposium on Visual Computing, pp. 521–533 (2021)
2. Blundell, C., Cornebise, J., Kavukcuoglu, K., et al.: Weight uncertainty in neural network. In: International Conference on Machine Learning, pp. 1613–1622 (2015)
3. Camps-Valls, G., Tuia, D., Zhu, X.X., Reichstein, M.: Deep Learning for the Earth Sciences: A Comprehensive Approach to Remote Sensing, Climate Science and Geosciences. Wiley (2021)
4. Cao, W., Wang, D., Li, J., Zhou, H., et al.: BRITS: bidirectional recurrent imputation for time series. Adv. Neural Inf. Process. Syst. **31** (2018)
5. Che, Z., Purushotham, S., Cho, K., Sontag, D., Liu, Y.: Recurrent neural networks for multivariate time series with missing values. Sci. Rep. **8**(1), 6085 (2018)
6. Chen, J., Jönsson, P., Tamura, M., Gu, Z., Matsushita, B., Eklundh, L.: A simple method for reconstructing a high-quality NDVI time-series data set based on the Savitzky-Golay filter. Remote Sens. Environ. **91**(3–4), 332–344 (2004)
7. Chen, S.: PM2.5 data of five Chinese cities. UCI Machine Learning Repo (2017)
8. Das, M., Ghosh, S.K.: A deep-learning-based forecasting ensemble to predict missing data for remote sensing analysis. IEEE J. Sel. Top. Appl. Earth Observ. Remote Sens. **10**(12), 5228–5236 (2017)
9. Diaconu, C.A., Saha, S., Günnemann, S., et al.: Understanding the role of weather data for earth surface forecasting using a ConvLSTM-based model. In: IEEE/CVF Conference on Computer Vision and Pattern Recognition, pp. 1362–1371 (2022)
10. Du, W., Côté, D., Liu, Y.: SAITS: self-attention-based imputation for time series. Expert Syst. Appl. **219**, 119619 (2023)
11. Ebel, P., Garnot, V.S.F., Schmitt, M., et al.: UnCRtainTS: uncertainty quantification for cloud removal in optical satellite time series. In: IEEE/CVF Conference on Computer Vision and Pattern Recognition, pp. 2086–2096 (2023)
12. Ferrari, F., Ferreira, M.P., Almeida, C.A., Feitosa, R.Q.: Fusing Sentinel-1 and Sentinel-2 images for deforestation detection in the Brazilian Amazon under diverse cloud conditions. IEEE Geosci. Remote Sens. Lett. **20**, 1–5 (2023)
13. Gal, Y., Ghahramani, Z.: Dropout as a Bayesian approximation: representing model uncertainty in deep learning. In: International Conference on Machine Learning, pp. 1050–1059 (2016)
14. Gal, Y., Hron, J., Kendall, A.: Concrete dropout. Adv. Neural Inf. Process. Syst. **30** (2017)
15. Garnot, V., Landrieu, L., Chehata, N.: Multi-modal temporal attention models for crop mapping from satellite time series. ISPRS J. Photogram. Remote. Sens. **187**, 294–305 (2022)
16. Gawlikowski, J., et al.: A survey of uncertainty in deep neural networks. Artif. Intell. Rev. **56**(Suppl 1), 1513–1589 (2023)

17. Guo, C., Pleiss, G., Sun, Y., Weinberger, K.Q.: On calibration of modern neural networks. In: International Conference on Machine Learning, pp. 1321–1330 (2017)
18. Helber, P., Bischke, B., Packbier, C., Habelitz, P., Seefeldt, F.: An operational approach to large-scale crop yield prediction with spatio-temporal machine learning models. In: IGARSS 2024-2024 IEEE International Geoscience and Remote Sensing Symposium, pp. 4299–4302. IEEE (2024)
19. Hüllermeier, E., Waegeman, W.: Aleatoric and epistemic uncertainty in machine learning: an introduction to concepts and methods. Mach. Learn. **110**, 457–506 (2021)
20. Inglada, J., Vincent, A., Arias, M., Marais-Sicre, C.: Improved early crop type identification by joint use of high temporal resolution SAR and optical image time series. Remote Sens. **8**(5), 362 (2016)
21. Kendall, A., Gal, Y.: What uncertainties do we need in Bayesian deep learning for computer vision? Adv. Neural Inf. Process. Syst. **30** (2017)
22. King, M.D., Platnick, S., Menzel, W.P., Ackerman, S.A., Hubanks, P.A.: Spatial and temporal distribution of clouds observed by MODIS onboard the Terra and Aqua satellites. IEEE Trans. Geosci. Remote Sens. **51**(7), 3826–3852 (2013)
23. Ma, Y., Zhang, Z., Kang, Y., Özdoğan, M.: Corn yield prediction and uncertainty analysis based on remotely sensed variables using a Bayesian neural network approach. Remote Sens. Environ. **259**, 112408 (2021)
24. Maddison, C.J., Mnih, A., Teh, Y.W.: The concrete distribution: a continuous relaxation of discrete random variables. In: International Conference on Learning Representations (2017)
25. Maimaitijiang, M., Sagan, V., Sidike, P., Hartling, S., Esposito, F., Fritschi, F.B.: Soybean yield prediction from UAV using multimodal data fusion and deep learning. Remote Sens. Environ. **237**, 111599 (2020)
26. Mena, F., Arenas, D., Dengel, A.: Increasing the robustness of model predictions to missing sensors in earth observation. arXiv preprint arXiv:2407.15512 (2024)
27. Mena, F., Arenas, D., Nuske, M., Dengel, A.: Common practices and taxonomy in deep multi-view fusion for remote sensing applications. IEEE J. Selected Top. Appl. Earth Observ. Remote Sens. 4797–4818 (2024)
28. Metzger, N., Turkoglu, M.O., D'Aronco, S., Wegner, J.D., Schindler, K.: Crop classification under varying cloud cover with neural ordinary differential equations. IEEE Trans. Geosci. Remote Sens. **60**, 1–12 (2021)
29. Miller, L., Pelletier, C., Webb, G.I.: Deep learning for satellite image time-series analysis: a review. IEEE Geosci. Remote Sens. Mag. (2024)
30. Miranda, M., Pathak, D., Nuske, M., Dengel, A.: Multi-modal fusion methods with local neighborhood information for crop yield prediction at field and subfield levels. In: IGARSS 2024-2024 IEEE International Geoscience and Remote Sensing Symposium, pp. 4307–4311. IEEE (2024)
31. Mobiny, A., Yuan, P., Moulik, S.K., Garg, N., Wu, C.C., Van Nguyen, H.: DropConnect is effective in modeling uncertainty of Bayesian deep networks. Sci. Rep. **11**(1), 5458 (2021)
32. Mohammadi Foumani, N., Miller, L., Tan, C.W., Webb, G.I., Forestier, G., Salehi, M.: Deep learning for time series classification and extrinsic regression: a current survey. ACM Comput. Surv. **56**(9), 1–45 (2024)
33. Nguyen, L.H., et al.: Spatial-temporal multi-task learning for within-field cotton yield prediction. In: Advances in Knowledge Discovery and Data Mining, pp. 343–354 (2019)

34. Nix, D.A., Weigend, A.S.: Estimating the mean and variance of the target probability distribution. In: IEEE International Conference on Neural Networks, vol. 1, pp. 55–60 (1994)
35. Ofori-Ampofo, S., Pelletier, C., Lang, S.: Crop type mapping from optical and radar time series using attention-based deep learning. Remote Sens. **13**(22), 4668 (2021)
36. Orfanidis, S.J.: Introduction to Signal Processing. Prentice-Hall, Inc. (1995)
37. Pathak, D., Miranda, M., Mena, F., Sanchez, C., et al.: Predicting crop yield with machine learning: an extensive analysis of input modalities and models on a field and sub-field level. In: IEEE International Geoscience and Remote Sensing Symposium, pp. 2767–2770 (2023)
38. Perich, G., et al.: Pixel-based yield mapping and prediction from Sentinel-2 using spectral indices and neural networks. Field Crop. Res. **292**, 108824 (2023)
39. Rao, K., Williams, A.P., Flefil, J.F., Konings, A.G.: SAR-enhanced mapping of live fuel moisture content. Remote Sens. Environ. **245**, 111797 (2020)
40. Requena-Mesa, C., Benson, V., Reichstein, M., Runge, J., Denzler, J.: Earthnet2021: a large-scale dataset and challenge for Earth surface forecasting as a guided video prediction task. In: IEEE/CVF Conference on Computer Vision and Pattern Recognition, pp. 1132–1142 (2021)
41. Sarukkai, V., Jain, A., Uzkent, B., Ermon, S.: Cloud removal from satellite images using spatiotemporal generator networks. In: IEEE/CVF Winter Conference on Applications of Computer Vision, pp. 1796–1805 (2020)
42. Scarpa, G., Gargiulo, M., Mazza, A., Gaetano, R.: A CNN-based fusion method for feature extraction from Sentinel data. Remote Sens. **10**(2), 236 (2018)
43. Shen, H., et al.: Missing information reconstruction of remote sensing data: a technical review. IEEE Geosci. Remote Sens. Mag. **3**(3), 61–85 (2015)
44. Shukla, S.N., Marlin, B.: Multi-time attention networks for irregularly sampled time series. In: International Conference on Learning Representations (2021)
45. Srivastava, N., Hinton, G., Krizhevsky, A., Sutskever, I., Salakhutdinov, R.: Dropout: a simple way to prevent neural networks from overfitting. J. Mach. Learn. Res. **15**(1), 1929–1958 (2014)
46. Tan, C.W., Bergmeir, C., Petitjean, F., Webb, G.I.: Time series extrinsic regression: predicting numeric values from time series data. Data Min. Knowl. Disc. **35**(3), 1032–1060 (2021)
47. Tseng, G., Cartuyvels, R., Zvonkov, I., Purohit, M., Rolnick, D., Kerner, H.R.: Lightweight, pre-trained transformers for remote sensing timeseries. In: NeurIPS 2023 Workshop on Tackling Climate Change with Machine Learning (2023)
48. Valdenegro-Toro, M.: I find your lack of uncertainty in computer vision disturbing. In: Proceedings of the IEEE/CVF Conference on Computer Vision and Pattern Recognition, pp. 1263–1272 (2021)
49. Valdenegro-Toro, M., de Jong, I.P., Zullich, M.: Unified uncertainties: combining input, data and model uncertainty into a single formulation. arXiv e-prints, pp. arXiv–2406 (2024)
50. Weilandt, F., et al.: Early crop classification via multi-modal satellite data fusion and temporal attention. Remote Sens. **15**(3), 799 (2023)

Performative Drift Resistant Classification Using Generative Domain Adversarial Networks

Maciej Makowski[(✉)], Brandon Gower-Winter, and Georg Krempl

Department of Information and Computing Sciences, Utrecht University, Heidelberglaan 8, 3584 CS Utrecht, The Netherlands
m.w.makowski@students.uu.nl, {b.gower-winter,g.m.krempl}@uu.nl

Abstract. Performative Drift is a special type of Concept Drift that occurs when a model's predictions influence the future instances the model will encounter. In these settings, retraining is not always feasible. In this work, we instead focus on drift understanding as a method for creating drift-resistant classifiers. To achieve this, we introduce the Generative Domain Adversarial Network (GDAN) which combines both Domain and Generative Adversarial Networks. Using GDAN, domain-invariant representations of incoming data are created and a generative network is used to reverse the effects of performative drift. Using semi-real and synthetic data generators, we empirically evaluate GDAN's ability to provide drift-resistant classification. Initial results are promising with GDAN limiting performance degradation over several timesteps. Additionally, GDAN's generative network can be used in tandem with other models to limit their performance degradation in the presence of performative drift. Lastly, we highlight the relationship between model retraining and the unpredictability of performative drift, providing deeper insights into the challenges faced when using traditional Concept Drift mitigation strategies in the performative setting.

Keywords: Concept Drift · Performative Prediction · Generative Neural Network · Domain Adversarial Neural Network · Drift Modeling

1 Introduction

Machine Learning has transformed decision-making across various domains. However, once deployed, models may encounter unforeseen factors that degrade performance by requiring the model to make predictions on data that originated from a distribution different to its training data. This phenomena is more broadly known as Concept Drift [6] and is most commonly associated with environmental factors that are intrinsic to the setting the predictive model is deployed in. Recently, a new type of drift has been identified that does not arise due to these

Supplementary Information The online version contains supplementary material available at https://doi.org/10.1007/978-3-031-91398-3_30.

G. Krempl et al. (Eds.): IDA 2025, LNCS 15669, pp. 403–416, 2025.
https://doi.org/10.1007/978-3-031-91398-3_30

intrinsic factors, but rather due to the presence of the predictive model embedded within the setting it is making predictions in. Such scenarios are described as Performative [22], and are identifiable by the causal relationship between the predictions a model makes and the drift the system experiences.

In such cases, performative drift can lead to dangerous feedback loops. For example, a company that struggles financially might face higher debt service costs (assigned by a predictive model) which, in turn, further engenders the company's financial struggles. To combat Concept Drift generally, methods like retraining with new data or online learning exist [10] but they are not always sustainable, especially in the performative setting where model updates are likely to introduce additional or different performative drift phenomena that must be perpetually reacted to. An alternative approach would be drift understanding [16] whereby predictive models may be imbued with the ability to not only understand the drift they are experiencing, but adapt to it.

To achieve this goal, we propose a novel approach to handling model deterioration caused by performative drift. Leveraging domain adversarial neural networks (DANN) [7] and generative adversarial networks (GAN) [4,8], our architecture, GDAN, extracts domain-invariant features, enabling more robust classification, regardless of whether the data originates from the original training distribution or one altered by model predictions.

Our main contributions are the following:

1. The leveraging of drift understanding as a method for addressing performance degradation caused by performative drift.
2. A novel architecture (GDAN) that is drift-resistant and capable of simulating distribution shifts caused by performative drift.
3. We demonstrate why performative drift is unique when compared to intrinsic drift, and illustrate why traditional methods for handling Concept Drift (such as incremental learning) may prove ineffective in performative settings.

Lastly, we make the source code and Supplementary Materials for this paper available at: https://tinyurl.com/erat3t47.

2 Background and Related Work

2.1 Performativity

If a setting is performative, a model's predictions affect the data that it must later predict on. Performative drift is a model-dependent form of Concept Drift and is distinct from traditional (intrinsic) concept drift which is model-independent. The term performativity was introduced described by Perdomo et al. [22] when they introduced the concept of performative risk, a loss function that describes the performance degradation a model will experience when evaluated on future data distributions it engenders. They also introduced *repeated risk minimisation* (RRM), a procedure which repeatedly finds a model that minimizes the performative risk. Under strict assumptions, retraining in RRM can cause performative risk to converge, thus eliminating the need for model retraining.

Performativity has been discussed under various names in different domains. In the context of classification in data streams, the concept drift caused by this phenomenon has earlier been described as prediction-induced drift [15,16]. Performativity has also been observed in Recommender Systems, which have to manage the performative effect recommendations have on the content consumed by their users [2,5,18]. Adversarial Learning is also inherently performative with adversaries updating their strategies in response to detectors which react and update their strategies in response to the adversaries [1,14].

2.2 Domain Adversarial Neural Networks (DANNs)

Domain Adversarial Neural Networks (DANNs) [7] address domain adaptation by learning domain-invariant features that generalize across different data distributions. DANNs consist of three components: a feature extractor (F), a label predictor (LC), and a domain classifier (DC). The feature extractor aims to confuse the domain classifier by minimizing the following objective function:

$$E = \mathcal{L}_{LC} - \lambda \cdot \mathcal{L}_{DC} \tag{1}$$

Training involves a min-max game where LC minimizes $\mathcal{L}_{LC}$, while DC tries to minimize $\mathcal{L}_{DC}$ and simultaneously maximize the overall objective.

2.3 Generative Adversarial Networks

Generative Adversarial Networks (GANs) [8] generate data resembling a given dataset by employing two neural networks: a generator (G) and a discriminator (D). The networks engage in a two-player min-max game, with the generator trying to produce fake samples $G(z)$ from noise $z \sim P(Z)$, while the discriminator tries to distinguish real data $x \sim P(X)$ from generated data. Training GANs can pose several challenges, such as training instability [17], undifferentiable outputs [13] and mode collapse. Standard GANs focus on distinguishing between real and fake data. Mirza et al. [19] enhanced this by conditioning the GAN on additional information, such as class labels, to generate more targeted outputs. Both the generator and discriminator receive extra inputs, enabling more controlled data generation. Building on this, Odena et al. [20] introduced the Auxiliary Classifier GAN (AC-GAN), where the discriminator outputs two probability distributions: one for real vs. fake classification and another for class prediction. Isola et al. [11] introduced a method for image-to-image translation using conditional GANs. The discriminator's input is extended by pairing real data x with either target data y or generator outputs $G(x)$, enabling the generator to learn how to produce target images from x. To improve results, the L1 distance metric is used between the translated image and the target image [21].

3 GDAN Generative Domain Adversarial Networks

This section presents GDAN, a performative drift resistant architecture that maps new distributions back to their original pre-drift distributions. The sys-

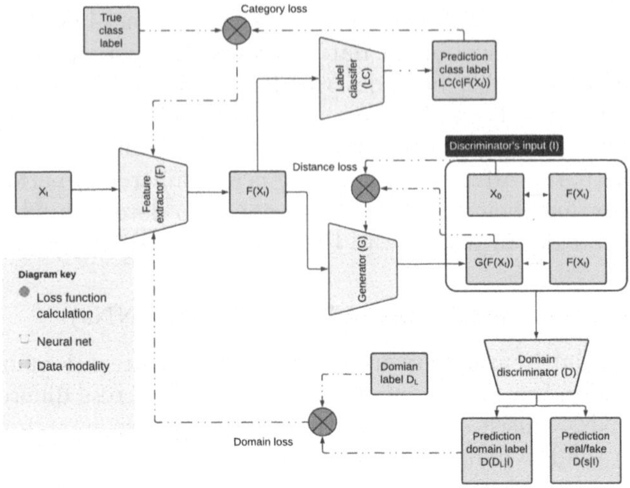

Fig. 1. GDAN architecture, composed of a domain adversarial network and a generative adversarial network. Single-sided arrows illustrate an input/output relationship, while double-sided arrows describe concatenation of data.

tem integrates Domain Adversarial Neural Networks and Generative Adversarial Networks, employing necessary modifications to combine these concepts.

Figure 1 illustrates the architecture of the model. Here, X_t refers to a data point at iteration t. Each sample is represented as a tuple (X_t, D_l, c). Where X_t is the set of attributes, D_l is the label of the distribution from which the point is sampled, and c is the class label. GDAN comprises of the following components:

- **Feature Extractor** (F): Extracts domain-invariant features from X_t, making the output $F(X_t)$ generic and domain-independent.
- **Label Classifier** (LC): Uses $F(X_t)$ to predict class c, ensuring robustness in classification despite distributional shifts caused by performative drift.
- **Generator** (G): Maps non-domain-specific representations $F(X_t)$ back to the original distribution. G learns the nature of the drift from X_0 to X_t
- **Discriminator** (D): Input to the discriminator I consists of real data points paired with their domain-invariant representations $(X_t, F(X_t))$ split equally with generated data points paired with the corresponding representations used to generate said data points $(G(F(X_t)), F(X_t))$. D outputs two probability distributions: one indicating the origin of the data point (D_l) and the other assessing whether the point is from the real or generated distribution.

3.1 Objective

GDAN comprises of several components, all of which need to be considered when developing an objective function. First, consider the discriminator D. The loss of D is two-fold: L_s denotes the segment responsible for predicting whether the instance is sampled from $P(X)$, and L_d denotes the log-likelihood of predicting the correct domain (i.e. which distribution was the point sampled from).

$$L_s = \mathbb{E}[log(P(S = 1|X_0, F(X_t))] + \mathbb{E}[log(P(S = 0|G(F(X_t)), F(X_t))] \quad (2)$$

$$L_d = \mathbb{E}[log(P(D_l = d_l|X_0, F(X_t))] + \mathbb{E}[log(P(D_l = d_l|G(F(X_t)), F(X_t))] \quad (3)$$

In Fig. 1, I symbolizes the concatenated inputs supplied to D. The output of D can then be written as $D(I) = (P(S|I), P(D_l|I))$. The goal of D is to maximise $\mathcal{L}(G, D, F) = L_s + L_d$ while the goal of the G is to maximise $L_d - L_s$. G has two tasks: to fool D and produce outputs which are similar to the target distribution. Therefore, G needs to be penalised based on the distance (dissimilarity) between its output and target instances: $\mathcal{L}_{L1}(G) = \mathbb{E}_{X_0,X_t}[||X_0 - G(F(X_t)||_1]$.

The task of selecting the optimal generator G^*, can be expressed as follows:

$$G^* = arg\,\min_G \max_D\,(\mathcal{L}(G, D, F) + \lambda\mathcal{L}_{L1}(G)) \quad (4)$$

Finally, let us consider the case of the label classifier LC and feature extractor F. Optimisation of LC follows the classic supervised learning paradigm, whereby LC aims to minimise its prediction loss $\mathcal{L}_{LC}(F, LC)$ by adjusting its parameters to accurately predict class labels based on features outputted by F. F is trained using $\mathcal{L}_{LC}$ and $\mathcal{L}_d$. F aims to maximise LC's accuracy while confusing D, ensuring the extracted features are generic and domain-invariant.

$$F^* = arg\,\min_F \max_D\,(\mathcal{L}_{LC} - \lambda \cdot \mathcal{L}_{P(D_l=d_l|X_0,F(X_t))}) \quad (5)$$

To summarize the approach, during training, there are two adversarial two-player games. The first one involves the feature extractor F and the part of the discriminator D responsible for predicting the domain label, and the second one is the generator G vs. the discriminator D. For further clarity, Pseudocode for training GDAN can be found in Appendix A.

4 Experimental Setup

This section describes our process for evaluating GDAN. By leveraging two data-generators developed by Perdomo et al. [22] and Izzo et al. [12], we design a suite of experiments to compare the accuracy (performance) of GDAN against two benchmark models. We also describe how we evaluate GDAN's understanding of performative drift. Results from a baseline experiment has been included in Table 5 and a full list of the network architectures and experimental parameters used in this work have been included in the Supplementary Materials (Fig. 2).

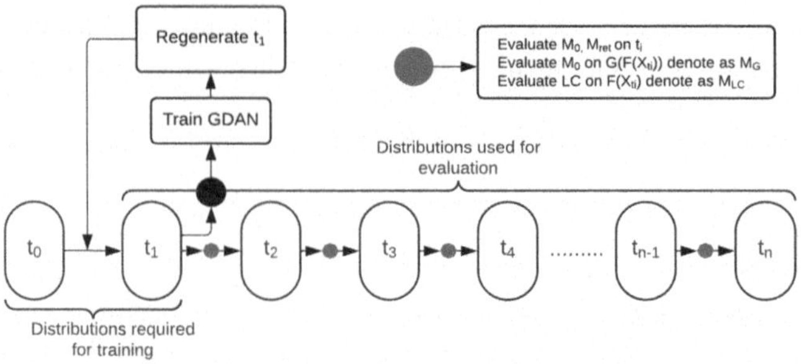

Fig. 2. Visualization of the flow of the simulation. For training of the GDAN architecture, the only necessary distributions are annotated baseline $t = 0$ and the first distribution when drift is detected $t = 1$. Iterations $1 \dots n$ are used for evaluation purposes. After a model is trained, $t = 1$ is regenerated and the testing starts on it. In each iteration, four models are evaluated. The diagram depicts a division, which is similar to a test/train split in a classic ML setup.

4.1 Simulation Design and Evaluation Metrics

To the best of our knowledge, there exists no publicly available dataset with known performative drift. We are therefore required to use the semi-real Perdomo [22] and synthetic Izzo [12] data generators. Additionally, our experiments require a dataset with ground truth labels, where each data point is associated with a specific distribution. These distributions are sequentially generated by inducing performative drift on the preceding one. For performance evaluation, access to the parameters of the drift-inducing model is mandatory.

A simulation run starts by generating data distribution X_0 using the chosen data generator. Then, a logistic regression model M_0 is trained on X_0 to establish performance benchmarks. M_0's parameters are then given to the data generator, influencing the data distribution of future instances, simulating performative drift. Each iteration (t) represents a performative drift event. At each iteration, four models are evaluated:

1. M_0: The original logistic regression model trained on data X_0 and evaluated on data X_i.
2. M_{ret}: A logistic regression model trained and evaluated on data X_i.
3. M_G: The original model M_0 evaluated on data generated by $G(F(X_t))$.
4. M_{LC}: GDAN's label classifier.

At each iteration, prediction accuracy is recorded to ascertain the performance of each model. Additionally, The L_1 norm between X_0 and X_i is compared to the L_1 norm between X_0 and $G(F(X_i))$. As a simulation run progresses, performative drift causes X_i to look less like data distribution X_0, which should be reflected in the L_1 norm. If GDAN successfully learned a domain-invariant representation of the incoming data, the L_1 norm between X_0 and $G(F(X_i))$ should remain as close to 0.0 as possible throughout the simulation run.

There is currently no framework for evaluating models in the performative setting. Thus, we treat the models M_0 and M_{ret} as our baselines with M_0 representing a lower-bound given that it undergoes no retraining, while M_{ret} acts like an upper-bound as it is continuously retrained after each drift event.

4.2 Data Generation

Perdomo Generator. We implement a slightly modified version of the data generation process, described with the following equation.

$$x_{t+1} = x_t - \epsilon B\theta \tag{6}$$

θ symbolizes the feature coefficients of the logistic regression, B indexes whether the feature is performative, i.e. it gets influenced by the drift. And ϵ informs about the strength of the performativity. X_0 is created by loading the data from *GiveMeSomeCredit* dataset [3]. For this experiment, we test two setups. One where θ is changed by retraining every iteration (Dynamic Drift), and second when θ is constant from $t = 0$ (Monotonous Drift).

Izzo Generator. For this generator we also slightly change the scheme of data generation, as we move it from $\mathbf{R}$ to $\mathbf{R^n}$. The main goal of this generator is to model a real-world spam detection situation, where only one class (the spammers) tries to modify their behaviour, to deceive the predictor.

$$x|y = 0, \theta \sim \mathcal{N}(\mu_0, \sigma_0^2) \qquad x|y = 1, \theta \sim \mathcal{N}(f(\theta), \sigma_1^2) \tag{7}$$

The function of moving the average of the distribution can be described with the following equation, where n is the index of the feature.

$$\mu_{i,n} = f(\theta_n) = \mu_1 - \epsilon \cdot \theta_n \cdot B_n \tag{8}$$

5 Results and Discussion

Table 1 presents the results of the experiments with the Perdomo data generator. The accuracies of M_{LC} and M_G are competitive in earlier iterations. Later they start diverging from the baseline, which is more rapid in the case where the drift is dynamic. Given that GDAN is trained on only two distributions of data, it only learns to replicate one direction of drift. Performance is better when the drift is monotonous as the direction of drift is constant. In the dynamic case, the direction of the drift is not constant which makes GDAN's task more complex. This is reflected in the larger standard deviations observed when the drift is dynamic. As the newly created distributions become increasingly different from the original distributions used for training, GDAN's performance degrades. However, the accuracy of M_{LC} in the first five (monotonous), and first three (dynamic) iterations show that it is possible to train a classifier whose performance is superior to a retrained predictor. Additionally, the performance

Table 1. Summary of experiments with Perdomo generator. Each experiment has been performed 10 times, and each time we sample 10000 points from the data generator and evolve them by inducing drift. The values presented in the table are means and standard deviations.

Method	Metric	1	2	3	4	5	6	7	8	9	10
Monotonous Drift	Acc M_0	**71.53** ±0.285	**71.04** ±0.311	**70.05** ±0.291	**68.92** ±0.274	**67.92** ±0.285	**66.91** ±0.303	**66.04** ±0.329	**65.13** ±0.334	**64.14** ±0.327	**63.22** ±0.303
	Acc M_G	**72.71** ±0.268	**72.26** ±0.206	**71.96** ±0.178	**71.88** ±0.207	**71.8** ±0.180	**71.56** ±0.167	**70.88** ±0.142	**70.06** ±0.236	**68.94** ±0.249	**67.53** ±0.341
	Acc M_{LC}	**76.02** ±0.302	**75.92** ±0.279	**74.91** ±0.198	**73.59** ±0.203	**71.91** ±0.177	**69.55** ±0.209	**66.63** ±0.337	**63.38** ±0.273	**59.94** ±0.269	**56.78** ±0.286
	Acc M_{ret}	**71.38** ±1.48	**71.38** ±1.485	**71.38** ±1.481	**71.38** ±1.487	**71.38** ±1.482	**71.38** ±1.484	**71.38** ±1.48	**71.38** ±1.482	**71.38** ±1.489	**71.38** ±1.485
	$L_1(X_0, X_i)$	**0.107** ±0.00	**0.213** ±0.00	**0.32** ±0.00	**0.427** ±0.00	**0.533** ±0.00	**0.64** ±0.00	**0.747** ±0.00	**0.853** ±0.00	**0.96** ±0.00	**1.067** ±0.00
	$L_1(X_0, G(F(X_i))$	**0.112** ±0.001	**0.119** ±0.001	**0.136** ±0.001	**0.158** ±0.001	**0.182** ±0.001	**0.204** ±0.001	**0.226** ±0.001	**0.247** ±0.001	**0.266** ±0.001	**0.285** ±0.001
Dynamic Drift	Acc M_0	**71.79** ±0.387	**71.2** ±0.431	**70.17** ±0.655	**69.07** ±0.774	**68.52** ±0.896	**66.96** ±0.953	**66.04** ±1.043	**65.12** ±1.252	**64.19** ±1.284	**63.38** ±1.255
	Acc M_G	**72.8** ±0.261	**72.29** ±0.42	**71.89** ±0.456	**71.5** ±0.474	**70.71** ±1.143	**69.98** ±1.874	**69.3** ±2.277	**68.3** ±2.672	**67.05** ±2.99	**65.83** ±3.292
	Acc M_{LC}	**75.99** ±0.264	**75.35** ±0.832	**72.32** ±3.793	**68.34** ±6.717	**64.3** ±8.599	**60.97** ±9.147	**58.21** ±9.122	**55.59** ±8.515	**53.35** ±7.729	**51.62** ±6.976
	Acc M_{ret}	**72.29** ±0.945	**72.29** ±0.945	**72.29** ±0.947	**72.29** ±0.947	**72.29** ±0.949	**72.29** ±0.947	**72.29** ±0.945	**72.29** ±0.938	**72.29** ±0.952	**72.29** ±0.944
	$L_1(X_0, X_i)$	**0.107** ±0.000	**0.232** ±0.022	**0.36** ±0.041	**0.488** ±0.061	**0.616** ±0.081	**0.744** ±0.01	**0.872** ±0.012	**1.0** ±0.141	**1.128** ±0.161	**1.255** ±0.180
	$L_1(X_0, G(F(X_i))$	**0.112** ±0.001	**0.127** ±0.008	**0.155** ±0.015	**0.186** ±0.021	**0.216** ±0.025	**0.245** ±0.031	**0.275** ±0.037	**0.304** ±0.046	**0.335** ±0.055	**0.367** ±0.065

of M_G indicates that GDAN can create domain-invariant representations which are understandable by models only trained on data from $t = 0$ (e.g. M_0).

Considering the distance (L_1) metrics, it is noticeable that the data generator with dynamic performative drift produces data with greater dissimilarity from iteration $t = 0$ to $t = i$ when compared to the data generator with monotonous performative drift. Additionally, GDAN's reconstructed distributions $G(F(X_i))$ are always closer to the original distribution X_0, indicating an ability to reverse the distribution shifts caused by the performative drift.

These results demonstrate one of the key characteristics of performative drift. Whenever a model is retrained, the performative drift's behaviour changes too. Retraining is a commonplace solution for handling intrinsic drift. Our results suggest that retraining would work in the performative setting if the deployed model has unfettered access to the labels of the instances it predicts on. If that is not case, our results indicate that less frequent retraining with a greater focus on drift understanding and drift reversal (using GDAN for example) may reduce the rate of performance degradation (See M_g compared to M_0 in Table 1 for example). This is because fewer retraining steps would provide more monotonous performative drift which our results clearly show is easier to manage.

Table 2 presents the results of the Izzo experiments. Interestingly, the nature of the Izzo generator's drift is cyclic. This can be seen by the oscillation of M_0

Table 2. Summary of experiments with Izzo data generator (20k datapoints sampled, averaged over 20 runs). M_{LC} and M_{ret} are excluded, as both achieve perfect accuracy. All values oscillate with the retraining of the logistic regressor (every 3 iterations), demonstrating the cyclic nature of the performative drift.

Metric/iter	1	2	3	4	5	6	7	8	9	10
Acc M_0	**50.3** $\pm$0.535	**50.21** $\pm$0.295	**50.02** $\pm$0.349	**50.21** $\pm$0.449	**60.03** $\pm$19.98	**60.11** $\pm$19.95	**59.88** $\pm$20.06	**49.99** $\pm$0.372	**50.21** $\pm$0.436	**49.94** $\pm$0.431
Acc M_G	**78.59** $\pm$18.36	**78.52** $\pm$18.34	**78.63** $\pm$18.28	**78.63** $\pm$18.21	**94.33** $\pm$12.54	**94.47** $\pm$12.26	**94.24** $\pm$12.74	**77.12** $\pm$18.72	**77.07** $\pm$18.78	**77.17** $\pm$18.68
$L_1(X_0, X_i)$	**3.78** $\pm$0.065	**3.78** $\pm$0.068	**3.78** $\pm$0.065	**3.78** $\pm$0.066	**3.67** $\pm$0.143	**3.67** $\pm$0.138	**3.66** $\pm$0.142	**3.82** $\pm$0.096	**3.82** $\pm$0.09	**3.82** $\pm$0.091
$L_1(X_0, G(F(X_i)))$	**3.39** $\pm$0.024	**3.39** $\pm$0.017	**3.39** $\pm$0.014	**3.39** $\pm$0.012	**3.41** $\pm$0.028	**3.39** $\pm$0.026	**3.39** $\pm$0.014	**3.4** $\pm$0.042	**3.4** $\pm$0.031	**3.39** $\pm$0.03

Metric/iter	11	12	13	14	15	16	17	18	19	20
Acc M_0	**65.1** $\pm$22.76	**64.88** $\pm$22.91	**64.73** $\pm$23.01	**49.83** $\pm$0.70	**50.15** $\pm$0.65	**50.14** $\pm$0.287	**63.66** $\pm$21.06	**63.74** $\pm$21.1	**63.57** $\pm$21.13	**50.13** $\pm$0.289
Acc M_G	**96.42** $\pm$10.73	**96.32** $\pm$11.02	**96.35** $\pm$10.94	**76.07** $\pm$19.05	76.57 $\pm$18.68	**76.44** $\pm$18.78	**99.56** $\pm$1.07	**99.53** $\pm$1.149	**99.56** $\pm$1.033	**74.54** $\pm$17.57
$L_1(X_0, X_i)$	**3.67** $\pm$0.133	**3.67** $\pm$0.132	**3.68** $\pm$0.134	**3.79** $\pm$0.03	**3.79** $\pm$0.033	**3.79** $\pm$0.035	**3.68** $\pm$0.139	**3.69** $\pm$0.137	**3.69** $\pm$0.13	**3.79** $\pm$0.037
$L_1(X_0, G(F(X_i)))$	**3.38** $\pm$0.021	**3.38** $\pm$0.021	**3.39** $\pm$0.019	**3.39** $\pm$0.019	**3.39** $\pm$0.019	**3.38** $\pm$0.026	**3.38** $\pm$0.019	**3.38** $\pm$0.036	**3.39** $\pm$0.024	**3.39** $\pm$0.015

and M_G's accuracies every three iterations. Additionally, the original logistic regression achieves greater accuracy on the mapped representations generated by GDAN (M_G). This indicates that despite GDAN's inability to completely minimize the dissimilarity of the drifted data points, GDAN is still capable of providing drift resistant classification. This is further supported by GDAN's label classifier which achieved perfect accuracy across all simulation runs.

Figures 3a and 3b illustrate how GDAN acts on on the performative features of the Perdomo data generator. The generated density region (green) does not overlap perfectly with the original one (red). However, GDAN's generator projects the mean of each distribution correctly. Despite GDAN's struggles to replicate outliers, the centres of the green and red density regions remain relatively close, suggesting alignment of central tendencies. A slight degradation occurs over the course of a simulation run, which is visible when analysing feature 8 where two spikes are visible on the green curve at iteration 0, while this same area is more uniform at iteration 9. This explains the decrease in performance observed for both the M_{LC} and M_G (Table 1) models.

Figure 4 presents the evolution of the PCA clusters in the Izzo experiment where the blue (drifted) clusters oscillate between the red (original) ones. Indicated by the inconsistent resemblance of the green (generated) clusters to the original (red) clusters, this oscillatory phenomena of the Izzo data generator impacts the results of GDAN's generator. This is because GDAN is only trained over a single iteration, making it incapable of understanding, and reversing the effects of multi-iteration drift phenomena.

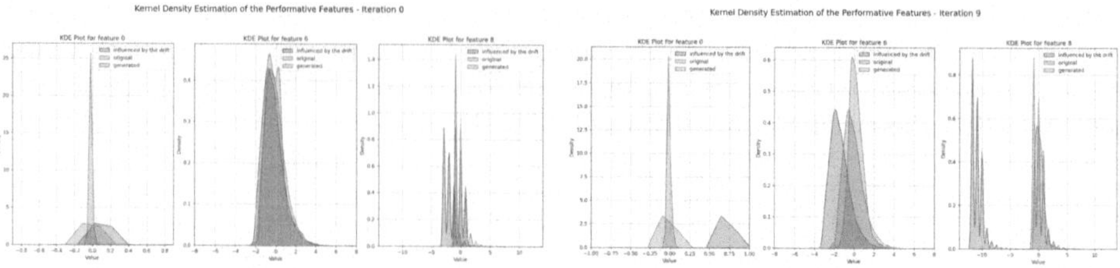

(a) Iteration 0 (b) Iteration 9

Fig. 3. Kernel Density Estimation of the Perdomo data generator's performative features. Using GDAN, the blue (drifted) data points are projected back to the red (original) data distribution. These projections are indicated by the green areas which show that GDAN is capable of reversing the effects of the drift, despite not being able to recreate the shape of the original distribution exactly.

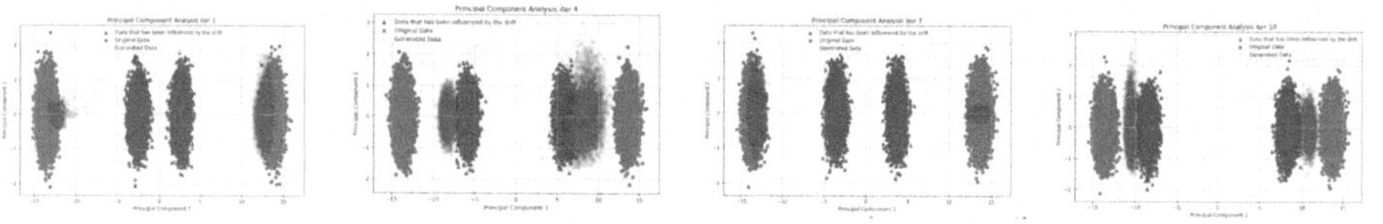

(a) Iteration 1 (b) Iteration 4 (c) Iteration 7 (d) Iteration 10

Fig. 4. PCA visualization of the Izzo data generator on iterations where the logistic regressor was retrained. The figure illustrates the cyclical nature of the drift, as the blue (drifted) clusters oscillate, moving to and from the stationary red (original data distribution) clusters. This dynamic significantly affects GDAN's ability to project the drifted data points as shown by the green clusters. (Color figure online)

5.1 Sensitivity of Dataset Size

Tables 3 and 4 present the results of additional experiments conducted to ascertain GDAN's sensitivity to the size of the training set. The experiment using the Perdomo generator demonstrated that a reduction in training data size negatively affects accuracy metrics. Models trained on $5k$ and $10k$ samples perform worse and performance decays more rapidly. With fewer data points, GDAN struggles to create a domain-invariant representation, which is visible when comparing the accuracies of M_{LC}. A spike in the performance of M_{LC} is observed when GDAN is trained on $10k$ samples. This spike is temporary and arises because the model fails to learn the correct direction of drift for one of the performative features (feature 6). The results of the Izzo generator are more straightforward. The original results (Table 2) were obtained using a model trained on $20k$ samples. The results in Table 4 are inferior to those. The model trained on $5k$ data points achieves an accuracy marginally above random guessing, indicating severe underfitting. The model trained on $40k$ data points exhibits clear signs of overfitting, failing to generalize effectively on drifted distributions. Overall, these

Table 3. Results of varying dataset sizes with the Perdomo data generator. In general, GDAN performs is incapable of creating a domain-invariant (drift-resistant) representation of the data with fewer training samples.

No data-points	Metric	1	2	3	4	5	6	7	8	9	10
5k	Acc M_g	**71.65** ±0.35	**71.67** ±0.45	**71.45** ±0.44	**70.98** ±1.20	**70.08** ±1.77	**68.58** ±1.97	**66.62** ±2.00	**65.05** ±1.71	**63.62** ±1.93	**62.18** ±3.12
	Acc M_{LC}	**61.16** ±0.44	**60.74** ±0.47	**60.47** ±0.55	**60.36** ±0.65	**60.54** ±0.97	**60.84** ±1.33	**60.80** ±1.50	**60.74** ±1.58	**60.69** ±1.49	**60.34** ±1.58
10k	Acc M_g	**71.61** ±0.15	**71.05** ±0.41	**69.88** ±0.55	**68.25** ±0.71	**66.38** ±1.09	**64.41** ±1.68	**62.64** ±1.93	**61.01** ±1.92	**59.81** ±2.01	**58.75** ±2.04
	Acc M_{LC}	**62.47** ±0.21	**62.82** ±0.33	**63.41** ±0.41	**63.74** ±0.40	**64.16** ±0.55	**65.02** ±0.85	**66.27** ±1.17	**67.52** ±1.52	**68.50** ±1.59	**69.26** ±1.67

Table 4. Results of varying dataset sizes with the Izzo data generator. In both cases, GDAN's performance is inferior exhibiting both over (40k) and underfitting (5k). M_{LC} and M_{ret} are omitted as they achieve perfect scores.

No data points	Metric/iter	1	2	3	4	5	6	7	8	9	10
5k	Acc M_0	**50.01** ±0.49	**49.91** ±0.35	**49.76** ±0.57	**50.03** ±0.57	**54.92** ±15.03	**54.96** ±15.02	**54.81** ±15.07	**49.94** ±0.40	**50.14** ±0.54	**49.94** ±0.42
	Acc M_g	**50.01** ±0.49	**49.91** ±0.35	**49.76** ±0.57	**50.03** ±0.57	**57.84** ±16.73	**57.85** ±16.75	**57.75** ±16.77	**49.95** ±0.40	**50.14** ±0.54	**49.94** ±0.42
40k	Acc M_0	**49.95** ±0.58	**49.99** ±0.52	**49.91** ±0.51	**49.84** ±0.47	**55.86** ±15.32	**55.79** ±15.32	**56.10** ±15.21	**50.00** ±0.48	**49.98** ±0.46	**50.00** ±0.39
	Acc M_g	**68.10** ±10.95	**68.25** ±10.92	**68.09** ±10.85	**68.16** ±10.96	**58.76** ±18.26	**58.75** ±18.28	**59.02** ±18.14	**68.63** ±10.89	**68.75** ±10.90	**68.80** ±10.89

results show that GDAN, like most ML-models, is susceptible to performance degradation as the number of training samples tends to zero.

6 Conclusions and Future Work

In this work we addressed the problem of performance degradation in settings with performative concept drift. That is, scenarios where a model's predictions influence the data it must later predict on. We developed GDAN, a novel architecture that combines Generative and Domain Adversarial Neural Networks. By creating domain-invariant representations of incoming data, GDAN is able to reverse the effects of performative drift and enable stable classification. Empirically evaluating GDAN on the Perdomo [22] and Izzo [12] data generators, results indicated that GDAN performs sufficiently well, consistently outperforming baseline models trained under the same conditions. Our results also indicated GDAN's performance degrades in scenarios where the direction of performative drift changes frequently. In such scenarios, GDAN can be inferior to simple retraining regimes. However, there are cases where retraining is not feasible, such as when new labels arrive with delay. We also showed that GDAN can model the direction of performative drift, and reverse its effects. This allows GDAN to provide deeper insights into the nature of performative drift, and allows other models to leverage GDAN to make themselves drift-resistant. Future work will aim to use real-world data to validate GDAN's performance as the data generators used in this work only approximate real-world phenomena. Furthermore, this work only investigated scenarios with performative drift whereas real world settings may have both performative and traditional (intrinsic) concept drift. Extending GDAN to include performative drift detection methods [9] would also be beneficial. This would address one the strongest assumptions in Performative Prediction literature which is that the presence of performative drift is known beforehand. Lastly, performative drift does not behave like intrinsic drift. Unlike intrinsic drift, the volatility of performative drift is tied to the rate at which deployed predictors are updated. Our results show that in settings that are performative, incremental learning strategies may be less effective as they result in more volatile drift phenomena. Future work will examine these claims in greater detail and seek to propose alternative learning strategies.

Disclosure of Interests. The authors have no competing interests to declare that are relevant to the content of this article.

A Training GDAN

Algorithm 1 highlights the process of training GDAN. θ denotes the weights of a model, with the subscript indicating the specific model. The parameter μ is the model specific learning rate. λ_1 and λ_2 are regularisation parameters, which control the degree of penalisation a model receives. The exact values for training GDAN in this work are included in the Supplementary Materials.

Table 5. Baseline experiment whereby data is influenced once and then continuously sampled from distribution t_1. In this scenario, no deterioration occurs and the performance of M_{LC} and M_g are superior to the accuracies achieved by M_0 and M_{ret}. This experiment serves as a sanity check for our proposed solution.

Metric	1	2	3	4	5	6	7	8	9	10
Acc M_0	**71.62** ±0.23	**71.62** ±0.23	**71.62** ±0.23	**71.62** ±0.23	**71.62** ±0.23	**71.62** ±0.23	**71.62** ±0.23	**71.62** ±0.23	**71.62** ±0.23	**71.62** ±0.23
Acc M_{ret}	**70.69** ±0.98	**70.69** ±0.98	**70.69** ±0.98	**70.69** ±0.98	**70.69** ±0.98	**70.69** ±0.98	**70.69** ±0.98	**70.69** ±0.98	**70.69** ±0.98	**70.69** ±0.98
Acc M_g	**72.76** ±0.20	**72.76** ±0.20	**72.76** ±0.20	**72.76** ±0.20	**72.76** ±0.20	**72.76** ±0.20	**72.76** ±0.20	**72.76** ±0.20	**72.76** ±0.20	**72.76** ±0.20
Acc M_{LC}	**76.00** ±0.20	**76.00** ±0.20	**76.00** ±0.20	**76.00** ±0.20	**76.00** ±0.20	**76.00** ±0.20	**76.00** ±0.20	**76.00** ±0.20	**76.00** ±0.20	**76.00** ±0.20
$L_1(X_0, X_i)$	**0.107** ±0.000	**0.107** ±0.000	**0.107** ±0.000	**0.107** ±0.000	**0.107** ±0.000	**0.107** ±0.000	**0.107** ±0.000	**0.107** ±0.000	**0.107** ±0.000	**0.107** ±0.000
$L_1(X_0, G(F(X_i)))$	**0.112** ±0.001	**0.112** ±0.001	**0.112** ±0.001	**0.112** ±0.001	**0.112** ±0.001	**0.112** ±0.001	**0.112** ±0.001	**0.112** ±0.001	**0.112** ±0.001	**0.112** ±0.001

Algorithm 1. Gradient Descent Training of GDAN

Input: Training iterations N, batch size m and k (the number of steps after which the generator starts to be updated and the feature extractor stops being updated)

Create $Z = X_0 \cup X_1$ with class labels C and domain labels D_l

for number of training iterations **do**

- Sample minibatch of $\frac{m}{2}m$ examples $\{x^{(1)}, \ldots, x^{(\frac{m}{2})}\}$ from Z
- Arbitrarily pair each sample with a point from distribution X_0
- Pass each sample through the network components and record their outputs.
- Generate another minibatch of $\frac{m}{2}$ samples by using the generator with $F(x^{(\frac{m}{2})})$ as input, and assign domain labels. The generator loss will only be calculated based on these generated instances (later denoted as m_1). The discriminator is trained on both real and generated samples.
- Calculate loss functions.
  - Update F, LC, D, and G:

$$\theta_F \leftarrow \theta_F - \mu_F(\nabla_{\theta_F}\mathscr{L}_{LC}^{m} - \lambda_1\nabla_{\theta_F}\mathscr{L}_{domain}^{\frac{1}{2}m})$$

$$\theta_{LC} \leftarrow \theta_{LC} - \mu_{LC}\nabla_{\theta_{LC}}\mathscr{L}_{LC}^{m}$$

$$\theta_D \leftarrow \theta_D - \mu_D\nabla_{\theta_D}(\mathscr{L}_{source}^{m} + \mathscr{L}_{domain}^{m})$$

$$\theta_G \leftarrow \theta_G - \mu_G\nabla_{\theta_G}(\mathscr{L}_{domain}^{m_1} + \mathscr{L}_{source}^{m_1} + \lambda_2\mathscr{L}_{L1})$$

end for

Output: Trained F, LC, G, and D

References

1. Ben-Eliezer, O., Jayaram, R., Woodruff, D.P., Yogev, E.: A framework for adversarially robust streaming algorithms. J. ACM (JACM) **69**(2), 1–33 (2022)
2. Chen, J., Dong, H., Wang, X., Feng, F., Wang, M., He, X.: Bias and debias in recommender system: a survey and future directions. Trans. Inf. Syst. **41**(3), 1–39 (2023)
3. Credit Fusion and W. Cukierski. Give me some credit (2011)
4. Creswell, A., White, T., Dumoulin, V., Arulkumaran, K., Sengupta, B., Bharath, A.A.: Generative adversarial networks: an overview. Signal Process. Mag. **35**(1), 53–65 (2018)
5. Epstein, R., Huang, Y., Megerdoomian, M., Zankich, V.R.: The "opinion matching effect" (OME): a subtle but powerful new form of influence that is apparently being used on the internet. PLoS ONE **19**(9), e0309897 (2024)
6. Gama, J., Žliobaitė, I., Bifet, A., Pechenizkiy, M., Bouchachia, H.: A survey on concept drift adaptation. ACM Comput. Surv. (CSUR) **46** (2014)
7. Ganin, Y.: Domain-adversarial training of neural networks. J. Mach. Learn. Res. **17**(59), 1–35 (2016)

8. Goodfellow, I., et al.: Generative adversarial nets. Adv. Neural Inf. Process. Syst. **27** (2014)
9. Gower-Winter, B., Krempl, G., Dragomiretskiy, S., Jelsma, T., Siebes, A.: Identifying predictions that influence the future: detecting performative concept drift in data streams. In: Proceedings of the AAAI Conference on Artificial Intelligence (2025)
10. Hoi, S., Sahoo, D., Lu, J., Zhao, P.: Online learning: a comprehensive survey. Neurocomputing **459**, 249–289 (2018)
11. Isola, P., Zhu, J.Y., Zhou, T., Efros, A.A: Image-to-image translation with conditional adversarial networks (2018)
12. Izzo, Z., Zou, J., Ying, L.: How to learn when data gradually reacts to your model. In: International Conference on Artificial Intelligence and Statistics, pp. 3998–4035. PMLR (2022)
13. Kodali, N., Abernethy, J.D., Hays, J., Kira, Z.: How to train your DRAGAN. arXiv, abs/1705.07215 (2017)
14. Korycki, Ł, Krawczyk, B.: Adversarial concept drift detection under poisoning attacks for robust data stream mining. Mach. Learn. **113**(5), 3303–3304 (2024)
15. Krempl, G., Bodnar, D., Hrubos, A.: When learning indeed changes the world: diagnosing prediction-induced drift. In: De Bie, T., Fromont, E., Leeuwen, M. (eds.) Advances in Intelligent Data Analysis XIV. LNCS, vol. 9385. Springer (2015)
16. Krempl, G., Hofer, V., Webb, G., Hüllermeier, E.: Beyond adaptation: understanding distributional changes (Dagstuhl Seminar 20372). In: Dagstuhl Reports, Vol. 10. Schloss Dagstuhl-Leibniz-Zentrum für Informatik (2021)
17. Luo, T., Zhu, Z., Chen, J., Zhu, J.: Stabilizing GANs' training with Brownian motion controller. In: International Conference on Machine Learning, pp. 23140–23156. PMLR (2023)
18. Mansoury, M., Abdollahpouri, H., Pechenizkiy, M., Mobasher, B., Burke, R.: Feedback loop and bias amplification in recommender systems. In: Proceedings of the 29th ACM International Conference on Information and Knowledge Management, pp. 2145–2148 (2020)
19. Mirza, M., Osindero, S.: Conditional generative adversarial nets. arXiv preprint arXiv:1411.1784 (2014)
20. Odena, A., Olah, C., Shlens, J.: Conditional image synthesis with auxiliary classifier GANs. In: Precup, D., Teh, Y.W. (eds.) In: Proceedings of the 34th International Conference on Machine Learning, vol. 70, pp. 2642–2651. PMLR (2017)
21. Pathak, D., Krahenbuhl, P., Donahue, J., Darrell, T., Efros, A.A.: Context encoders: feature learning by inpainting. In: Proceedings of the IEEE Conference on Computer Vision And Pattern Recognition, pp. 2536–2544 (2016)
22. Perdomo, J., Zrnic, T., Mendler-Dünner, C., Hardt, M.: Performative prediction. In: Daumé, H., III Singh, A. (eds.) Proceedings of the 37th International Conference on Machine Learning, vol. 119, pp. 7599–7609. PMLR (2020)

Explainable and Interpretable Data Science

Extracting Moore Machines from Transformers Using Queries and Counterexamples

Rik Adriaensen(✉) and Jaron Maene

KU Leuven, Departement of Computer Science, Leuven, Belgium
{rik.adriaensen,jaron.maene}@kuleuven.be

Abstract. Fuelled by the popularity of the transformer architecture in deep learning, several works have investigated what formal languages a transformer can learn from data. Nonetheless, existing results remain hard to compare due to methodological differences. To address this, we construct finite state automata as high-level abstractions of transformers trained on regular languages using queries and counterexamples. Concretely, we extract Moore machines, as many training tasks used in literature can be mapped onto them. We demonstrate the usefulness of this approach by studying positive-only learning and the sequence accuracy measure in detail.

Keywords: Interpretability · Finite State Machines · Transformers

1 Introduction

Transformers [25] have become a dominant architecture in natural language processing and deep learning more generally. This explains the increase of interest in characterising the expressive power of this architecture. Recently, significant effort has been devoted to proving what formal languages can or cannot be *expressed* by the transformer architecture [1,19,22].

The *trainability*—what formal languages transformers can learn from data—remains less well understood. Indeed, the fact that a transformer can express a formal language does not necessarily mean it can also learn said language from data. Consider the Parity language, for example, which consists of all binary sequences with an odd number of ones. Although Chiang et al. [10] proved that soft attention transformers can express this language, experimental studies found that transformers fail to learn the Parity language from data [5,11,18].

Empirically studying the trainability of transformers is challenging, however. Many possible choices for the specific training setup make it hard to draw consistent conclusions [8,18,29]. Moreover, small changes in the transformer architecture can significantly impact what languages can or cannot be expressed [22]. For example, transformers can no longer express Parity when masked hard attention is used instead of soft attention [28].

G. Krempl et al. (Eds.): IDA 2025, LNCS 15669, pp. 419–431, 2025.
https://doi.org/10.1007/978-3-031-91398-3_31

To study the trainability of regular languages, we propose a method to *extract finite state automata from transformers.* Inspired by prior work that extracts deterministic finite automata from recurrent neural networks [26], our method uses an extension of the classical L^* algorithm [2] to recover the automaton simulated by a transformer. Experimentally, we show that the extracted automata faithfully model the behaviour of transformers that recognise a regular language.

We demonstrate the use of our extraction method by studying the influence of the choice of training task on trainability. Our extraction method accommodates these different training tasks, as they can be seen as simulating a specific Moore machine. In our experiments, we find that transformers do not model garbage states when trained on purely positive examples and that the prevalent sequence accuracy measure can report a perfect score when the transformer has learnt the wrong language.

In summary, the contribution of this paper is: (1) a method for extracting Moore machines from transformers and (2) applying this method to identify two common flaws in trainability studies.

2 Preliminaries

We assume basic familiarity with transformers [25]. In this section, we briefly introduce regular languages and finite state automata. We refer to Hopcroft et al. [16] for a more thorough introduction.

We write Σ or Γ for a finite set of symbols, also called an *alphabet*. The *Kleene closure* Σ^* of an alphabet Σ is the set of all finite sequences using the symbols of Σ. For example, the binary alphabet $\Sigma = \{0, 1\}$ has as closure $\Sigma^* = \{\epsilon, 0, 1, 00, 01, 10, 11, \dots\}$. We denote the empty sequence with ϵ and the empty set with $\emptyset$. A *language* L over the alphabet Σ is a subset of Σ^*.

Definition 1. *[16] The* regular languages *are those subsets of Σ obtained by closing the empty set and the singleton sets containing a symbol from $\Sigma \cup \{\epsilon\}$ over union, concatenation, and the Kleene operator.*

The star-free languages are a subset of the regular languages constructable without using Kleene closure.

Definition 2. *[7] The* star-free regular languages *are those subsets of Σ obtained by closing the empty set and the singleton sets containing a symbol from $\Sigma \cup \{\epsilon\}$ over union, concatenation, and complement.*

For example, the language containing no consecutive zeros can be constructed as $\overline{\Sigma^* 00 \Sigma^*}$ and, as $\Sigma^* = \overline{\emptyset}$, is therefore star-free.

Definition 3. *A* deterministic finite automaton *(DFA) is a tuple $(Q, \Sigma, \delta, q_0, F)$, such that Q is a finite set of states, Σ is the input alphabet, $\delta : Q \times \Sigma \rightarrow Q$ is the transition function, $q_0 \in Q$ is the start state, and $F \subset Q$ is the set of final states.*

Given a sequence $s \in \Sigma^*$, a symbol $\sigma \in \Sigma$, and a state $q \in Q$, we can define the extended transition function $\hat{\delta} : Q \times \Sigma^* \to Q$ as $\hat{\delta}(q, \sigma.s) = \hat{\delta}(\delta(q, \sigma), s)$, with $\hat{\delta}(q, \epsilon) = q$ in the base case. We say that a DFA *accepts* a sequence s if and only if $\hat{\delta}(q_0, s) \in F$. Each DFA $\mathcal{D}$ hence represents a language $L(\mathcal{D}) = \{s \mid \hat{\delta}(q_0, s) \in F\}$. Famously, the class of languages that a DFA can accept is precisely the set of regular languages.

Moore machines generalise DFAs by replacing the set of accepting states by an output function $\gamma : Q \to \Gamma$, with Γ the output alphabet. Whereas DFAs are limited to $\Gamma = \{\text{Reject}, \text{Accept}\}$, Moore machines allow an arbitrary number of labels, analogous to the distinction between binary and multiclass classification.

Example Languages We consider the following regular languages defined over the binary alphabet $\Sigma = \{0, 1\}$. Figure 1 shows some of their corresponding DFAs. All these languages are star-free except for Parity.

- **Ones**: The language Ones (1^*) contains all sequences consisting of only ones.
- **First**: The language First $(1(0 \mid 1)^*)$ contains all sequences starting in one.
- **Depth-bounded Dyck** $(\mathcal{D}_i)$: The depth-bounded Dyck languages contain all sequences where the two symbols are correctly balanced and the maximal nesting depth is bounded. We denote the Dyck language with maximum depth i as $\mathcal{D}_i$. For example, $\mathcal{D}_1 = (01)^*$ and $\mathcal{D}_2 = (0(01)^*1)^*$.
- **Grid** $(\mathcal{G}_n)$: The DFAs accepting the Grid languages model a one-dimensional world with n squares, see Fig. 1. The two symbols are interpreted as moving one square/state to the left or right, such that trying to move beyond the boundaries results in staying on that square.
- **Parity**: Parity contains all sequences with an odd number of ones.

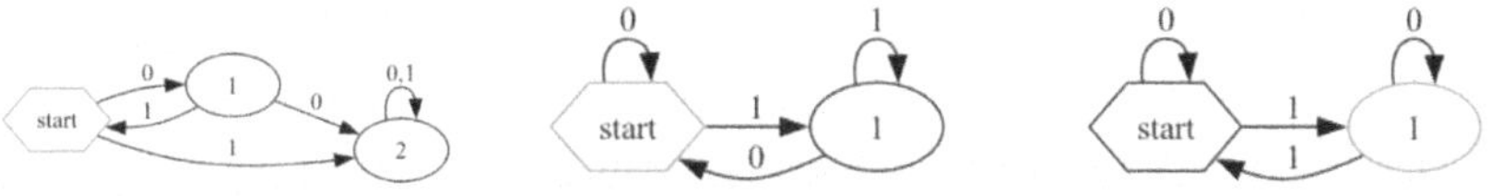

(a) $\mathcal{D}_1$ (b) $\mathcal{G}_2$ (c) Parity

Fig. 1. The DFAs accepting $\mathcal{D}_1$, $\mathcal{G}_2$ and Parity.

3 Training Tasks

Various training tasks have been used to study the trainability of regular languages. As such, the aim is often not to predict membership to a language directly. Rather it is to predict some task-specific labelling over the states of the automaton recognising that language. Such a labelling over states is equivalent to a Moore machine. We briefly introduce some common tasks.

In *state prediction* [18], the transformer predicts the automaton's state after processing a sequence s. In other words, it predicts the generalised transition function $\hat{\delta}(q_0, s)$. Directly accepting a language, or *membership prediction* [29], comes down to predicting a binary label for each state indicating whether it is accepting.

A third example is *character prediction* [8], where the aim is to predict which next symbols are valid continuations of a sequence. A symbol $\sigma \in \Sigma$ is a valid continuation of a sequence s if the extended sequence $s.\sigma$ has an accepting suffix, that is, $\exists s' \in \Sigma^* : s.\sigma.s' \in L$. Additionally, the special end-of-sequence symbol ω is a valid continuation of any accepting sequence. As the transformer must decide the validity of $|\Sigma \cup \{\omega\}|$ symbols, this is equivalent to predicting one of $2^{|\Sigma|+1}$ possible labels for each state. Figure 2 shows the labellings associated to these three tasks for the language $\mathcal{D}_1$.

Since state prediction corresponds to a unique labelling of each state, it corresponds to the fully observable case of predicting the automaton's state. In contrast, the other two tasks allow states to share a label, making them partially observable variants of this problem. In the end, they all boil down to the general task of simulating a Moore machine specific to the task and target language.

In addition to the training task, we highlight two other common choices in the training setup. Often *sequence accuracy* is used in combination with character prediction [8,12,24]. This metric counts correctly accepted sequences and considers a sequence accepted if all symbols in it are predicted to be valid and ω is valid in the final position. Furthermore, some studies train only on positive sequences [8,24], resembling the language modelling setting.

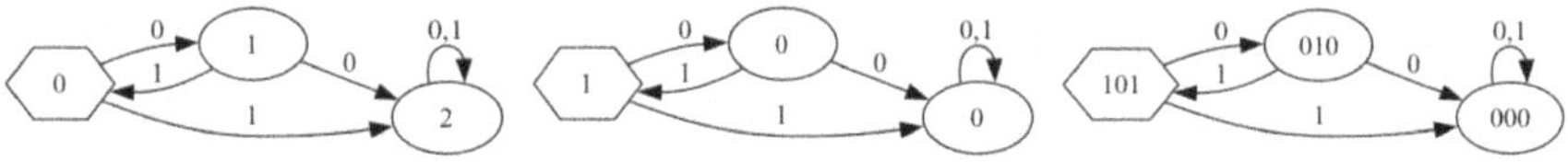

(a) State prediction (b) Membership prediction (c) Character prediction

Fig. 2. Labellings of $\mathcal{D}_1$ corresponding to the three tasks. The character prediction labels are encoded in binary. The three bits, from left to right, indicate the validity of 0, 1 and ω.

4 Extracting State Machines

Assuming a neural network faithfully implements an automaton, it must be possible to extract said automaton. Weiss et al. [26] proposed a method for extracting DFAs from recurrent neural networks (RNN) using the exact learning algorithm L^* [2]. We briefly summarise their method below.

The L^* algorithm learns a DFA by asking queries to a teacher, here the RNN. These queries are either membership queries, asking whether the teacher accepts a sequence, or equivalence queries, asking whether a proposed DFA implements the teacher's language and returning a counterexample should they differ.

Learning a minimal DFA consistent with a set of positive and negative examples is NP-hard in general [14]. However, L^* can learn any minimal DFA in time polynomial in the number of states of that DFA from a teacher answering only membership and equivalence queries [2].

Asking membership queries to a neural network is trivial, given that they accept the query as input and are trained as a classifier. Answering equivalence queries is generally intractable, however. Weiss et al. [26] deal with this intractability by checking equivalence against a discrete approximation of the RNN. They partition the hidden activation space such that each partition represents a state and state transitions are defined by a single sample in each partition [13]. In other words, they keep track of two DFAs: the DFA $\mathcal{D}$ being extracted by L^*, and the DFA $\mathcal{D}'$ specified by the partitioning of the activation space.

To answer the equivalence query, the states of $\mathcal{D}$ and $\mathcal{D}'$ are traversed breadth-first in parallel. Two types of conflicts may occur. First, the sequences accepted by $\mathcal{D}$ may differ from those accepted by the RNN, which produces a counterexample. Second, one state of $\mathcal{D}'$ may be associated with two different states of $\mathcal{D}$, in which case the partitioning must be refined.

A decision tree keeps track of the partitioning, where internal nodes are support vector machines and leaves are states of $\mathcal{D}'$. A new support vector machine splits a leaf in this tree when a refinement is needed.

A key strength of this partitioning approach is that it aligns the automaton's states with discrete regions of the RNN's state space, ensuring the automaton reflects the RNN's continuous state representations.

4.1 Extracting from Transformers

Importantly, the method of Weiss et al. [26] is not specific to RNNs but can be used on any *neural membership predictor*, provided a suitable state space is chosen. In a transformer, we propose using the activations just before unembedding. Similar to the state vectors of the RNNs in the original study, the output is computed from these activations with a single affine transformation followed by a softmax.

This transformation from the state space to the output space represents the output function of the simulated DFA. For RNNs, the transition function of the DFA is correspondingly implemented in its recurrent connections and input function. However, for a transformer, this cannot hold as there are no recurrent connections. Instead, it is more accurate to think of the transformer's forward projection from a sequence to the chosen state space as simulating the *generalised* transition function. This does make sense, as all previous symbols may impact the state directly through the attention mechanism.

4.2 Extracting Moore Machines

On transformers not trained as membership predictors, we cannot directly apply the extraction method previously described. However, as covered in Sect. 3, some of the alternative training tasks map to a labelling over the automaton's states. As a Moore machine is the formal description of such a labelling we cover all these tasks by extending the method to extract Moore machines.

To extract Moore machines instead of DFAs, we use a generalisation of L^*. This generalisation is a special case of an existing extension of L^* that learns product Moore machines [20]. We briefly describe this special case here.

Internally, the learner keeps track of an observation table $\mathcal{O}$, where answers to membership queries are stored and from which proposals are constructed.

Definition 4. *(Observation Table) An observation table over alphabet Σ is a triple $\mathcal{O} = (S, E, T)$, with:*

- *S a non-empty finite set of sequences, where $\forall w \in S : prefix(w) \in S$. S is said to be prefix-closed.*
- *E a non-empty finite set of sequences, where $\forall w \in E : suffix(w) \in E$. E is said to be suffix-closed.*
- *T a function mapping $((S \cup S.\Sigma).E)$ to Γ.*

The observation table can be thought of as a table where the rows are labelled with elements from $(S \cup S.\Sigma)$ and the columns with elements from E. The entry for row s and column e is then $T(s.e)$. The observation table defines a Moore machine $M(S, E, T)$ as $Q = \{row(s) \mid s \in S\}$, $\delta(row(s), \sigma) = row(s.\sigma)$, $q_0 = row(\epsilon)$, and $\gamma(s) = T(s)$.

An observation table is called *closed* if, for each $t \in S.\Sigma$, there is an $s \in S$ such that $row(t) = row(s)$. An observation table is called *consistent* if for all $s_1, s_2 \in S$ and $\sigma \in \Sigma$, $row(s_1) = row(s_2) \Rightarrow row(s_1.\sigma) = row(s_2.\sigma)$. The following property follows from Moerman [20].

Theorem 1. *If (S, E, T) is a closed, consistent observation table, the Moore machine $M(S, E, T)$ is consistent with the finite function T.*

Finally, the complete algorithm consists of the following steps.

1. Ask membership queries to make the (initially empty) observation table closed and consistent.
2. Query the equivalence of $M(S, E, T)$ and the teacher's language.
3. If equivalent, return $M(S, E, T)$; otherwise, extend T with the returned counterexample and return to step 1.

5 Related Work

Following the L^* algorithm [2] for learning DFAs, more algorithms based on Angluin's exact learning framework [3] of using queries and counterexamples have been developed to learn other concepts, such as weighted finite automata [6] and horn theories [4]. Similar to Weiss et al. [26], these have been applied to construct a high-level abstraction of neural networks [9,27]. Our work contributes to this line of research by extracting Moore machines from transformers.

Concurrently to our work, Zhang et al. [30] attempted to extract DFAs from transformers. As they claim transformers are inherently stateless, Zhang et al. [30] first train an RNN to simulate the transformer. Next, they apply the original extraction method by Weiss et al. [26] on that RNN. As now the discretisation is performed on the internal state of this intermediate RNN, the extracted DFA is not aligned with the original transformer but with that of the intermediate RNN. As discussed in Sect. 4.1, transformers must simulate a state if they are to successfully recognise a regular language, which they can do by modelling the *generalised* transition function, making the intermediate RNN redundant.

6 Experiments

In this section, we apply our extraction method to transformers trained on regular languages and analyse the resulting Moore machines to answer the following two research questions. All code to replicate the experiments is available on https://github.com/Rik-A3/Extracting-Moore.

RQ1. *Do the extracted state machines faithfully model the language learnt by the transformer?* There are two reasons why the extracted automaton could differ from the target language. Either because the transformer did not learn the correct language or because the extracted automaton does not correctly model the transformer. For the extraction method to be faithful, the latter reason should not occur.

RQ2. *Can automata extraction help explain the impact of different training setups?* Two aspects of the training setup are studied more closely using the extraction technique. First, we investigate whether transformers can learn a language using only positive examples. Second, we look at sequence accuracy commonly used in combination with character prediction.

6.1 Experiment Setup

Datasets. The training datasets contain 10 000 sequences of length 32 labelled using a target Moore machine. The validation and test sets contain 1 000 sequences of length 100. Each sequence is prepended with a unique beginning-of-sequence symbol, allowing the transformer to learn an output for the start state. Unless stated otherwise, the sequences are generated by randomly traversing the target Moore machine. To prevent overly skewed class labels due to garbage states, the probability $p(q, \sigma)$ of picking the next symbol σ in the state q is taken as $p(q, 0) = \frac{c(\delta(q,1))}{c(\delta(q,0))+c(\delta(q,1))}$ for $\sigma = 0$ and similarly for $\sigma = 1$, where $c(q)$ is the number of times state q has been visited during the generation of the dataset.

Architecture. We use one-layer encoder-only transformers with soft attention [25], pre-norm layer normalisation [21], and rotary positional encodings [23]. The latter two techniques have been demonstrated to enhance the original transformer architecture in terms of training efficiency and length generalisation, respectively. We use a residual stream width of 16, a single-layer MLP block of width 64, and 4 attention heads. This is a subset of the transformer configurations studied in Bhattamishra et al. [8]. We do not compare different transformer configurations, as this has been extensively done before [8,11].

Training. We train with the Adam optimiser [17] using a learning rate of $3 \cdot 10^{-4}$, no learning rate schedule, and a batch size of 32. We employ early stopping on the validation loss, with a patience of 5 epochs. Each experiment is repeated three times with different seeds. Training can be replicated in under six hours on a single NVIDIA GeForce RTX 3080 Ti GPU.

Extraction. Weiss et al. [26] employ three tactics to improve practical performance, which we also adopt. First, we impose a time limit after which the algorithm is stopped and the latest automaton proposed during an equivalence query is returned. Second, we use an aggressive initial splitting strategy on the first set of conflicting activation vectors by splitting along the dimensions where these vectors differ the most. The number of dimensions d to initially split on is a hyperparameter called the *initial split depth*, which we keep at 10. Third, all possible sequences of length three are supplied as starting examples for the L^* learner. These last two tactics prevent the algorithm from terminating prematurely on an automaton with a single state.

6.2 Are the Extracted State Machines Faithful?

To evaluate our approach, we trained transformers to recognise each of the example languages described in Sect. 2 under the state prediction task, as this is the fully observable case. We measure the prediction accuracy on the final ten positions of sequences in the test set, emphasising generalisation beyond the training length. We also report the *average correct length* (ACL), meaning the average length up to which the transformer makes no mistakes.

From each of these transformers, we extract a Moore machine. Again we report the prediction accuracy on the last ten positions and average correct length, where now the predictions of the transformer serve as the ground truth, which the extracted Moore machine must predict. All results are in Table 1. When the extracted automaton is exactly the target, all mistakes are due to the transformer imperfectly modelling the language. This is why we include how often the exact target machine was extracted in the last column.

The languages First, Ones, $\mathcal{G}_2$, $\mathcal{G}_3$, $\mathcal{D}_1$ and $\mathcal{D}_2$ are all learnable by our transformers as they achieve high accuracy and generalise well beyond the training length of 32. A single transformer failed to learn the First language well, potentially due to suboptimal hyperparameters. For Parity, generalisation does not occur. Only sequences shorter than the training length are predicted accurately.

Table 1. We report the accuracy and average correct length of transformers on a test set of length 100 as well as the accuracy and average correct length of the extracted Moore machine compared to predictions of the transformer on the same test set. Results are averaged over 3 runs. We report the extraction time, where T/O indicates a time-out of 30 s, and how often the target Moore machine was extracted.

	Test		Extraction			
Name	Accuracy	ACL	Accuracy	ACL	Time	Target?
First	0.650	59.292	0.650	59.667	0.15, T/O, 0.19	2/3
Ones	1.000	101.000	1.000	101.000	0.10, 0.11, 0.13	3/3
$\mathcal{G}_2$	1.000	101.000	1.000	101.000	0.16, 0.15, 0.16	3/3
$\mathcal{G}_3$	0.950	77.125	0.950	77.125	0.21, 0.17, 0.20	3/3
$\mathcal{D}_1$	0.996	88.125	0.996	88.125	0.20, 0.25, 0.19	3/3
$\mathcal{D}_2$	1.000	44.958	1.000	44.958	0.22, 0.23, 0.30	3/3
Parity	0.458	17.083	0.496	17.917	0.24, 0.12, T/O	2/3

The extraction results show that the target automaton is often recovered from the transformer. This is what we expect when the transformer has perfectly learnt the language beyond the length checked during the breadth-first search. We thus conclude our method finds a small and faithful automaton resembling the language accepted by the transformer. However, when the transformer's modelling does not generalise well towards longer sequences, the abstraction inevitably breaks down, and our method returns a large automaton when the time limit is reached. This intolerance to noise is typical for exact learning algorithms such as L^*. Although large, the returned intermediate Moore machine does represent the transformer slightly better than the target machine, as shown by the increased accuracy and average correct length on the test set.

RQ1. Our results show that the correct automata is extracted in all cases where the transformer successfully learnt the language.

6.3 Can Automata Extraction Explain Experimental Differences?

Similar to the previous section, we train transformers on the example languages, apply the extraction method, and report the average correct length as well as the accuracy on the last ten positions of the test set. However, we now consider two common variations on the training setup.

Positive-Only Learning. First, we train the transformers on a dataset containing only positive examples, similar to Bhattamishra et al. [8]. The dataset is generated by traversing the target DFA at random after removing the garbage state for efficiency and only keeping accepted sequences. The results are shown in Table 2. For languages $\mathcal{G}_2$, $\mathcal{G}_3$ and Parity, the results are as before. The other languages are learnt poorly. As we will explain, this is because of their garbage state.

Table 2. We report the accuracy and average correct length of positive-only trained transformers on a test set of length 100 as well as the accuracy and average correct length of the extracted Moore machine compared to predictions of the transformer on the same test set. Results are averaged over 3 runs.

	Test		Extraction			
Name	Accuracy	ACL	Accuracy	ACL	Time	Target?
First	0.417	42.667	1.000	101.000	0.18, 0.15, 0.15	0/3
Ones	0.000	10.167	1.000	101.000	0.06, 0.05, 0.06	0/3
$\mathcal{G}_2$	1.000	101.000	1.000	101.000	0.15, 0.27, 0.13	3/3
$\mathcal{G}_3$	0.992	93.750	0.992	93.750	0.19, 0.16, 0.16	3/3
$\mathcal{D}_1$	0.000	17.000	1.000	97.000	0.10, 0.11, T/O	2/3
$\mathcal{D}_2$	0.000	10.583	0.621	35.333	T/O, T/O, T/O	0/3
Parity	0.471	15.458	0.517	15.917	T/O, 0.12, 0.11	2/3

Since the gradient update of a transformer is not only based on the prediction of the last symbol of the sequence but of all symbols in the sequence, the transformer is essentially trained on all prefixes of the examples in the training dataset. When there is no garbage state, a positive-only dataset containing accepting examples of a certain length can have in its prefix set all sequences—including non-accepting ones—of a length slightly smaller than the dataset length. When there is a garbage state, however, sequences visiting it never appear as the prefix of a positive example and are, therefore, never seen by the transformer.

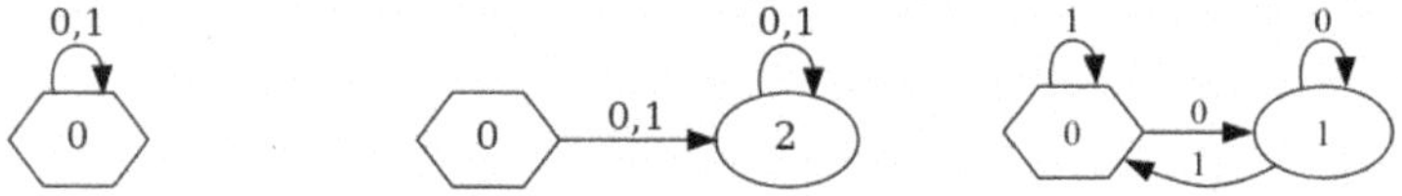

(a) Ones (b) First (c) $\mathcal{D}_1$

Fig. 3. Moore machines extracted from transformers trained on Ones, First, and $\mathcal{D}_1$.

We thus expect that garbage states will not be learnt and only languages containing a garbage state will be affected. This explains the decreased scores on a test set containing both positive and negative examples. Nonetheless, the extraction method finds small, faithful Moore machines for Ones, First, and $\mathcal{D}_1$. These automata, shown in Fig. 3, are exactly the target automata where the garbage state is not modelled.

This observation is related to recent work [15] showing encoder-only transformers are biased towards learning *functions of low sensitivity*, that is, functions where flipping a single bit in the input often does not cause a change in the output. All three alternative languages the transformer has learnt are less sensitive than the expected languages with a garbage state, while still accepting all examples in the dataset. We leave further investigations on the sensitivity of languages with or without garbage states to future work.

Character Prediction and Sequence Accuracy. This observation sparks a discussion on the use of another common training setup: character prediction and sequence accuracy. In the character prediction task, a symbol in a sequence is invalid if from that point no accepting suffixes exist, meaning the sequence is in a garbage state. Since, according to sequence accuracy, a sequence is rejected when a single symbol is predicted to be invalid, all predictions of the symbols after it are irrelevant. Therefore, transformers can achieve perfect sequence accuracy by correctly labelling non-garbage states without explicitly modelling the garbage state. Indeed, the automaton for language $\mathcal{G}_n$, given the correct labelling, would score a perfect sequence accuracy on sequences labelled using $\mathcal{D}_{n-1}$.

RQ2: Using our approach, we have shown transformers do not learn garbage states when trained on only positive examples and observe that the sequence accuracy can be high even though a language different from the target has been learnt. These examples highlight that judging whether a language has been learnt based on simple metrics can lead to misleading results, which our method prevents by extracting said language.

7 Conclusion

Inspired by prior work leveraging Angluin's exact learning framework to learn high-level abstractions from neural networks, we extracted Moore machines from encoder-only transformers trained on regular languages using an extension of L^*. Our method accommodates a variety of training tasks used in literature since, as we have shown, many of these tasks reduce to simulating Moore machines. We demonstrated that our extraction technique finds small automata that faithfully describe the transformer, given that the transformer learnt the language. Finally, we found that transformers do not model garbage states when trained on only positive examples and that the common sequence accuracy can report a perfect score even though the transformer has learnt a language different from the target.

Acknowledgments. This research received funding from the Flemish Government (AI Research Program), the Flanders Research Foundation (FWO) under project G097720N, and the European Research Council (ERC) under the European Union's Horizon 2020 research and innovation program (Advanced Grant DeepLog No. 101142702). We are grateful to Luc De Raedt for his helpful feedback.

Disclosure of Interests. The authors have no competing interests to declare that are relevant to the content of this article.

References

1. Ackerman, J., Cybenko, G.: A survey of neural networks and formal languages. arXiv preprint arXiv:2006.01338 (2020)
2. Angluin, D.: Learning regular sets from queries and counterexamples. Inf. Comput. **75**(2), 87–106 (1987)
3. Angluin, D.: Queries and concept learning. Mach. Learn. **2**(4), 319–342 (1988). https://doi.org/10.1023/A:1022821128753
4. Angluin, D., Frazier, M., Pitt, L.: Learning conjunctions of horn clauses. Mach. Learn. **9**(2), 147–164 (1992)
5. Anil, C., et al.: Exploring length generalization in large language models. Adv. Neural. Inf. Process. Syst. **35**, 38546–38556 (2022)
6. Balle, B., Mohri, M.: Learning weighted automata, pp. 1–21 (2015)
7. Barrington, D., Compton, K., Straubing, H., Thérien, D.: Regular languages in nc1. J. Comput. Syst. Sci. **44**(3), 478–499 (1992)
8. Bhattamishra, S., Ahuja, K., Goyal, N.: On the ability and limitations of transformers to recognize formal languages. In: Proceedings of the 2020 Conference on Empirical Methods in Natural Language Processing (EMNLP), pp. 7096–7116. Association for Computational Linguistics (2020)
9. Blum, S., Koudijs, R., Ozaki, A., Touileb, S.: Learning horn envelopes via queries from language models. Int. J. Approximate Reasoning **171**, 109026 (2024). synergies between Machine Learning and Reasoning
10. Chiang, D., Cholak, P.: Overcoming a theoretical limitation of self-attention. In: Proceedings of the 60th Annual Meeting of the Association for Computational Linguistics (2022)
11. Delétang, G., et al.: Neural networks and the chomsky hierarchy. In: The Eleventh International Conference on Learning Representations (2022)
12. Gers, F., Schmidhuber, E.: Lstm recurrent networks learn simple context-free and context-sensitive languages. IEEE Trans. Neural Netw. **12**(6), 1333–1340 (2001)
13. Giles, C.L., Miller, C.B., Chen, D., Sun, G.Z., Chen, H.H., Lee, Y.C.: Extracting and learning an unknown grammar with recurrent neural networks. In: Moody, J., Hanson, S., Lippmann, R. (eds.) Advances in Neural Information Processing Systems, vol. 4. Morgan-Kaufmann (1991)
14. Gold, E.M.: Complexity of automaton identification from given data. Inf. Control **37**(3), 302–320 (1978). https://doi.org/10.1016/S0019-9958(78)90562-4
15. Hahn, M., Rofin, M.: Why are sensitive functions hard for transformers? . In: Ku, L.W., Martins, A., Srikumar, V. (eds.) Proceedings of the 62nd Annual Meeting of the Association for Computational Linguistics (Volume 1: Long Papers), pp. 14973–15008. Association for Computational Linguistics, Bangkok, Thailand (2024). 10.18653/v1/2024.acl-long.800
16. Hopcroft, J.E., Motwani, R., Ullman, J.D.: Introduction to Automata Theory, Languages, and Computation, 3rd edn. Addison-Wesley Longman Publishing Co., Inc, USA (2006)
17. Kingma, D.P., Ba, J.L.: Adam: a method for stochastic gradient descent. In: International Conference on Learning Representations, pp. 1–15 (2015)

18. Liu, B., Ash, J.T., Goel, S., Krishnamurthy, A., Zhang, C.: Transformers learn shortcuts to automata. In: The Eleventh International Conference on Learning Representations (2023)
19. Merrill, W.: Formal language theory meets modern nlp. arXiv preprint arXiv:2102.10094 (2021)
20. Moerman, J.: Learning product automata. In: Unold, O., Dyrka, W., Wieczorek, W. (eds.) Proceedings of Machine Learning Research, vol. 93, pp. 54–66. PMLR (2019)
21. Nguyen, T.Q., Salazar, J.: Transformers without tears: Improving the normalization of self-attention. In: Niehues, J., et al., (eds.) Proceedings of the 16th International Conference on Spoken Language Translation. Association for Computational Linguistics, Hong Kong (2019)
22. Strobl, L., Merrill, W., Weiss, G., Chiang, D., Angluin, D.: Transformers as recognizers of formal languages: a survey on expressivity (2023)
23. Su, J., Ahmed, M., Lu, Y., Pan, S., Bo, W., Liu, Y.: Roformer: enhanced transformer with rotary position embedding. Neurocomputing **568**, 127063 (2024)
24. Suzgun, M., Belinkov, Y., Shieber, S.M.: On evaluating the generalization of lstm models in formal languages, pp. 277–286 (2019)
25. Vaswani, A., et al.: Attention is all you need. In: Guyon, I., et al., (eds.) Advances in Neural Information Processing Systems, vol. 30. Curran Associates, Inc. (2017)
26. Weiss, G., Goldberg, Y., Yahav, E.: Extracting automata from recurrent neural networks using queries and counterexamples. In: International Conference on Machine Learning, pp. 5247–5256. PMLR (2018)
27. Weiss, G., Goldberg, Y., Yahav, E.: Learning deterministic weighted automata with queries and counterexamples. In: Wallach, H., Larochelle, H., Beygelzimer, A., d' Alché-Buc, F., Fox, E., Garnett, R. (eds.) Advances in Neural Information Processing Systems, vol. 32. Curran Associates, Inc. (2019)
28. Yang, A., Chiang, D., Angluin, D.: Masked hard-attention transformers recognize exactly the star-free languages (2023)
29. Yao, S., Peng, B., Papadimitriou, C., Narasimhan, K.: Self-attention networks can process bounded hierarchical languages. In: Zong, C., Xia, F., Li, W., Navigli, R. (eds.) Proceedings of the 59th Annual Meeting of the Association for Computational Linguistics and the 11th International Joint Conference on Natural Language Processing, vol. 1, pp. 3770–3785. Association for Computational Linguistics, Online (2021)
30. Zhang, Y., Wei, Z., Sun, M.: Automata extraction from transformers (2024)

Obtaining Example-Based Explanations from Deep Neural Networks

Genghua Dong[1(✉)], Henrik Boström[1], Michalis Vazirgiannis[1,2], and Roman Bresson[1]

[1] KTH Royal Institute of Technology, Stockholm, Sweden
{genghua,bostromh,mvaz,bresson}@kth.se
[2] LIX, École Polytechnique, IP Paris, Paris, France
mvazirg@lix.polytechnique.fr

Abstract. Most techniques for explainable machine learning focus on feature attribution, i.e., values are assigned to the features such that their sum equals the prediction. Example attribution is another form of explanation that assigns weights to the training examples, such that their scalar product with the labels equals the prediction. The latter may provide valuable complementary information to feature attribution, in particular in cases where the features are not easily interpretable. Current example-based explanation techniques have targeted a few model types only, such as k-nearest neighbors and random forests. In this work, a technique for obtaining example-based explanations from deep neural networks (EBE-DNN) is proposed. The basic idea is to use the deep neural network to obtain an embedding, which is employed by a k-nearest neighbor classifier to form a prediction; the example attribution can hence straightforwardly be derived from the latter. Results from an empirical investigation show that EBE-DNN can provide highly concentrated example attributions, i.e., the predictions can be explained with few training examples, without reducing accuracy compared to the original deep neural network. Another important finding from the empirical investigation is that the choice of layer to use for the embeddings may have a large impact on the resulting accuracy.

Keywords: Explainable AI · Example-based explanations · Deep neural networks

1 Introduction

Deep neural networks (DNN) play an indispensable role in modern society, including in areas such as autonomous driving, AI-assisted medical diagnosis, and financial fraud prevention. However, due to their 'black box' nature, there has been growing concern about the inner workings of these networks [4]. Explainable artificial intelligence (XAI)[1] aims to produce models for which the

[1] The term XAI is in this work used as a synonym to explainable machine learning.

G. Krempl et al. (Eds.): IDA 2025, LNCS 15669, pp. 432–443, 2025.
https://doi.org/10.1007/978-3-031-91398-3_32

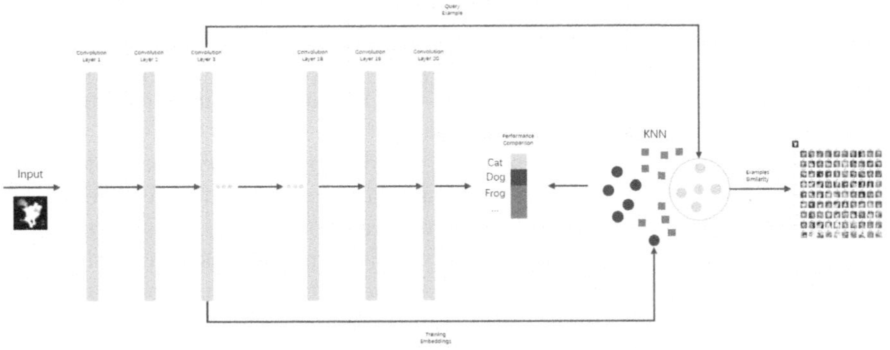

Fig. 1. The structure of EBE-DNN

predictions can be explained in a way that is understandable to humans and that may provide insights into the inner workings of the models, with the end goal of allowing users to trust and manage them effectively. XAI provides transparency into how decisions are made by a machine learning model, identifying the key features or examples contributing to its predictions [1,4].

Some machine learning models are by design interpretable, e.g., logistic regression, decision trees, decision rules, and k-nearest neighbors (KNN). Others, such as DNN, can be interpreted by employing post-hoc analysis techniques, using e.g., feature-based, concept-based, rule-based, and example-based explanation methods [11]. Feature-based explanations (FBE) describe how input features contribute to the model output. This class includes the well-known model-agnostic techniques LIME [16] and SHAP [10], as well as the model-specific Grad-CAM [18], which targets Convolutional Neural Network (CNN)-based models for computer vision tasks. Since enumerating all feature values involved in a prediction is often not useful, concept-based approaches, such as TCAV [7], address these limitations by finding the concept, e.g., a color and an object, embedded within the latent space learned by DNN. Anchors [17] is a representative from the class of rule-based explanation method, which explains individual predictions of any black box classification model by finding a decision rule that 'anchors' the prediction sufficiently.

In contrast to the aforementioned techniques, example-based explanations (EBE) enable an understanding of the model behavior through specific examples in the training data [11]. Central to this is the concept of example attribution, which refers to identifying and quantifying the influence of training examples on a model's prediction. Example attribution provides a direct connection between the model's predictions and its training data, offering insights into how specific examples guide the model's decision-making. A straightforward method for example attribution is the KNN algorithm, which bases predictions on the most common class among the nearest neighbors of an example. However, its reliance on raw data without feature extraction limits its performance in tasks requiring high-level feature understanding, such as CIFAR-10, where features like edges or textures are critical. Beyond KNN, methods for example attribution in other models, such as random forests, assign weights to training examples based on their contributions to predictions [2]. However, research exploring example attribution in the context of DNN remains sparse.

Our focus is on forming the example attribution of a test example, which involves identifying and presenting training examples that influence the model's prediction. Recent surveys provide different categorizations of EBE methods, see e.g., [15]. However, these methods often face limitations in practical deep learning tasks. For instance, the prototype and criticisms approach [6] focuses on representing the data distribution by selecting prototypes that are representative of all the data points and criticisms that are not well represented by the prototypes. However, this method does not directly provide example-level attributions for individual predictions, limiting its utility for explaining specific test examples. Similarly, counterfactual explanations [3] consist of slight modifications to the test instance such that the prediction is affected. While it may provide some insights, this technique relies on the generation of semantically meaningful counterfactuals, which is challenging and often requires prior knowledge.

To address these limitations, we propose EBE-DNN, a method that leverages the strengths of both KNN and DNN to form example attributions for test examples. As illustrated in Fig. 1, embeddings from a trained DNN are used as the input for KNN, enabling the retrieval of training examples that are most common to the test image in terms of the DNN's learned feature representation. By combining the interpretability of KNN-based example attributions with the feature extraction ability of DNN, EBE-DNN bridges the gap between explanation and model performance. It maintains the accuracy of the underlying DNN while providing example-based explanations. This approach ensures that example attributions are not only aligned with the model's decision-making process but also provide support for the predictions in the form of examples, enabling trust and transparency in deep learning models. Our method differs from previous work such as Deep k-Nearest Neighbors (DkNN) [12], which employs KNN in multiple DNN layers to estimate prediction credibility and detect out-of-distribution samples. In contrast, EBE-DNN is explicitly designed to improve interpretability by retrieving specific training examples that contribute to a test example's prediction, aiming for highly concentrated example attributions rather than global estimates of prediction confidence.

The remainder of this paper is structured as follows. In Sect. 2, we introduce details of EBE-DNN. In Sect. 3, we illustrate and evaluate EBE-DNN on image classification tasks. Finally, we summarize the main conclusions and discuss future work in Sect. 4.

2 EBE-DNN

The proposed approach, EBE-DNN, which is outlined in algorithm 1, operates by leveraging embeddings extracted from a specific layer of a deep neural network (DNN) to generate example attributions and predict the label of a test example. Given a set of training examples X with their corresponding labels y, and a deep neural network D, embeddings are generated with respect to a specified layer l, upper bounded by D_L (the number of layers in D), resulting in a transformed set of examples X'. The transformation step aims to convert the high-dimensional input examples to a more useful feature space, reflecting both low-level and high-level properties. Similarly, the test example X_{n+1} is converted into its embedding X'_{n+1} using the same transformation function. The k nearest neighbors of the test instance in the transformed space are retrieved, using some distance metric (δ). From these examples, which in the algorithm are denoted as $X'(1), \ldots, X'(k)$, where each (i) corresponds to the index of the example with rank i, the resulting example attribution for the test instance are obtained, i.e., $N = (X_{(1)}, y_{(1)}), \ldots, (X_{(k)}, y_{(k)})$. Finally, the predicted label for the test instance is given by the mode of the labels from the retrieved examples. The algorithm implicitly assumes uniform weighting of the k returned examples, i.e., each weight is $1/k$, but alternative weighting schemes are possible.

Algorithm 1. EBE-DNN

Require:
- $X = \{X_1, \ldots, X_n\}$ ▷ objects
- $y = \{y_1, \ldots, y_n\}$ ▷ labels
- D ▷ deep neural network
- $0 < k \leq n$ ▷ no. of examples
- $0 < l \leq D_L$ ▷ layer no.
- δ ▷ distance metric (e.g.,cosine)
- X_{n+1} ▷ test object

1: $X' \leftarrow Transform(X, D, l)$
2: $X'_{n+1} \leftarrow Transform(X_{n+1}, D, l)$
3: $\{X'_{(1)}, \ldots, X'_{(k)}\} \leftarrow KNN(X'_{n+1}, X', k, \delta)$
4: $N \leftarrow \{(X_{(1)}, y_{(1)}), \ldots, (X_{(k)}, y_{(k)})\}$
5: $\hat{y} \leftarrow Mode(\{y_{(1)}, \ldots, y_{(k)}\})$
6: **return** $\hat{y}, N$

3 Empirical Investigation

In this section, we investigate the example attributions generated by EBE-DNN and evaluate its predictive performance across three widely studied image classification tasks. Our primary objective is to demonstrate that EBE-DNN effectively provides example attributions for test images without compromising the predictive accuracy of the employed DNN.

3.1 Experimental Setup

We evaluate EBE-DNN using three widely adopted datasets: MNIST [9], Fashion-MNIST [19], and CIFAR-10 [8]. MNIST is a dataset of grayscale 28×28

images of handwritten digits (0–9), while Fashion-MNIST consists of grayscale 28×28 images of clothing items, such as T-shirts, pants, and shoes. CIFAR-10 is a more diverse dataset containing grayscale 36×36 images of 10 classes, including animals (e.g., cats, dogs, horses) and vehicles (e.g., cars, trucks, airplanes). MNIST and Fashion-MNIST each have 60 000 training images and 10 000 test images, while CIFAR-10 contains 50 000 training images and 10 000 test images.

We select ResNet18 [5] as the backbone model for EBE-DNN due to its proven effectiveness in image processing tasks. ResNet18 consists of 4 blocks and 20 convolutional layers, which are sequentially numbered from shallow to deep layers for reference. Among these, layers 8, 13, and 18 perform downsampling, enabling the network to reduce spatial dimensions. After training ResNet18 on each dataset using PyTorch [13] with a batch size of 128, a learning rate of 0.001, and 60 epochs, the model weights are frozen to ensure consistency during embedding extraction.

To generate example attributions, embeddings are obtained from each convolutional layer of ResNet18 for both training and test images. These embeddings, representing the semantic features learned by ResNet18, differ across layers due to the hierarchical nature of feature abstraction in networks. To investigate the impact of the choice of layer, we apply EBE-DNN to each layer, i.e., $l \in \{1, \ldots, 20\}$. The function KNN in the algorithm is using the Scikit-learn implementation [14], with cosine similarity as the distance metric. This metric was chosen as it is particularly well-suited for high-dimensional spaces and is robust to magnitude scaling introduced by operations like convolution, pooling, and batch normalization.

For illustration, we first present some example attributions provided by EBE-DNN. To investigate whether the example attribution comes at a cost in terms of predictive performance (accuracy), EBE-DNN is compared to the underlying DNN (ResNet18).

3.2 Illustration of Example Attributions

In this section, we present example attributions generated by EBE-DNN for two CIFAR-10 test images, a cat and an automobile, demonstrating how the method identifies training examples that contribute to the model's predictions, here using $k = 10$ neighbors and embeddings from layer 14 of ResNet18. In addition, we explore how example attributions vary across different layers for the automobile test image, showing how the supporting training examples may change with the choice of convolutional layer.

Example Attribution for a Cat Image. In Fig. 2a, the example attribution for a test image of a cat is shown, with all the training examples that contribute to a prediction being ranked. By inspecting the retrieved examples, some key properties such as fur patterns, facial structure, and ear shape, can be identified, e.g., most of the retrieved images show cats with a white patch of fur around their mouths or chests. Additionally, the round facial shapes in the examples

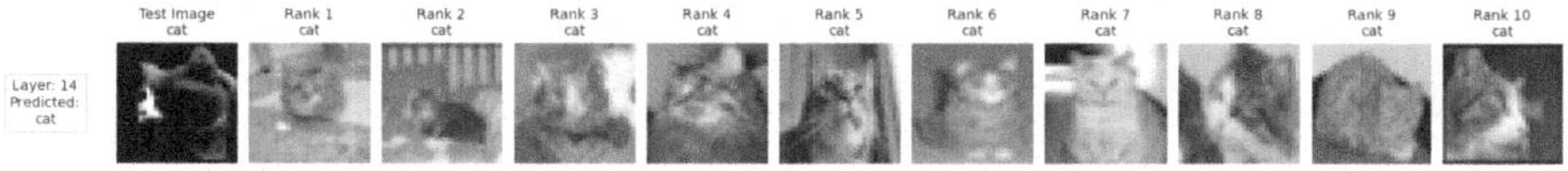

(a) Example attribution of a cat image using convolutional layer 14

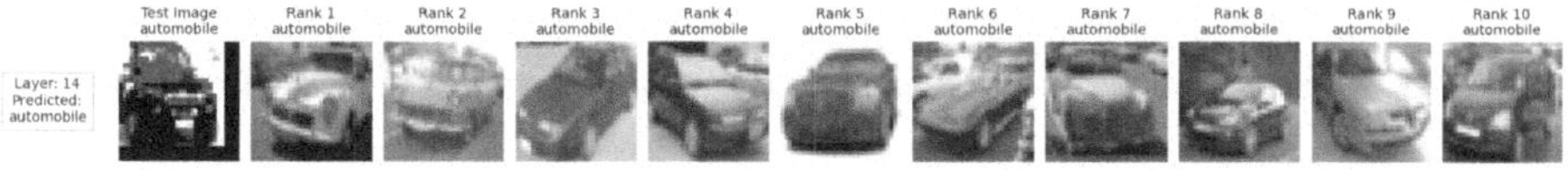

(b) Example attribution of an automobile image using convolutional layer 14

Fig. 2. Example attributions of cat and automobile test images

suggest that the model relies on structural similarity in facial features to classify the image. The triangular ear shapes observed across all examples further underscore their importance as a distinguishing property. While ranks 4 and 9 deviate slightly by lacking prominent white fur, other properties seem to be quite consistent across all ten examples.

Example Attribution for an Automobile Image. In Fig. 2b, the example attribution for a test image of an automobile is shown. The retrieved examples indicate that the test image is classified as an automobile based on properties such as the shape of the car's front face, the visibility of tires, and color similarity. A majority of the retrieved examples capture automobiles with a front face structure similar to that of the test image, with headlights positioned symmetrically and contours of the car hood reflecting the design of the test automobile. Although the examples ranked 2, 3, 10 deviate slightly with smaller car hoods influenced by shooting angles, they still maintain structural consistency with the test image. The presence of one or two visible tires across all examples further indicate their role as a critical property for classifying automobiles. Additionally, six of the retrieved images share a blue color similar to the test image, suggesting that color consistency contributes to classification but plays a secondary role compared to structural properties. This example attribution gives an indication of the ability of ResNet18 to capture visual and structural properties essential for classifying automobiles.

Example Attribution Across Layers. In Fig. 3, example attributions for the above automobile test image are presented, using embeddings from different convolutional layers of ResNet18.

In the top three layers, the retrieved images primarily reflect low-level properties, such as color and texture. For instance, many of the retrieved examples

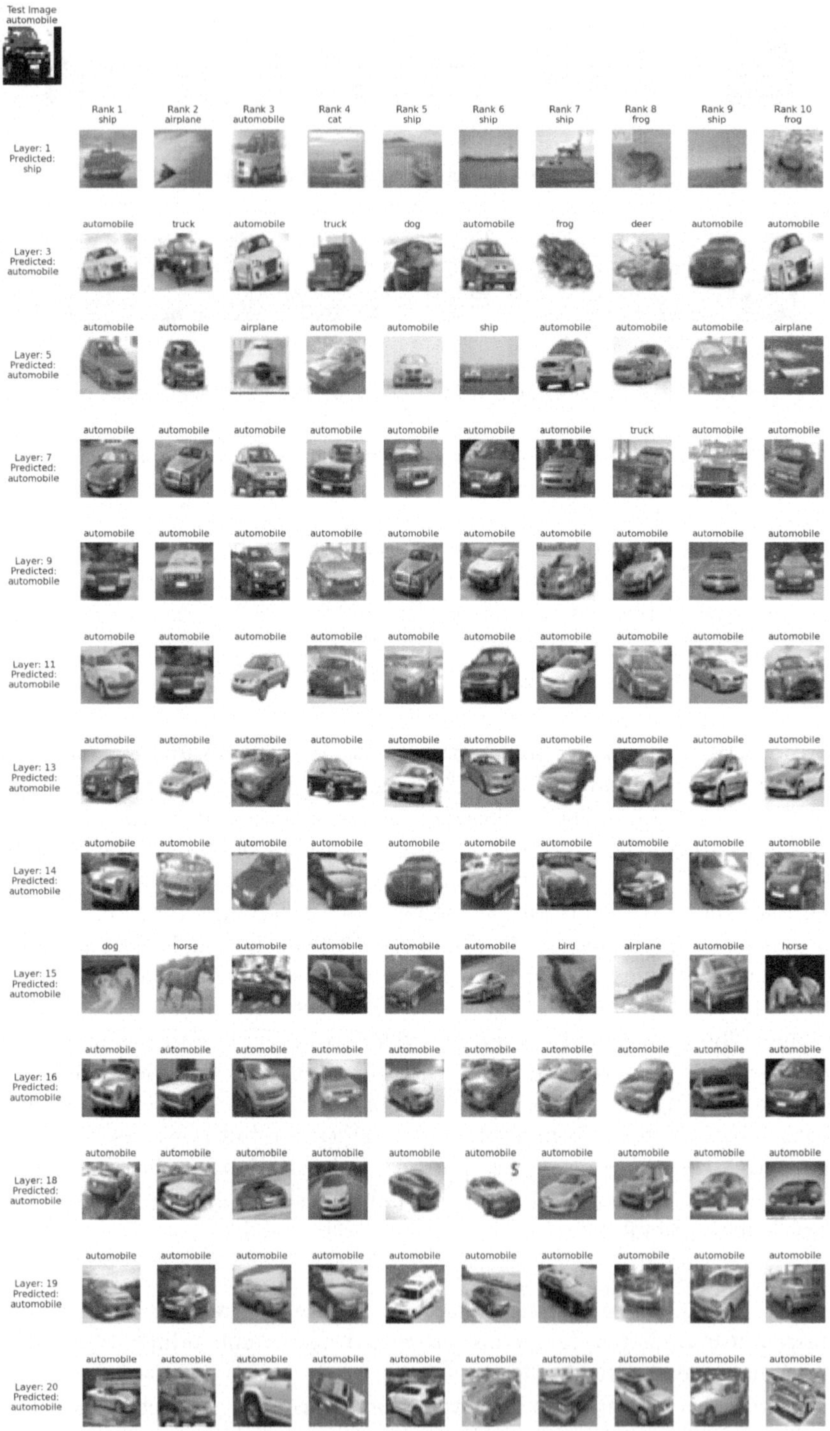

Fig. 3. Example attributions of an automobile test image from CIFAR-10

share the blue-and-white color scheme of the test image, which corresponds to the car's blue body and white background. Additionally, texture patterns, such as the smooth surfaces of the car body, are captured prominently. However, the retrieved images exhibit minimal structural consistency, as the embeddings at these layers focus on localized patterns rather than overall object structures. Notably, higher-ranked images (e.g., rank 1 and 2) generally show better alignment with the test image's color distribution and texture patterns compared to lower-ranked images, which display greater variability in these properties.

As the embeddings progress through middle layers, such as layers 7, 9, and 14, the retrieved examples begin to incorporate more high-level and discriminative properties. Properties like the car hood's shape, the placement of headlights, and the presence of visible tires become more prominent. These layers achieve a balance by preserving low-level details, such as color and texture, while simultaneously capturing structural features critical for classification. The rank order in this stage reflects the relevance of structural consistency: higher-ranked images exhibit more similar frontal shapes and component placements to the test image, while lower-ranked images still align structurally but may vary slightly in less significant features, such as the angle of the car hood or the positioning of tires.

In the deeper layers, such as layers 18, 19, and 20, the example attributions generalize further, capturing more abstract and category-level properties. While higher-ranked images maintain a closer similarity to the test image in terms of critical features like wheels and overall vehicle structure, lower-ranked images deviate more, often representing generalized characteristics of the automobile class. For instance, some lower-ranked images, such as layer 19 rank 8 and layer 20 rank 6, appear visually dissimilar to the test image, yet they retain essential discriminative properties, such as the presence of wheels and a land vehicle's structural form. This suggests that deeper layers prioritize semantic alignment over instance-specific details.

Through these observations of example attributions generated by EBE-DNN, we arrive at a conclusion consistent with [20]: deeper layers seem to generally produce more discriminative features. This affinity demonstrates that the EBE-DNN method through the example attributions offer some insights into ResNet18's classification mechanism and that it provides a transparent view of the decision-making process.

3.3 Predictive Performance

In this section, we compare the predictive performance of EBE-DNN to the original DNN, and investigate in particular the effect of the layer to use for the embeddings and the number of examples used in the explanations.

Results for MNIST and Fashion-MNIST. The predictive performance on the MNIST and Fashion-MNIST datasets are shown in Fig. 4 and Fig. 5 for different values of k (number of examples used in the attributions) and the choice of convolutional layers. Notably, despite limiting the number of training

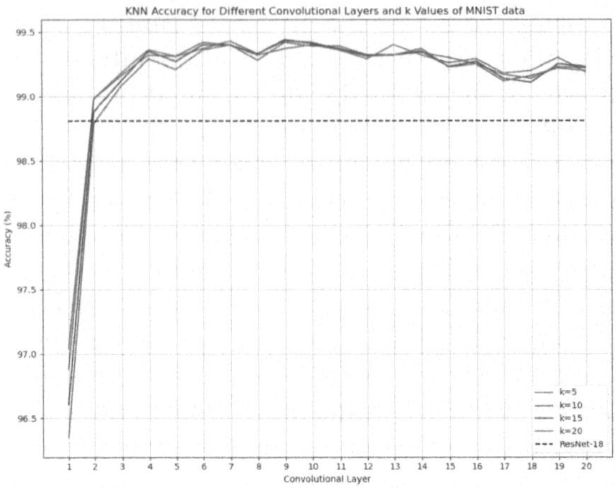

Fig. 4. Accuracy on MNIST for different layers and number of examples

examples contributing to a prediction, EBE-DNN does not seem to generally degrade performance compared to the original model. On the contrary, the predictive performance is improved in several cases. For MNIST, improvements are observed when using any layer, except for the top-most, independently of the choice of k. For Fashion-MNIST, similar results are observed for layers 9 and below. Somewhat surprisingly, the highest accuracies of are observed for when the embeddings are extracted from the middle convolutional layers of ResNet18, rather than the deepest layers, as one could have been expecting. Additionally, the choice of examples (k) has only a minor influence on accuracy compared to the significant impact of the embedding layer selection.

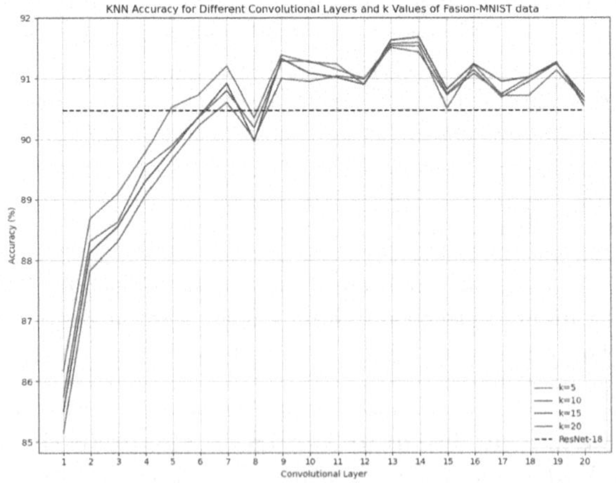

Fig. 5. Accuracy on Fashion-MNIST for different layers and number of examples

Results for CIFAR-10. The results for the CIFAR-10 dataset are shown in Fig. 6. EBE-DNN achieves accuracy comparable to ResNet18 when the embeddings are extracted from layers 16 and 19 of ResNet18. These layers are particularly effective because their embeddings capture class-specific properties, as demonstrated by the example attributions; most retrieved images belong to the same class as the test image. Interestingly, at layer 15, predictive performance drops significantly, consistent with the attributions presented in Fig. 3, where the attributions for this layer included images from multiple other classes. This suggests that embeddings obtained from layer 15 generalize across classes, reducing both attribution quality and accuracy. The results also show that the choice of k has less impact on accuracy compared to layer selection. The results indicate that choosing $k = 10$ for this dataset seems to give the best balance of predictive performance and interpretability (as measured by the number of training examples included in the explanations).

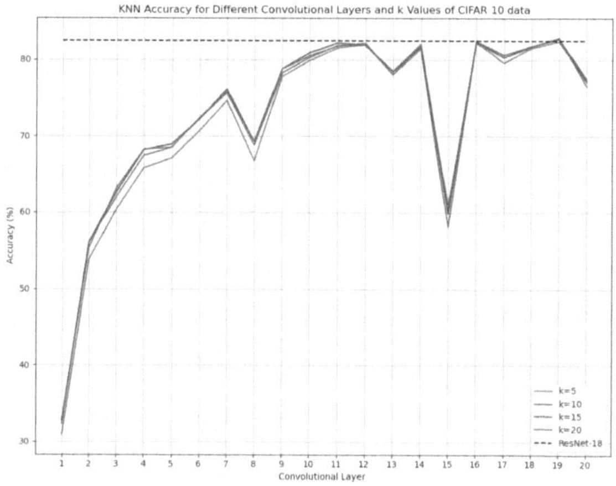

Fig. 6. Accuracy on CIFAR-10 for different layers and number of examples

4 Concluding Remarks

A novel approach to obtaining example attributions from deep neural networks, called EBE-DNN, has been introduced, which combines KNN with the feature extraction capabilities of DNN to generate example attributions that are both interpretable and aligned with model predictions. By associating test examples with the training data that influence their predictions, EBE-DNN offers a transparent and interpretable framework for emulating the predictions of deep learning models, while maintaining competitive predictive performance.

Our investigation has highlighted the progression of properties captured across ResNet18 layers. Early layers focus on low-level properties like color and texture, which provide surface-level similarities but lack structural information. Middle layers extract distinctive structural features critical for classification, achieving a balance between preserving low-level details and capturing high-level abstractions. Deeper layers generalize to category-level properties, enabling the model to recognize diverse instances within a class but often at the expense of fine-grained details. These insights are consistent with prior work indicating that higher layers produce more discriminative features, demonstrating the interpretability and effectiveness of EBE-DNN in forming example attributions.

In terms of predictive performance, EBE-DNN achieves accuracy comparable to or exceeding ResNet18 across datasets, with middle layers delivering the best results. By leveraging embeddings that capture structural features critical for classification, EBE-DNN balances interpretability and predictive performance. The method performs consistently across different settings of k (number of examples to include in the attributions), with $k = 10$ providing a reasonable trade-off between capturing sufficient training examples for accurate predictions and maintaining interpretability by avoiding excessively large sets of retrieved examples.

While effective, EBE-DNN presents potential opportunities for further improvement. One area for future work is addressing the dependence on the choice of convolutional layers, which currently acts as a hyperparameter requiring tuning. Developing automated or adaptive strategies for selecting the most effective layers for attributions and predictions could enhance usability and efficiency. Additionally, while KNN provides interpretable example attributions, it can become computationally expensive for high-dimensional embeddings or large-scale datasets. Future research could explore scalable implementations or alternative similarity-based approaches to reduce computational cost. Other directions concern investigating alternative distance metrics and forming predictions by weighted voting, e.g., using the distances. Another direction for future research is to complement the example attributions with feature attributions, to more clearly determine the factors influencing the model's predictions. Another direction is a more formal identification and measurement of the concepts used by the model at each layer to make two pictures closer or farther in the latent space. Finally, further exploration is needed to determine whether the EBE-DNN framework is equally effective in other domains, such as text or time-series data, potentially broadening its applicability across machine learning tasks.

Acknowledgments. This project was funded by the Wallenberg AI, Autonomous Systems, and Software Program (WASP).

References

1. Adadi, A., Berrada, M.: Peeking inside the black-box: a survey on explainable artificial intelligence (xai). IEEE Access **6**, 52138–52160 (2018)
2. Boström, H.: Example-based explanations of random forest predictions. In: International Symposium on Intelligent Data Analysis, pp. 185–196. Springer (2024)

3. Guidotti, R.: Counterfactual explanations and how to find them: literature review and benchmarking. Data Min. Knowl. Disc. **38**(5), 2770–2824 (2024)
4. Gunning, D., Aha, D.: Darpa's explainable artificial intelligence (xai) program. AI Mag. **40**(2), 44–58 (2019)
5. He, K., Zhang, X., Ren, S., Sun, J.: Deep residual learning for image recognition. In: Proceedings of the IEEE Conference on Computer Vision and Pattern Recognition, pp. 770–778 (2016)
6. Kim, B., Khanna, R., Koyejo, O.O.: Examples are not enough, learn to criticize! criticism for interpretability. Adv. Neural Inform. Process. Syst. **29** (2016)
7. Kim, B., Wattenberg, M., Gilmer, J., Cai, C., Wexler, J., Viegas, F., et al.: Interpretability beyond feature attribution: quantitative testing with concept activation vectors (tcav). In: International Conference on Machine Learning, pp. 2668–2677. PMLR (2018)
8. Krizhevsky, A., Hinton, G., et al.: Learning multiple layers of features from tiny images (2009)
9. LeCun, Y.: The mnist database of handwritten digits. http://yannlecun.com/exdb/mnist/ (1998)
10. Lundberg, S.: A unified approach to interpreting model predictions. arXiv preprint arXiv:1705.07874 (2017)
11. Molnar, C.: Interpretable machine learning. Lulu. com (2020)
12. Papernot, N., McDaniel, P.: Deep k-nearest neighbors: towards confident, interpretable and robust deep learning. arXiv preprint arXiv:1803.04765 (2018)
13. Paszke, A., et al.: Pytorch: an imperative style, high-performance deep learning library. Adv. Neural Inform. Process. Syst. **32** (2019)
14. Pedregosa, F.: Scikit-learn: machine learning in python fabian. J. Mach. Learn. Res. **12**, 2825 (2011)
15. Poché, A., Hervier, L., Bakkay, M.C.: Natural example-based explainability: a survey. In: World Conference on EXplainable Artificial Intelligence, pp. 24–47. Springer (2023)
16. Ribeiro, M.T., Singh, S., Guestrin, C.: " Why should i trust you?" explaining the predictions of any classifier. In: Proceedings of the 22nd ACM SIGKDD International Conference on Knowledge Discovery and Data Mining, pp. 1135–1144 (2016)
17. Ribeiro, M.T., Singh, S., Guestrin, C.: Anchors: high-precision model-agnostic explanations. In: Proceedings of the AAAI Conference on Artificial Intelligence, vol. 32 (2018)
18. Selvaraju, R.R., Cogswell, M., Das, A., Vedantam, R., Parikh, D., Batra, D.: Grad-cam: visual explanations from deep networks via gradient-based localization. Int. J. Comput. Vision **128**, 336–359 (2020)
19. Xiao, H., Rasul, K., Vollgraf, R.: Fashion-mnist: a novel image dataset for benchmarking machine learning algorithms. arXiv preprint arXiv:1708.07747 (2017)
20. Zeiler, M.: Visualizing and understanding convolutional networks. In: European Conference on Computer Vision/arXiv, vol. 1311 (2014)

Relevance-Aware Algorithmic Recourse

Dongwhi Kim(✉) and Nuno Moniz

University of Notre Dame, Notre Dame, IN, USA
{dkim37,nuno.moniz}@nd.edu

Abstract. As machine learning continues to gain prominence, transparency and explainability are increasingly critical. Without an understanding of these models, they can replicate and worsen human bias, adversely affecting marginalized communities. Algorithmic recourse emerges as a tool for clarifying decisions made by predictive models, providing actionable insights to alter outcomes. They answer, "What do I have to change?" to achieve the desired result. Despite their importance, current algorithmic recourse methods treat all domain values equally, which is unrealistic in real-world settings. While algorithmic recourse is extensively studied in classification tasks, its application to regression tasks remain scarce. In this paper, we propose a novel framework, Relevance-Aware Algorithmic Recourse (RAAR), that leverages the concept of relevance in applying algorithmic recourse to regression tasks. Relevance can supplement other algorithmic recourse methods, but we focus on Bayesian optimization-based methods as a baseline in this paper. We conducted multiple experiments on 15 datasets to outline how relevance influences recourses. Results demonstrate that our approach is comparable to well-known baselines while achieving greater efficiency, measured by shorter computation times and fewer iterations, and lower relative costs, indicated by more minor modifications required to achieve desired outcomes. The results, datasets, and code for replicating this study are available on GitHub.

Keywords: Algorithmic Recourse · Fairness · Domain Relevance

1 Introduction

The increased adoption of machine learning has drawn attention from governments and large organizations in regulation [1]. Evidenced by the "right to explanation" introduced by the EU General Data Protection Regulation (GDPR), it seeks to safeguard individuals adversely affected by semi-autonomous decisions, giving them the right to understand to understand the decision-making process [2]. Similarly, within the United States, the "right to explanation" is enforced in credit scoring, ensuring that creditors explain the key factors that influence an individual's credit score [3]. Consequently, there is a growing emphasis on developing programs and analyses that provide avenues for recourse to those impacted by semi-autonomous decisions [4]. Regulation is necessary to manage the associated risks associated with AI while promoting its development.

G. Krempl et al. (Eds.): IDA 2025, LNCS 15669, pp. 444–455, 2025.
https://doi.org/10.1007/978-3-031-91398-3_33

Counterfactual explanations are excellent tools to mitigate risks, promote the ethical use of AI, and abide by regulations [5]. They outline the necessary conditions to achieve a desired outcome, offering insights into complex models. In addition, they reduce the regulatory burden by providing practical recourse options and align with the GDPR's "right to explanation" [4].

In recent literature, multiple approaches exist for algorithmic recourse, i.e., counterfactuals, from gradient-based methods to integer programming-based approaches and probabilistic optimizations [6–8]. These methods primarily target classification tasks, while regression-specific solutions remain scarce. For instance, one method introduces a latent-space disentanglement framework for high-dimensional regression models, focusing on maintaining the proximity of recourses. Yet, these works overlook domain relevance [9].

Motivation Suppose an individual arrives at a bank seeking a $100,000 loan but receives only $50,000. In such a scenario, algorithmic recourse is only valuable if it offers actionable and realistic counterfactuals, guiding the applicant on precise, feasible adjustments to become eligible. As such, disregarding user preferences can lead to useless and impractical explanations. Standard approaches lack consideration for user preferences over the domain, i.e., relevance, and generally assume all values in a domain are equally crucial in obtaining an accurate prediction [10]. This issue is especially prevalent in regression tasks, where the target is not a single class but a continuum of values. To address this, we introduce the concept of relevance, which helps define the importance of outcomes that matter most from the user's perspective, such as a desired loan amount spanning from $100,000 to $120,000. Additionally, relevance is especially useful for imbalanced regression tasks when predicting particular values [10,11]. For instance, if we had a small group of individuals who could get a $100,000,000 loan, a counterfactual would help us understand what changes a typical individual could make to be part of such a group or help the loan model prevent these extreme cases.

Contribution. In this paper, we introduce the first framework for regression algorithmic recourse, Relevance-Aware Algorithmic Recourse (RAAR), which leverages domain relevance to generate accurate and realistic counterfactuals [10]. RAAR enhances decision-making by providing actionable explanations that maximize outcomes relative to a user's unique constraints and preferences. We implement RAAR using Bayesian optimization because it is globally convergent, model-agnostic, and sample efficient [12,13]. Our experiments demonstrate that RAAR is comparable to existing baselines in validity and sparsity while achieving faster convergence and fewer changes. As a result, we enable users to act on recourses more easily.

2 Related Works

Algorithmic recourse provides actionable recourse to individuals affected by semi-autonomous decisions [14]. It answers: **how can I change my situa-**

tion [15,16]? As mentioned previously, we explore relevance's impact on algorithmic recourse. Therefore, although there are many other approaches papers have outlined, from enforcement of sparsity in counterfactuals to plausibility constraints to making models robust to dataset shifts in the real world, we concentrate on the use of domain preferences, i.e., relevance [14,17–19]. This paper focuses on the scope of Bayesian optimization methods on regression tasks to generate recourses.

2.1 Bayesian Optimization

Bayesian optimization is a global optimization problem to solve:

$$x^* = \max_{x \epsilon X} f(x), \quad (1)$$

where $f(x)$ is the objective function, X is the set of features, x is a feature, and x^* is the feature for the global maximum of the objective function [20]. When $f(x)$ is unknown, Bayesian optimization treats it as a black-box function, a scenario commonly encountered in algorithmic recourse [21]. The methodology revolves around three primary components: objective function (black box), surrogate model (Gaussian process), and acquisition function [22].

Algorithm 1. Bayesian optimization pseudo code

```
function BayesOpt(f, M_0, N, A)
    H ← φ
    for n ← 1 to N do
        x* ← max_x A(x, M_{n-1}),
        evaluate f(x*)
        H ← H ∪ (x*, f(x*)),
        Fit a new model M_n to H
    end for

    return H
end function
```

Pseudo Code of the Algorithm. Where H is the observation history of the pair (feature, value), N is the max number of iterations, f is the true objective function (black box), M is the surrogate function (Gaussian process), A is the Acquisition function, x^* is the next feature chosen to evaluate [23].

Surrogate Model (Gaussian Process). It uses the model's history to construct multivariate Gaussian distributions to predict the objective function. The *mean* and *covariance function* decide where to steer the acquisition function.

In our paper, we use the *matern (covariance) kernel function* along with this. Matern kernels are a generalization of power exponential kernels where the additional parameter ν controls the smoothness of the function. The larger ν is the smoother the function until $\nu \rightarrow \infty$, where the matern kernel becomes a power exponential kernel [21]. A 2.5 matern kernel has $\nu = 2.5$ for twice differentiable functions. This kernel is given by:

$$k(x_i, x_j) = \frac{1}{\Gamma(\nu)2^{\nu-1}} \left(\frac{\sqrt{2\nu}}{l} d(x_i, x_j) \right)^{\nu} K_{\nu} \left(\frac{\sqrt{2\nu}}{l} d(x_i, x_j) \right), \tag{2}$$

where k is the kernel, $d(x_i, x_j)$ is the euclidean distance between x_i and x_j, ν is the smoothness of the function, $K_{\nu}()$ is the modified Bessel function, and $\Gamma(\nu)$ is the gamma function [22].

Acquisition Function. This is optimized to balance the exploration and exploitation of the objective function. We use the *Upper Confidence Bound* which is a simple and common function, given by:

$$a(x; \lambda) = \mu(x) + \lambda\sigma(x), \tag{3}$$

where $\mu(x)$ is the expected performance of our surrogate model that controls the exploration, and $\sigma = \sqrt{k(x_i, x_j}$ which adds uncertainty through the standard deviation (root of the kernel), in effect controlling the exploitation [22]. The larger λ is, the more this function is rewarded for searching unknown spaces, leading to more exploration [20].

2.2 Relevance

The concept of relevance is defined by Ribeiro and Moniz [10] as a function given by:

$$\phi(Y) : \mathcal{Y} \rightarrow [0, 1], \tag{4}$$

where ϕ is the continuous relevance function, displaying the bias of domain $\mathcal{Y}$ by mapping it to the continuous range $[0, 1]$ with 0 being the minimum and 1 being the maximum relevance [24]. To calculate the relevance of all points, we require:

a) ***control points*** which are points with a known relevance
b) ***interpolation function*** that can determine the relevance of other points [10].

Control points are required for interpolation, with a minimum of three: the minimum, median, and maximum points for relevance. The control points can be defined by the equation given by:

$$S = \{\langle y_k, \varphi(y_k), \varphi'(y_k)\rangle\}_{k=1}^{s}, \tag{5}$$

where y_k is the control point, $\varphi(y_k)$ is the relevance at y_k, and $\varphi'(y_k)$ is the derivative of the relevance function at that point. By default, $\varphi'(y_k)$ will be zero because the control points are assumed to be the minimums and maximums of the function [10]. Default generation of a relevance function follows data distribution, making common values less relevant and rare values more relevant. If we have domain knowledge, we can set the relevance of control points to reflect their real-world significance. Concerning interpolation, [10] proposes the use of *pchip* (Piecewise Cubic Hermite Interpolating Polynomials) to achieve the desirable conditions, such as continuity in generated points.

Relevance is also an important concept in addressing imbalanced regression tasks, where the goal is to accurately predict high-importance value ranges. For instance, in stock market forecasting, we'd want to accurately predict stock market crashes or surges to better understand the market [10,11,25]. Generating counterfactuals for these rare cases can help users understand their situation better and highlight bias in the model.

3 Our Framework: Relevance-Aware Algorithmic Recourse

This section details our framework, Relevance-Aware Algorithmic Recourse (RAAR). First, we introduce background information and details about our algorithmic recourse problem. Then, we introduce our objective function and its optimization.

3.1 Preliminaries

Through Bayesian optimization, we optimize for a particular outcome by applying changes to x, where $f(x + \epsilon)$ is the new outcome with a cost of ϵ. Unlike most studies with algorithmic recourse, we set bounds for the features given by $\epsilon \in [-cx, cx]$ where c constrains the search space and ensures the recourse is actionable. We explore the solution space through the Gaussian process, generating a counterfactual distribution. Formally, we have a regression model $f : \chi \rightarrow \mathbb{R}$ with the Gaussian Process surrogate Model, $GP(\mu_n, K_n)$. This surrogate model accepts a mean function $\mu_n : \chi \rightarrow \mathbb{R}$ and a covariance function $K_n : \chi * \chi \rightarrow \mathbb{R}$, with an Upper Confidence Bound acquisition function [21,26].

3.2 Formulating and Optimizing Our Objective

To better explain the formulation and optimization of our objective functions, let us revisit the example of the individual applying for a \$100,000 loan at a bank but being offered only \$50,000. In classification tasks, most approaches would minimize the feature costs to change the class label from \$50,000 to \$100,000. These approaches also assume all domain values are equal, which fails to account for the varying significance of loan amounts to the individual. As a result, we propose and compare two approaches since this is a regression task. For our baseline, we use Bayesian optimization to find the counterfactual for the y that the individual wants ($100,000). Here, the domain values are treated as being of equal relevance.

Our second approach, relevance-focused optimization, instead introduces a mapping between the loan amount (y) and a relevance score, emphasizing outcomes that align more closely with the target. By incorporating relevance, we prioritize solutions that approach the target value, treating domain values nearer to the target as significantly more important and allowing their importance to decrease exponentially as they move farther away. This approach reflects more realistic modeling of decision-making processes where minor improvements near a critical threshold have a disproportionately higher impact than similar improvements far from the threshold.

Beyond target optimizations, we may also look for the maximum improvement we could achieve, given minor changes to our features. Unlike traditional algorithmic recourse, which attempts to find the small feature costs to change the class label, we focus on finding the outcome. As a result, to control how actionable these resources are, we set feature bounds. We use ±5% for maximum loan optimizations to focus on actionable recourses with minimum changes and ±100% for a loan target ($100,000). However, relevance-focused optimizations introduce a pseudo-bound by plateauing at a relevance of one beyond a certain threshold, further increasing their actionability. We compare these two optimizations of our objective functions to analyze the impact of relevance:

1) ***Y optimizations*** (baseline): Bayesian optimization to find the counterfactual that maximizes the loan amount (outcome) or one that is closest to a specific target (e.g. $100,000).
2) ***Relevance optimizations***: Bayesian optimization to find the counterfactual that maximizes the relevance (i.e., 1). Changing the relevance function will change whether we want to maximize or find a specific target (e.g.$100,000).

Y Optimization. More formally, for y, our maximum optimizations use the objective function as follows:

$$x^* = \arg\max_{x^* \epsilon X} \frac{f(x^*) - f(x)}{|f(x)|} \tag{6}$$

Target optimizations optimize for the point that achieves the closest value to our target value.

$$x^* = \arg\max_{x^* \epsilon X} \frac{-|f(x^*) - y_{target}|}{|y_{target} - f(x)|} \tag{7}$$

Relevance Optimization. We optimize for relevance through using $\phi(f(x))$ instead of $f(x)$, where $\phi(y)$ is the relevance function. The relevance function further pseudo-bounds the optimization process to provide a realistic and relevance-bound recourse. We use the equation below for both maximum and target optimizations.

$$x^* = \arg\max_{x^* \epsilon X} \phi(f(x^*)) \tag{8}$$

The above ϕ function can be generated by changing whether we optimize for a maximum relevance value or a target y value. We generate an automated (distribution-based) relevance function with only the right extreme for the maximum relevance value. We use the training set to optimize for its most extreme values, investigating how using the relevance function for optimization compares with target value optimizations. We generate a unique relevance function for target values with both extremes for every target value we want to reach (local). To define a unique relevance function for a target value, we define its control points as follows:

$$S = \{\langle y_{min}, 0, 0\rangle, \langle y_{original}, 0.5, 0\rangle, \langle y_{target}, 1, 0\rangle, \langle y_{max}, 0, 0\rangle, \} \tag{9}$$

Setting these control points as above ensures that the optimization direction is always at maximum relevance or y_{target}.

4 Experiments

This paper aims to analyze if Relevance-Aware Algorithmic Recourse (RAAR) generates better counterfactuals. Here, we discuss the experimental process of evaluating our novel framework Fig. 1. We investigate through two methods of generating a recourse, with two different optimization objectives and two regression models each. The following describes how we leverage the well-known learning algorithms Random Forest and LGBM in our optimization [27,28].

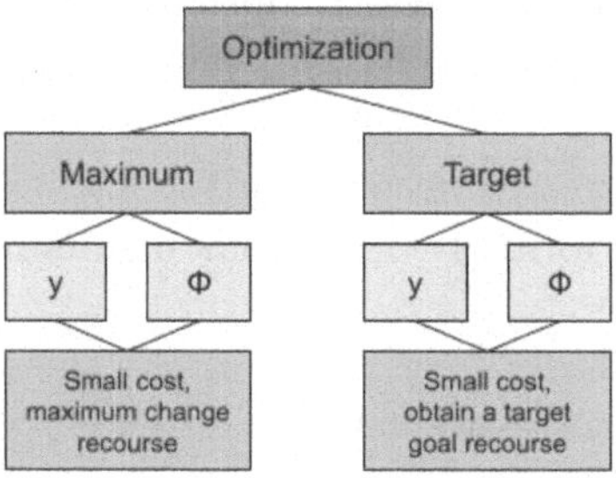

Fig. 1. Experimental Process

4.1 Experimental Setup

We compiled 15 datasets to run the process above, ranging from insurance expenses and mortgages to wine ratings, which can be accessed on Github. Each dataset consists of regression values that are used for RAAR. During every optimization process, we use Bayesian optimization to reach our target. For (1), the target is the maximum positive difference we can get while minimizing the cost, which is the change to features (e.g., the maximum loan with the minimum feature change). For (2), the target is any value we provide (e.g., $100,000), and we used four different percentage (e.g., +50%) increases to analyze how close we can get to a target value to replicate real-world scenarios. The results of these experiments answer how relevance affects recourses and how the two methods compare.

4.2 Maximum Optimization Results

Before comparing y and relevance-based recourses for maximum optimizations, it is essential to note that the relevance function introduces a pseudo-bound by plateauing at a relevance of 1 beyond a certain threshold. This results in more realistic, actionable recourses, as the function stops prioritizing extreme values that may be much farther away. In contrast, optimizing directly for y allows for continuous maximization, often resulting in less actionable recourses Table 1.

Table 1. Maximum Optimization Data Results. Larger $\Delta y\%$ and ϕ is better. Δd is the distance difference between the y and relevance optimization models w.r.t. relevance. Negative Δd means the relevance model produces cheaper counterfactuals. Boldface denotes the best performance for each metric.

Model	Optimization	$\Delta y\%$ Mean	$\Delta y\%$ SD	ϕ Mean	ϕ SD	Δd
RF	**Y**	**30.2**	61.74	0.18	0.23	−0.07%
	ϕ	25.36	60.98	0.18	0.23	
LGBM	**Y**	**41.23**	65.37	0.24	0.26	−0.12%
	ϕ	36.62	63.8	0.24	0.26	

Overall, y-optimization recourses achieve better outcomes than relevance-based models but at a higher cost. Relevance-based recourses, while less optimal in outcome, incur lower costs (i.e., shorter distances) due to the pseudo-bound discussed above. This bounded optimization is evident as both approaches reached the same relevance level, confirming that relevance functions effectively constrain extreme cases. This effect occurs because relevance functions are automatically generated to prioritize larger, rarer outcomes without additional guidance. In the context of the previous example, a y-based optimization might suggest a recourse of $200,000 but require an additional 30 h of work per week. In contrast, a relevance-based optimization might yield a recourse of $150,000, requiring only 20 additional hours per week.

4.3 Target Optimization Results

However, in many scenarios, individuals are more interested in understanding what adjustments they can make to achieve a specific outcome rather than simply maximizing it. We focused on optimizing for a target value instead of a maximum to address this. We then compared the y-difference model with the relevance model for predictions at 10%, 20%, 50%, and 100% above the original outcome. We manually designed relevance functions for the relevance model to reflect each of these target values.

Table 2. Target Optimization Data Results. ΔIt is the difference in iterations required to reach their best recourse between the y and relevance optimization models w.r.t relevance. Negative ΔIt means relevance model converges quicker. Δd is the distance difference between the y and relevance optimization models w.r.t. relevance. Negative Δd means the relevance model produces cheaper counterfactuals. Boldface denotes the best performance for each metric.

Model	Δy_{target}%	Opt.	Δy%	Y SD	ϕ Mean	ϕ SD	ΔIt	Δd
RF	10%	**Y**	**9.57%**	1.25	0.94	0.13	−6.24%	−2.15%
		ϕ	13.14%	1.52	0.96	0.10		
	20%	Y	18.02%	0.74	0.90	0.17	−3.48%	−1.13%
		$\boldsymbol{\phi}$	**20.78%**	0.89	0.93	0.13		
	50%	Y	41.39%	0.41	0.87	0.18	−1.19%	−0.33%
		$\boldsymbol{\phi}$	**42.63%**	0.48	0.90	0.13		
	100%	**Y**	**55.87%**	0.27	0.65	0.19	−5.71%	−0.82%
		ϕ	48.83%	0.31	0.74	0.12		
LGBM	10%	**Y**	**9.61%**	0.44	0.96	0.12	−4.98%	−1.91%
		ϕ	13.09%	0.82	0.98	0.08		
	20%	Y	18.71%	0.27	0.94	0.13	−5.89%	−2.17%
		$\boldsymbol{\phi}$	**21.08%**	0.50	0.96	0.10		
	50%	**Y**	**44.03%**	0.22	0.90	0.16	−6.15%	−1.76%
		ϕ	43.40%	0.32	0.91	0.12		
	100%	**Y**	**64.12%**	0.2	0.66	0.18	−8.38%	−2.04%
		ϕ	53.32%	0.24	0.76	0.12		

We evaluate the effectiveness of the two optimization processes in reaching a target value. In the table, the y-optimization model achieves recourses close to the target approximately 60% of the time. However, as shown in Table 2, Fig. 2a, and Fig. 2b, relevance optimization models reach similar targets more quickly and at lower cost, as indicated by the negative percentage.

However, relevance models beat the y model 100% of the time when comparing the mean relevance obtained. Additionally, when given a moderate target value such as 20% or 50%, relevance optimizations reach closer, more actionable recourses in quicker time than y-optimizations. When the target increases to 100%, this devolves into a maximum optimization which relevance optimizations perform worse as shown in the previous experiment.

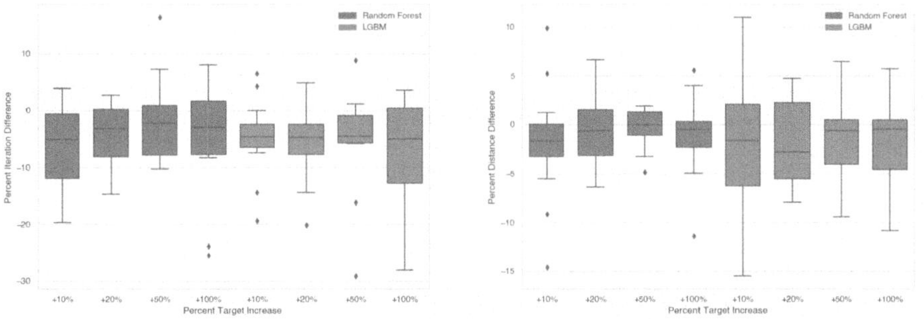

(a) **Percent Optimization Iteration Diff.** (b) **Percent Optimization Distance Diff.**

Fig. 2. Optimization Iteration and Distance Comparisons

Overall, while relevance optimizations produce outcomes comparable to y-optimizations, they do so more efficiently by reducing the number of iterations required and minimizing the distance between the recourses and the original points. This introduces a trade-off, balancing improved efficiency with a minor decrease in accuracy and precision. In the context of the previous example, both y-based and relevance-based optimizations reach $100,000 loan, but the relevance-based optimization requires less change in our current situation.

These differences in performance stem from the underlying structure of the objective functions. Optimizing for y is more straightforward and will attempt to get as close as possible to the desired value. However, since the relevance model beats the y model in relevance obtained every time, this suggests that a better result could be achieved with more domain knowledge.

5 Conclusion and Future Work

In this paper, we introduced a novel framework for relevance-aware algorithmic recourse (RAAR) to address the limitations of uniformly weighted values, which is often impractical in real-world scenarios by supplementing existing methods. By incorporating relevance, a concept that assigns variable importance to different values, we enabled the model to prioritize changes that better align with practical goals. Utilizing Bayesian optimization, we leveraged objective functions, a Gaussian process, and an acquisition function to generate recourses that maximize impact with minimal cost or meet targeted outcomes. Results across 15 datasets reveal that relevance-based recourses are faster and more cost-effective than y-based recourses, often achieving comparable results, thereby underscoring their potential in real-world applications. The results, datasets, and code for replicating this study are available on https://github.com/JohndKim/Relevance-Aware-Algorithmic-Recourse.

This work opens pathways for further development in relevance-based recourse strategies. Future research could introduce new cost functions that further focus on balancing feature costs and desired outcomes, or investigate how varying shapes of relevance functions might improve recourse accuracy and precision. Additionally, integrating relevance-based recourse with other approaches other than Bayesian optimization may provide further insights into how relevance can enhance the effectiveness of other recourse solutions in diverse contexts.

References

1. Black, J., Murray, A.D.:"Regulating AI and machine learning: setting the regulatory agenda". Eur. J. Law Technol. **10** (2019)
2. July 2018
3. Code of federal regulation. https://www.ecfr.gov/current/title-12/chapter-X/part-1002/subpart-A/section-1002.9. Accessed 28-Jan-2024
4. Voigt, P., Bussche, A.: The eu general data protection regulation (gdpr): a practical guide (2017)
5. Sokol, K., Flach, P.: Counterfactual explanations of machine learning predictions: opportunities and challenges for AI safety. In: 2019 aaai workshop on Artificial Intelligence Safety, Safeai (2019) (CEUR Workshop Proceedings, 2019)
6. Wachter, S., Mittelstadt, B.D., Russell, C.: Counterfactual explanations without opening the black box: automated decisions and the GDPR. CoRR, abs/1711.00399 (2017)
7. Ustun, B., Spangher, A., Liu, Y.: Actionable recourse in linear classification. In: Proceedings of the Conference on Fairness, Accountability, and Transparency, FAT* '19, pp. 10–19 (2019)
8. Karimi, A.-H., Von Kügelgen, J., Schölkopf, B., Valera, I.: Algorithmic recourse under imperfect causal knowledge: a probabilistic approach. Adv. Neural. Inf. Process. Syst. **33**, 265–277 (2020)
9. Zhao, X., Broelemann, K., Kasneci, G.: Counterfactual explanation for regression via disentanglement in latent space (2023)
10. Ribeiro, R.P., Moniz, N.: Imbalanced regression and extreme value prediction. Mach. Learn. **109**, 1803–1835 (2020)
11. Japkowicz, N., Stephen, S.: The class imbalance problem: a systematic study. Intell. Data Anal. **6**, 429–449 (2002)
12. Wang, X., Jin, Y., Schmitt, S., Olhofer, M.: Recent advances in bayesian optimization (2022)
13. Jones, D.R., Schonlau, M., Welch, W.J.: Efficient global optimization of expensive black-box functions. J. Glob. Optim. **13**, 455–492 (1998)
14. Karimi, A.-H., Barthe, G., Schölkopf, B., Valera, I.: A survey of algorithmic recourse: definitions, formulations, solutions, and prospects. arXiv preprint arXiv:2010.04050 (2020)
15. Venkatasubramanian, S., Alfano, M.: The philosophical basis of algorithmic recourse. In: Proceedings of the 2020 Conference on Fairness, Accountability, and Transparency, FAT* '20, pp. 284–293 (2020)
16. Joshi, S., Koyejo, O., Vijitbenjaronk, W., Kim, B., Ghosh, J.: Towards realistic individual recourse and actionable explanations in black-box decision making systems (2019)
17. Barocas, S., Selbst, A.D., Raghavan, M.: The hidden assumptions behind counterfactual explanations and principal reasons. In: Proceedings of the 2020 Conference on Fairness, Accountability, and Transparency, FAT* '20, pp. 80–89 (2020)

18. Karimi, A.-H., Schölkopf, B., Valera, I.: Algorithmic Recourse: from Counterfactual Explanations to Interventions. Virtual Event, Canada (2021)
19. Upadhyay, S., Joshi, S., Lakkaraju, H.: Towards robust and reliable algorithmic recourse. Adv. Neural. Inf. Process. Syst. **34**, 16926–16937 (2021)
20. Frazier, P.I.: A tutorial on bayesian optimization. arXiv preprint arXiv:1807.02811 (2018)
21. Snoek, J., Larochelle, H., Adams, R.P.: Practical bayesian optimization of machine learning algorithms. Adv. Neural Inform. Process. Syst. **25** (2012)
22. Schulz, E., Speekenbrink, M., Krause, A.: A tutorial on gaussian process regression: modelling, exploring, and exploiting functions. J. Math. Psychol. **85**, 1–16 (2018)
23. Bergstra, J., Bardenet, R., Bengio, Y., Kégl, B.: Algorithms for hyperparameter optimization. Adv. Neural Inform. Process. Syst. **24** (2011)
24. Torgo, L., Branco, P., Ribeiro, R.P., Pfahringer, B.: Resampling strategies for regression. Expert. Syst. **32**, 465–476 (2015)
25. Akbilgic, O., Bozdogan, H., Balaban, M.E.: A novel hybrid rbf neural networks model as a forecaster. Stat. Comput. **24**, 365–375 (2014)
26. Spooner, T., Dervovic, D., Long, J., Shepard, J., Chen, J., Magazzeni, D.:"Counterfactual explanations for arbitrary regression models". arXiv preprint arXiv:2106.15212 (2021)
27. Rigatti, S.J.: Random forest. J. Insur. Med. **47**, 31–39 (2017)
28. Sibindi, R., Mwangi, R.W., Waititu, A.G.: A boosting ensemble learning based hybrid light gradient boosting machine and extreme gradient boosting model for predicting house prices. Eng. Rep. **5**, e12599 (2023)

Expanding Polynomial Kernels for Global and Local Explanations of Support Vector Machines

Rikard Vinge[1(✉)], Stefan Byttner[2], and Jens Lundström[2]

[1] Munich, Germany
research@rikardvinge.se
[2] Center for Applied Intelligent Systems Research in Health (CASIR Health), Halmstad University, Halmstad, Sweden
{stefan.byttner,jens.r.lundstrom}@hh.se

Abstract. Researchers and practitioners of machine learning nowadays rarely overlook the potential of using explainable AI methods to understand models and their predictions. These explainable AI methods mainly focus on the importance of individual input features. However, as important as the input features themselves, are the interactions between them. Methods such as the model-agnostic but computationally expensive Friedman's H-statistic and SHAP investigate and estimate the impact of interactions between the features. Due to computational constraints, the investigation is often limited to second-order interactions. In this paper, we present a novel, model-specific method to explain the impact of feature interactions in SVM classifiers with polynomial kernels. The method is computationally frugal and calculates the interaction importance exactly for any order of interaction. Explainability is achieved by mathematical transformation to a linear model with full fidelity to the original model. Further, we show how the model provides for both global and local explanations, and facilitates post-hoc feature selection. We demonstrate the method on two datasets; one is an artificial dataset where H-statistics requires extra care to provide useful interpretation; and one on the real-world scenario of the Wisconsin Breast Cancer dataset. Our experiments show that the method provides reasonable, easy to interpret and fast to compute explanations of the trained model.

Keywords: SVM · Explainability · Trustworthy AI

1 Introduction

Support Vector Machines (SVMs) can be seen as *geometrical* classifiers able to be trained to maximize the margin that separates two target classes. In low dimensions of the input space and using a linear kernel trained models are naturally prone to interpretation. However, often classification and regression problems

R. Vinge—Independent researcher.

G. Krempl et al. (Eds.): IDA 2025, LNCS 15669, pp. 456–468, 2025.
https://doi.org/10.1007/978-3-031-91398-3_34

handled by SVMs benefit in generalization performance by transforming input features to a higher-dimensional feature space. As class categories becomes separable in higher-dimensional feature space the model becomes more complex for direct human interpretation, even at low number of input features. This paper proposes a method to mathematically reformulate SVM models trained with polynomial kernels into linear models that improve the interpretability of the model, both how model predictions can be explained in average, denoted global explanations, and how the model can be explained in individual predictions, denoted local explanations[1]. The proposed method therefore gives explanations for both individual features as well as their interactions with other features.

There are many explainable AI methods available for studying the importance of individual features, which put less focus on studying feature interactions. Understanding interactions can, however, lead to increased prediction performance by selecting relevant features, deeper understanding of the data, and more fair ML models. Despite the age of the SVM method it is still a popular machine learning technique to consider when the number of training data samples are low where other more data-craving models would not be suitable. SVMs have been shown to perform well for a number of more recent classification tasks [11,12], are frequently used in anomaly detection in the form of one-class SVMs [19], and method developments continue to be an active area of research [13,17,23].

The remainder of this paper is structured as follows. In Sect. 2, we discuss related work and further describes our method in Sect. 3. In Sect. 4, we test our method on simulated and real data and compare the explanations with state-of-the-art methods, and finally conclude the results in Sect. 5.

2 Related Work

Interpretability in SVMs has been studied extensively from a multitude of approaches. Early methods generally applicable to all SVMs focused on rule extraction from pre-trained SVM models [1,2,5,20]. By identifying prototype observations, if-then rules are constructed after model training either globally [9] or locally [25]. Methods for identifying representative observations for explanations using PCA [21] in the kernel space is suggested in [26], and an open-source library for identifying the most influential support vectors is provided in [15]. In Biological sciences, interpretation methods utilizing knowledge on the structure of the kernels have been suggested in [24,27].

Support vector machine-specific visualization using nomograms is proposed in [4,29] for the RBF kernel and in [3] for the RBF and polynomial kernels. These methods utilize Taylor expansion to approximate the SVM models into a polynomial model, truncated to a second-order polynomial.

The aforementioned methods either exclude interactions between input features in their explanations, or only include interactions up to the second order. Interactions in the higher order can, however, have strong impact on the model.

[1] https://github.com/rikardvinge/explainpolysvm.

Model-agnostic methods such as Friedman's H-statistic [8] can be extended to an arbitrary order, but to the cost of an exponentially increasing computational burden. Following the importance of explaining higher order interactions, the method proposed in this paper is designed to explain SVMs with polynomial kernels of arbitrary degree. The method is based on the mathematical reformulations of the polynomial kernel SVM introduced in [28], where the decision function is shown to be related to the statistical moments of the support vectors.

3 Method

The proposed method aims to provide both global and local explanations to the polynomial SVM without the loss of certainty from approximations nor restrictions in the dimension of the polynomial kernel. This chapter defines the problem and introduce the theory behind the method.

To achieve explainability the SVM decision function,

$$f(x) = \sum_{i=1}^{N} \alpha_i y_i K(x_i, x) + b_0, \tag{1}$$

where (x_i, y_i), $x_i \in \mathbb{R}^p$, $y_i \in \{-1, +1\} \forall i = 1 \dots N$ are the training observations and their respective class labels, x is the observation to classify, $\alpha_i > 0$ are the dual coefficients, b_0 the intercept and $K(\cdot, \cdot)$ is the kernel function, is reformulated as an equivalent linear function

$$f(x) = w^T \phi(x) + b_0'.$$

Here, ϕ is a transformation of the original features into a space where the each dimension correspond to a single interaction and w is a weight vector constituting the contribution of each interaction to the decision function. The kernel is assumed to be the common polynomial kernel,

$$K(x, y|D, r, \gamma) = \left(r + \gamma \cdot x^T y\right)^D, \tag{2}$$

where r, γ and D are the *independent term*, *kernel coefficient* and *degree* respectively. The reformulation of the decision function is, however, applicable to any polynomial kernel or power series of a kernel.

As detailed in the next section, interactions can appear multiple times in (2). For an interpretable linear model, each interaction should appear only once, and the magnitude of the interaction should correspond to the total contribution from all its repeats.

Our method constitute two parts: in the first, the trained polynomial SVM model is transformed into a linear model containing repeated interactions; and in the second, the redundant interactions are removed to produce a linear model of only unique interactions.

3.1 Expanding Polynomial SVMs to Linear Models

To turn the SVM into a linear model, the kernel (2) is expanded and (1) is rewritten in the equivalent form

$$f(x) = \sum_{d=0}^{D} \binom{D}{d} r^{D-d} \gamma^d \sum_{i=1}^{N} \alpha_i y_i (x_i^T x)^d + b_0. \tag{3}$$

Next, we utilize that

$$\left(x_i^T x\right)^d = \left(x_i^{\otimes d}\right)^T \left(x^{\otimes d}\right) \tag{4}$$

where $x^{\otimes d} = x \otimes x \otimes \ldots \otimes x$ with d Kronecker products, e.g. $x^{\otimes 3} = x \otimes x \otimes x$. This space constitute multiplications of, or interaction between, the input features.

With (3) and (4) the SVM can express a linear combination of linear models in a transformed space,

$$f(x) = \sum_{d=0}^{D} \binom{D}{d} r^{D-d} \gamma^d \sum_{i=1}^{N} \alpha_i y_i \phi(x_i|d)^T \phi(x|d) + b_0, \tag{5}$$

with the transformation $\phi(x|d) = x^{\otimes d}$. By treating the outer summation as a dot product, we arrive at an equivalent linear model of the polynomial SVM,

$$f(x) = \begin{bmatrix} D r^{D-1} \gamma \sum_{i=1}^{N} \alpha_i y_i \phi(x_i|1) \\ \binom{D}{2} r^{D-2} \gamma^2 \sum_{i=1}^{N} \alpha_i y_i \phi(x_i|2) \\ \vdots \\ \gamma^D \sum_{i=1}^{N} \alpha_i y_i \phi(x_i|D) \end{bmatrix}^T \begin{bmatrix} \phi(x|1) \\ \phi(x|2) \\ \vdots \\ \phi(x|D) \end{bmatrix} + b_0. \tag{6}$$

The 0th order term is omitted due to the Karush-Kuhn-Tucker condition that $\sum_{i=1}^{N} \alpha_i y_i = 0$.

3.2 Model Compression

The linear model in (6) contains a high degree of redundancy [10]. In fact, $\phi(x|d)$ contains

$$\frac{(d+p-1)!}{d!(p-1)!} \tag{7}$$

unique interactions, compared to the total number of elements p^d. Thus, a polynomial kernel of degree D can be completely described by $\mathcal{O}\left(p^D/D!\right)$ terms, compared to the $\mathcal{O}\left(p^D\right)$ used in (6). The number of repeats, or permutations, of an interaction of degree d with n_i occurrences of original feature i, $i = 1 \ldots p$, and $\sum_{i=1}^{p} n_i = d$, is

$$\frac{d!}{n_1! n_2! \ldots n_p!}. \tag{8}$$

For example, the interaction $x_0 x_1^2$ have $d = 3$, $n_0 = 1$ and $n_1 = 2$ and occurs 3 times in $\phi(x|3)$.

To create a compressed linear model, a modified transformation $\phi'(x|d)$ is defined such that it only contains the unique interactions of $\phi(x|d)$. Additionally, we define an occurrence vector p_d whose elements are the number of repeats (8) of each unique interaction in $\phi(x|d)$, given the order in $\phi'(x|d)$. With $\phi'(x|d)$ and p_d, a linear model equivalent to (6) can be created without repeated interactions,

$$f(x) = \begin{bmatrix} Dr^{D-1}\gamma p_1 \odot \sum_{i=1}^{N} \alpha_i y_i \phi'(x_i|1) \\ \binom{D}{2} r^{D-2}\gamma^2 p_2 \odot \sum_{i=1}^{N} \alpha_i y_i \phi'(x_i|2) \\ \vdots \\ \gamma^D p_D \odot \sum_{i=1}^{N} \alpha_i y_i \phi'(x_i|D) \end{bmatrix}^T \begin{bmatrix} \phi'(x|1) \\ \phi'(x|2) \\ \vdots \\ \phi'(x|D) \end{bmatrix} + b_0. \quad (9)$$

Here, $\odot$ denotes the Hadamard product. From (9), both global and local explanations can be extracted.

3.3 Global Explanations and Feature Selection by Feature Importance

The compressed linear model (9) contains a single element for each interaction in the polynomial SVM. These elements, or coefficients, are calculated from the support vectors. Global explanations of the model are given by the relative magnitude of the elements of the weight vector in (9),

$$\begin{bmatrix} Dr^{D-1}\gamma p_1 \odot \sum_{i=1}^{N} \alpha_i y_i \phi'(x_i|1) \\ \binom{D}{2} r^{D-2}\gamma^2 p_2 \odot \sum_{i=1}^{N} \alpha_i y_i \phi'(x_i|2) \\ \vdots \\ \gamma^D p_D \odot \sum_{i=1}^{N} \alpha_i y_i \phi'(x_i|D) \end{bmatrix} \quad (10)$$

The global explanations shall be interpreted as the change in the decision function by a unit change in one of the constituting features in the interaction, all other features held constant. If a feature appears in multiple interactions, a unit change in the feature changes the decision function by the sum of the interaction that involves the feature. If all input features have the same range, the magnitude of the weight vector elements can be compared directly.

Feature Selection. One use for exlainable AI methods is feature selection, where features with low impact on the output of a trained model are removed and the model trained with the remaining subset of features [14]. The compressed linear model (9) allows for a different approach to reduce the dimensionality of the model, namely to suppress individual features or interactions directly in the decision function of the trained model. To this end, we define mask vectors

m_d, $d = 1 \dots D$ with ones on elements corresponding to interactions in $\phi'(x|d)$ that should be kept, and zero elsewhere. Then, a masked model $\hat{f}(x)$ is given by

$$\hat{f}(x) = \begin{bmatrix} Dr^{D-1}\gamma p_1 \odot m_1 \odot \sum_{i=1}^{N} \alpha_i y_i \phi'(x_i|1) \\ \binom{D}{2} r^{D-2}\gamma^2 p_2 \odot m_2 \odot \sum_{i=1}^{N} \alpha_i y_i \phi'(x_i|2) \\ \vdots \\ \gamma^D p_D \odot m_D \odot \sum_{i=1}^{N} \alpha_i y_i \phi'(x_i|D) \end{bmatrix}^T \begin{bmatrix} \phi'(x|1) \\ \phi'(x|2) \\ \vdots \\ \phi'(x|D) \end{bmatrix} + b_0. \quad (11)$$

This feature masking is performed post-hoc with low additional computational cost. In Sect. 4.2 we show that masking of the decision function can improve the performance of a model. The exploration of and identification of guidelines for masking of the decision function is left to future work.

3.4 Local Explanations

Local explanations are achieved for an observation x by a partial evaluation of the dot product in (9), namely the element-wise multiplication,

$$\begin{bmatrix} Dr^{D-1}\gamma p_1 \odot \sum_{i=1}^{N} \alpha_i y_i \phi'(x_i|1) \\ \binom{D}{2} r^{D-2}\gamma^2 p_2 \odot \sum_{i=1}^{N} \alpha_i y_i \phi'(x_i|2) \\ \vdots \\ \gamma^D p_D \odot \sum_{i=1}^{N} \alpha_i y_i \phi'(x_i|D) \end{bmatrix} \odot \begin{bmatrix} \phi'(x|1) \\ \phi'(x|2) \\ \vdots \\ \phi'(x|D) \end{bmatrix}. \quad (12)$$

Each element of the resulting vector is the contribution to the decision function of the corresponding interaction, such that the sum of all contributions and the intercept add up to the latent variable.

4 Experiments

In this section, we present two example uses of the feature and interaction importance method presented in this paper. The first example is based on an artificial dataset and compares the global explanations from our method with those from Friedman's H-statistic. The second example is based on the Wisconsin breast cancer (WBC) dataset [6] with the aim of showing how our method provides easy-to-interpret local explanation. All calculations are performed on a consumer laptop with 16 GB RAM and Intel i5-12500H CPU. For the sake of simplicity, we denote all terms, both linear terms and interactions between features, as interactions.

4.1 Global Explanation Comparison with H-Statistic

In this example we present a comparison of global explanations from our method and from H-statistic. The SVM is trained using scikit-learn [22] and scikit-explain

[7] is used for H-statistic and partial dependence (PD). We focus this experiment on feature interactions and, therefore, do not analyze the single-feature PDs nor apply other methods to explain the input feature importance.

H-statistic is sensitive to the relative magnitude of the PD of the single variable and multiple variables [18], particularly when the individual PDs of two features are small but the total effect is primarily interaction. To showcase this and to compare explanation of the interactions by our method, we create two datasets. Both datasets have five features, each feature created by uniform sampling within the range $[-5, 5]$. The samples are assigned to a class according to

$$0.5x_0x_1 - 0.5x_0x_2 - 0.15x_3^2 + a\sum_{i_1}^{5} x_i + \epsilon \underset{c=-1}{\overset{c=1}{\gtrless}} -1 \tag{13}$$

where $\epsilon \sim \mathcal{N}(0, \sigma)$. The first dataset, denoted *weak interaction*, has $a = 1$ and $\sigma = 1$ to create a scenario where the model should be primarily linear and only weakly dependent on interactions. The second dataset, denoted *strong interaction*, has $a = 0.1$ and $\sigma = 0.4$ such that the model should depend mainly on interactions. The value of σ is set to create the same amount of overlap between the classes; in both datasets approximately 4% class labels are switched due to noise.

An SVM is trained on 2000 observation using parameters: kernel = 'poly', $C = 1$, $d = 3$, $r = \sqrt{2}$, and $\gamma =$ 'scale'. The accuracy of the trained model on a held-put test set is 95.5% on the weak interaction dataset and 95.7% on the strong interaction dataset.

The global explanations (10) for the 15 most important features are presented in Fig. 1.

For both datasets, our method indicate that the trained models have learned to approximate (13), both in terms of relative magnitude and sign of the polynomial coefficients. Clearly, the interactions are emphasized for the strongly interacting dataset at the bottom of Fig. 1 as opposed to the weaker interactions in the top where the underlying function (13) used $a = 1$, making the individual terms dominant.

The H-statistic are significantly different for the models trained on the two datasets, as shown in Fig. 2. For the weak interaction dataset the H-statistics are informative and indicate that the interactions x_0x_1 and x_0x_2 plays the strongest role in the model of all the interactions. However, the other interactions are also indicated to play a significant role. There exists, currently, no test for significance for H-statistics [18] to distinguish between real and spurious interactions. H-statistics also does not provide any information on the sign of an interaction.

The sensitivity to strong interactions but weak individual PDs is evident for the strong interaction dataset. We find that half of the interactions exceed 1 and are therefore difficult to interpret. In this case, the unnormalized H-statistic, shown in Fig. 3, provide extra evidence that only the x_0x_1 and x_0x_2 interactions are of interest. The same conclusions can not be drawn from the unnormalized H-statistic for the model of the weak interaction dataset.

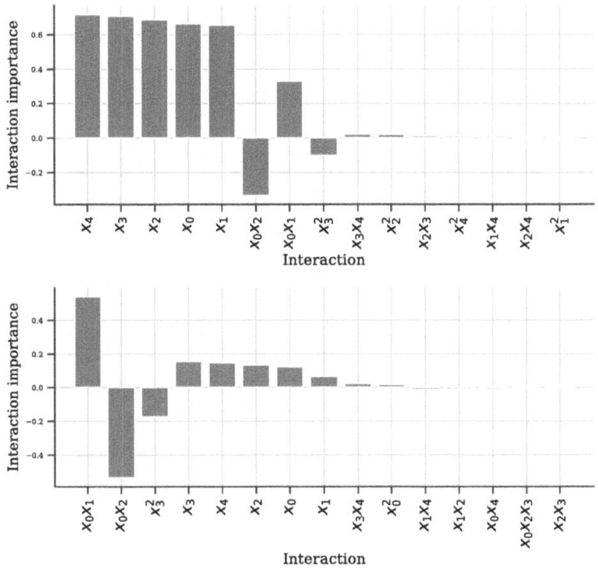

Fig. 1. Signed interaction importance from the proposed approach for the top 15 interactions in the third-order polynomial kernel SVM trained on (top) the weak interaction dataset and (bottom) the strong interaction dataset. The interaction importance should be interpreted as the change in the decision value due to a unit change in one of the interactions.

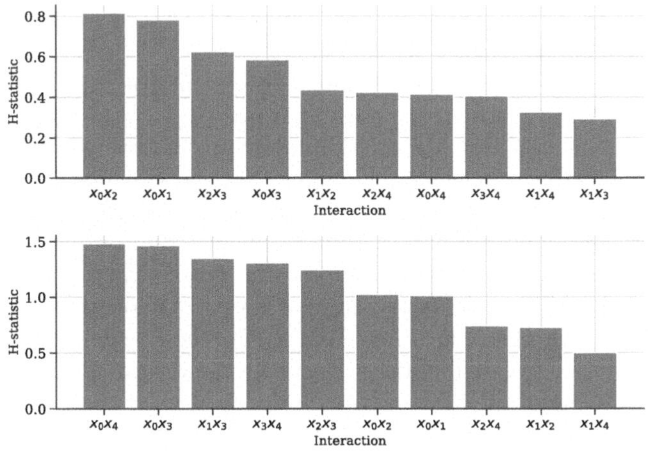

Fig. 2. Friedman's H-statistics for the model trained on (top) the weak interaction dataset and (bottom) the strong interaction dataset.

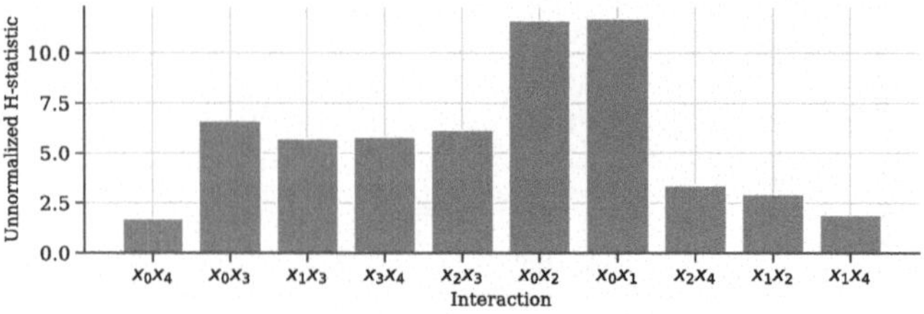

Fig. 3. Unnormalized H-statistic for the model trained on the strong interaction dataset.

For H-statistic, the total computation time, calculating one- and two-dimensional partial dependence without sub-sampling the training set, is 80 s. The corresponding calculation time with our method, including third-order interactions, is 1 ms, four orders of magnitude faster.

4.2 Local Explanations and Feature Selection on the WBC Dataset

The WBC data set [6] constitutes 30 features and 357 benign and 212 malignant observations. A hold-out test set is created by randomly sampling 20% of the observations. Feature values are normalized by computing the z-scores per feature. An SVM is tuned through 5-fold cross-validation on the normalized samples, and achieves a test-accuracy of 97% with the SVM parameters: kernel = 'poly', $C = 0.5$, $d = 2$, $r = 2.5$, and $\gamma =$ 'scale'. Global explanations from our model are compared with H-statistic, and an example of local explanations is compared with SHAP [16].

In Fig. 4 we show the global explanations computed using the proposed method for the 40 most important interactions.

In this case, there is not a clear dominance by a few features. The global explanations indicate that the model is mainly linear. Calculating H-statistic on all 435 quadratic interactions shows that 250 of the interactions exceed 1 and the remainder shows an approximately uniform distribution between 0.3 and 1. Neither method provides for easily interpretable global explanations of the interactions in this case. The calculation time of H-statistic on the 235 interactions from 426 training samples is 6.5 min while our method requires 2.5 ms.

Even though the cross-validation selected a quadratic kernel, all input features except for 'smoothness error' are more important than the interactions between two features. This indicates that a linear SVM could suffice. In fact, a linear SVM trained on the same observations achieve a performance of 96%, comparable to 97% for the quadratic kernel. Further, masking the trained quadratic SVM with (11) to only include linear terms increases the performance to 98.7%.

Local explanations for the most influential interactions are shown in Fig. 5 for a sample from the negative class. The local explanations are presented as waterfall plots where the cumulative contributions of interactions of diminishing importance are shown for our method and SHAP.

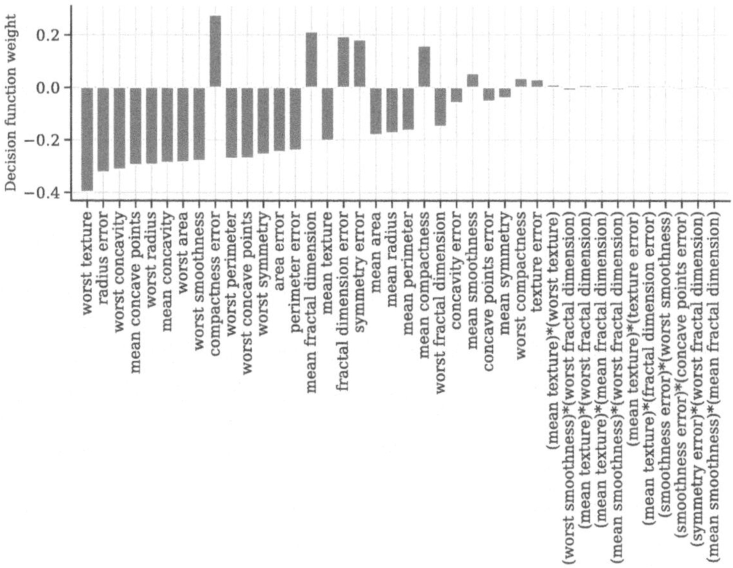

Fig. 4. Importance for the 40 most important interactions in the second-order polynomial kernel SVM trained on the Wisconsin breast cancer data set (using the proposed approach).

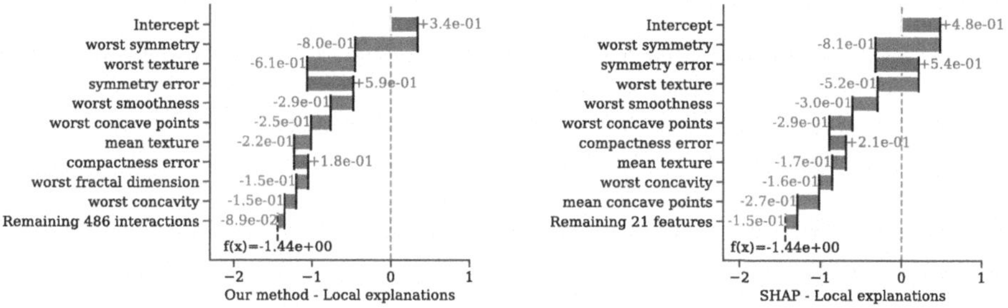

Fig. 5. Local explanations for a sample from the negative class in the WBC dataset. (Left) shows our method and (right) the SHAP values. The plots show the intercept and the 9 features or interactions most impactful on the decision value, ordered by magnitude. The length of the bars and the values next to the bars indicate the magnitude and sign of the contribution to the decision values. The contributions are additive and sum to the decision value of -1.44. Red and blue bars denote a contribution toward a positive and negative decision value, respectively. The number of terms not shown differ between the plots due to our method calculate the impact from all features and interactions, while the SHAP values are calculated for the input features only. (Color figure online)

Our method and SHAP are in strong agreement on the magnitude and sign of the contributions from the features. We want to note that SHAP displays the accumulated importance of a feature: the feature itself and all interactions it is part of. Contrastingly, our method shows the importance of a single feature or a single interaction.

5 Conclusions

We have presented a novel method to extract, without approximation, both global and local explanations from SVM models trained with polynomial kernels of arbitrary degree. The method is intrinsic and requires only information from the trained model and the support vectors. It allows for feature selection by removing specific, unwanted, interactions in the trained model. From our experiments, performed both on artificial and real data sets, we conclude that the global explanations given by importance of interactions match the expected interactions also in scenarios where H-statistic can be difficult to interpret. The method reproduce with high similarity the local explanations of SHAP, as demonstrated for an individual sample. Moreover, the proposed method requires a fraction of the computation time used by H-statistics to compute the global explanation, 2.5 ms as compared to 6.5 min in our experiments on the Wisconsin Breast Cancer dataset.

In the presented work, we have limited the scope to the most common polynomial kernel. The method, however, can be extended to other polynomial kernels. Promising future work involve expanding the method to truncated power series of the RBF kernel and to other types of SVMs, e.g. Support Vector Regression and one-class SVM. We also see interesting topics for further research into clarifying and explaining the impact from dependence between features in the original space as well as dependent interactions, e.g. two interactions that cancel the effects of each other. We further want to develop the method to provide insight into inductive bias and the degree the polynomial kernel should have for a specific problem. We have already seen that the explanations can guide the machine learning designer to a suitable kernel degree, but the conclusions here are so far vague and a proper framework for gauging the suitability of a polynomial kernel would benefit many applications.

Acknowledgments. We would like to thank Tomas McKelvey for the fruitful discussions that led up to the presented work. R.V. gratefully acknowledge the support by Volvo Group, Robert Valton and Adam Stahl.

Disclosure of Interests. The authors have no competing interests to declare that are relevant to the content of this article.

References

1. Barakat, N.H., Bradley, A.P.: Rule extraction from support vector machines: a review. Neurocomputing **74**, 178–190 (2010)

2. Barbado, A., Corcho, Ó.: Rule extraction in unsupervised anomaly detection for model explainability: application to oneclass SVM. Expert Syst. Appl. **189**, 116100 (2019)
3. van Belle, V., van Calster, B., van Huffel, S., Suykens, J.A.K., Lisboa, P.J.G.: Explaining support vector machines: a color based nomogram. PLoS ONE **11** (2016)
4. van Belle, V., Lisboa, P.J.G.: Automated selection of interaction effects in sparse kernel methods to predict pregnancy viability. In: 2013 IEEE Symposium on Computational Intelligence and Data Mining (CIDM), pp. 26–31 (2013)
5. Diederich, J.: Rule Extraction from Support Vector Machines: An Introduction, pp. 3–31. Springer, Heidelberg (2008). https://doi.org/10.1007/978-3-540-75390-2_1
6. Dua, D., Graff, C.: UCI machine learning repository (2017). http://archive.ics.uci.edu/ml
7. Flora, M.L., Potvin, C.K., McGovern, A., Handler, S.: scikit-explain (2022). https://github.com/monte-flora/scikit-explain
8. Friedman, J.H., Popescu, B.E.: Predictive learning via rule ensembles. Ann. Appl. Stat. **2**(3), 916–954 (2008). https://doi.org/10.1214/07-AOAS148
9. Gurumoorthy, K.S., Dhurandhar, A., Cecchi, G.A., Aggarwal, C.C.: Efficient data representation by selecting prototypes with importance weights. In: 2019 IEEE International Conference on Data Mining (ICDM), pp. 260–269 (2017)
10. Herbrich, R.: Learning Kernel Classifiers: Theory and Algorithms. The MIT Press (2001). https://doi.org/10.7551/mitpress/4170.001.0001
11. Khairandish, M.O., Sharma, M., Jain, V., Chatterjee, J.M., Jhanjhi, N.: A hybrid CNN-SVM threshold segmentation approach for tumor detection and classification of MRI brain images. IRBM **43**(4), 290–299 (2022)
12. Kurani, A., Doshi, P., Vakharia, A., Shah, M.: A comprehensive comparative study of artificial neural network (ANN) and support vector machines (SVM) on stock forecasting. Ann. Data Sci. **10**(1), 183–208 (2023)
13. Lee, J., Lee, H., Hwang, K., Kwak, N.: Deep support vectors. In: Globerson, A., et al. (eds.) Advances in Neural Information Processing Systems, vol. 37, pp. 118155–118181. Curran Associates, Inc. (2024). https://proceedings.neurips.cc/paper_files/paper/2024/file/d60e14c19cd6e0fc38556ad29ac8fbc9-Paper-Conference.pdf
14. Li, J., et al.: Feature selection: a data perspective. ACM Comput. Surv. **50**(6) (2017). https://doi.org/10.1145/3136625
15. Loor, M., Tapia-Rosero, A., Tré, G.D.: An open-source software library for explainable support vector machine classification. In: 2022 IEEE International Conference on Fuzzy Systems (FUZZ-IEEE), pp. 1–7 (2022)
16. Lundberg, S.M., Lee, S.I.: A unified approach to interpreting model predictions. In: Guyon, I., et al. (eds.) Advances in Neural Information Processing Systems, vol. 30, pp. 4765–4774. Curran Associates, Inc. (2017). http://papers.nips.cc/paper/7062-a-unified-approach-to-interpreting-model-predictions.pdf
17. Marchetti, F., Perracchione, E.: Local-to-global support vector machines (LGSVMs). Pattern Recogn. **132**, 108920 (2022). https://doi.org/10.1016/j.patcog.2022.108920
18. Molnar, C.: Interpretable Machine Learning, 2 edn. (2022). https://christophm.github.io/interpretable-ml-book
19. Nassif, A.B., Talib, M.A., Nasir, Q., Dakalbab, F.M.: Machine learning for anomaly detection: a systematic review. IEEE Access **9**, 78658–78700 (2021). https://doi.org/10.1109/ACCESS.2021.3083060

20. Núñez, H., Angulo, C., Català, A.: Rule Extraction from Support Vector Machines, pp. 237–248. Springer, Heidelberg (2005). https://doi.org/10.1007/3-540-28803-1_10
21. Pearson, K.: On lines and planes of closest fit to systems of points in space. Philos. Mag. Ser. **1**(2), 559–572 (1901)
22. Pedregosa, F., et al.: Scikit-learn: machine learning in Python. J. Mach. Learn. Res. **12**, 2825–2830 (2011)
23. Que, Z., Lin, C.J.: One-class SVM probabilistic outputs. IEEE Trans. Neural Netw. Learn. Syst. 1–13 (2024). https://doi.org/10.1109/TNNLS.2024.3395148
24. Rätsch, G., Sonnenburg, S., Schäfer, C.: Learning interpretable SVMs for biological sequence classification. BMC Bioinform. **7** (2006)
25. Ribeiro, M.T., Singh, S., Guestrin, C.: Anchors: high-precision model-agnostic explanations. In: Proceedings of the AAAI Conference on Artificial Intelligence, vol. 32 (2018). https://doi.org/10.1609/aaai.v32i1.11491
26. Schölkopf, B., et al.: Input space versus feature space in kernel-based methods. IEEE Trans. Neural Networks **10**(5), 1000–17 (1999)
27. Sonnenburg, S., Zien, A., Philips, P., Rätsch, G.: POIMs: positional oligomer importance matrices-understanding support vector machine-based signal detectors. Bioinformatics **24**, i6–i14 (2008)
28. Vinge, R., McKelvey, T.: Understanding support vector machines with polynomial kernels. In: 2019 27th European Signal Processing Conference (EUSIPCO), pp. 1–5 (2019)
29. Walters, B., Ortega-Martorell, S., Olier, I., Lisboa, P.J.G.: The partial response SVM. In: ESANN 2021 proceedings (2021)

A Constrained Declarative Based Approach for Explainable Clustering

Mathieu Guilbert(✉), Christel Vrain, and Thi-Bich-Hanh Dao

Université d'Orléans, INSA CVL, LIFO UR 4022, Orléans, France
{mathieu.guilbert,christel.vrain,thi-bich-hanh.dao}@univ-orleans.fr

Abstract. Clustering is an unsupervised exploratory task that helps experts understanding the structure of their data. Constraints based on expert knowledge can be introduced, but obtaining them remains challenging, making the explanation of results essential for adjusting parameters and uncovering new insights. We address explainable clustering by modeling the data in two spaces: one for clustering and one for explanation. Our method ECS (Explainability-driven Cluster Selection) aims to produce a high-quality clustering while ensuring interpretability through patterns that cover most instances in a cluster and distinguish them from others. It relies on ensemble clustering and a new constraint programming (CP) model for selecting the clusters and their explanations.

Keywords: Clustering · Explainability · Ensemble method

1 Introduction

Clustering is an unsupervised learning task that aims at finding a structure of data into groups (clusters) such that observations within the same cluster are similar while differing from those in other clusters. It is an exploratory task allowing experts to better understand the underlying structure of their data. However, its outcome is influenced by multiple factors, making explanations essential for adjusting the parameters. Expert knowledge, expressed as constraints, has been introduced during the clustering process to get results closer to the expert expectations (see for instance [8]). However, eliciting expert knowledge is a challenging task, highlighting the necessity of explanations not only for parameter setting but also for assisting the expert in uncovering new knowledge.

As an illustration, let us consider a medical application where patients are assigned a severity score reflecting the progression of their disease associated with clinical features such as age, stress level, ...: it could be useful to cluster patients according to their scores with an interpretation on their clinical features, thus highlighting relations between the disease severity and the clinical properties. In this paper we address the problem of explainable clustering under the assumption that data is described in two (possibly identical) representation spaces, a feature space for clustering and a Boolean property space for explanations. Introducing two representation spaces allows to propose a general framework encapsulating

G. Krempl et al. (Eds.): IDA 2025, LNCS 15669, pp. 469–483, 2025.
https://doi.org/10.1007/978-3-031-91398-3_35

two situations: when features are used both for building the clustering and its explanation, and when another set of descriptors may be interesting for building explanations, thus highlighting new relations between these spaces.

Most clustering approaches, such as K-Means, are based on a distance between observations. However, there exists a less known family of methods called conceptual clustering, based on the notion of concept. First introduced in [22], current approaches often rely on Formal Concept Analysis and closed pattern mining. It is of high interest for explainable clustering, since it is based on the idea that a cluster must be a concept, that is a set of observations coupled with a set of properties exclusively satisfied by the observations of the cluster. However, this notion of concept is often too strong for real applications and must be relaxed as proposed in [6]. Our framework based on two representation spaces (clustering space/explanation space) bridges the gap between distance-based and conceptual clustering, thus benefiting from advantages of both approaches. The aim is to find an explainable clustering defined as follows: a cluster is composed of a set of observations which are similar in the clustering space, along with its explanation consisting of a set of patterns (conjunctions of Boolean descriptors from the representation space) that are characteristic (satisfied by most observations of this cluster), discriminant w.r.t. observations not in this cluster, and concise. Explanations are built during the clustering process (*explanation by design*), as done for instance in [6,11].

Our work differs from [6], where an explanation is a set of Boolean descriptors and which does not explicitly consider the notion of discrimination between clusters. Moreover, we propose a new method for simultaneously building a clustering and an explanation for each cluster. We rely on declarative frameworks, allowing to easily represent expert knowledge and more particularly on Constraint Programming to take advantage of its ability to reason in different algebras.

Our contributions are as follows:

- a generic framework for explainable clustering based on two representation spaces (clustering/explanation),
- a new explainable clustering method ECS (Explainability driven Cluster Selection) and based on two main steps that are building a set of clusters and their descriptions and selecting the best ones to satisfy the constraints.
- a CP model for cluster and explanation selection,
- experiments on one artificial and multiple real datasets, including a case study in education [5].

The paper is organized as follows: state of the art (Sect. 2), our formulation of explainable clustering (Sect. 3), presentation of the overall method (Sect. 4), the CP model (Sect. 5), experimental results (Sect. 6) and conclusion (Sect. 7).

2 Related Work

The field of Explainable AI (XAI) has attracted a lot of attention in recent years. In clustering, many works rely on decision trees (see for example [16] for a discussion of their interpretability). Some of these trees are constructed a posteriori to approximate the results of clustering algorithms such as K-Means [9,12,17,20]. Intrinsic explainable approaches (*explainable-by-design*) have also been introduced with separation criteria such as cluster compactness [19] or Silhouette and Dunn indices [2].

Other approaches focus on delimiting clusters using encompassing shapes, and each cluster is then described by a set of rules delimiting its boundaries. These shapes can be hyper-rectangles [4] with contours parallel to the axes, or polytopes [18] with oblique contours.

A third family of techniques aims to integrate distance-based clustering and conceptual clustering in the same framework [6,27,33]. It was introduced in [6] under the term descriptive clustering: data are described by two modalities, one to build compact clusters and the other (composed of semantic descriptors) to form interpretable descriptions. We introduce a new descriptive clustering framework that, unlike these approaches, takes into account more complex explanations in the form of patterns, i.e. conjunctions of Boolean descriptors, and incorporates the notion of discrimination between clusters.

Our work differs from redescription mining [13] and concept learning (see for instance [22]) since groups are not known beforehand and the explanation of a cluster depend on the other clusters.

Many constrained clustering approaches construct a partition by assigning points to clusters, while satisfying the constraints. Some works [29] first generate a set of partitions and then select the best one based on its ability to satisfy the constraints. To our knowledge, only [3,24,25] create a partition from a given set of clusters while satisfying user-defined constraints. With respect to constrained clustering, we introduce several new ideas: the formalization of explainability and its integration during clustering construction, the introduction of discrimination constraints, the combination of distance-based and conceptual clustering, and, inspired by ensemble methods, the integration of diverse clustering techniques during the generation of base partitions.

Our approach, called ECS for Explainability-driven Cluster Selection, is related to ensemble-based clustering methods that generate many base partitions and build a single final partition, often called the consensus partition. In our work, the final clustering is built by selecting clusters so as to satisfy the constraints. While several methods have integrated expert knowledge into ensemble clustering [14,31,32], none addressed the selection of interpretable clusters.

3 Explainable Clustering

In our framework we suppose that data is described by two views: one for clustering ($\mathbb{F}$) and one for explanation ($\mathbb{B}$), which can be built from two different representation spaces or from the same one. The data is therefore a set $\mathcal{O}$ of n objects, described by a set $\mathbb{F}$ of f (typically numerical) features, and a set $\mathbb{B}$ of r Boolean descriptors. This is represented by a $n \times f$ matrix, where $\mathbb{F}_{oa}$ denotes the value of object o for feature a, and a $n \times r$ Boolean matrix where $\mathbb{B}_{op} = 1$ indicates that object o satisfies property p.

A *pattern* is defined as a conjunction of Boolean descriptors, while an *explanation* for a cluster is a set of patterns. A set of patterns explaining a cluster can be interpreted as a disjunction of its patterns, representing alternative combinations of descriptors that collectively explain the cluster. We consider that a good explanation of a cluster C must be *characteristic* (cover most of the objects of C), *discriminant* (satisfied by only a few objects outside the cluster) and *concise* (patterns must be quite short and the number of patterns associated to a cluster should not be too large).

To formalize these notions, we introduce the predicate *cover* between a pattern π and an object o as: $cover(\pi, o) \stackrel{def}{=} \forall t \in \pi, \mathbb{B}_{o,t} = 1$. The expected properties for an explanation are as follows: (# denotes the cardinality of a set and θ, θ', ϱ, ϕ are predefined parameters in $[0, 1]$)

- Coverage of a cluster C by a pattern π:

$$cover(\pi, C) \stackrel{def}{=} \#\{o \in C | cover(\pi, o)\} \geq \theta.\#C$$

- Coverage of a cluster C by a set of patterns $\Pi = \{\pi_1, \ldots, \pi_j\}$:

$$cover(\Pi, C) \stackrel{def}{=} \#\{o \in C | \exists l \in [1, j], cover(\pi_l, o)\} \geq \theta'.\#C$$

- Discrimination of a pattern π:
  - Dataset-wise: π should not cover more than a ratio ϱ of $o \notin C$:

$$disc\_global(\pi, C) \stackrel{def}{=} \#\{o \in \mathcal{O} \backslash C \mid cover(\pi, o)\} \leq \varrho.\#(\mathcal{O} \backslash C)$$

  - Cluster-wise: for all clusters $C' \neq C$, π should not cover more than a ratio ϕ of objects of C':

$$disc\_cluster(\pi, C, C') \stackrel{def}{=} \#\{o \in C' \mid cover(\pi, o)\} \leq \phi.\#C'$$

Patterns inside the explanation of a cluster can be linked by a generality relation: $subsumes(\pi_1, \pi_2) \stackrel{def}{=} \pi_2 \subseteq \pi_1$. A more general pattern provides a better coverage within a cluster but tends to have a lower discriminative power. Our framework allows expert preferences to be incorporated through a constraint stating that if two patterns π_1 and π_2 explain the same cluster and satisfy $subsumes(\pi_1, \pi_2)$, then the most general (or the most specific depending on the expert preference) must be selected. Additionally, domain knowledge on Boolean descriptors (for instance a generality relation) can also be integrated. For a list of such constraints, refer to [8].

4 Proposed Method

Our method aims at finding a clustering and explanations for each cluster satisfying conditions both in terms of clustering requirements (as for instance forming a partition, allowing overlapping between clusters, limiting the diameter of the clusters) and in terms of coverage and discrimination for the explanations, according to Sect. 3. Unlike previous approaches [6,7] that build a clustering by searching an assignment of points to clusters, our method takes its inspiration from ensemble clustering [30]: it constructs a set of candidate cluster and selects a subset satisfying all the requirements. On the theoretical side, the clustering problem is NP-Hard [1] as well as clustering under constraints [10]. On the practical side, one of the advantages of our method in comparison to the previous methods is a control of complexity: instead of finding an assignment of n data points to clusters, the problem is reformulated as *selecting* a set of k clusters among a set of c candidates, where c can be significantly smaller than n. The problem is still NP-Hard, since the set covering problem can be reduced to it.

The clustering and explanation requirements are expressed by constraints, which can be classified into three categories:

- **IC** Constraints on individual cluster composition, as for instance cluster size forbidding too large or too small clusters, cluster diameter restricting the spread of the clusters, or must-link/cannot-link constraints requiring that two objects must/cannot be in the same cluster.
- **ID** Constraints on individual cluster description, as for instance on the size, on the coverage or on the dataset-wise discrimination of patterns that can explain a cluster. If a constraint on the length of patterns is set to 1, then a pattern corresponds to a single descriptor.
- **AC** Across cluster/pattern constraints: either explanation-based constraints, as for instance cluster-wise discrimination of clusters, or clustering-based constraints such as restriction on the cluster overlapping.

The framework of our method is illustrated in Fig. 1. Each step is described below and ensures a different type of constraints.

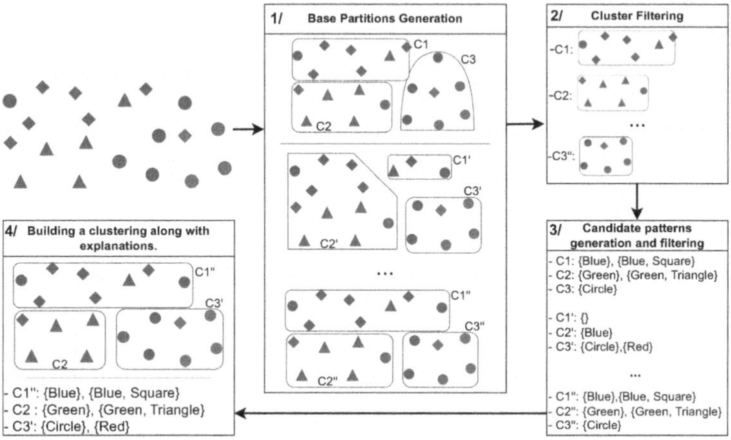

Fig. 1. Overall ECS approach.

Building Cluster Candidates. Inspired by ensemble clustering, several clustering algorithms such as K-means, hierarchical clustering, density-based clustering... are applied to generate various partitions, whose clusters constitute a set of cluster candidates (Step 1, Fig. 1). The aim is to ensure a diversity of candidates with clusters of different shapes (globular with K-means, elongated with single-link hierarchical or dense with density-based clustering). Constraints on individual cluster composition **IC** are then enforced to keep only the clusters satisfying them (Step 2).

Building Pattern Candidates. For each cluster candidate C, the set of frequent closed patterns in C is generated using LCM [28], with the frequency set to the coverage threshold. **ID** constraints are enforced to retain only the patterns satisfying them. Moreover, only candidate clusters with at least one remaining pattern are kept (Step 3).

Building a Clustering Along with Explanations. Cluster candidates are selected to form a clustering, and for each selected cluster C some patterns are selected to explain C according to the **AC** constraints (Step 4). It is worth to notice that cluster and pattern selection is achieved at the same time. We call this problem Explainable Cluster Selection (ECS), which is formulated as a Constrained Optimization Problem (COP). A model using Constraint Programming (CP) for ECS is presented in Sect. 5 which can be solved by a CP solver [26].

5 Explainable Cluster Selection Model

As input, we have the set $\mathcal{O}$ of N data points, $\mathbf{C}$ the set of cluster candidates, each cluster $c \in \mathbf{C}$ is associated with a set $\mathbf{D}_c$ of pattern candidates. The aim is to select a subset of clusters in $\mathbf{C}$ such that the clustering and pattern constraints **AC** are satisfied and an objective function representing the clustering/explanation quality is optimized. In order to model this problem, CP is chosen rather than Integer Linear Programming since many constraints are written as logical implications, which are more naturally expressed in CP. Moreover, non linear constraints on clusters can be added. Compared to SAT, constraints are expressed in a higher language level, making the model much more readable than by decomposing them into CNF formulas. The selection problem is defined by the following variables:

- for each cluster $c \in \mathbf{C}$, let $C_c \in \{0,1\}$ be a Boolean variable expressing whether the cluster c is selected;
- for each cluster $c \in \mathbf{C}$, for each pattern $d \in \mathbf{D}_c$ let $D_{cd} \in \{0,1\}$ be a Boolean variable expressing whether the pattern d is selected to explain c;
- for each $x \in \mathcal{O}$, let $A_x \in \mathbb{N}$ be an integer variable counting the number of clusters point x is assigned to;
- for each $x \in \mathcal{O}$, let $P_x \in \{0,1\}$ be a Boolean variable expressing whether point x is explained by a selected pattern.

The constraints in the model present the **AC** constraints, clustering constraints or user-defined constraints. Links with A_x and P_x are expressed by constraints ($\mathbf{1}(\cdot)$ the indicator function):

$$\forall x \in \mathcal{O},\ A_x = \sum\nolimits_{c \in \mathbf{C}} C_c \mathbf{1}(x \in c)$$

$$\forall x \in \mathcal{O},\ P_x = \sum\nolimits_{c \in \mathbf{C}, d \in \mathbf{D}_c} D_{cd} \mathbf{1}(x \in d \wedge x \in c)$$

Constraints on Explanation Size. For instance, *non-empty explanations*[1]: $\forall c \in \mathbf{C},\ C_c = 1 \iff \sum_{d \in \mathbf{D}_c} D_{cd} \geq 1$ or *explanation concision* [23]: $\forall c \in \mathbf{C}$, $\sum_{d \in \mathbf{D}_c} D_{cd} \leq 5$

Constraints on Explanation Discrimination. A pattern d explaining a selected cluster c should not be used for more than a ratio η of all the selected clusters:

$$\forall c \in \mathbf{C}, \forall d \in \mathbf{D}_c, \quad D_{cd} = 1 \Longrightarrow \sum\nolimits_{c' \in \mathbf{C}, d \in \mathbf{D}_{c'}} D_{c'd} \ \leq \ \eta \sum\nolimits_{c' \in \mathbf{C}} C_{c'}$$

A pattern d explaining a cluster c should not cover any other selected cluster c' more than a ratio ϕ:

$$\forall c \in \mathbf{C}, \forall d \in \mathbf{D}_c, \quad D_{cd} = 1 \Longrightarrow \bigwedge\nolimits_{c' \in \mathbf{C}, c' \neq c,\ freq(d,c') \geq \phi \# c'} (C_{c'} = 0)$$

Constraints on Explanation Coverage. Most of the points, up to a ration α, are covered by a selected explanation pattern: $\alpha N \leq \sum_{x \in \mathcal{O}} P_x$.

Clustering Constraints. The number of selected clusters can be bounded: $K_{min} \ \leq \ \sum_{c \in \mathbf{C}} C_c \ \leq \ K_{max}$, as well as the number of clusters containing each point:

$$\forall x \in \mathcal{O},\ nOverlap_{min} \ \leq \ A_x \ \leq \ nOverlap_{max}$$

and the overall cluster overlapping:

$$\sum\nolimits_{x \in \mathcal{O}} \mathbf{1}(A_x > 1) \leq nOverlapAll$$

User-Defined Constraints. Constraints between selected clusters could be introduced. For instance two clusters c, c' cannot be selected together: $C_c + C_{c'} < 2$.

[1] This constraint is expressed by an equivalence $\iff$, implying that if a cluster is not selected then none of its candidate patterns can be selected.

Objectives. Several objective functions on clustering composition can be used, such as minimizing the sum of squared errors or the cluster diameter. We introduce new objectives: minimizing the number of unassigned points, maximizing the number of points assigned to a unique cluster or maximizing the number of points covered by at least one pattern. In the experiments we consider an objective aiming at assigning and explaining a maximum of points while limiting the overlapping:

$$\arg\max \sum_{x \in \mathcal{O}} (P_x + \mathbf{1}(A_x = 1)) \quad (1)$$

Complexity Analysis. The number of variables in the model heavily depends on the number of candidate clusters and patterns. The number of candidate clusters depends in part on the number of generated base partitions, and is also influenced by the coverage ratio θ which directly affects the amount of candidate patterns. As mentioned in Sect. 4, we filter out the clusters that do not have any covering candidate patterns. Lowering the coverage increases the number of potential patterns found for each cluster, and thus raises the number of non-filtered out candidate clusters.

By adjusting the number of candidate clusters and patterns (via the number of base partition, the coverage parameter or any other constraints during the filtering step) the model complexity can be effectively managed, balancing computational feasibility and potential solution quality. A smaller set of candidates reduces the search space, leading to faster computations, while a larger set increases the ability to satisfy the constraints and optimize the objective criterion.

The discrimination ratio and the unassignement/overlapping settings also influence the combinatorial complexity of ECS, since they impact the number of configurations to be evaluated by the model. Similarly, higher pattern discrimination is more restrictive, thus lowering the number of possibilities.

6 Experiments

We introduce three new evaluation measures for explanations; their values are between 0 and 1 and they have to be maximized. For a pattern p and a cluster C, PCR (resp. IPC) measures the ratio of observations covered (resp. not covered) by p within (resp. outside) C. For each cluster, we compute the average PCR of the patterns appearing in its explanation; tables in the experiments give the averages of these values over all clusters. For an explanation D of C, EC measures the number of points in C covered by at least one pattern in D; tables in the experiments give the average of EC over all clusters.

$$\text{Pattern Coverage Rate } \mathbf{PCR}(p, C)\text{: } \frac{\#\{o \in C : cover(p,o)\}}{\#C}$$

$$\text{Explanation Coverage } \mathbf{EC}(D, C)\text{: } \frac{\#\{o \in C \mid \exists p \in D\ cover(p,o)\}}{\#C}$$

$$\text{Inverse Pattern Contrastivity } \mathbf{IPC}(p, C)\text{: } \frac{1}{K-1} \sum_{C' \neq C} 1 - \frac{\#\{o \in C' \mid cover(p,o)\}}{\#C'}$$

The approach is implemented in Python 3 and is available on Github[2]; Pattern generation is performed using LCM in Python, and the CP model is developed with CPMPY [15]. Experiments were performed on a computer equipped with an 11th Gen Intel® Core™ i7-1165G7 @ 2.80 GHz × 8 processor and 31.0 GB memory.

In the tables presented in this section, patterns are written between braces and separated by ||. The character "-" indicates that no solution has been found. Various colors are used to enhance readability.

Illustration on an Artificial Dataset. To illustrate our method, we present results on an artificial dataset visualized in Fig. 2. Descriptive properties are instance colors, shapes, and sizes.

Clusters were generated using random Gaussian distributions: C1 and C2 correspond to single Gaussians, while C3 consists of four adjacent smaller Gaussians. The significant descriptive property distributions are as follows; Cluster 1: 80% blue, 60% square, 70% large; Cluster 2: 70% red, 50% triangle, 60% small; Cluster 3: 80% green, 70% round, 70% large.

We applied ECS using base partitions generated by K-means and Spectral Clustering. Results were evaluated using ARI and our explanation measures.

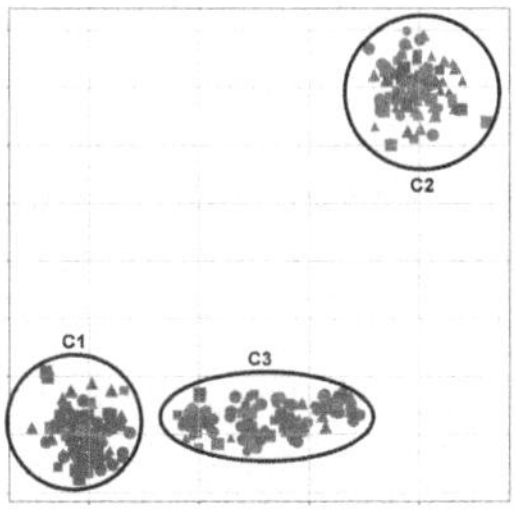

Fig. 2. Artificial dataset. (Color figure online)

With a high coverage parameter, we expected clusters to be explained primarily by colors, the most characteristic properties ($\geq$ 70% in each cluster). This was confirmed, as shown in Table 1. Conversely, with a lower coverage and a lower discrimination thresholds, we expected more specific combinations of properties, such as *red triangle* for Cluster 2 or *blue square* for Cluster 1. This was also verified, as presented in Table 2. The differences in the execution time is mainly due to the number of variables given to the model; With a lower coverage parameter, Table 1's model had more possible candidate clusters and patterns to consider, thus making it more time consuming. In both cases, the obtained clustering matches the ground truth with ARIs equal to 1.

[2] https://github.com/MathieuGuilbert/ECS.

Table 1. Artificial dataset, pattern of any length, K = 3, $\theta = 0.7$, $\phi = 0.5$. Time: 14 s.

C	Explanation	PCR	EC	IPC
0	{green}	0.8	0.8	0.94
1	{red}	0.8	0.8	0.88
2	{blue}	0.7	0.7	0.88

Table 2. Artificial dataset, pattern of any length, K = 3, $\theta = 0.3$, $\phi = 0.1$. Time: 84 s.

C	Explanation	PCR	EC	IPC
0	{green} ||{green, large} ||{circle, large}	0.58	0.85	0.97
1	{red, small} ||{red, triangle}	0.46	0.69	0.96
2	{blue, large} ||{blue, square}	0.58	0.78	0.92

Use Case on Education. The dataset NC-1014 is composed of 1014 programs written in 2020 by 60 students from the University of New Caledonia, answering 8 different exercises from a Python learning module [5]. The feature space for clustering $\mathbb{F}$ is a 20-dimensional embedding computed using the *sec2vec* method [21] on the raw programs. No information was given to the model about the program authors, nor about the exercises they were supposed to solve. The $\mathbb{B}$ explanation space is made up of descriptors indicating the presence of important tokens such as *if, else, while, for*, the exercise number, the user, and certain numerical values discretized according to the median (e.g. the number of returns).

NC-1014 - Complete Data Set. Our method was applied by maximizing the number of points covered and uniquely assigned (Eq. 1). We observed that most clusters are partially explained by an exercise, suggesting that programs responding to the same exercise have similar embeddings.

NC-1014 - First Exercise. The method was applied to programs corresponding to the first exercise with patterns of length 1 (a single descriptor) and with patterns of any length. Table 3 shows that, overall, the clusters correspond to different students. The results differ somewhat when using patterns of any length, where the explanations are made up of patterns containing information about the students and their coding style. Recall that the identity of the program authors was not taken into account when calculating the embedding. This suggests that writing style is induced. An interesting discovery is thus that even when clusters are generated only on the embedding space, they can be explained by the coding style of some students.

Table 3. NC-1014, 1st exercise, pat.length = 1, $\theta = 0.5, \phi = 0.3$, Time = 475 s

C	Explanation	Size	PCR	EC	IPC
0	{user4} || {user12}	24	0.35	0.71	0.99
1	{user16} || {user13}	14	0.46	0.93	1.0
2	{user39}	15	0.73	0.73	1.0
3	{user19}	6	1.0	1.0	1.0
4	{user28}	15	0.33	0.33	0.99
5	{user50}	7	0.71	0.71	1.0
6	{user36}	19	0.37	0.37	1.0
7	{user18}	5	1.0	1.0	1.0

Table 4. Automobile, pattern of size 1, K = 3, $\theta = 0.7$, $\phi = 0.5$, Time=55 s

C	Explanation	Size	EC	IPC	Price
0	{mpfi-fuel-injection}	42	0.88	0.68	17090
1	{turbo} || {five-cylinders} || {mercedes-benz}	7	0.86	0.93	29251
2	{curb-weight $\leq$ 2340} || {city-mpg >26} || {highway-mpg >32} || {front-wheel-traction}	110	0.91	0.93	8157

Automobile Dataset. The UCI Automobile dataset contains 159 cars (after the removal of 55 incomplete observations) characterized by 24 descriptive properties and their prices. Clustering was performed based on price, while the remaining attributes were used to explain the resulting clusters. Numerical attributes were discretized according to the median. Base partitions (Step 1) were generated by K-means and Spectral Clustering with a number of clusters ranging from 2 to 15. Table 4 shows results with patterns of length 1. The clusters represent 3 price categories with cheap, medium and expensive cars, each characterized by distinctive intrinsic properties. For example, cluster 1 contains the 7 most expensive cars featuring a five-cylinder turbo engine and/or the Mercedes-Benz brand. Cluster 0 is characterized by the presence of MFPI injection systems.

Comparison to Post-Hoc Decision Trees. The most widely used explainable clustering methods are decision trees (see Sect. 2). We now compare our method to explanations derived from post-hoc decision trees, computed using DecisionTreeClassifier[3] on the clusters obtained by our method. They are constructed with the default parameters only varying the complexity parameter *ccp_alpha*. This parameter (Cost Complexity Pruning Alpha) controls how much pruning is applied to the tree. It determines a trade-off between model complexity and generalization by penalizing tree size. The higher the ccp_alpha value, the more nodes are removed, leading to a more compact structure. Conversely, a lower ccp_alpha value results in deeper trees, potentially leading to overfitting as it creates very specific paths. As shown in Table 5 that compares decision tree with different pruning parameters to ECS for the Automobile and Artificial datasets, the higher *ccp_alpha*, the higher PCR but the lower EC and IPC.

[3] https://scikit-learn.org.

Table 5. Decision Tree Performance with different ccp_alpha values and ECS.

	Automobile					Artif. Dataset			
ccp_alpha	0.0	0.01	0.02	0.03	**ECS**	0.0	0.01	0.02	**ECS**
PCR	0.20	0.35	0.47	0.52	0.77	0.13	0.50	0.75	0.54
EC	0.99	0.93	0.86	0.81	1.0	0.77	0.74	0.74	0.78
IPC	1.0	0.99	0.98	0.95	0.77	0.98	0.92	0.87	0.95
Number of patterns	6.67	3.67	2.33	1.67	3	6.00	1.67	1.00	2.33
Pattern size	4.83	3.86	3.00	2.67	4.5	4.00	2.67	1.67	1.83

Table 6 presents ECS results on the Automobile dataset, with a high ϕ, thus a low constraint on discrimination. Table 7 contains the paths generated by the tree (cp_alpha = 0.02). Decision trees are built to be discriminant. ECS is able to find competitive discriminant explanations, covering all data objects with patterns covering a significant amount of cluster objects.

Comparison with Other Works. We consider two competing approaches: the first applies K-Means and then searches for patterns of length 1 (*K-Means*) or patterns of any length (*K-Means–LCM*); the second, considering only patterns of length 1 (single descriptor), is presented in [6]. This latter one aims to find a Pareto front of partitions maximizing cluster diameter and the MMCTA criterion (number of tags Boolean descriptors shared by all instances in a cluster). Our results were compared with 3 results from their Pareto front: the first one maximizing MMCTA, the last one maximizing cluster diameter and the median (denoted *central solution*).

The [6] descriptions maximize the PCR and EC measures, but are less discriminative than our model. Their high coverage values, due to the MMCTA objective, leads to low IPC values, as discrimination is not taken into account. In comparison, K-Means-based methods produce results with comparable or higher IPC but lower coverage. Overall, ECS generates more discriminating explanations with only small decreases in coverage measures (Tables 8 and 9).

Table 6. Automobile, pattern of any length, K = 3, $\theta = 0.7$, $\phi = 0.5$

C	Explanation	PCR	EC	IPC
0	{curb-weight>2340.0, horsepower>88.0, rwd, front} || {wheel-base>96.9, length>172.4, height>54.1, curb-weight>2340.0, horsepower>88.0, highway-mpg<=32.0, front}	0.77	1.0	0.79
1	{gas, fwd, front, ohc, four} || {city-mpg>26.0, gas, std, front, four} || {highway-mpg>32.0, front} || {compression-ratio>9.0, std, front, four}	0.78	1.0	0.81
2	{length>172.4, horsepower>88.0, front, four} || {width>65.4, gas, front, four} || {wheel-base>96.9, length>172.4, width>65.4, gas, std, front}	0.77	1.0	0.7

Table 7. Decision tree paths on Automobile. cp _alpha = 0.02.

C	Explanation	PCR	EC	IPC
0	{curb-weight>2340.0, not rwd, mpfi, not symboling<=1.0} || {curb-weight>2340.0, rwd}	0.5	1.0	0.89
1	{not curb-weight>2340.0} || {curb-weight>2340.0, not rwd, not mpfi}	0.47	0.94	0.96
2	{curb-weight>2340.0, not rwd, mpfi, symboling<=1.0}	0.42	0.42	0.98

Table 8. K-means baseline and ECS results ($\theta = 0.7$, $\phi = 0.5$)

	Data	Auto	Student	NC
	K	3	3	8
K-Means	PCR	0.87	0.83	0.89
	EC	1.0	1.0	1.0
	IPC	0.48	0.21	0.4
K-Means (LCM)	PCR	0.8	0.78	0.84
	EC	1.0	1.0	1.0
	IPC	0.77	0.28	0.63
ECS	PCR	0.86	-	0.83
	EC	0.94	-	0.83
	IPC	0.82	-	0,97
ECS (LCM)	PCR	0.81	0.74	0.83
	EC	0.99	0.89	0.83
	IPC	0.89	0.61	0.98

Table 9. Dao *et al.* 2018. results. Note: Student's Pareto front has only two solutions.

	Data	Auto	Student	NC
	K	3	3	8
Dao *et al.* 2018, max MMCTA	PCR	1.0	1.0	1.0
	EC	1.0	1.0	1.0
	IPC	0.42	0.14	0,38
Dao *et al.* 2018, central solution	PCR	1.0		1.0
	EC	1.0		1.0
	IPC	0.25		0.25
Dao *et al.* 2018, max diameter	PCR	1.0	0.0	1.0
	EC	1.0	0.0	1.0
	IPC	0.19	0.0	0.15

7 Conclusion

We propose a new *explainable by design* clustering method, where an explanation is defined as a set of patterns covering the majority of instances in a cluster and discriminating them from others. These notions are expressed by constraints, enabling experts to incorporate domain knowledge and expectations. Two representation spaces are used: one for clustering and another, Boolean, for explanations. Although the explanation space may be derived from the clustering space, we have demonstrated the benefits of this separation on two use cases (in education and automotive).

Future work includes improving the model itself by optimizing its formulation and computational efficiency, and expanding its functionality by incorporating additional constraints and objective criteria. Additionally, developing a method to automatically determine optimal coverage and discrimination parameters would further enhance the adaptability and usability of our approach. Another promising extension of our method is its application to chemoinformatics where molecules need to be explained by structural characteristics, such as pharmacophoric properties, in order to justify their biological activities when encountering a certain type of proteins.

Acknowledgements. This work was financed by the InvolvD ANR project Interactive constraint elicitation for unsupervised and semi-supervised data mining (ANR-20-CE23-0023).

Disclosure of Interests. The authors have no competing interests to declare that are relevant to the content of this article.

References

1. Aloise, D., Deshpande, A., Hansen, P., Popat, P.: Np-hardness of euclidean sum-of-squares clustering. Mach. Learn. **75**(2), 245–248 (2009)
2. Bertsimas, D., Orfanoudaki, A., Wiberg, H.: Interpretable clustering: an optimization approach. Mach. Learn. **110**(1), 89–138 (2021)
3. Chabert, M., Solnon, C.: Constraint programming for multi-criteria conceptual clustering. In: CP, pp. 460–476 (2017)
4. Chen, J., et al.: Interpretable clustering via discriminative rectangle mixture model. In: ICDM, pp. 823–828. IEEE (2016)
5. Cleuziou, G., Flouvat, F.: Learning student program embeddings using abstract execution traces. In: 14th International Conference on Education DM, pp. 252–262 (2021)
6. Dao, T.B.H., Kuo, C.T., Ravi, S., Vrain, C., Davidson, I.: Descriptive clustering: ILP and CP formulations with applications. In: 27th IJCAI, pp. 1263–1269 (2018)
7. Dao, T.B.H., Lesaint, W., Vrain, C.: Clustering conceptuel et relationnel en programmation par contraintes. JFPC 11 (2015)
8. Dao, T., Vrain, C.: A review on declarative approaches for constrained clustering. Int. J. Approx. Reason. **171**, 109–135 (2024)
9. Dasgupta, S., Frost, N., Moshkovitz, M., Rashtchian, C.: Explainable k-means and k-medians clustering. In: ICML, pp. 12–18 (2020)
10. Davidson, I., Ravi, S.S.: The complexity of non-hierarchical clustering with instance and cluster level constraints. Data Mining Knowl. Discov. (2007)
11. De Raedt, L., Blockeel, H.: Using logical decision trees for clustering. In: Lavrač, N., Džeroski, S. (eds.) ILP 1997. LNCS, vol. 1297, pp. 133–140. Springer, Heidelberg (1997). https://doi.org/10.1007/3540635149_41
12. Frost, N., Moshkovitz, M., Rashtchian, C.: Exkmc: expanding explainable k-means clustering. arXiv preprint arXiv:2006.02399 (2020)
13. Galbrun, E., Miettinen, P.: What is Redescription Mining. Springer (2017)
14. Guilbert, M., Vrain, C., de Souto, M.C., et al.: Anchored constrained clustering ensemble. In: IJCNN, pp. 1–8. IEEE (2022)
15. Guns, T.: Increasing modeling language convenience with a universal n-dimensional array, cppy as python-embedded example. In: 18th Modref Workshop at CP (2019)
16. Izza, Y., Ignatiev, A., Marques-Silva, J.: On tackling explanation redundancy in decision trees. J. Artif. Intell. Res. **75**, 261–321 (2022)
17. Laber, E., Murtinho, L., Oliveira, F.: Shallow decision trees for explainable k-means clustering. Pattern Recogn. **137**, 109239 (2023)

18. Lawless, C., Kalagnanam, J., Nguyen, L.M., Phan, D., Reddy, C.: Interpretable clustering via multi-polytope machines. In: AAAI, vol. 36, pp. 7309–7316 (2022)
19. Loyola-Gonzalez, O., et al.: An explainable artificial intelligence model for clustering numerical databases. IEEE Access **8**, 52370–52384 (2020)
20. Makarychev, K., Shan, L.: Near-optimal algorithms for explainable k-medians and k-means. In: ICML, pp. 7358–7367 (2021)
21. Martinet, T., Cleuziou, G., Exbrayat, M., Flouvat, F.: From document to program embeddings: can distributional hypothesis really be used on programming languages? In: ECAI 2024, pp. 2138–2145. IOS Press (2024)
22. Michalski, R.S., Stepp, R.E.: Automated construction of classifications: conceptual clustering versus numerical taxonomy. IEEE Trans. Pattern Anal. Mach. Intell. (1983)
23. Miller, G.A.: The magical number seven, plus or minus two: some limits on our capacity for processing information. Psychol. Rev. **63**(2), 81 (1956)
24. Mueller, M., Kramer, S.: Integer linear programming models for constrained clustering. In: DS, pp. 159–173 (2010)
25. Ouali, A., Loudni, S., Lebbah, Y., Boizumault, P., Zimmermann, A., Loukil, L.: Efficiently finding conceptual clustering models with integer linear programming. In: IJCAI, pp. 647–654 (2016)
26. Rossi, F., van Beek, P., Walsh, T.: Handbook of Constraint Programming. Elsevier B.V, Foundations of Artificial Intelligence (2006)
27. Sambaturu, P., Gupta, A., Davidson, I., Ravi, S., Vullikanti, A., Warren, A.: Efficient algorithms for generating provably near-optimal cluster descriptors for explainability. In: AAAI, vol. 34, pp. 1636–1643 (2020)
28. Uno, T., Asai, T., Uchida, Y., Arimura, H.: LCM: an efficient algorithm for enumerating frequent closed item sets. In: FIMI, vol. 90 (2003)
29. Van Craenendonck, T., Blockeel, H.: Constraint-based clustering selection. Mach. Learn. **1**, 1497–1521 (2017). https://doi.org/10.1007/s10994-017-5643-7
30. Vega-Pons, S., Ruiz-Shulcloper, J.: A survey of clustering ensemble algorithms. Int. J. Pattern Recognit. Artif. Intell. **25**(03), 337–372 (2011)
31. Yang, J., Sun, L., Wu, Q.: Constraint projections for semi-supervised spectral clustering ensemble. Pract. Exp. Concurr. Comput. **31**(20), e5359 (2019)
32. Yu, Z., et al.: Semi-supervised ensemble clustering based on selected constraint projection. IEEE TKDE (2018)
33. Zhang, H., Davidson, I.: Deep descriptive clustering. In: IJCAI (2021)

Author Index

A
Adriaensen, Rik 419
Ahavi, Paul 28

B
Barandas, Marília 343
Baratchi, Mitra 67, 181
Bicego, Manuele 154
Boisaubert, Hugo 373
Boström, Henrik 432
Bresson, Roman 432
Buurke, Jaap H. 17
Byttner, Stefan 456

C
Cerqueira, Vitor 343
Cestnik, Bojan 261
Clavier, Ianis 373
Cohen-Boulakia, Sarah 274
Cornuéjols, Antoine 28
Cuissart, Bertrand 167
Cule, Boris 305

D
Dang, Vu Minh Hoang 288
Dao, Thi-Bich-Hanh 469
Davis, Jesse 113
de Campos, Cassio 209
Dengel, Andreas 389
Djamai, Abdallah Alaa-Eddine 246
Dong, Genghua 432
Duivesteijn, Wouter 195
Dupont, Pierre 234

F
Faulon, Jean-Loup 28
Ferret, Olivier 274
Fink, Jérôme 54
Frénay, Benoît 54
Fromont, Elisa 81

G
Ghandi, Soroush 209
Gower-Winter, Brandon 403
Gross-Amblard, David 3
Guglielmo, Gianluca 305
Guilbert, Mathieu 469
Gunasekara, Nuwan 358

H
Heinrich, Carl Vico 195

I
Inácio, Ricardo 343
Ittermann, Till 41

K
Kaneko, Kunitake 141
Kastner, Lise 167
Kastrin, Andrej 261
Kenmogne, Edith Bélise 54
Keraghel, Imed 246
Khalal, Noor 246
Kim, Dongwhi 444
Kindji, G. Charbel N. 81
Klincewicz, Michal 305
Kok, Joost N. 17
Koloski, Boshko 261
Koutamanis, Alexander 67
Kramer, Stefan 318
Krempl, Georg 403
Krijthe, Jesse H. 154

L
Lamotte, Jean-Luc 167
Lavrač, Nada 261
Lejus-Bourdeau, Corinne 373
Lombarts, Tommie 195
Loog, Marco 154
Lundström, Jens 456
Lutton, Evelyne 28

G. Krempl et al. (Eds.): IDA 2025, LNCS 15669, pp. 485–486, 2025.
https://doi.org/10.1007/978-3-031-91398-3

M

Madjoukeng, Ariel Basso 54
Maene, Jaron 419
Makowski, Maciej 403
Mallens, Jules 195
Mena, Francisco 389
Miranda, Miro 389
Mollet, Bastien 28
Moniz, Nuno 444

N

Nadif, Mohamed 246
Nasri, Maedeh 67
Nelke, Martin 318
Névéol, Aurélie 274
Nijssen, Siegfried 234
Nowaczyk, Sławomir 97, 127, 330, 358
Nuyts, Loren 113

O

Ozen, Canberk 97

P

Pashami, Sepideh 97, 358
Persson, Daniel 330
Picarougne, Fabien 373
Poitier, Pierre 54

Q

Quost, Benjamin 209

R

Reenalda, Jasper 17
Rieffe, Carolien 67
Rojas-Barahona, Lina M. 81
Ronval, Benoît 234
Rother, Anne 41
Rotman, Stijn J. 305

S

Sassi, Dorra 3
Scheltinga, Bouke L. 17
Schneider, Johannes 223
Schröder, Sietse 181
Sebe, Clémence 274
Sinoquet, Christine 373
Soares, Carlos 343
Spiliopoulou, Myra 41

T

Thierry, Constance 3
Tiwari, Prayag 97
Tonda, Alberto 28
Tortike, Luc 195

U

Urvoy, Tanguy 81

V

van Rijn, Jan N. 181
Vazirgiannis, Michalis 432
Verma, Rakesh M. 288
Vettoruzzo, Anna 330
Vexler, Julian 318
Victor, Quentin 373
Vieten, Björn 318
Vinge, Rikard 456
Vrain, Christel 469

W

Wahlberg, William 330
Wolf, David 195

Y

Yamashita, Tsuyoshi 141

Z

Żarski, Mateusz 127

The manufacturer's authorised representative in the EU is Springer
Nature Customer Service Centre GmbH, Europaplatz 3, 69115 Heidelberg,
Germany. If you have any concerns regarding our products, please
contact ProductSafety@springernature.com

Printed and bound by CPI Group (UK) Ltd, Croydon, CR0 4YY
01/10/2026
02225908-0017